Estates Gazette Law Reports

1981

Volume 1

Series edited by
Barry Denyer-Green LLM PhD FRICS
of the Middle Temple, barrister

Cases edited by
J Muir Watt OBE MA
of the Inner Temple, barrister

Audrey Boyle BA
Publisher's editor

Estates Gazette

Routledge
Taylor & Francis Group
LONDON AND NEW YORK

EG Books is an imprint of Routledge
2 Park Square, Milton Park, Abingdon, Oxon OX14 4RN
711 Third Avenue, New York, NY 10017, USA
Routledge is an imprint of the Taylor & Francis Group, an informa business

ISBN for complete set of 2 volumes: 0 7282 0233 6
ISBN for this volume 0 7282 0234 4
ISSN 0951-9289

For information on all EG Books publications
visit our website www.routledge.com

CONTENTS

TABLE OF CASES

Titles of cases shown in bold type
Names of parties reversed shown in ordinary type

INDEX OF SUBJECT-MATTER

LANDLORD AND TENANT — *continued*
Landlord and Tenant Act 1954 Part II — *continued*
Contracting-out — continued

Landlords accept tenant's offer to surrender — Held, agreement to surrender void under s38(1) 1954 Act — Covenant itself did not preclude tenant making application or request under Part II of 1954 Act — Therefore, no declaration that tenants entitled to assign without offering to surrender to landlords.
***Allnat London Properties Ltd* v *Newton (Ch)*, 49**

Covenant
Sublessees agree with freeholder to observe covenants from date of assignment "and thenceforth during the residue of the term granted by the underlease" — Whether covenant expired with contractual term or continued during time subtenants held over under Part II of 1954 Act — Held, effect of s24(1) was to make term granted by tenancy continue to subsist — Accordingly, agreement to be construed so that obligations continued so long as subtenants held over.
***GMS Syndicate Ltd* v *Gary Elliott Ltd (Ch)*, 37**

Landlord's opposition to new tenancy
S30(1)(*f*) — Condition attached to landlord's planning permission to build supermarket required access to be built on land occupied by tenant — Landlord had not appointed building contractor and had not formalised building lease — Held, no doubt that landlord had moved into "valley of decision" — Landlord prevented from proceeding by delaying tactics of tenant.
***AJA Smith Transport Ltd* v *British Railways Board (CA)*, 54**

S30(1)(*g*) — Lease of vacant site used as car park — Landlord intends to develop site to expand MOT testing business — Tenants argue that landlord's plans to build on site would create a new holding and would not constitute an occupation of "the holding" — Held, tenant's argument rejected — Landlord's plans would result in occupation of vacant site, which was holding, and something else, the buildings — *Nursey* v *P Currie (Dartford) Ltd* distinguished.
***Cam Gears Ltd* v *Cunningham (CA)*, 63**

New tenancy term
Ordered that new tenancy "duration should be three years" — Landlords contend term unreasonable on various grounds — Insufficient account of fact that landlords could not rely on s30(1)(*g*) as just outside five-year period — Improper to take account of offer by tenants to move elsewhere in building — Held, discretion reasonably exercised — Guidance that orders for new tenancy should express date on which tenancy to end.
***Chipperfield* v *Shell UK Ltd (CA)*, 51**

Notice to terminate tenancy
S30(1)(*f*) — Notice stated that landlord intended "to carry out substantial work of redecoration" as opposed to s30(1)(*f*) of 1954 Act which reads "work of construction" — Whether valid notice to terminate — Held, notice contained sufficient statement of one of statutory grounds — Grounds for notice, when read in conjunction with notes on back of notice, would have left tenant in no doubt landlord aiming at subpara (*f*).
***Philipson-Stow* v *Trevor Square Ltd (Ch)*, 56**

S57 certificate
Appeal against refusal to judicially review certificate made under s57(1) 1954 Act — Various grounds, including *mala fides*, insufficient evidence, refusal of Divisional Court to admit fresh evidence and misinterpretation of "requisite" in s57(1) — Held, finding that no *mala fides* not to be disturbed by way of judicial review — Fresh evidence inadmissible — Evidence was adequate — "Requisite" interpreted and properly so, by Secretary of State, as "reasonably necessary".
***R* v *Secretary of State for the Environment*, ex parte *Powis (CA)*, 58**

Lease or licence
Possession claim
Properties purchased with previous owner's gardener in occupation of one of them — New owner employs gardener and supplies him with statement of terms of employment which is signed by gardener — Whether gardener held protected tenancy or licence — Held, signing of statement effected surrender of any protected tenancy and created bare licence.
***Scrimgeour* v *Waller (CA)*, 68**

Leasehold enfranchisement
"House"
Premises part of small parade of shops each with living accommodation above — 75% of premises, including backyard, used as shop and associated storage — Remainder, living room and two bedrooms — Whether "reasonably so-called" a house — Held, not house under s2(1) Leasehold Reform Act.
***Tandon* v *Trustees of Spurgeons Homes (CA)*, 66**

Qualifying tenant
Regulated tenant of part of house acquires residue of long lease of whole house — Whether qualifying tenant — Construction of s1(1)(*b*) Leasehold Reform Act 1967 — Held, did not satisfy conditions of subsection — Had not been tenant under long tenancy at low rent "for the last five years or for periods amounting to five years in the last ten years".
***Harris* v *Plentex Ltd (Ch)*, 65**

Rent Acts
Demolition and closing orders
Closing order under Housing Act 1957 becomes operative before fair rent objection decided by rent assessment committee — Whether committee bound to decide fair rent as nil or nominal — Held, no rule of law that nil rent must be assessed after coming into force of closing order — Discretion under s72 Rent Act 1977 to fix effective date of registration significant.
***Williams* v *Khan (CA)*, 79**

Holiday letting
Property occupied by students — One student held tenancy — Prior to tenancy agreement forms filled-in giving personal details of occupants — Under agreement tenant to have "right to occupy and to use for a holiday" — Whether fell within s9 Rent Act 1977 — Held, proper approach to look at words of written agreement and, thereafter, to consider extrinsic evidence as to whether in reality holiday letting — Clear that parties knew that flat was to be occupied for purposes of work as students — Protected tenancy.
***R* v *Rent Officer for London Borough of Camden*, ex parte *Plant (QB)*, 73**

Rent assessment
Appellant landlords registered housing association of which tenants members — Committee determine service element in rent fixed element — Held, on analysis of constitution and rules of association tenants did have rights to see audited service charge costs and apportionment — Order that service element of rent variable.
***Re Heathview Tenants' Co-operative Ltd (QB)*, 82**

Fair rent — Proceedings by landlords under s141 Rent Act 1977 alleging holiday letting — Consultation with rent officer and parties where former decides protected tenancy — Later consultation on level of rent — Rent officer refuses to determine and register rent pending county court decision — Held, once officer is satisfied he has jurisdiction under duty to proceed — *Mandamus* to determine and register fair rent.
***R* v *Rent Officer for London Borough of Camden*, ex parte *Ebiri (QB)*, 71**

Rent assessment committee
Complaint of inadequacy of reasons — Allegations that committee failed to state sufficiently their opinion on landlord's submissions on capital value comparables balancing scarcity and amenity value — Held, provided reasons were intelligible and substantial points were addressed decision was good — No rule that committee must state methods of valuation it adopted nor explain final figure when own knowledge and experience relied upon — Application dismissed.
***Guppys Properties Ltd* v *Knott (No 3) (QB)*, 85**

Restricted contract
Room let at rent with services — Whether letting was regulated tenancy or restricted contract under s19 Rent Act 1977 — Interpretation of s19(5)(*a*) — Held. Act provides that if tenancy is regulated tenancy it is not restricted tenancy and not the reverse.
***Baldock* v *Murray (CA)*, 70**

Rent arrears
Estoppel
Rent fixed, on three occasions, by rent assessment committee on basis that landlords liable for repairs under covenant implied by s32 Housing Act 1961 — Landlords accept enhanced rent — Commenced proceedings for arrears on that basis — Landlords inconsistently dispute set-off for repairs carried out by tenant on basis that no covenant implied by law — Held, representation of fact that landlords liable for repairs — Estopped from denying liability for repairs until such time as fair rent corrected to reflect true position that no implied covenant.
***Brikom Investments Ltd* v *Seaford (CA)*, 30**

Rent review clause
Arbitration
Option exercised for new lease — Executed lease makes provision for rent review — Does not state that in absence of agreement rent to be determined by arbitrator — Landlord's unilateral mistake that such provision made — Tenants do not draw omission to landlord's attention — Omission contrary to landlord's interests — Whether lease to be rectified to include arbitration provision — Held, conditions for rectification satisfied — Arbitrator to determine rent as what would have been reasonable for parties to have agreed and not market rent.
***Thomas Bates & Son* v *Wyndham's (Lingerie) Ltd (CA)*, 91**

Powers of arbitrator — Clause for renewal of tenancy for further 21 years — Rent to be "at the then prevailing market rent" to be determined, in default of agreement, by arbitration — In other respects new lease to be identical to old — Whether arbitrator had power to direct periodic rent reviews — Held, had no such power to introduce variations — In effect, parties had agreed on lease of 42 years with rent review after 21 years.
***National Westminster Bank Ltd* v *BSC Footwear Ltd (CA)*, 89**

Construction: valuation
Increased rent to be determined by reference to "rack rent market value" — Whether rent to be fixed by reference to vacant possession

ficient evidence that caused loss — No negligence giving rise to vandalism.
***Cedar Transport Group Ltd v First Wyvern Properties Trustees Co Ltd (QB),* 154**

Completion date
Variation
Completion date passed by as parties to contract for the sale of land negotiate variations — Whether agreement varied so as to leave completion date open and contract void for uncertainty — Held, completion date agreed — *Obiter*, where completion to take place on date to be agreed court would imply reasonable time in default of agreement.
***Walters v Roberts (Ch),* 151**

AGRICULTURE

QUEEN'S BENCH DIVISION

October 17 1980

(Before Mr Justice WOOLF)

KELLETT AND ANOTHER v ALEXANDER

KELLETT AND ANOTHER v CADY AND ANOTHER

Estates Gazette February 7 1981

(1980) 257 EG 494

Agricultural holdings — Agriculture (Miscellaneous Provisions) Act 1976 — Succession provisions — Whether an application to an agricultural land tribunal under section 20 for a direction entitling the applicant to a tenancy was invalid because of a failure to give notice of the application to the landlord — Effect of section 20(14) of Act and Rule 4(1) of the rules in the schedule to the Agricultural Land Tribunals (Succession to Agricultural Tenancies) Order 1976 — No notice given to landlord on Form 3, although Form 1 had contained the statement that the applicant would notify the landlord — Landlord was in fact notified of the application by service upon her of a copy by the secretary of the tribunal — Tribunal had held that the failure to give a formal notice to the landlord was a fatal defect invalidating the application — Held on appeal that the tribunal were wrong — Section 20(14) merely required an order to be made containing the procedural steps mentioned in the order and the relevant rule, Rule 4(1), was a procedural rule failure to comply with which did not automatically invalidate an application — The tribunal had a discretion to extend time and not to regard as fatal a failure by the applicant to comply with the requirement to give the landlord formal notice of the application — The chairman had in addition the specific powers given in Rule 37 and Rule 38 of the "principal rules" currently in Schedule 1 to the Agricultural Land Tribunals (Rules) Order 1978 to enlarge the time for taking any step and to disregard a failure to comply with the 1976 rules — Both appeals allowed.

These were two appeals by case stated from decisions of the Agricultural Land Tribunal for the Northern Area. The applicants in both cases were the same, Harold Kellett and John Edward Kellett, who had made joint applications to the tribunal under sections 20(1) and 21(2) of the Agriculture (Miscellaneous Provisions) Act 1976 for tenancies on the death of their father. The landlords, respondents to the applications, were in the one case Noral Alexander and in the other Mary Elizabeth Cady and Edith Margaret Ellershaw. As both cases raised the same point the judge dealt with the first case for the purpose of the appeal.

R J Furber (instructed by Boxall & Boxall, agents for Kendal & Fisher, of Dalton in Furness) appeared on behalf of the appellants (applicants) in both cases; H W P Eccles (instructed by Wedlake, Bell) represented the respondents.

Giving judgment, WOOLF J said: There are before me two appeals, both by way of case stated, under section 6(1) of the Agriculture (Miscellaneous Provisions) Act 1954 by the Agricultural Land Tribunal for the Northern Area.

For the purposes of this appeal I do not propose to make any distinction between the two appeals because the point which arises is identical on both appeals; but for the facts I will refer to the papers in the matter of Kellett and Kellett v Alexander. So everything that I have to say about that appeal is equally applicable to the other appeal.

The question which was posed by the tribunal at the end of the case was in these terms: whether an applicant's failure to give notice of his application to the landlord is fatal, or whether the tribunal have a discretion to waive service on the landlord in circumstances where the landlord is not prejudiced? In order to understand the question which has been stated by the tribunal it is necessary to look at certain provisions of the code of statutory provisions dealing with agricultural holdings. It is convenient to start with the Agricultural Holdings (Notices to Quit) Act 1977.

Section 2 of that Act contains restrictions on the operation of a notice to quit agricultural holdings. The result of those restrictions is that the notice to quit does not take effect unless the tribunal consents to its operation. However, the restrictions do not apply to certain cases which are set out in section 2(2). One of the cases so set out is Case G, which provides that they shall not apply where:

> There had died within three months before the date of the giving of the notice to quit a person who immediately before his death was the sole (or sole surviving) tenant under the contract of tenancy, and it is stated in the notice that it is given by reason of the matter aforesaid.

The exclusion of what I will call the statutory protection provided by section 2 of the 1977 Act in such a case was modified by the provisions of the Agriculture (Miscellaneous Provisions) Act 1976, which introduced a provision dealing with succession on the death of a tenant.

Section 18(1) provides:

> Where after the passing of this Act the sole (or sole surviving) tenant of an agricultural holding dies and is survived by any of the following persons . . .

then the categories are set out

> the following provisions of . . . the Act . . . shall apply unless excluded by subsection (4) of that subsection.

I do not need to trouble, for the purposes of this case, with subsection (4) and I can proceed to the subsequent sections of the Act starting with section 19, which provides:

> A notice to quit the holding given to the tenant thereof by reason of the death of the deceased and falling within

what was to become Case G under the 1977 Act

> shall not have effect unless either

and these words are important

> (a) no application to become the tenant of the holding is made under section 20 of this Act within the relevant period; or (b) one or more such applications having been made within that period, either —
> (i) none of the applicants is determined by the Tribunal to be in their opinion a suitable person to become the tenant of the holding; or
> (ii) the Tribunal consent under section 22 of this Act to the operation of the notice to quit.

I stress the wording so far as the requirement of an application is concerned because it is quite clear from that wording that, unless there is an application to become the tenant of the holding under section 20 of the Act, the restriction on the operation of the notice to quit in section 19 does not apply at all. Section 20(1) provides:

> Any eligible person may within the relevant period apply to the Tribunal for a direction entitling him to a tenancy of the holding.

Section 20(14) provides:

> Provision shall be made by order under section 73(3) of the Agriculture Act 1947 . . . for requiring any person making an application to such a tribunal under this or the following section to give notice of the application to the landlord of the agricultural holding to which the application relates and to take such steps as the order may require for bringing the application to the notice of other persons interested in the outcome of the application.

So far as subsection (14) is concerned, it will be seen that it specifically provides that there should be an order requiring any person making the application to give notice of the application to the landlord. In this case, for reasons which appear in the case stated, no notice of the application was in fact given to the landlord.

Section 21 contains provisions which enable a person who is not a fully eligible person to be treated as eligible; and section 22 provides an opportunity for the landlord to obtain the tribunal's consent to the operation of the notice to quit; and that links up with the provisions of section 20(1) because section 20(1) enables an eligible person who seeks a tenancy to apply for a direction entitling him to it.

Section 22 enables the obverse application,. by the landlord, to come before the tribunal so that he can obtain the consent to the operation of the notice to quit notwithstanding the fact that there is somebody who contends that they are entitled to the benefit of the provisions contained in Part II of the Act of 1976 (which deals with the provision for succession on death). As one would expect, there is a provision which enables the two applications to be heard at the same time.

The specific rules which were made governing the procedure for dealing with the applications under Part II of the 1976 Act are the Agricultural Land Tribunals (Succession to Agricultural Tenancies) Order 1976 (SI 1976 No. 2183). Rule 2 specifically lays down that the "application to the tribunal under section 20 of the 1976 Act for a direction entitling the applicant to a tenancy . . . shall be made in Form 1." That form is in the schedule to the order and says, among other things, who the landlord of the holding is and, in paragraph 19, sets out under the heading "The persons whom I shall notify of this application . . . are: (a) The landlord of the holding whose name and address are: . . ."

I draw attention to the words "I shall notify" which look to the future.

Rule 2(3) says:

> An application made under this rule shall not be entertained by the tribunal if it is not made within the relevant period.

and the relevant period is, where the deceased died on or after January 1 1977, the period of three months beginning with the day after the date of death.

Rule 2(3) reflects the requirement of the Act, contained in section 20(1), that an eligible person may within the relevant period apply to the tribunal for a direction entitling him to a tenancy of the holding; the relevant period in the rules and the Act are the same.

Rule 4(1) is the rule which complies with the requirement contained in section 20(14) of the Act. It provides:

> An applicant shall at the time of making his application serve notice of the application in Form 3 on the landlord and on any person who, to the knowledge of the applicant, has made or may be able to make an application under rule 2(1), and shall inform the tribunal in his application of the name and address of every person to be notified by him.

Again, it is relevant to look at Form 3 because that form says, "I . . . hereby give you notice that I applied on . . ." and then there is to be inserted the date of his application which seems to presuppose that when the form is completed the application has already been made.

I made the reference to Form 3 and to Form 1 to contrast the language in the forms with the words of Rule 4(1), which state that the "applicant shall at the time of making his application serve notice of the application in Form 3 on the landlord." Quite clearly those words have not to be applied absolutely literally. It would be impossible at the same time as making the application to actually, there and then, give notice, and what, it seems to me, is intended by the wording of Rule 4(1) is that, having made the application, the applicant shall promptly give notice. What will be compliance with that requirement to give notice promptly will have to be considered in the circumstances of the particular case. Sometimes, exercising reasonable diligence, a certain period will have to elapse before it is possible for the notice to be given, but as long as the applicant concerned acts with reasonable promptitude in the circumstances of the particular application, the letter of that rule will be complied with.

In addition to the specific rules there are what are described as principal rules which apply to proceedings before tribunals generally. Those principal rules were linked to the specific rules, to which I have just made reference, by Rule 19 of the specific rules.

The current principal rules are set out in the Agricultural Land Tribunals (Rules) Order 1978 (SI 1978 No. 259). Those rules replaced earlier rules and, in Schedule 2, amended Rule 19 of the specific rules so that the rules of 1976 then read as follows:

> 19(1) Rules 28 to 31 and 33 to 38 of the principal rules shall apply with the necessary modifications to proceedings under these Rules as they apply to other proceedings before the tribunal.
> (2) For the purposes of rules 31, 33 and 34 of the principal rules, any dismissal of an application under these Rules shall be a decision, and all such decisions, and the reasons for them, may be given in a single document at the conclusion of the proceedings unless the Chairman otherwise decides.

Turning therefore in particular to Rules 33 to 38 of the principal rules, we find that: Rule 33 deals with references to the High Court; Rule 34 deals with the modification of tribunals' decisions; Rule 35 contains general rules as to service; Rule 36 deals with substituted service; Rule 37 says "The time appointed by or under these Rules for doing any act or taking any step in connection with any proceedings may be extended by the chairman on such terms and conditions, if any, as appear to him just"; Rule 38 says, "Any failure on the part of any person to comply with the provisions of these Rules shall not render the proceedings, or anything done in pursuance thereof, invalid unless the chairman or the tribunal so direct."

Having referred to the relevant statutory provisons contained in the Acts and the subordinate legislation, it is convenient to refer shortly to the case stated. That recites the following facts:

The father of the two applicants had died on April 2 1979 and the time by which the applications under section 20(1) and section 21(2) had to be made was July 2 1979. The applicants made a joint application on June 27 1979 and, by letter dated June 29 indicated that the applications were to be treated as joint; and then the case goes on,

> The applications had contained a statement that the landlord had been notified that the applications had been made but no notice in Form 3 had been served on the landlord.

The recital in the case that the applications had contained a statement that the landlord had been notified is not entirely accurate, because the applications were in the statutory form and, in accordance with the statutory form, they did not say that the landlord had been notified but indicated that the landlord was someone whom the applicant "shall notify of the application".

The case goes on to set out the contention of the landlord that the applications were invalid on two counts; the first of which is not relevant to this appeal; the second being that the landlord contended that the applications were not valid because no notice in Form 3 had been served on the landlord as required by Rule 4(1) of the 1976 rules.

In respect of that contention the applicants conceded that, in the haste of compiling and submitting the applications before the time expired, the requirement that Form 3 be served on the landlord had been overlooked, but they contended this was merely a procedural fault which had not prejudiced the landlord who was notified of the application by the service on her by the secretary to the tribunal. There is no dispute that that in fact happened, there being a provision in the rules which requires the secretary of the tribunal to notify the landlord of the application. That was a requirement in addition to the requirement placed upon the applicant to give notice as required by Rule 4(1).

The tribunal decided that the failure to serve notice of the application on the landlord was not a procedural error. They took the

A view that section 20(14) of the Act of 1976 made it a provision of the Act (and they underline the word "Act") that the applicants give notice of the applications to their landlord; and the tribunal did not consider that the discretion provided by the rules to waive the requirements of the rules could be extended by the tribunal to waive a requirement contained in the Act. They therefore found the failure to serve notice of the applications on the landlord to be a fatal error and dismissed the applications. The tribunal considered, therefore, that the consequence of the mistake which was made, in not giving the notice required by Rule 4(1), was that the application failed *in limine* and there was no jurisdiction in the tribunal to consider the application.

B Before me the applicants contend that this is a complete misinterpretation of the statutory provisions. First of all, the very question framed by the tribunal seems to underline the view of the tribunal that, because there was no notice given of the application to the landlord by the applicants, this was a case where an express requirement of the Act had been disobeyed by the applicants. I do not read the provisions of section 20(14) in that way. The express requirement of section 20(14) of the 1976 Act is to make an order which complies with the terms of that subsection. The subsection itself does not set out any specific provision as to giving notice. The form of notice required and the time within which the notice had to be given is the requirement of the order made pursuant to the
C subsection and therefore, in order to decide whether or not there is any power to overcome the admitted failure to comply with Rule 4(1), does not depend, in my view, on the terms of section 20(14) but on the terms of Rule 4(1) and the other rules which are relevant, construed in the light of the provisions contained in section 20(14).

The position has to be contrasted with that contained in the Act, which requires an application to be made within the relevant period. That is a requirement of the Act and it is conceded before me, and in my view rightly conceded, that if there had been no application made here within the relevant time then this would be a situation where in fact there would be no power for the tribunal to consider the matter because an imperative provision of the Act had
D been disobeyed.

The position if no application had been made would be identical to that indicated by the Court of Appeal in the case of *Howard v Secretary of State for the Environment* reported *inter alia* in [1974] 1 All ER 644, as to the requirement to make an application within the prescribed time. That conclusion is equally applicable to the wording of the order of 1976 which reiterates the requirement of the Act that an application should be made within the relevant period. Therefore, so far as Rule 2(3) is concerned, which says that the application shall not be entertained if it is not made within the relevant period, I would readily come to the conclusion that that is a requirement which cannot be waived by the tribunal, and that Rule
E 37 and Rule 38 have no application. I do not think that on their normal reading those rules are appropriate to deal with an application which fails *in limine*. Such an application, which is made out of time, appears to be outside the express words of Rule 37 and Rule 38, because until a proper application has been made there are no proceedings to which those rules can apply. However, so far as the failure to give notice is concerned, it is strongly argued on behalf of the applicants that Rule 37 and Rule 38 should apply.

Before considering Rule 37 and Rule 38 it is convenient to consider what would be the situation assuming those principal rules were not ones which were relevant to a failure to comply with Rule 4(1) of the 1976 Order. It is my view that looking at that rule in its
F context it is the sort of procedural rule which you would not normally expect to be regarded as an imperative requirement and normally, even if there were no Rule 37 or 38, the approach to the interpretation of that rule would be to regard it as one which, while it should be complied with, the failure to comply with it would not necessarily be fatal to an application. It would be a procedural matter which the tribunal as the master of its own procedure would be entitled to take into account and decide according to the relevant circumstances whether to regard it as something which should prevent the application proceeding or not. Normally if there was no prejudice to the other parties, the view of a tribunal would be that such a failure should not be fatal.

It follows from what I have said that section 20(14), which G
required such a rule, was doing no more than indicating that it should be a procedural requirement that notice of the application should be given, and placing a statutory obligation upon those responsible for making the order to see that a procedural requirement was included in the order.

The fact that I have come to this conclusion about the terms of Rule 4(1) means that the decision of the tribunal must be wrong, because if the tribunal had a discretion to extend time and not to regard as fatal the failure to comply with Rule 4(1), clearly the tribunal should have adopted a different approach from that which they adopted at the hearing of this application.

It is, however, still important to consider whether Rule 37 and H
Rule 38 of the principal rules apply to this particular failure to comply with the requirements of the 1976 Order. I did at the outset have some concern as to whether Rule 37 and Rule 38 were applicable. It seems to me that it would be possible to construe those rules as having a more limited ambit. The side note to the amended Rule 19 contained in the schedule to the 1978 rules refers to the application of principal rules relating to evidence, decisions, etc, and it would be possible to limit those rules so that they did not apply to the giving of notice to the initial application.

However, although I had reservations about this matter I have been persuaded that both Rule 37 and Rule 38 apply to the failure to give notice, and the proper approach in this application for the J
tribunal to have adopted would be to consider those rules in deciding whether to exercise a discretion to waive the requirement of giving notice. If, as is my view, there existed a power to extend time, then quite clearly that, in an appropriate case, would be equivalent to a power to waive the requirement to give notice. It would have been pointless of the tribunal, on the hearing of this application, to decide that the time should be extended so that notice of the application should be given there and then. Once the tribunal's secretary had given notice of the application, then any purpose in giving a second notice disappeared, and the more satisfactory way of dealing with an extension of time was to waive the requirement. Furthermore, if Rule 38 applies, then even without K
any specific application being made under Rule 37, it was open to the tribunal to disregard the failure to comply with the notice requirement and to treat the proceedings as being valid notwithstanding its absence.

In practice, of course, the application of Rule 37 and Rule 38 would probably not result in a different consequence than if the position were one where Rule 37 and Rule 38 did not apply. The only distinction is that Rule 38 does say that it shall not be invalid unless the chairman or the tribunal so directs. That merely indicates that there has to be a consideration by the tribunal of the issues as to validity, and the fact that Rule 38 is worded in that way will not have any material effect in practice. L

Accordingly, it follows that this appeal, and the appeal in the second case, will be allowed and the question which is posed by the tribunal will be answered by saying that the applicants' failure to give notice to the landlord is not fatal and the tribunal have a discretion to waive service on the landlord in an appropriate case.

The appeals were allowed with costs in the High Court. The costs of the previous hearing before the tribunal were reserved to the tribunal to decide.

COURT OF APPEAL M

February 1 1980

(Before Lord Justice STEPHENSON, Lord Justice WALLER and Dame Elizabeth LANE)

STONE v WHITCOMBE

Estates Gazette March 7 1981

(1980) 257 EG 929

Agricultural Holdings Act 1948 — Whether proviso to section 2(1) applied on the ground that there was an agreement "in contemplation of the use of the land only for grazing or mowing during some

specified period of the year" — Court of Appeal's decision (dismissing appeal from county court judge) constitutes a development of the law governing the interpretation of section 2(1) and the proviso — Difficulty caused by judge's finding that the agreements between the parties for the sale of grazing rights were from Lady Day to Lady Day in each of two successive years — Whether it was open to judge, having found that the agreement covered a full 12 months in each case, to hold that it was nevertheless in contemplation of the use of the land only for a specified period of the year — Held that the judge was justified on the evidence in holding that, although the expressed duration of the agreement was from Lady Day to Lady Day (namely a year certain), the licence to occupy was made in contemplation only of its use for grazing and mowing during the grazing season, April to October — Held also that the grazing season was a sufficiently "specified period of the year" (agreeing with the Scottish Court of Session decision in Mackenzie v Laird) — Decision of Court of Appeal likely to give rise to discussion

This was an appeal from a decision of Judge Best at Bridgwater County Court in an action between John Neville Stone, plaintiff, and David Roy Whitcombe, defendant, concerning a piece of land, of over 17 acres, at Catcott Moor in the Bridgwater Plain. The plaintiff claimed that, as a result of section 2(1) of the Agricultural Holdings Act 1948, he was entitled to a tenancy from year to year of the land; the defendant, the owner of the land, submitted that the proviso to section 2(1) applied and that accordingly the plaintiff had not acquired such a tenancy. The judge decided in favour of the defendant and the plaintiff appealed to the Court of Appeal.

James Wigmore (instructed by Reed & Reed, agents for Ash, Clifford & Co, of Bridgwater) appeared on behalf of the appellant; Hubert Dunn (instructed by Gregory Rowcliffe & Co, agents for Pardoe, David & Shaw, of Bridgwater) represented the respondent.

Giving the first judgment at the invitation of Stephenson LJ, WALLER LJ said: This is an appeal from a decision of Judge Best sitting in the Bridgwater County Court and delivered on October 5 1978. It arises out of a dispute over 17 acres of land at Catcott Moor in the Bridgwater Plain. The land was owned for many years by Mr Burgoyne, but in 1976 he sold it to the defendant Mr Whitcombe. In April 1974 the plaintiff enquired from Mr Burgoyne whether he could have the land. Mr Burgoyne said that he could either purchase it or rent it or have the grazing. An agreement was reached and as a result the plaintiff was able to put his stock on the land in the summer of 1974. The plaintiff paid £55 cash to Mr Burgoyne on June 6, at a time when Mr Burgoyne was ill in bed. In 1975, in circumstances to which I shall refer later, the arrangements made in 1974 were renewed and the plaintiff put his cattle on the land. In March 1976 Mr Burgoyne conveyed the land to the defendant and when the plaintiff put his cattle on the land the defendant turned them off. There is a dispute between the plaintiff and the defendant as to the terms on which the plaintiff had rights over the land. The issue is, was it a tenancy as the plaintiff alleges, in which case by section 2(1) of the Agricultural Holdings Act 1948 the tenancy would become a tenancy from year to year; or was it merely the selling of grazing and mowing as the defendant submits, in which case the proviso to section 2(1) of the Agricultural Holdings Act may take this land out of the provisions of section 2(1)?

The learned judge found for the defendant and the plaintiff appeals to this court on two main grounds: firstly, that the judge was in error in finding the facts in the way in which he did and should have preferred the plaintiff's evidence and found that there was a tenancy. Alternatively, if it is not possible to reverse the judge's findings of fact, the plaintiff submits that his conclusion that on the evidence the agreement was in contemplation of the use of the land only for a specified period of the year was wrong.

I deal first, briefly, with the submissions relating to the finding of fact. The original arrangements were made at a meeting in the 5-acre field, which is part of the land in question. The plaintiff's evidence was that he agreed to rent the land and not buy it, and that he was not interested in the grass keep. Mr Burgoyne's evidence was that he sold the grass for 12 months and the following year sold it again. The plaintiff's evidence was supported to some extent by his sons, who were present for part of the meeting at the 5-acre field, and also Mr Burgoyne was later paid £55 in cash and he signed a receipt. The receipt was in Mrs Stone's, the plaintiff's wife's, writing, and it described it as a receipt for rent. Mr Burgoyne had not read the terms of the receipt. He received the £55 in cash and signed the receipt without reading it. He had been ill in bed.

There was no direct support of the evidence of Mr Burgoyne, but there was other evidence to which the judge paid considerable attention. Before I come to this I should relate some of the other events which took place. The plaintiff used the land during the grazing season of 1974 and then, at a meeting sometime in early 1975, Mr Burgoyne said that he agreed to sell the grass keep again for 1975, and a cheque for £55 was paid on April 15 when Mr Burgoyne called on the plaintiff to ask for the payment.

It appears from the judge's judgment that he was disposed to prefer the evidence of Mr Burgoyne to that of the plaintiff, but because Mr Burgoyne was ill and in hospital (his evidence was taken in hospital; the judge going to the hospital) the judge felt reluctant to prefer his evidence in the absence of other evidence which tended to support it. It appears from the judge's judgment that there were three matters supporting Mr Burgoyne, one of which was decisive in his mind. That was this:

First of all, the evidence of Mr Chapman. Mr Chapman was an adjoining landowner and he gave evidence that in 1975 (he had originally given a different year) the plaintiff asked him if he could do the same as Mr Burgoyne and told Mr Chapman that he had bought the grass on the Burgoyne land and had the option of buying the land. He wanted to use the land for 12 months to see how it shaped up. The judge said about this:

> This is a vital piece of evidence. This ties in with what Mr Burgoyne said. I find it incredible that Mr Stone, who is a man who knows well the difference between grass keep and a tenancy, should make an error about this. If I was in any doubt about Mr Burgoyne's evidence, it is made firm by the account Mr Chapman gives about what he was told later.

("later" refers to a later date when the conversation took place). This evidence the judge regarded as confirming his view that Mr Burgoyne was right and that of the plaintiff was not to be accepted.

There were two other matters which the judge did not regard as important but should be mentioned, namely, firstly that the drainage board rate, which is levied annually on the occupier of the land, was paid in the first year by Mr Burgoyne, in the second year by Mr Whitcombe and never by the plaintiff. Secondly, the payment in advance of £55 was more consistent with the sale of grass keep than with a tenancy.

Mr Wigmore submitted that Mr Chapman's evidence was inherently improbable and that the judge was wrong to pay attention to it. In my view the attack on the judge's findings of fact is really almost unarguable. This case was one which depended very much on the impression that the witnesses made: the judge has made clear his reasoning in arriving at his conclusion and I do not think it is possible to change it in any way. I have dealt with this shortly because I entirely agree with the way in which the judge has set out the facts under his findings.

I come next to the plaintiff's alternative submission. Mr Wigmore was asked by my Lord if he would summarise on paper his main contention, and this is his contention:

> Whether upon the finding of the judge that there was an express agreement to grant to the plaintiff the right to occupy land for grazing or mowing for 12 months, it was open to the judge upon a true construction of the proviso to section 2(1) of the Agricultural Holdings Act 1948, and upon the evidence, to hold that the agreement was in contemplation of the use of the land only for a specified period of the year.

Then the proviso, which, of course, is of the essence in this case:

> Provided that this subsection shall not have effect in relation to an agreement for the letting of land, or the granting of a licence to occupy land, made (whether or not the agreement expressly so provides) in contemplation of the use of the land only for grazing or mowing during some specified period of the year, . . .

The critical finding on which this ground of appeal is based is contained in the learned judge's judgment where he says:

> I find that the agreement between the plaintiff and Mr Burgoyne was that Mr Burgoyne agreed to sell the plaintiff grazing rights to the grass from Lady Day to Lady Day, 1974 to 1975. By a fresh agreement he gave further grazing rights over the said land from March 25 1975 to March 25 1976.

Mr Wigmore's submission is that the judge, having made that finding, it was not possible to say the proviso applied.

I read also another part of the judge's judgment in which he enlarges upon that finding. He said this:

> I have come to the conclusion that the effect of the proviso is this: where the agreement or grant of licence is made in contemplation of using the land only for grazing or mowing during some specified part of the year, this applies only to the period of grazing or mowing and not to use ancillary to grazing or mowing.

The submission on behalf of the appellant is that this finding was not open to the learned judge because the period ran from Lady Day to Lady Day, and from March 25 to March 25 and that the second period was consecutive to the first. Mr Wigmore submitted that although the grazing only covered the period from March to October, this was not a specified period because it would vary from year to year, and furthermore there was a right to go on the land at other times for purposes ancillary to grazing: therefore he submitted that in fact it was a licence for 12 months, which would be converted into a tenancy from year to year because the proviso would not then apply.

It is necessary first to consider the evidence relating to the grazing season. The plaintiff said, when asked about the use of the land:

> Land would be used for grazing or mowing only from about April to November as impractical to use any other time. When took land realised would use it grazing, mowing only and during grazing season.

There is also evidence from both surveyors. Mr M J Garrett [FRICS] agreed that land could be used only for grazing during the fairly short season. He had said that payment for grass keep could be in April. Then Mr L B Sellick [FRICS] both in evidence-in-chief and in cross-examination, said: "Would normally sell grass April to October. No one could expect to stock land after November." The plaintiff's son, who was also called, said:

> I think in 1974 we only stocked it once for two to three weeks. Stock off in October to November.

There was no evidence to the contrary of those witnesses.

The judge amplified his view of the agreement in two other passages, one when he was resolving the factual dispute, he having said:

> The two accounts are contradictory, for a tenancy is a very different thing from an agreement to sell grass. I find that the agreement to sell grass or keep in this area was the right to graze cattle on the land during the limited season in which cattle are normally grazed from April to October.

He also said:

> The land could only be used during the customary period of the year, and that accordingly, comes within the proviso.

He went on:

> Mr Burgoyne's attitude towards Mr Stone is still one of friendship; one has to read into the agreements between the parties that it was in their contemplation that the land would be used only during the season of the year. They knew what the ordinary custom was.

Unless it can be said that "specified" in the section requires a firm date, the submission that the period was not specified is, in my opinion, not made out. The evidence was, in fact, all one way. The period was specified as the period of grazing or mowing (that was the judge's finding); the judge quoted the plaintiff and said: "... it is plain from the plaintiff's answers that this land can only be used during a limited period of the year (April to October or early November) for cattle." The judge found that the plaintiff and Mr Burgoyne were well aware of the difference and said: "... I am sure the plaintiff knew exactly what the difference was": and there is the passage which I have already read, to the same effect.

When considering the various arguments which have been put forward, it is necessary first to consider what is the nature of the licence which can be converted into a tenancy from year to year. Lord Diplock, in *Bahamas International Trust Co Ltd v Threadgold* [1974] 1 WLR 1514, said at p 1527 when dealing with this particular problem:

> To come within section 2(1) of the Agricultural Holdings Act 1948 the licence granted under an agreement must be a "licence to *occupy* land for use as agricultural land". To satisfy this requirement the right of occupation for agricultural purposes must be an exclusive right under which the grantee is entitled to prevent the grantor and any other person authorised by the grantor from making any use of the land, at any rate for agricultural purposes, during the period of the grant.

So one has to consider whether or not there was a licence giving an exclusive right to the land for the relevant period. Provisionally it seems that there was an exclusive right to go on to the land for the purpose of feeding cattle and mowing during the season which both parties recognise as being from April to October or November, but there is no evidence that I can see that there was any exclusive right apart from that period; I should, however, also refer to the judgment of Parker LJ in *Scene Estate Ltd v Amos* [1957] 2 QB 2095, at p 212, where he was considering the meaning of this section and proviso. He said this:

> The words are not altogether easy to construe, but this at any rate, I think, is clear: first, that "specified period of the year" is equivalent to "specified part of the year" — that, I think, is the effect of the decision of this court in *Reid v Dawson* [1955] 1QB 214; secondly, that the "contemplation" must be not only that the use of the land shall be only for grazing or mowing but that the period of use shall be only a period which is less than a year. It seems to me that, as a matter of grammar and ordinary reading, the word "only" must apply not only to the use but to the period; and indeed, if it were not so, it seems to me that the words "some specified period of the year" were quite unnecessary, because this is a proviso to a section which is only dealing with tenancies which are for part of a year. Thirdly, it seems to me that the "contemplation" must be mutual, both on the part of the landlord and of the tenant. Fourthly, that one judges the contemplation as at the beginning of the agreement, if there is but one agreement. . . .

The test is, taking both of those authorities: was it in contemplation that the plaintiff should have exclusive use of the land for the whole of the period from Lady Day to Lady Day, or was it only in contemplation that he should have exclusive use during the period when he could provide feed for his cattle or mowing.

In my judgment it is clear from the understanding of the two parties, although the words used included the period Lady Day to Lady Day, that they contemplated use only during the season when it was possible to graze cattle, or to mow. But there is a question — raised again by Mr Wigmore — whether "specified" means specified in terms and that it is not sufficient to take a period which everybody understands but which may vary from year to year because the season starts early or finishes late, as the case may be, but in *Mackenzie v Laird* 1959 SLT 268, the court had to consider whether, if there was seasonal letting of the land for grazing, that would be "during some specified period of the year". It is unnecessary for me to cite the words of the Lord Justice-Clerk or others of their Lordships in the judgments which they gave. Suffice it to say that they held that letting during the grazing season is letting for a specified period of the year and it would be in that particular case within the contemplation of the parties that use would be made of the land during the grazing season. If I had felt any difficulty in coming to the conclusion that the period could be specified without there being specific dates, that would remove any doubt that I had.

The plaintiff also submits that use for purposes ancillary to grazing during the non-grazing months would result in a continuous right to possession on the part of the plaintiff. The only evidence of such use was that in March 1975 the plaintiff did go on the land to rotovate. In my opinion, the specified period was for grazing and it is not possible to add to that other periods, for example, a period for rotovating.

In the case of *Avon County Council v Clothier* (1977) 242 EG 1048 the question arose as to whether certain buildings which were adjacent to the grazing were covered by the proviso, and it was held in this court that as they were ancillary to the grazing they were covered. I take it that it would follow from that, that if work were being done during the season — work which was not in itself grazing or mowing — that would be ancillary to the grazing or mowing: for instance, if rotovating or ditching were being done. Whether or not there was a right to do anything outside the period is, in my view, much more doubtful.

Mr Burgoyne was asked for the grass again in 1975. There would be no point in the plaintiff having the right to rotovate after, say, November 1974, unless he had already had the promise of the grazing and mowing in 1975; but even if there were a right to go on the land for some purpose ancillary to grazing or mowing, it would not, in my view, be a right to exclusive possession and, therefore, would not fulfil the test set out by Lord Diplock in *Bahamas International Trust Co Ltd v Threadgold* [1974] 1WLR 1514 at p. 1527, mentioned above.

For these reasons I have come to the conclusion that the judge was correct in the judgment which he gave. The agreement was for a sale of grazing rights. It was the grant of a licence made in contemplation of the use of the land only for grazing or mowing during some specified period of the year and, therefore, he was correct in finding for the defendants. I would dismiss this appeal.

Agreeing, DAME ELIZABETH LANE said: I would add only this on one submission made by Mr Wigmore; and that is, that this is a case where it was clearly shown and found what was in the contemplation of the parties, and it was not a mere case of expectation, or expectancy, which on authority is something less than is required to bring the agreement within the proviso of this subsection.

Also agreeing, STEPHENSON LJ said: The judge was, in my opinion, entitled to prefer the evidence of Mr Burgoyne, supported as it was by the evidence of Mr Chapman, and to reject the plaintiff's evidence that the agreements of 1974 and 1975 were for a tenancy. The learned judge balanced all the relevant pointers in both directions, with characteristic care and common sense, including the use of the word "rent". The significance of that word is even less when used in receipts of the plaintiff's own making, one of which the judge found, on the acceptable evidence of an independent witness, to be a deliberate misrepresentation of the nature of the receipted payment.

The judge's careful understatement that "the plaintiff was not as careful of other people's rights as he should have been" was justified by the evidence and led inevitably to the failure of the plaintiff's pleaded case. The judge was accordingly entitled to find, as he did, in a passage already read by Waller LJ: ". . . that the agreement between the plaintiff and Mr Burgoyne was that Mr Burgoyne agreed to sell the plaintiff grazing rights to the grass from Lady Day to Lady Day, 1974 to 1975. By a fresh agreement he gave further grazing rights over the said land from March 25 1975 to March 25 1976." So the plaintiff was granted a licence to occupy Mr Burgoyne's land for use as agricultural land, which section 2(1) of the Agricultural Holdings Act 1948 made to take effect as a tenancy from year to year unless the granting of the licence was made in contemplation of the use of the land only for grazing or mowing during some specified period of the year.

The judge found, in another passage which my Lord has read: "that the agreement to sell grass or keep in this area was the right to graze cattle on the land during the limited season in which cattle are normally grazed from April to October". I think that Mr Dunn, for the defendant, is right in regarding that as a finding that the grant was made, both in 1974 and in 1975, in contemplation of the use of the land only for grazing or mowing, and only for that purpose during some specified part of each of those years — to adopt the interpretation put upon the words of the proviso by Parker LJ in the passage from *Scene Estate Ltd v Amos* which my Lord has read.

Was the judge entitled to find that that was the use contemplated by the grant and the agreement?

The period of the year, or limited season of contemplated use, was not exactly defined, but it does not have to be. . . . "if the period of the year is one which is capable of reasonably clear ascertainment, the language of the proviso is satisfied", said Lord Patrick in *Mackenzie v Laird* 1959 SLT 268 at p 272, with which I respectfully agree. The learned judge found the period to be April to October. The plaintiff called it April to November and was supported by his son, Clive. The plaintiff's expert, Mr Garrett, called the season "fairly short". The defendant's expert plumped for April to October, which could be extended to November.

In my judgment the judge was entitled to find that the season was April to October, or November. Was he also entitled to find that the use for grazing and mowing during that period was all that was contemplated by the plaintiff and Mr Burgoyne, or was that finding inconsistent with the grant of grazing rights to the grass from Lady Day to Lady Day?

First, that use was the actual use which the plaintiff made of the land with the exception of the rotovation of the soil on the 5 acres in March 1975, and possibly the putting of artificials on the soil and chain harrowing, or rolling it at an unspecified date. Next, that was the only use to which the land could be put, according to the evidence of the plaintiff and both expert witnesses. Finally, and most important of all, that was the plaintiff's own view of what was contemplated, as is abundantly plain from the passage which has been read from his evidence: "The land would be used for grazing or mowing only from about April to November as impractical to use any other time. When took land realised would use it grazing, mowing only and during grazing season."

The difficulty in the way of accepting that evidence at its face value, and that its effect is to bring this agreement within the words of the proviso, lies in the expressed duration of the agreement as for a year certain from Lady Day to Lady Day. That, in the opinion of the defendant's expert, made it a surprising agreement. Why was that surprising feature admittedly expressed in the agreement unless the parties contemplated a permitted use for grazing or mowing, not for part of a year but for a whole year? My answer to that is that it was expressed because it was suggested by the plaintiff, with the intention of making sure that he would get some security — perhaps the security of an agricultural tenant — and it was accepted by Mr Burgoyne because he believed it would not have that effect. Does it have that effect? Mr Wigmore has powerfully submitted that it does.

After some vacillation of mind I conclude that it does not. I respectfully agree with what Denning LJ said in *Reid v Dawson* [1955] 1 QB 214 at p 217, that a simple letting of grazing for one whole year would be caught by the Act without escape through the proviso. So would a simple licence to occupy for grazing and mowing for a whole year. Hence the device of letting for 364 days or licensing grasskeep for eight months only. But the question here is whether the licence to occupy for a whole year was made in contemplation only of its use for a particular purpose for a period of less than a year. The language of the proviso, taking agreements out of the subsection, uses ordinary words, nowhere better interpreted, if I may respectfully say so, than by Denning LJ in *Scene Estate Ltd v Amos* [1957] 2 QB 205 at p 211. There, the learned Lord Justice said:

> In my opinion the object of the word "contemplation" in the proviso is to protect a landlord who has not expressly inserted a provision that it is for grazing only, or for mowing only, or that it is for a specified part of a year; but, nevertheless, both parties know that that is what is contemplated.

There the Lord Justice appears to have had in mind an agreement into which the contemplated use could be implied from its expression in a previous agreement, and he expresses the opinion that in the ordinary way it is from the terms of the agreement alone that the contemplation of the parties can be gathered. But it seems to me, on principle and authority, to be a question of fact, to be decided on the express terms of the agreement and on extrinsic evidence, whether a landowner and the party he licenses to graze or mow his land both contemplate only a specified part of the year and only for that grazing and mowing, even when the agreement is to run for a year certain. Was it contemplated that though there were to be grazing rights to the grass for 12 months, those rights of grazing and mowing were only to be exercised during what the plaintiff himself called "the grazing season"?

Parker LJ said in *Scene Estate Ltd v Amos, supra*, at p 213:

> Prima facie it seems to me that an agreement is not made in contemplation of something being done under it unless there is power and obligation to do that very thing.

A And it might be said that under this agreement the plaintiff had the power to graze and mow, and Mr Burgoyne was under an obligation to let him graze and mow, during the rest of the year outside the grazing and mowing season. Grazing rights included such rights ancillary or incidental to taking the grass for grazing stock, or for mowing and sale, by fencing, rotovating or fertilising, which would be exercisable throughout the year.

The decision of this court on the ancillary use of buildings during the period of contemplated use in *Avon County Council v Clothier*, to which Waller LJ has referred, is relied upon by Mr Wigmore. But Parker LJ's words were spoken in a case where the court was asked to assume that the parties contemplated an extension of the time
B provided for by the agreement, not a reduction of the time. They were prefaced by the phrase "prima facie" and ended with a reference to "that very thing", which was "use for grazing purposes only" and were not concerned with ancillary purposes. If Parker LJ's words impose a different interpretation from Denning LJ's on the language of the proviso, and could apply to this case, I would prefer to follow Denning LJ and ask: "Was there adequate evidence that the plaintiff and Mr Burgoyne contemplated physically using this land only for mowing and grazing and only for the seven or eight months when that was practical?"

If "grazing or mowing" includes incidental or ancillary activities, there was admittedly no specific evidence of ancillary activities
C being contemplated by either parties. Mr Burgoyne gave no evidence himself of what was in his contemplation; but the plaintiff did, and I have quoted the passage in which he says what that contemplation was. The learned judge was entitled to infer from his evidence and all the rest of the evidence that what was in contemplation was the use of this land for grazing and mowing during the short season. So I would answer that question, "yes". That may mean that references in the evidence to selling the grass for 12 months, and from Lady Day to Lady Day, were ineffective to confer any rights beyond that of occupation for use as agricultural land, perhaps so as to exclude anybody else from using it out of the grazing season for agricultural purposes: see what Lord Diplock
D said in the passage from *Bahamas International Trust Co Ltd v Threadgold* [1974] 1 WLR 1514 at p 1527 which Waller LJ has quoted.

What was contemplated when these agreements were made was, in my judgment, the exercise of grazing rights, or rights of grazing and mowing, or rights to the grass, within the grazing season only. The plaintiff's suggestion, adopted by Mr Burgoyne, may have given the plaintiff the right to stop others using or doing things on this land throughout the year, but it did not, in my judgment, give him the right to exercise his rights to the grass outside the grazing season or show that both he and Mr Burgoyne contemplated the exercise of those rights outside that season.

E I would therefore agree in dismissing this appeal.

The appeal was dismissed with costs.

QUEEN'S BENCH DIVISION

December 1 1980

(Before His Honour Judge Edgar FAY QC, sitting as a High Court judge)

LITTLEWOOD v ROLFE AND ANOTHER

Estates Gazette April 11 1981

(1980) 258 EG 168

F **Agriculture (Miscellaneous Provisions) Act 1976 — Application of section 21 in the case of an applicant who is not fully eligible because the condition specified in paragraph (b) of the definition of "eligible person" in section 18(2) of the Act is not fully satisfied — To be "treated" as eligible in such a case the condition in paragraph (b) must be satisfied "to a material extent" — Condition in paragraph (b) requires the applicant's only or principal source of livelihood for five years to have been derived from his or her agricultural work on the holding — Analysis of tribunal decisions which included interpretations of "material extent" ranging from "only just failing to qualify as eligible" to "substantial in terms of time and important in terms of value" — Obiter dictum by Lord Russell in Jackson v Hall that section 21 applied to a person "who does not quite fulfil"** G **the eligibility requirement explained — The tribunal in the present case considered that an applicant must "almost fulfil" the condition in paragraph (b) of the definition of eligibility — On this principle they rejected the applicant, part of whose livelihood had been derived from nursing — Held that the tribunal had taken too narrow a view — The judge adopted the formulation of the tribunal in Dagg v Lovett, "substantial in terms of time and important in terms of value" — Case remitted to tribunal with a direction to find that the applicant was a person to whom section 21 applied and to consider whether it would be fair and reasonable for her to be able to apply under section 20**

This was an appeal by case stated from the East Midland Agricul- H tural Land Tribunal by Mrs Edna Ada Littlewood. She challenged the dismissal by the tribunal of her application to be treated as an "eligible person" under section 21 of the Agriculture (Miscellaneous Provisions) Act 1976. The respondents were the landlords, Thomas Donald Rolfe and Rita Rolfe, and the holding which was the subject of the application consisted of five fields, amounting to 26.774 acres, at North Leverton with Habblesthorpe and South Leverton in the County of Nottingham.

H Richards (instructed by Hancock & Willis, agents for Hodgkinson & Tallents, of Newark) appeared on behalf of the applicant; C B Priday (instructed by Collyer-Bristow, agents for Hayes, Son & J Richmond, of Gainsborough, Lincs) represented the respondents.

Giving judgment, JUDGE EDGAR FAY said: This is an appeal by way of case stated from the Agricultural Land Tribunal for the East Midland Area. The appellant, Mrs Edna Ada Littlewood, had applied to that tribunal for a declaration that she should be treated as an eligible person to apply to succeed to the tenancy of an agricultural holding formerly held by her husband, who had died on June 13 1979.

The tribunal heard her application on April 2 1980 and dismissed it. The appeal requires this court to pronounce upon the interpretation of a section of the Agriculture Acts which has caused controversy and has led to differing decisions by different tribunals.

The matter arises in this way. Section 20 of the Agriculture K (Miscellaneous Provisions) Act 1976 confers upon certain relatives of a deceased tenant of an agricultural holding the right to apply for a declaration entitling him to a tenancy of the holding. If the applicant is what the Act calls an "eligible person" the tribunal has to decide whether he is, in their opinion, a suitable person to become a tenant of the holding. If they decide that he is, then, subject to certain qualifications irrelevant to this decision, they give a declaration entitling him to a tenancy. The applicant has thus to be both eligible and suitable.

Eligibility is defined in section 18(2) of the Act. It requires the satisfaction of three conditions. The first is relationship to the L deceased (the list of relatives includes the widow). The third is that the applicant is not the occupant of a commercial unit of agricultural land; this is not in question here and it is agreed that the first and third conditions are satisfied. The second condition is that:

> (b) in the seven years ending with the date of death his only or principal source of livelihood throughout a continuous period of not less than five years, or two or more discontinuous periods together amounting to not less than five years, derived from his agricultural work on the holding or on an agricultural unit of which the holding forms part;

This condition has two elements. One, that the applicant's only or principal source of livelihood has been derived from his work on the M holding; the other, that this derivation of livelihood has enured for five years.

In the present case this condition is not satisfied. There is a shortfall in the period and there may be a difficulty over the principal source of livelihood point. But Parliament has provided for cases where the second condition is not fully satisfied; this is by section 21 of the Act. Its somewhat convoluted provisions provide that on application under the section the tribunal has power to determine that the applicant is an eligible person. They must exercise this power in the applicant's favour if they are satisfied that he is a person to whom the section applies and also that in all the circumstances it would be fair and reasonable for the applicant to be able to apply under section 20 of the Act for a declaration entitling him to a tenancy of the holding. Subsection (1) enacts that the applicant is a person to whom the section applies if the first and third conditions of the definition of eligible person in section 18 are satisfied and if:

> (b) the condition specified in paragraph (b) of that definition, though not fully satisfied, is satisfied to a material extent.

The result of all this verbiage is that one who fails to pass the second condition may, nevertheless, be deemed to pass. But if he does, a further hurdle is placed in the way, namely, the need to show that it is fair and reasonable to make the order. Now this deemed eligibility depends upon whether the second condition is "satisfied to a material extent" and what I have to determine in this case is what those deceptively simple words mean.

Let me now go back to the facts on the findings. Attached to the tribunal's stated case is the tribunal's written decision and scheduled to that decision is the tribunal's written findings and reasons. In paragraph 1 of this schedule the tribunal sets out its findings of fact in 11 numbered subparagraphs, as follows:

> i. the applicant is a woman aged 40. Michael Jack Littlewood ("the deceased") was her husband and died on June 13 1979 aged 39;
> ii. Thomas Donald Rolfe and Rita Rolfe ("the landlords") granted the deceased a tenancy of five fields at North Leverton with Habblesthorpe and South Leverton in the County of Nottingham ("the holding"). The holding was 26.774 acres and the tenancy commenced on October 1 1975;
> iii. the deceased or the applicant

it should be "the deceased *and* the applicant"

> also owned a further almost 25 acres of land in the same area. That land was an agricultural unit of which the holding forms part. The holding and that agricultural unit are collectively called "the farm";
> iv. the deceased and the applicant together with two children they then had moved to the area from London in the summer of 1975. The deceased then worked full time on the farm until the latter part of 1978 when he obtained employment as a nurse. The applicant had worked as a pupil nurse at the Royal Free Hospital in London training to become a state enrolled nurse;
> v. throughout the period between October 1 1975 and June 13 1979, some 44½ months, the applicant did agricultural work on the farm;
> vi. between October 1 1975 and June 13 1977, some 20½ months, the applicant was working either full time or part time at Worksop Victoria and Kilton Hospital ("the hospital");
> vii. between October 1 1975 and October 31 1976 the applicant worked full time either as a pupil nurse or as a state enrolled nurse. She was earning between October 1 1975 and July 31 1976 at the gross annual rate of £1,953, and for a very short period between August 1 1976 and August 20 1976 at the gross annual rate of £2,052, and between August 20 1976 and October 31 1976 at the gross annual rate of £2,184. The applicant ceased being a pupil nurse in May 1976. The tribunal presume that was at the end of May, a period of eight months since October 1 1975;
> viii. between November 1 1976 and June 13 1977 the applicant worked part time (half the normal working hours) at the hospital earning at the gross annual rate of £1,092;
> ix. for exactly two years as it turned out the applicant worked on the farm without having earnings from any other employment;
> x. the deceased's drawings from the farm were as follows:
>
> | | |
> |---|---|
> | i. year ended March 31 1976 | £1,565.34 |
> | ii. year ended March 31 1977 | £1,918.46 |
> | iii. year ended March 31 1978 | £2,009.97 |
> | iv. year ended July 31 1979 | £1,608.85 |
>
> xi. the applicant is a courageous and industrious woman, who much impressed the tribunal as a witness. She had worked hard on the farm despite her other employment as a nurse, despite the birth of her third child, David, born on September 24 1977, and despite the untimely and sudden death of her husband, the deceased.

Pausing there; it is clear that the period during which the applicant derived her livelihood wholly or partly from the farm was 44½ months as against the 60 months required by the second condition of eligibility. Throughout that period she is found, as a fact, to have worked hard on the farm. She and her husband derived their livelihood partly from the farm and partly from nursing. The livelihood aspects may be divided into three periods. Firstly, from October 1 1975 to June 13 1977, when the husband is working full time on the farm and the wife is working part time on the farm. Secondly, from June 13 1977 to what is called "the latter part of 1978", when both were working full time on the farm, and, thirdly, from the latter part of 1978 down to June 13 1979, the date of the husband's death, when the wife was working full time and the husband part time on the farm. The joint livelihood of husband and wife was derived wholly from the farm during the second period. During the first period it would seem to have been derived almost equally from the farm and from the wife's nursing. We are not told what the husband earned when nursing for the last six or seven months of his life, but the proportions derived from the two pursuits are likely to have been similar to the period when it was the wife who was nursing.

The case stated makes, in paragraph 7, a further finding, namely, that "livelihood" might include other means of support such as the provision of food, clothing and shelter and that that other means of support was present only to a very small and limited extent. This I take to refer to the shelter of the farmhouse and to such foodstuffs as came off the farm.

Upon these facts the tribunal decided by a majority that the applicant's principal source of livelihood did not come from her work on the farm during any period in which she worked full or part time as a nurse. They went on to consider, under section 21, whether the second condition had been satisfied to a material extent and they said this, in paragraph 6 of their findings:

> 6. The decision of the tribunal by a majority was:
> i. That the specified condition could be satisfied under section 21(1)(b) of the Act only if the applicant almost fulfilled its requirements. The tribunal was referred to the speech of Lord Russell in *Jackson v Hall* [1980] 1 All ER 177 at 189, where he said this about section 21:
> "Section 21 in effect enables the tribunal to treat as an eligible person 'any survivor of the deceased' who does not quite fulfil the earlier requirements for eligibility related to the activities of the applicant during the seven years preceding the death of the tenant";
> ii. the test of satisfaction of the specified condition to a material extent under section 21(1)(b) of the Act must be applied by considering:
> a. the length of the continuous period during which the applicant's principal source of livelihood derived from her agricultural work on the farm;
> b. the proportion of the applicant's livelihood derived from her agricultural work on the farm and derived, on the other hand, from her work elsewhere during any period when her principal source of livelihood was *not* derived from her agricultural work on the farm;
> iii. that even if the applicant's principal source of livelihood derived from agricultural work on the farm throughout the whole period from October 1 1975 until June 13 1979, nevertheless, as that period was less than 75 per cent of the required period of five years the specified condition was not satisfied to a material extent;
> iv. that on the facts found the only period during which the applicant's principal source of livelihood derived from her agricultural work on the farm was the period of two years from June 13 1977 until June 13 1979.

The crucial part of this finding is the holding that "satisfied to a material extent" means "almost fulfilled". If that is its meaning, their decision must clearly be right. But, Mr Hesketh Richards, for the appellant, contends that a much wider meaning should be attached to the phrase. Both he and Mr Christopher Priday, for the respondents, helpfully put before me certain decisions of other agricultural land tribunals touching on the interpretation. These are, of course, not binding on me, but I find it helpful to see how other minds have approached this question.

Before I come to them, however, I would deal with the tribunal's reference to what Lord Russell of Killowen said in *Jackson v Hall*. That was a case which went to appeal on the meaning of the third of the three conditions of eligibility together with the question of the date at which the applicant had to be eligible. It was in no way concerned with the interpretation of section 21. Lord Russell, whose words on the subject are quoted in full in the passage I have read from the tribunal's findings, was referring to the section by way of what he described as a summary of the statutory provisions. Lord Fraser of Tullybelton did the same at p 185, where he summarised the second condition and added:

> . . . though that condition may be relaxed by the tribunal under section 21(1)(b) if it has been not fully satisfied but satisfied to a material extent.

Lord Russell's words can have no binding effect in these circumstances. The words themselves are not open to controversy if they

A bear a literal meaning. The tribunal has gone further and found in them an implied assertion that one who goes beyond a "not quite" test is outwith the section. Lord Russell did not say that, although it may be that this was in his mind on first impression and without argument upon the point.

Tribunals have, however, been wrestling with section 21 and hearing argument upon it for some time. I was referred to four decisions. In *Judge v Umpleby (Deceased)* in July 1978, the Yorkshire and Lancashire Area Tribunal had this to say in paragraph 22:

> . . . It was accepted by Mr Maguire that the provision in section 21(1)(b) of the Act that this condition need only be satisfied to a material extent applied
> B not only to the *period* during which the applicant's source of livelihood had derived from her work on the holding, but also to the *extent* to which her livelihood had so derived. In other words, he accepted that it would be sufficient for the applicant to show that throughout the period in question — or some material part thereof — a material source of her livelihood had derived from her agricultural work on the holding, although it fell short of being her only or principal source of livelihood. In making this concession, Mr Maguire said that he did so in the light of section 21(6) of the Act. In the view of the tribunal this concession was rightly and properly made, and represented a correct interpretation of the relevant law.

In paragraph 31:

> C 31. The tribunal did not consider it possible or realistic to make any precise mathematical calculation of the extent to which any of these other benefits could be said to have been attributable to the applicant's work on the holding. Accordingly it was in our view impossible to say that we were satisfied that for at least five out of the seven years ending with her father's death the applicant had derived her *principal* source of livelihood from her agricultural work on the holding. However, we were quite satisfied that for at least five of these years she had derived a *material* source of livelihood from her agricultural work on the holding or (to put it in a slightly different way) had derived her livelihood from such work to a material extent. In other words, but for the rewards in cash and kind that she received for her agricultural work on the holding, she would have been materially worse off throughout this period.

D

Next is *Dagg v Lovett*, in February 1979, where the Northern Area Tribunal said, in paragraph 29 of the decision:

> 29. This section

that is to say section 21(2)

> of the 1976 Act provides for applicants to be treated as eligible where they cannot show that they fully satisfy the requirements of section 18(2)(b) but can show that they satisfy these requirements to a "material" extent. What then is to be taken to be the interpretation of the word "material" in this context? Mr Price suggested that it was to be taken to mean "more than
> E minimal"; "there was", he said, "salt in seawater to a material extent in the ordinary meaning of the word". The tribunal did not take this view. It surely could not be the intention of the Act that applicants who clearly fail to pass the first hurdle should be admitted under this section. If it were, what would be the value of the test at 18(2)(b)? Its effect would be stultified. This tribunal take the view that the meaning to be given to the word "material" in this section is "substantial in terms of time and important in terms of value".

In *Colson v Midwood*, in June 1979, the Eastern Area Tribunal adopted a narrower construction. At paragraph 14(4) of their decision they said:

> "The condition though not fully satisfied is satisfied to a material extent"
> F (section 21(1)(b)). This provision is applicable both to cases where there is a failure to satisfy section 18(2)(b) as to proportion of livelihood coming from agricultural work and to the period. There have been conflicting decisions but in our opinion section 21(6) makes the matter clear. As to what is a material extent — either as to amount or period — we do not think that this means that the source of livelihood must be shown to be only just below 51 per cent or the period to be very nearly 60 months. But on the other hand we do not think it is enough for an applicant to show merely that a "substantial" proportion of his (or her) livelihood has come from agricultural work or, as the case may be, that such has been the position for a substantial part of the full five-year period. If this had been all that was required the section could, and indeed surely would, have been quite differently and much more simply worded. Again hard and fast lines cannot be drawn but in our opinion though an applicant need not show that he (or she) has come "within a hair's G
> breadth" of full compliance it must be shown on the facts that there has not been a really large failure.

Finally, in *Clark v Bernard*, in March 1980, the South Eastern Area Tribunal said:

> The personal aspects of hardship that may ensue from the decision were not, in the tribunal's view, relevant to this application and therefore attention was devoted to deciding whether the conditions specified in section 18(2)(b) of the Act were satisfied to a material extent. This was not easy. It was felt that it should not be done solely on the basis of any mathematical percentage or proportion. It was also felt that it would be unwise to do anything other than to consider the meaning of the word "material". That is H
> the word in the Act and if Parliament had wished to substitute some other word it would have done so. It is not proposed to enter into any lengthy discussion on what "material", in the context in which it is used, could otherwise mean. Certainly other meanings could be attempted. Definitions in the *Oxford Dictionary* include: "of much consequence" and "important". If this is the right approach, it has to be considered, on the facts of this case, whether Mr Clark has brought himself, to the tribunal's satisfaction, within the conditions of section 18(2)(b), *to a material extent*.

There has thus been a whole gamut of interpretation, from the strict "only just failing" view of the tribunal in the present case, to the wide interpretation accorded in *Dagg v Lovett*. This underlines the difficulty of the task of interpretation, and when I find the late J
Lord Dilhorne, at p 183 in *Jackson v Hall*, saying: "My Lords, it is to be regretted that this lengthy and no doubt expensive litigation has been brought about the inadequacy of the drafting of this Act", I feel disposed, respectfully, to echo his sentiments.

"Satisfied to a material extent". It is a strange phrase. "Extent" presents no difficulty; nor does "satisfied" in the context although it is a misuse of the word to apply it to a condition which, *ex hypothesi*, is not satisfied. The crux of the question is the word "material". It is an adjective frequently used by lawyers — they speak of material facts, of material considerations. In this sense it means relevant to the matter in hand, but it can hardly mean relevant here. It has another sense — of something substantial, more than minimal; that K
seems to have been the path followed by the Northern and Yorkshire and Lancashire Tribunals.

Mr Priday argues that because this litigation takes away a landlord's right to possession it ought to be narrowly construed. On the other hand the policy of Part II of the Act is clearly to create new rights in deserving persons with a view, I would think, to preserving that family continuity which has long been a feature of English farming life. It can equally well be argued that a wide construction is needed to give effect to the policy of the Act.

I do not find either of these approaches compelling. I do, however, find some assistance in subsection 2(b) of section 21, the L
paragraph setting up the further hurdle to the successor tenant of whether the order is fair and reasonable. I think this points to a wide interpretation of the "material extent". If the "not quite" view were right and the let-out consisted of what may be called "the hard-luck cases" (where the applicant has 59 months instead of 60; or 49 per cent in livelihood instead of the 51 per cent which would make it the principal source) then it is difficult to see how any tribunal could find that it was not fair and reasonable to make the order bearing in mind that the applicant still has the final hurdle of suitability to surmount. On the other hand, if jurisdiction is given over a wide spectrum of facts by a beneficial interpretation of the word "material" the tribunal will have a real task in deciding where M
lies fairness and reasonableness.

I therefore find that the tribunal erred in law in following the narrow interpretation. But I must go beyond this and endeavour to afford some guidance for those who have to solve these difficult problems. I agree wholeheartedly with the view that no mathematical formula can be laid down. Percentages of fulfilment, when worked out, are a useful guide to put the facts of finance or of time in perspective and to help judge their weight, but I would think it wrong to try to impose a mathematical cut-off point to what is material. After considerable casting about I do not feel that I can do better by way of definition than to adopt what the Northern Area

A Tribunal said in *Dagg v Lovett*, namely, that "material" means "substantial in terms of time and important in terms of value".

The case stated in the present appeal asks me to say whether an applicant can satisfy the requirements "to a material extent" if the period of the tenancy itself was less than 75 per cent of the period of five years. It was in fact 44½ months against the 45 months which would be 75 per cent. My answer to that question is: "Yes". I can envisage that 50 per cent could well, in some circumstances, be satisfaction to an extent that was material in the sense of substantial. Here a young couple took on a farm — no doubt hoping to make their permanent home there and bring their children up there. These are relevant facts when considering whether the
B period is deemed to be satisfied.

The tribunal pose a similar question as regards principal source of livelihood. I do not think one can approach the question of source of livelihood of husband and wife in two separate compartments. C
Husbands and wives usually pool their income. Here the husband and wife had two sources of income between them for part of the time and one source only, namely the farm, for another part. The figures in this case show that an important part of the joint income at all times was the farm and I entertain no doubt that, applying the interpretation I have to the word "material", this applicant has brought herself within section 21.

I accordingly remit the case to the tribunal with the direction that they find that the applicant is a person to whom section 21 applies and with the further direction that they do now proceed to consider the question whether in all the circumstances it would be fair and reasonable for the applicant to be able to apply under section 20 of D
the Act for a direction entitling her to a tenancy of the holding.

Judgment was given in favour of the applicant with costs.

COURT OF APPEAL

July 4 1980

(Before Lord DENNING MR, Lord Justice WALLER and Lord Justice DUNN)

PARKER v SECRETARY OF STATE FOR THE ENVIRONMENT AND ANOTHER

Estates Gazette February 21 1981

(1980) 257 EG 718

Application to quash compulsory purchase order — Properties in clearance area — Secretary of State's decision letter confirming compulsory purchase order challenged on two main grounds — Submission that inspector's report of inquiry and decision letter failed to come to a definite conclusion on a matter much canvassed at inquiry, namely, that on the basis of comparative costs the local authority's plans were considerably less economic than an alternative scheme — Further submission that, contrary to natural justice, the inspector took into account a factor with which the objectors had not had an opportunity to deal, namely, the effect of redevelopment in smaller, separate parcels in producing fewer dwellings and thus adversely affecting the authority's ability to deal with housing demands elsewhere in their area — Criticisms rejected by Court of Appeal — Lord Denning's warning that an inspector's report and a minister's decision letter should not be gone through like a statute, in search of semantic objections — As long as they were broadly correct, and no injustice had been done, the resulting order should not be upset by the courts — Appeal from decision of Willis J dismissed

This was an appeal by Mrs Joan Parker, of 5 Woldcarr Villas, Hull, from a decision of Willis J upholding the order of the Secretary of State for the Environment confirming a compulsory purchase order affecting a number of houses in Hull. The order was made by Kingston upon Hull City Council in pursuance of a redevelopment scheme affecting land designated as a clearance area.

D Keane and W Bowring (instructed by Breeze, Benton & Co) appeared on behalf of the appellant; D Latham (instructed by the Treasury Solicitor) represented the first respondent, the Secretary of State; B Ash (instructed by Sharpe, Pritchard & Co) represented the second respondent, the Kingston upon Hull City Council.

Giving judgment, LORD DENNING MR said: This case concerns an area of 20 acres of land in Kingston upon Hull. It has been designated a slum clearance area. As far back as 1977 the Hull Council made a compulsory purchase order for most of the area. Under the provisions of the Housing Act 1957 they decided to acquire compulsorily a large number of houses and land and redevelop the area. A number of the residents have objected to the order. The question is whether or not they are justified in so doing.

Before going into details, I will relate the history of this matter. Within the area of 20 acres there are 12.54 acres of land on which there are 430 houses which are either completely unfit for habitation and ought to be demolished or in such a condition that they could be rehabilitated only at very considerable expense. On the remaining 6.95 acres there are 105 properties which are not unfit: but they are on what is called "added land". That is land taken to be included in the redevelopment scheme: because — by so including that land — the area will be of a suitable size for comprehensive development. We have been shown a plan in which the clearance areas are in pink and the added lands are in grey.

When the compulsory purchase order was made the minister ordered that there should be a public inquiry. It was held by an experienced inspector appointed by the minister. He heard it over five days in September 1978. He made a long report of 71 pages, putting forward the observations of the council on the one side and the objectors on the other. Then, subject to some minor variations, he confirmed the whole of the compulsory purchase order. That was confirmed by the minister. Then the minister gave his decision letter on August 12 1979.

Under the provisions of the statute, within six weeks one or more of the residents could make application to the court to have the order quashed on the ground that the requirements of the statute had not been complied with. That is what happened in this case. Mrs Parker was one of the objectors. She occupies a house at 5 Woldcarr Villas. She — or her advisers — having studied the inspector's report and the decision letter complained that the statutory requirements had not been complied with. The judge found that the objections were not valid. He upheld the order. Now there is an appeal to this court.

The statutory requirements with regard to a slum clearance area are contained in the Housing Act 1957. Under section 42, where a local authority are satisfied as respects any area in their district that the houses are unfit for human habitation, and the most satisfactory method of dealing with them is the demolition of all the buildings in the area, they can declare it a clearance area. In addition, under section 43, as soon as they have declared it a clearance area, they can make an order for the demolition of the buildings in the area and can purchase the lands adjacent to it which are necessary for the purpose of securing a cleared area of convenient shape and dimensions. The powers to object — after an inquiry — are contained in the Fourth Schedule to that Act, whereby a person can come to the High Court within six weeks and complain of the order.

In passing I would refer to a point which was mentioned by Mr Latham and Mr Keane, namely that this inquiry is a statutory inquiry within the Tribunals and Inquiries Act 1971. Section 12(1) of that Act says of this and other inquiries:

> . . . it shall be the duty of the tribunal or Minister to furnish a statement, either written or oral, of the reasons for the decision if requested . . .

Implicitly in this case reasons were requested: and they were, in fact, given. Having received the reasons, the objector can bring them in as part of his argument, or in pursuance of it, so as to make good his case, if he can, under the Fourth Schedule to the Housing Act 1957. As we held in *Ashbridge Investments Ltd v Minister of Housing and Local Government* [1965] 1 WLR 1320 he can complain if the minister has taken into account something he ought not to have done, or failed to take into account something he ought to have done: or he has misdirected himself in law: or he has given reasons which, on the facts, cannot stand. That is the outline of the legal position in regard to this case.

Now I will turn to the facts. The inspector held a very careful inquiry. He went through the facts at length, and made his recommendations. But he failed to deal with three points which are raised by Mr Keane by way of objection to the order. I ought to say firstly that the operative order is contained in the decision letter of the minister. But, as he virtually incorporated the inspector's report into that letter, one is thrown back to that report to see whether that

A can be complained of in any way in pursuance of the principles which I have mentioned.

The first matter of complaint was about what has been called "the comparative cost method". That arises in this way. Before the inspector a great deal of argument was put forward on behalf of the objectors. They said that it would not be economically sensible to pull down 500 houses on this site: and then have to go to the expense of building new ones. That would cost £6,140,000. They said the best thing would be to demolish only half of them. New dwellings could be built in place of those 250 houses. But, as regards 200 others, what has been called "a housing action area" could be operated. They said that, if that were done, it would cost
B only £4,040,000. The residents' association, which was very well-advised, had gone into the matter carefully. They said, using the strategy which they suggested, the proposed redevelopment would cost £2,100,000 less than if the council's plan were adopted.

Mr Keane, on behalf of the objectors, said that that was very widely canvassed at the inquiry. But the inspector's report and the minister's letter—although they set out all the contentions—did not come to a definite conclusion upon it.

Having read through the inspector's report, it seems to me implicit that he did come to a conclusion upon it. What he virtually said was this: "Accepting that the residents' plan and strategy would be cheaper, nevertheless, viewing the matter on a much
C wider basis, the best course of action would be to pull down all the houses and build new dwellings in their place. New houses would last much longer: perhaps 50 or 60 years. Whereas, if half of the houses were renovated, they might last only for about 15 years". Although he did not spell it out in detail, it seems to me that the inspector's decision upon those contentions is virtually brought in in paragraph 364 of his report when he considered which would be the most satisfactory method generally. He said:

> Linking together the unfit conditions of the houses in the clearance areas, the general agreement that about half the properties in the area will have to be demolished no matter which course of action is selected to deal with the rest, and the considerations set out in paragraph 363 c to i above, it is
> D apparent that whereas the improvement to a limited standard of some of the dwellings is feasible there is no certainty that the scattered half which would remain would be improved to a standard which would provide good housing for many years to come . . .

I need not go further into it. It seems to me that he was taking into account the comparative cost. He was saying that it would be feasible to do what the residents suggested — and feasible to do it cheaper. But, taking into account the wider considerations, he said at paragraph 365:

> In general the most satisfactory method of dealing with the conditions which exist is by demolishing not only the conceptual half in respect of which there is no dispute, but also most of the other houses and with them most of the added lands.

E

So it seems to me that Mr Keane's first point is not good: and I would agree with Willis J in holding that that objections fails.

I need not go into subsidiary objections which were raised, but the other main objection in effect was this: The inspector took on himself an additional reason which the objectors had not had a chance of dealing with. They ought to have had a chance of dealing with it. As it was not afforded to them, there has been a want of natural justice in that regard. It is said that he took into account the effect on other areas in Hull — such as West Hull. The paragraph which gave rise to this objection is 363(f). It said:

> Further, the redevelopment of smaller, separate parcels of land almost
> F invariably results in fewer dwellings being built than if those areas had been amalgamated, and the joint site redeveloped. Although the council's ability to suitably rehouse the families displaced by clearance has been questioned, I accept their submission that they can do so in respect of this area; but building fewer dwellings here would be likely to affect adversely their ability to deal satisfactorily with demands arising out of any future clearance action in West Hull.

The judge said that he found some difficulty with this part of the case: but, on the whole, he thought that the questions which were being canvassed there were peripheral matters. He said that, if they had been further canvassed, they would not have affected his mind or that of the Secretary of State: and that no breach of the rules of natural justice arose.

It seems to me that the observations of the inspector were made G
using ordinary commonsense: and it was not necessary for him to invite further observations from the objectors. Therefore there was no breach of the rules of natural justice. In other words, it seems to me that the objections taken to the inspector's report were almost in the nature of semantic objections. They were objections as to the way one part or the other of it was phrased.

On this point, I would like to say that it is a mistake to go through an inspector's report, or a decision letter, as if one were going through a statute to see if one can find a hole in it. These inquiries are conducted by laymen — experience, good laymen — and they set out their reasons (in this case exceedingly well and in detail) for
coming to their decision. As long as it is broadly correct, and no H
injustice has been done, then the order should not be upset by the courts. Looking at the report of the inspector and the decision letter of the minister, it seems to me that no reasonable objection could be made to them. No statutory requirement has been omitted as far as I can see. The reasons are not such that they can be upset on any of the submitted grounds which I stated earlier on in this judgment.

So, on the whole, I find myself in agreement with Willis J and would dismiss the appeal.

Agreeing, WALLER LJ said: The inspector's report was a very full report. Mr Keane, on behalf of the appellant here, makes two main complaints. At the inquiry the objectors had sought to say that if
their plan were carried out it would cost £4m whereas the corpora- J
tion's plan would cost £6m. Mr Keane's complaint is that that issue was not dealt with by the inspector.

It appears from his report — which, as I say, was a very full one — that that consideration was certainly in his mind. He quoted in paragraph 39 the difficulty which the council had had on another occasion in persuading owners to obtain improvement grants and do the necessary improvements. There had been quoted to him two houses which had been improved out of 12 when an order of that kind was made; and the inspector quoted evidence from the local authority of the average cost of a full improvement grant standard to one of these houses set out in detail. He quoted the reply of the
council to the association's submissions at paragraph 78 in this way: K
"On the question of costs of improvement, there is no great difference between the submissions of the council and those of the WRA; because it is evident that the cost specified by the latter relate only to improvement to 'intermediate' standard rather than the full improvement standard required by the Council." Having quoted those, there is then paragraph 364, which my Lord has already quoted, but perhaps I could take the liberty of mentioning one particular sentence: ". . . it is apparent that whereas the improvement to a limited standard of some of the dwellings is feasible there is no certainty that the scattered half which would remain would be improved to a standard which would provide good housing for
many years to come." Then at paragraph 372 he again refers to the L
alternative way, showing that he has given full consideration to it. In my judgment, I agree both with Willis J and my Lord that it is quite clear that the inspector had those considerations fully in mind, and I do not think for myself that he can be faulted for not having said more about it.

The second main complaint which Mr Keane makes is the point about natural justice arising from two subparagraphs in the inspector's report, paragraph 363(e) and (f). I would just quote (e) because severe criticism of (e) was made by Mr Keane to support what Willis J described as "the cart before the horse." Subpara-
graph (e) says this: M

> A decision to exclude half the unfit properties from the order could result in the exclusion also of more than half the added lands, even if all the unfit properties were in one compact area.

Pausing there, the inspector is simply quoting the effect of the association's objection. They wish to have half the unfit properties excluded from the order, and he is stating, as it seems to me, pure commonsense. Then it goes on:

> . . . but where, as in the submission for the WRA, the excluded properties are in scattered smaller areas, the proportion and the extent of added lands which could be justified as reasonably necessary for the satisfactory development or use of the cleared areas would be that much less, with a corresponding loss in the number of housing units which could be built

A and then he quotes an example. Again, as it seems to me, that is a statement of simple commonsense. If you exclude one large area of houses, there will be in most cases an incidental result of having to exclude some added land (I use the phrase from the section of the Housing Act), but if you take an area — in this case something like 20 acres — and you, so to speak, take spots over that 20 acres, the effect on the whole plan is enormous, and the resultant number of houses which can be built will be greatly reduced. I am not sure how many individual areas the objectors have, but at any rate it is a considerable number with a number of islands in the area of 20 acres, and the effect would be much greater than the same amount of land in one spot. That is all that the inspector is saying; and he B quotes an example, which has been drawn to our attention, which shows in one small corner that there would be a very considerable effect in reducing the amount of added land and accordingly the amount of land which would be available for redevelopment.

One could state the proposition in other words. If you have an area of 10 acres with no obstructions in it, you would be able to build a certain number of houses. If you halved it directly down the middle, you would halve the number of houses which could be built in it. But, if you subtracted from that area of 10 acres 10 different portions of half an acre (that is the same as reducing it by half), the reduction in the number of houses would be very much greater. As it seems to me, that is all the inspector was saying in that paragraph. C So I do not find for myself that there was any injustice. Indeed I do not even share with Willis J any of the reservations which he expressed about it.

Then with regard to the next subparagraph (f), similarly my Lord has already quoted it, but I would just come to the last sentence:

> . . . but building fewer dwellings here would be likely to affect adversely their ability to deal satisfactorily with demands arising out of any future clearance action in West Hull.

Again that is a matter of commonsense. Mr Keane very fairly agreed that if the word "West" had been missed out he could not really complain about that sentence. Therefore it was perhaps unfortunate that the inspector happened to put it in, but he was D simply quoting an example that, if you build fewer houses, the ability to rehouse citizens from other parts of the town will be thereby reduced.

In my judgment, no complaint can properly be made about the inspector's report or about the minister's letter, and I agree that this appeal should be dismissed.

DUNN LJ also agreed.

The appeal was dismissed with one set of costs only, the order for costs not to be enforced without leave of the court.

E

COURT OF APPEAL

July 29 1980

(Before Lord Justice MEGAW, Lord Justice EVELEIGH and Lord Justice OLIVER)

ACKERMAN AND ANOTHER v SECRETARY OF STATE FOR THE ENVIRONMENT AND ANOTHER

Estates Gazette March 14 1981

(1980) 257 EG 1037

Compulsory purchase order — Application to quash order on ground of alleged breach of rules of natural justice — Negotiations between applicants and acquiring authority before inquiry with a F view to the compulsory purchase order being withdrawn in respect of applicants' property — Applicants believed until day when inquiry opened that they had satisfied conditions for withdrawal and that the inquiry would not take place — On the morning when the inquiry opened a solicitor employed by the authority telephoned one of the applicants, on the instructions of the inspector, to ask whether he proposed to attend the inquiry — There was a conflict of evidence as to the telephone conversation, but it was made clear to the applicants that the inquiry was taking place — The applicants did not attend, the inquiry proceeded, the inspector recommended confirmation of the order and this was done subsequently — The judge held that even if there was a reasonable basis for the applicants G believing up to the opening day that the inquiry would not take place, there was no such basis after the telephone conversation — The applicants had only themselves to blame for the case being heard in absentia — It was not reasonable for the applicants or one of them to fail to attend the inquiry if they wished to be heard — Held, upholding the decision of Willis J, that there had been no denial of natural justice and that the appeal must be dismissed

This was an appeal from a decision of Willis J rejecting an application by Jack Pesach Ackerman and Joseph Ackerman to quash a compulsory purchase order made by the second respondents, the London Borough of Camden and confirmed by the first respondent, the Secretary of State for the Environment. The order was in respect of premises at 72 Messina Avenue, London NW6. H

Matthew Horton (instructed by Joseph Ackerman) appeared on behalf of the appellants; Simon Brown (instructed by the Treasury Solicitor) represented the first respondent; J Sullivan (instructed by F Nickson, Solicitor to the London Borough of Camden) represented the second respondents.

Giving the first judgment at the invitation of Megaw LJ, EVELEIGH LJ said: This is an appeal from the judgment of Willis J when he refused the appellants' application to quash a compulsory purchase order made by the second respondents, the London Borough of Camden.

A compulsory order was made under section 32 of the Housing Act of 1969 in relation to premises, 72 Messina Avenue, London J NW6, owned by the two appellants, one of whom is a solicitor. There were negotiations between the council that made the order and the appellants, and finally the council indicated that they would be willing to cancel the compulsory purchase order if these appellants agreed to take as tenants three people nominated by the council after necessary work had been done to improve the premises. Reference was made in the course of these negotiations to a housing grant in respect of the alterations or improvements. The council intimated that they would be prepared to arrange for the postponement of a public inquiry into the question whether or not the compulsory purchase order should be confirmed by the Minister. K

The inquiry was ultimately held. The appellants did not attend. There was some conversation over the telephone between them and the solicitor for the second respondents who was asked to telephone the appellants by the inspector, and as a result of a report being made back to the inspector of the nature of that conversation on the telephone the inspector decided to carry on with the inquiry. It is said on behalf of the appellants that this the inspector ought not to have done, and his decision to proceed was in breach of the rules of natural justice.

It is necessary to look at the facts and the negotiations between the parties in rather greater detail. On October 27 1976, following the oral negotiations between the parties, the council wrote as L follows:

> Further to your telephone conversation of Thursday 21st October 1976 I would confirm that you wish to proceed with the conversion of the premises into four self-contained units of accommodation and further that you are now prepared to accept that the council rehouse three tenants and in return is entitled to nominate tenants to three of the units after completion of the works.

Then there came this passage:

> In the meantime, I have asked the Town Clerk to have their Public Inquiry adjourned for six months pending the completion of the necessary documents etc. As soon as the necessary paper work has been completed and M suitable proposals approved by the council, the C.P.O. can be formally withdrawn.

Then on March 1 1977 there came the appellants' reply to that letter: "Further to your recent letter, after consultations with our surveyors and legal representatives, we return the draft agreement with our slight amendment in blue which we trust will be acceptable in view of the extensive work to be carried out at the premises."

On March 3 1977 the appellants were notified that a date for the inquiry had been fixed, namely May 3 1977. On March 9 the second respondents, the London Borough of Camden, wrote acknowledging receipt of the amended agreement, which is called in the corres-

A pondence the nomination agreement, regarding the property. Then on April 5 the appellants wrote: "With reference to your letter of 9th ultimo, on several occasions we have spoken to you on the telephone and it was finally agreed last week that we accepted amendments in the agreement and that you would be within the course of the next few days from our conversation sending an agreement for signature." Then there came this sentence:

> We would ask you if you would submit this to us by return of post so that it may be fully tied up and that the pending compulsory purchase hearing may be postponed.

On April 6 the appellants returned the agreement signed, and said
B "Please confirm to us that the pending compulsory purchase hearing is now cancelled".

On April 14 there is an important letter written by the council to the appellants:

> With reference to your letter of the 6th April 1977 I acknowledge receipt of the nomination agreement which will be sent to the Town Clerk for stamping and signature in the immediate future. I shall be glad to receive your full grant application for the conversion of the premises into self-contained units when it will be possible to deal with the second point in your letter.

In other words, the council were saying "We will deal with the second point, that is to say the cancellation of the hearing, when we
C receive your full grant application".

The full grant application was not sent by the time May 3 had arrived, and the inspector appeared to conduct the inquiry, but there is a contemporaneous letter of that date, May 3 1977, written by the appellants to the house improvements manager, London Borough of Camden: "Further to our letter of the 27th ultimo, we now enclose the completed Form no 1, application for improvement grant, together also with a copy of the plans and a specification. The builders' cost will follow fairly shortly." Then there is a postscript: "Since dictating this letter, we have received the sealed copy of the nomination agreement dated April 26 1977", that is to say the finally agreed document bearing the council's signature.

D So that at that stage the nomination agreement had been accepted and formally signed. The grant had not yet been decided and application for the grant was not in the council's hands.

The facts that I shall now recite are taken from the affidavit of Mr Jack Ackerman, and in that affidavit he says in paragraph 8, at the end: "It was understood by my brother and myself that once this agreement" — that is to say, the nomination agreement — "was executed by both parties the said compulsory purchase order would be withdrawn." By paragraph 10 he says: "So far as my brother and I were concerned, upon the receipt at the end of April 1977 of Camden's part of the nomination agreement duly sealed, we assumed that the compulsory purchase order would be withdrawn,
E and that the May 3 1977 hearing would not be effective." That sentence in the affidavit leads me to think that Mr Ackerman is there asserting that he had received the council's completed and engrossed agreement at the end of April, and so substantially before May 3 was entitled to the view, which was in fact held, that the inquiry would not take place on May 3, which date had not yet arrived. It is difficult to reconcile that sentence with the postscript of the letter of May 3.

Be that as it may, the affidavit continues: "On May 3 1977 I received a telephone call at about 10 am at my office, from a person who did not identify himself but whom I took to be a clerk or other person in the employ of Camden. He asked me if it was my intention
F to attend the public inquiry which he told me was taking place that day. I expressed surprise at this, and pointed out that it had been my understanding that upon the nomination agreement being executed and exchanged the compulsory purchase order would be withdrawn. This had been the clear impression given to me both from the correspondence and from my various conversations from time to time with Mr Dawkins."

In paragraph 17 he says: "I dispute that I informed the representative of Camden who telephoned me that it was not our intention to appear but that we wished to sustain our objection. I have deposed to the terms of this telephone conversation in paragraph 12 hereinabove, and I repeat that I was left with the clear impression and understanding that the public inquiry would not be G proceeded with on that day."

I now turn to the affidavit of Mr Dawkins, who is the house improvements manager in the housing department of the council. Mr Dawkins refers to the letter of October 27 and to the particular sentence "I have asked the Town Clerk to have the Public Inquiry adjourned for six months pending the completion of the necessary documents, etc. As soon as the necessary paper work has been completed and suitable proposals approved by the council the CPO can be formally withdrawn", and he then deposes: "I therefore expected both the signing of a nomination agreement *and* a grant application to be made and approved in this six month period", and in his affidavit he emphasised the point that that second requirement was also expected. H

Mr Cunliffe, the senior assistant solicitor to the London Borough of Camden, in his affidavit in paragraph 5, having referred to the request of the inspector to telephone the appellant, says: "I am certain that I identified myself to Mr Ackerman, said that I was telephoning from the inquiry chamber, or possibly 'from the inquiry', and that I asked him if he intended to appear at the inquiry. He stated that he did not intend to appear. I asked him whether he wished to withdraw his objection to the order and he replied that he did not. I therefore told the inspector that the objector wished to sustain his objection. I am positive that I did not state that the inquiry would not proceed on that date although I do J not remember saying in terms that the inquiry would proceed in his absence."

Mr Cunliffe reported back to the inspector, and the inspector records that fact in paragraph 11 of his report in these words: "In the absence of the objectors the council's representative telephoned Mr J P Ackerman, and stated that the objectors did not intend to appear but that they wished to sustain their objection. I announced that I would take into account in my report the written submissions that had been made." The inspector in fact held the inquiry and recommended that the compulsory purchase order should be confirmed, and it was in July 1977.

Now as there was a conflict in these two affidavits in relation to K what took place on the telephone conversation, the learned judge allowed cross-examination of the two deponents in relation to that matter, and reading from his judgment he said this:

> As a result

— that is to say, as a result of the cross-examination —

> it is apparent that Mr Ackerman was very surprised and agitated to learn that the inquiry was taking place. There was reference both to the nomination agreements and to the grant applications, and to Mr Dawkins as the person who could confirm Mr Ackerman's understanding of the position. I am satisfied that Mr Cunliffe did not tell Mr Ackerman that the inquiry would not proceed on that day, nor did he tell him that it would proceed in L his absence if he did not appear. In his evidence Mr Cunliffe remained clear that Mr Ackerman did say he was not proposing to attend, although he did not say that the objections could be dealt with on the basis of the written representation. That was the way the inspector expressed himself on being told that the applicants were not attending, but did not wish to withdraw their objections. Mr Ackerman, while agreeing that he did not answer directly the questions put to him as to his attendance, was at a loss to understand how Mr Cunliffe could have got the impression that he was not coming to the inquiry but wished to sustain his objections. I have, therefore, to resolve this important matter of dispute as best I can and it is right that I should say that both counsel expressly disclaim any suggestion that either Mr Cunliffe or Mr Ackerman was not giving his honest recollection of what occurred during this telephone conversation. M

The learned judge went on to say later in his judgment: "Even if there was a genuine misunderstanding, there was no reasonable basis for it, viewed objectively, and therefore they failed to attend the inquiry without reasonable cause." That was the gist of the judge's finding in this case which he specifically made later, that there was no reasonable basis for the failure to attend the inquiry without reasonable cause.

The learned judge then said that he accepted Mr Cunliffe's evidence that Mr Ackerman was astonished that the inquiry was being held, and then went on to say that in a conversation which was largely one-sided Mr Ackerman spoke of the various matters relat-

A ing to the situation thus far. He gave Mr Cunliffe the impression that he believed that he had done all that was necessary to meet the council's condition for adjourning or cancelling the inquiry.

Later the learned judge says:

> I regret to say that I can find no reasonable basis for Mr Ackerman's non-attendance at the inquiry: it had originally been postponed for 6 months in order to help him and in accordance with council policy. He was aware that May 3 had been fixed by the ministry. He had received no intimation from either the ministry or the council that that date had been vacated, either temporarily or permanently. He had made no enquiry following the equivocal letter of April 14, in answer to his letter of April 6. If he did believe, up to May 3, that the inquiry would not take place following
> B the nomination agreement, and I find it unnecessary to make a finding as to that, such a belief had, in my view, no reasonable basis. But even if it had, after the telephone conversation on May 3 it seems to me impossible to find any reasonable basis for his non-attendance. I am afraid that I must conclude that the applicants have only themselves to blame for the case being heard *in absentia*.

So that the learned judge was saying that while Mr Ackerman contended that the letters were in such terms as to lead him to believe that all he had to do was to complete the nomination agreement and the compulsory purchase order would be cancelled and the inquiry cancelled or postponed, the learned judge did not regard that as a reasonable interpretation of the letters. But on the
C basis that Mr Ackerman so understood it, none the less, in view of the telephone call of May 3, it was not reasonable for Mr Ackerman to fail to attend the inquiry if he wished to be heard.

Now the procedure in these inquiries, so far as is relevant for the present case, is governed by the Compulsory Purchase by Local Authorities (Inquiry Procedure) Rules 1962, SI 1962 no 1424, and the particular rule that is relevant to the present argument is rule 7, subrule 6:

> If any statutory objector does not appear at the inquiry, the appointed person may at his discretion proceed with the inquiry, and, if he does so, shall (subject to disclosure thereof at the inquiry) take into account any previous written representations of such objector in so far as the same
> D appear to him to be proper and relevant to the matters in issue.

and it was no doubt by virtue of that rule that the inspector continued with the inquiry.

Counsel neither here nor below have been able to sustain, and indeed here have not presented, any argument to the effect that the inspector was in breach of specific rules; but it is said that compliance with the rules is not an end of the matter, and that the principles of natural justice must be observed. So far I would not quarrel with counsel's submission. He said that no breach was alleged, but the rule does not lay down how the inspector's decision to proceed should be arrived at, and there is, he submits, a general
E principle that the inspector must be satisfied that the party concerned is requesting him to continue in his absence; and in the course of developing the argument to sustain that principle Mr Horton went on to say that in the case of doubt (I am not sure that he was not saying as a routine anyway) the inspector himself should phone the party to satisfy himself that the party was requesting him to proceed in his absence. Alternatively, if he should not himself phone, he should have a formula to be used by someone on his behalf, so that the question could be specifically put to the party concerned, and thus the inspector could be assured that he was being requested to continue in his absence.

He referred to the case of *Priddle v Fisher & Sons* [1968] 1 WLR
F 1478, and he relied particularly upon a passage in the judgment of Lord Parker CJ, at p 1481, which reads as follows:

> The matter can be put in many ways, but the way in which it appeals to me is that a tribunal is acting wrongly in law if, knowing that an appellant has all along intended to attend and give evidence in support of his claim, and being satisfied, as they must have been, that he was unable for one reason or another to attend, they refuse to adjourn merely because he had not asked expressly for an adjournment. Before deciding to continue the tribunal should be satisfied that he was inviting them to continue in his absence.

That was an industrial tribunal case, a claim by an employee against an employer, but the principle would be applicable to other hearings also.

It seems to me that in considering the present case we are concerned to consider one question only, namely: Was the inspector G acting in a way that denied natural justice to the appellant? This case has centred upon that very narrow point. We have not been concerned to decide what would be the proper answer if an application had been made to reopen the inquiry. We are not concerned with a case where the court has received a request to postpone an inquiry where the very request would of itself indicate that the party was not willing for the hearing to proceed and where the very request may, in certain circumstances, require investigation into the facts in order for the president of the tribunal, whatever that tribunal might be, to conclude whether or not it would be right to continue with the inquiry. As I said, we are concerned with that H straightforward question as to whether or not the rules of natural justice had been broken by the inspector.

It seems to me, as the argument progressed, that Mr Horton was constrained to bring his argument to that very narrow point and say "The rules of natural justice in a case like this require the inspector to take more steps than the inspector here took". In the quotation from *Priddle v Fisher*, which I have just read, I would emphasise those words "knowing that an appellant has all along intended to attend and give evidence in support of his claim, and" — and here are the important words — "being satisfied, as they must have been, that he was unable for one reason or another to attend".

In the present case I think it is right to say that the inspector had J been under the impression that the appellants would attend, and it was for that reason that he asked the solicitor to make the telephone call. But after that call had been made and after there had been no positive approach by the appellants to anyone or in any way whatsoever to postpone the hearing, it seems to me that it cannot be said that thereafter the inspector was satisfied that they wanted to attend or were unable to attend for one reason or another. Certainly as to their ability to attend it is nowhere contended that they were unable to attend.

Reference was also made to the case of *Maxwell v Keun* [1928] 1 KB 645 and to various dicta in that case; but that case was dealing with a request for an adjournment and on that basis, if no other, is K to be distinguished from the present case.

So, I approach this problem by asking myself: Did the inspector do anything he ought not to have done, or did he omit to do anything which he ought to have done in pursuance of the basic principle which governs this case, namely affording the party the right to be heard? The court cannot compel a party to be heard, or is under no obligation to. In the circumstances of this case the tribunal certainly had no positive duty to persuade the party to be heard; its duty was to afford an opportunity to be heard. It seems to me that by asking the solicitor in effect to ensure that Mr Ackerman knew that the inquiry was being held that day, and he, the inspector, was there to hold it, was granting the opportunity to Mr Ackerman. If L there had been a request for an adjournment, or something like that, that had been communicated to the inspector and he had refused, it may be in all the circumstances of the case — and I do not say it must be — that there would be ground for complaint. But that is not the position in this case. The position here was that Mr Ackerman, not knowing precisely who was making the call to him, and certainly from his affidavit not assuming that this was a person in high authority, did not take any further step to see that the inquiry was postponed, if postponed it had to be for some inability on the part of Mr Ackerman to attend, or to approach the person on the council with whom he had been dealing to see that steps were taken by the council to cancel the order. I cannot, for myself, see in M this case that the learned judge's conclusion can be faulted when he says that "I am afraid that I must conclude that the applicants have only themselves to blame for the case being heard *in absentia*".

That being the position, I see no breach of the rules of natural justice here. I agree with the conclusion at which the learned judge arrived, and I would dismiss this appeal in relation to that ground.

There is one further ground of appeal which ought to be dealt with specifically, and that has been put before the court by a notice of further grounds of appeal. There it is said the learned judge erred in law in holding that there was a heavy burden on the appellants to establish a breach of the rules of natural justice. That ground results

A from the following passage in the judgment. Having said that no breach of the rules could be established, the learned judge said this: "No doubt an applicant who cannot establish a breach of the procedural rules, which are designed to ensure the fairness of local inquiry proceedings, faces a heavy burden when claiming a denial of natural justice in relation to the procedure at an inquiry governed by such rules." He referred to the *Lake District Special Planning Board v Secretary of State for the Environment* (1975) 236 EG 417. But I do not understand the learned judge to be approaching this case upon the basis that there is some special test that he is by law compelled to apply. He went on to say: "But notwithstanding that there was no breach of the rules in this case, Mr Sales submits that
B Mr Ackerman's absence from the inquiry is capable of a reasonable explanation and he should be allowed the chance therefore of putting his case to the minister for decision after hearing both sides: *Grimshaw v Dunbar* [1953] 1 QB 408". The learned judge then considered all of the arguments in relation to the natural justice point.

I regard the learned judge in the passage just referred to as stating what is in truth the common sense of the matter. Where it is necessary to prove that a person is unfair, he has in the very nature of things a somewhat difficult task when elaborate rules have been laid down and followed to ensure fairness. It just is a matter of fact that it is, and must be, in such a circumstance, a difficult task. It does
C not mean, and I do not read the learned judge as saying so, that there is some special kind of burden of proof attaching to the appellants in this case.

I, therefore, would reject that ground of appeal also and would dismiss this appeal.

Agreeing, OLIVER LJ said: I am prepared to accept, for my part, that up to the time of the telephone call on May 3 1977 the applicants believed, whether reasonably or not, that the inquiry was to be postponed or cancelled; but that picture changed entirely, in my judgment, on May 3 when the applicants were told of the inspector's presence at the town hall for the very purpose of holding
D the inquiry. Mr Ackerman's own evidence was that this information was retailed to him by an informant of whose identity and status he was unaware, and I am bound to say that I share in full measure the difficulty which the learned judge felt about this. I do not think that the rules of natural justice require an assumption that people of full age and understanding are going to behave otherwise than in a reasonable manner, and in the last analysis, in my judgment, the appellants' absence from the inquiry was due not to any non-observance by the inspector of the rules of natural justice but to the failure of the applicants to behave as reasonable people once they had been told that the inspector had arrived to hold the inquiry.

The learned judge in his judgment put it in this way:

E Mr Ackerman's surprise and agitation was, in the circumstances, understandable: but would not a reasonable man in his position, realising that his understanding was evidently not shared by the department or the council and that the inquiry was under way, have taken positive steps to protect his position? Would he not have asked to speak to the inspector if possible, or asked for an adjournment, or go at once to the town hall and explain the situation in relation to himself and his expert, and do his best to ensure that he was given a hearing then or on some other day? He did none of those things. He did not know who was speaking on the telephone, or with what authority he spoke. And yet he says, having asked his unknown interlocutor to speak to Mr Dawkins and sort things out, he assumed that that would be an end of the matter: in other words, that the council would drop the order, the inspector would pack up and go home, and he need not therefore attend.

F The learned judge stated his conclusion in the words which my Lord Eveleigh LJ has already read: "I am afraid that I must conclude that the applicants have only themselves to blame for the case being heard *in absentia*."

With every word of that I agree, and I therefore agree that this appeal must be dismissed.

Also agreeing MEGAW LJ said: In a recent decision of the House of Lords, *Bushell v Secretary of State for the Environment* [1980] 3 WLR 22, Lord Diplock, in the course of his speech dealing with the duties of an inspector in an inquiry held under the orders of the Secretary of State for the Environment in respect of schemes for
motorways, said this in relation to the discretion of the minister or G
the inspector appointed by him, at p 27 letter G:

In exercising that discretion, as in exercising any other administrative function, they owe a constitutional duty to perform it fairly and honestly and to the best of their ability, as Lord Greene MR pointed out in his neglected but luminous analysis of the quasi-judicial and administrative functions of a minister as confirming authority of a compulsory purchase order made by a local authority which is to be found in *B Johnson & Co (Builders) Ltd v Minister of Health* [1947] 2 All ER 395, at pp 399 to 400.

I break off from the citation of Lord Diplock's speech there to say that one of the matters stressed by Lord Greene in that judgment was that in matters of this sort there is a public interest which is an
interest that has to be taken into account independent of the inter- H
est of those who might be described as being the parties to a particular list or law suit.

Lord Diplock goes on:

That judgment contains a salutary warning against applying to procedures involved in the making of administrative decisions concepts that are appropriate to the conduct of ordinary civil litigation between private parties. So rather than use such phrases as "natural justice" which may suggest that the prototype is only to be found in procedures followed by English courts of law, I prefer to put it that in the absence of any rules made under the Tribunals and Inquiries Act 1971, the only requirement . . . as to the procedure to be followed at a local inquiry . . . is that it must be fair to all
those who have an interest in the decision that will follow it. J

I am quite unable to see, despite Mr Horton's interesting and ingenious submissions, that it can possibly be said that the inspector in the present case failed in any way in a duty of fairness in what he did in relation to this inquiry and the absence of the applicant when the inspector came to open the inquiry.

There is another possible aspect of this appeal, even though any charge of unfairness against the inspector is wholly misconceived, as in my judgment it is. It might still be possible to submit that this court would have jurisdiction to interfere and should interfere if the true view were that the absence of the applicant from the inspector's inquiry was caused by some fault, some negligent act or K
omission, on the part of a representative of the local authority. We do not, as I see it, need to consider further what the position would be in those circumstances. It may well be that there would be difficulties even so in the way of the applicant; but it does not fall to be considered in this case because it is clear from the conclusions which the learned judge reached on the facts — and his conclusions on facts included his consideration of oral evidence which he had heard and which we have not seen — that it is impossible to say that the absence of the applicant from the inquiry was due to any fault on the part of the representative of the local authority. The most that can be said in favour of the applicant on the facts as found by the judge is that the representative of the local authority who tele- L
phoned and spoke to Mr Ackerman may have misunderstood what Mr Ackerman intended to say. But on the judge's findings it is clear that that was not due to any fault on the part of the representative of the local authority. If the position were that Mr Ackerman intended to make it clear that he thought that the local inquiry was not going to proceed, he failed to do so. As his failure to do so was a failure on his own part and was not due to any mistake or misunderstanding for which the other party to the conversation was responsible, this cannot be a case in which that misunderstanding by him could entitle this court to interfere. I agree that the appeal must be dismissed.

The appeal was dismissed with costs. M

A **COURT OF APPEAL**

November 25 1980

(Before Lord DENNING MR, Lord Justice ORMROD and Lord Justice O'CONNOR)

COHEN v LONDON BOROUGH OF HARINGEY

Estates Gazette April 11 1981

(1980) 258 EG 165

Compulsory purchase — Question as to validity of notice to treat and notice of entry — Acquiring authority served two invalid notices to treat (invalid because served on previous owners who had
B sold the subject property) — Third notice to treat served on the correct person, the present appellant, but after authority had already taken possession of the land — Both notices, although served on June 24 1975, were back-dated to April 14 1975, the notice of entry being expressed to expire at midnight on April 27, the council having taken possession on April 28 — Appellant claimed that the notices were invalid, that the council had occupied the premises unlawfully and were trespassers — Held that the notice to treat, despite being back-dated, was valid at date of service and that the notice of entry was also valid, although served when the authority were already in possession — Notice of entry operated to legalise entry 14 days after service, mesne profits being payable by the
C authority for the period before the entry became legalised — Appellant's claim that he was entitled to have the land returned to him (together with two houses which the authority had built upon it) rejected — Appeal dismissed, but on grounds different from those in the judge's decision

This was an appeal by the plaintiff, Stephen Cohen, from a decision of Judge Stabb QC, sitting as a High Court judge in the Queen's Bench Division, in an action by the plaintiff against the London Borough of Haringey, the present respondents. The plaintiff claimed that, as a result of the alleged invalidity of a notice to treat and a notice of entry, the authority had entered on his land
D unlawfully. The judge held that the notices were bad, but decided against the plaintiff on the ground of equitable estoppel. The property which was the subject of the proceedings was a house at 29 Antill Road, Tottenham.

Philip Goodenay (instructed by Nicholas & Co) appeared on behalf of the appellant; Michael Rich QC and C J Lockhart-Mummery (instructed by T F Neville, chief solicitor, Haringey Borough) represented the respondents.

Giving judgment, Lord Denning MR said: In compulsory purchase the notice to treat is vital. *Halsbury*, vol 8, 4th ed, paragraph 136 says:

> The effect of serving a notice to treat is to establish a relation analogous in
> E some respects to that of vendor and purchaser, a relation which binds the owner of the land to give up the land subject to his being paid compensation, and which binds the undertakers or acquiring authority to take the land, but there is no contract of sale until the price is ascertained and the land remains the property of the landowner. Both parties have the right to have the price ascertained and the purchase completed in manner provided by the Lands Clauses Acts or the Compulsory Purchase Act 1965 or any Acts modifying those Acts. The rights and obligations created by the service of the notice to treat are legal as distinct from equitable, and they bind all persons claiming under the owner, whether with notice of the service or not.

If a notice to treat is valid, it binds the owner. If it is invalid, or it is not served on the proper persons, it does not bind the owner. The
F acquiring authority cannot proceed with the acquisition.

In the present case the acquiring authority, the London Borough of Haringey, served two notices to treat in succession. Each was bad because it was not served on the right persons. Then it served a third notice. This was served on the right person. But it was back-dated. The question is: Does that make it invalid? It was served at a time when the acquiring authority was already in possession of the land. Does that make the notice to treat invalid?

In 1972 there were slums in the Borough of Haringey. Many houses were unfit for human habitation. The borough council decided to clear the area under the slum clearance provisions. They made a list of the houses. They gave the preliminary notices to the owners — as at that time in 1972. In particular, they gave notice to G
the owners of 29 Antill Road, Tottenham. The owners at the time — in 1972 — were a Mr and Mrs Berkovitz. The house was occupied by a widow Mrs Thomas. She was a protected tenant, who paid £3.50 a week rent.

Two years passed. The compulsory purchase order was confirmed on October 23 1974. But meanwhile, since 1972, there had been dealings with the house. In September 1973 Mr and Mrs Berkovitz sold the house to a Mr Alexandros Papaioannou. In January 1974 Mr Papaioannou sold it to a Mr Stephen Cohen. Those transfers were notified to the rating department of the council. They demanded rates from those new owners: and they were duly paid. Also the agents for Mr Cohen notified the valuation H
department of the council that the property had been transferred to him. So the council knew perfectly well, from various notices which had been given, that the owner now was Mr Cohen.

But apparently other departments of the council (especially those concerned with notices to treat) did not realise this. They did not know of the change of ownership. They went by the 1972 list. In 1974, after the compulsory purchase order was made, the council's solicitor first served a notice to treat on Mr and Mrs Berkovitz. That was on December 9 1974. Mr and Mrs Berkovitz sent it on to their solicitors, who wrote to the council's solicitor, saying, "Our clients have sold the property to Mr Alexandros Papaioannou". So the council's solicitor served a new notice to treat on Mr Papaioan- J
nou — thinking he was the owner. But he was not. Mr Papaioannou did not reply. So the council's solicitor did not get to know that Mr Papaioannou had sold the house to Mr Cohen. The council's solicitor thought the notices he had served on Mr Papaioannou were all right. The council determined to enter into possession. They gave notice to Mrs Thomas that she must go. They were ready to rehouse her. She left on April 26 1975.

Mrs Thomas knew that her landlord was Mr Cohen. She told him or his agents that she was going. Then, for the first time, Mr Cohen got to know that the procedure for compulsory purchase was going forward. But he had never been served with a notice to treat or a notice of entry. So his solicitors took the point up with the council K
solicitor. They said that they were astonished and amazed that the council had gone into possession of Mr Cohen's house in this way. They found out in the first week of May 1975 that the premises had been boarded up by the Borough of Haringey. Mr Cohen could not get in to do repairs. So Mr Cohen's solicitors took the point that he had not been served with any notices at all.

When the council looked into the matter, they realised they had not served the proper notices. They tried to put it right. On June 24 1975 the council tried to put everything in order. They served notice to treat and notice of entry so that they could operate thereafter the compulsory purchase procedure. I will read the important parts of the letter of June 24 1975. The borough secretary L
and solicitor wrote to Mr Cohen, saying:

> . . . As your solicitors had affirmed that you are the proper owner of the property, it appears that the position should be regularised accordingly. I therefore enclose herewith a copy of the above compulsory purchase order as confirmed . . . together with a notice to treat in respect of the above property.

I will stop there. The notice to treat, which was enclosed, was a perfectly proper notice to treat to Mr Cohen saying, "the council are willing to treat with you for the purchase of your estate and interest in the said lands . . ." At the bottom the date was given "April 14 1975". That was wrong. It was back-dated. The true date was June 24 1975. Does that make it bad? I think not. The body of M
the notice is what matters. The operative part required Mr Cohen within 21 days *of the service thereof* to give the particulars of his property.

The letter continues:

> I also enclose three forms of claim, two copies of which should be completed and signed and returned to me within 21 days. I also enclose a formal notice of entry. . . .

The notice of entry was enclosed, saying that the council, in exercise of the power conferred on them by section 11 of the Compulsory Purchase Act 1965, were giving notice that upon the expiration of 14 days from the *service* of the notice, they were going to enter the premises. Now they had already entered the premises. Does that make it bad? Again I think not. That notice also was back-dated April 14. That does not matter either. In addition to the notice of entry, three forms of claim were enclosed. Mr Cohen was asked to set out the particulars of the property, the figure at which he valued it, and so forth. The letter goes on:

> Please note that I have back-dated both notices, so that in respect of the notice of entry, it will be deemed to have expired at midnight on April 27, the council having taken possession for the purposes of rehousing the tenant and subsequent demolition on April 28 last. Thus, from the date of such rehousing, the council will be responsible for the whole of the property actually taken and will be liable to pay interest at the appropriate rate on the compensation finally agreed from the date of entry up to the date of actual completion.

In the circumstances, I look upon that part of the letter as an offer: whereby — if Mr Cohen so wished — they back-dated everything to April (the date on which they took possession), and from that date he would get interest on the purchase money. That was a very important letter. Mr Cohen's solicitors did not reply to it until September 8. They did not take any objection to it at that stage. They said:

> Our client is prepared to dispose of his interest without prejudice in the sum of £8,000.00.

The borough secretary and solicitor wrote back on September 12 saying:

> It will facilitate negotiations if full particulars of the nature of your client's interest in the property and certain other particulars are submitted in the form of claim.

Then comes the important letter of September 18 1975 from Mr Cohen's solicitors, enclosing the completed form of claim but insisting that the notice to treat had been served not in April but on June 24. They said:

> We now enclose herewith claim form in answer to the notice to treat which incidentally was only served on our client on or about June 24 1975 when you have already taken possession.

The form of claim, which had been filled in, also stated distinctly that the notice to treat had been served on or about June 24. They gave all the particulars of the property, and put its value at £8,000.

It seems to me that Mr Cohen's solicitors were acting very properly. They treated the notice to treat and the notice of entry as valid but insisted that they should be regarded as served on June 24 — not on April 14. In short they ignored the back-dating. They treated it as a perfectly good notice to treat served on June 24. So from 14 days after the service — that is, July 9 — the possession would be lawful.

Unfortunately, thereafter there was disagreement. The parties could not agree upon the price. Then, in February 1976, the solicitors for Mr Cohen turned round and said that the council had occupied the premises unlawfully and were guilty of a trespass. In March 1976 Mr Cohen's solicitors issued a writ against the council, saying that they had been guilty of trespass.

Meanwhile the council have demolished the house and have built two new houses on the land. If Mr Cohen is right — and the property is still his — it means that, not only have the council been trespassing on the land, but that he is entitled to have the land returned to him, presumably with the houses upon it.

That is the issue in the case. The judge found the notices to be bad. He said:

> They (the council) sought, however, to regularise the position as from April 28 by back-dating the notices. I do not think that notices in that form could be said to comply with the Act because a notice to treat must be served before entry into possession, and it was not within the defendants' power to serve a notice to treat while in possession.

So the judge held that the notices were invalid. But he went on to hold, on the ground of equitable estoppel, that Mr Cohen had so acted that he was estopped from taking advantage of the bad notices.

I take a different view from the judge. In my judgment, the notice to treat was perfectly valid as at the date it was *served*. The date it bore at the bottom does not matter in the least. The council, on the date of *service*, demanded particulars of the estate and interest. They said they were willing to treat with him for the purchase of the estate, and gave him 21 days after the *service* of the notice. The service was on June 24. It seems to me that, although the date was April 14, nevertheless it was an effective notice to treat as at the date of its service on June 24.

Then I turn to the notice of entry. Again it seems to me that that was perfectly valid. It is true that the council were already in possession on June 24. But the effect of section 11 of the 1965 Act is that, after the 14 days have expired, the possession thereafter is perfectly lawful. It would be absurd if the council had to go through the form of going out of possession for a moment on June 24 and then immediately afterwards serving a fresh notice of entry. It seems to me that a notice of entry can be served while the acquiring authority is in possession. It may be — as in this case — that the possession at that time was unlawful. Nevertheless, on the notice of entry being given, it became lawful 14 days after service. In short, section 11 protects the acquiring authority against any claim for trespass after the 14 days have expired.

So it seems to me that the notice of entry was valid, and operated to legalise this entry as from July 9. Mesne profits for trespass are payable up to that date — but not thereafter. That is, in effect, the order which the judge made. It seems to me, therefore, on that ground his judgment should be upheld.

It is unnecessary in these circumstances to go into the question of estoppel. If it had been necessary, I should have been disposed to agree with what the judge said, namely, having regard to the dealings which had taken place between the parties, it would be inequitable to allow Mr Cohen now to tear the whole thing up and claim the land and the properties which have been built on it. He has treated the notices as good. The council have acted on them. Having regard to the principles we stated in *Crabb v Arun District Council* [1976] Ch 179, equity would restore the position to what it would have been if valid notices had been served on June 24. But it is quite unnecessary to go into all that because in my opinion the notice to treat was good and the notice of entry was good as at June 24.

I would dismiss the appeal accordingly.

Agreeing Ormrod LJ said: I think it is important in this case to set the facts in the correct key. This is not a case of a local authority endeavouring to acquire property by manipulating its compulsory purchase powers at a price much below its true value and to take advantage of the owner of the property. On the contrary, it is precisely the opposite. The key to this case is that it was the owner of the property, the plaintiff, who has been trying to take advantage since 1975 of a mistake made by the local authority in their paperwork in connection with the acquisition of this property; a mistake for which they are of course responsible; a mistake which they should not have made, but in the circumstances it is not very difficult to understand how it came to be made.

For that purpose one has to look back a little and try to keep in mind the fundamental fact of this case, which is that there was a compulsory purchase order in force covering this property, and many other properties, at all material times in this case. The council had made the original compulsory purchase order on November 27 1972; and it had been confirmed by the Secretary of State on October 23 1974. Meanwhile this property, and I dare say other properties subject to the compulsory purchase order, were changing hands in the most surprising fashion, as they were sold and bought and sold and bought. It is a remarkable fact that Mr and Mrs Berkovitz bought this property, already plainly subject to the compulsory purchase order, on April 16 1973. They got rid of it as quickly as they could and sold it to Mr Papaioannou on September 14 1973 for the same price as they paid for it; so they got out without burning their fingers. Mr Papaioannou sold it to Mr Cohen, the plaintiff in this case, in September 1974 so far as we know. But Mr Papaioannou did rather better; he got £2,900 for this property.

It may be that the plaintiff did not know there was a compulsory purchase order in force. That is his affair. But, with property

A changing hands at this rate in these circumstances, it is not altogether surprising that the local authority's solicitor's department sometimes failed to keep up with the changes. That they ought to have done is plain, but they did not. The result was, as my Lord has said, that they served their necessary notice to treat under section 5 of the Compulsory Purchase Act 1965 and the notice of entry under section 11, firstly, on the wrong people.

It is quite right to say that it seems as if Mr Cohen himself did not know that the local authority had taken possession of his property until he went one day to look at it with a builder and found that it had been boarded up to keep the vandals out because the tenant had gone. Then his solicitors wrote — perfectly sensibly — to the
B council protesting, as he thought, at their behaviour; and he received a letter back from the council, with which my Lord has dealt in detail and I will not repeat, which, like my Lord, I regard as a perfectly reasonable sensible letter suggesting a way of sorting out this mistake in such a way as to hurt Mr Cohen as little as possible. The object of the suggestion of back-dating the notice to treat and the notice of entry was to ante-date the date from which interest would run. The only person that would help would be Mr Cohen. Apart from that, I should have thought, speaking for myself, that it was as fair and as reasonable a way of dealing with this mistake as anyone could devise. But unfortunately Mr Cohen neither accepted the proposal nor rejected it; he did nothing for two months. Then,
C in a desultory kind of way he appears to have followed the procedure as if the notice to treat had been properly served. He never made any claim against the council for the possession of his property. There was not a hint of it until he finally issued a writ in March 1976 claiming possession.

So we reach the absurd situation that today, if he is right, he would be entitled to an order for possession of this land on which the council have now built at least one and possibly two houses — or it may be that one house is on one bit of the land and the other house is on another bit of the land. It would produce a ridiculous situation in which Mr Cohen would be able to bargain with the council in the strongest possible way to get the maximum
D amount of money out of them. The normal compensation to which he would be entitled presumably, as with the other properties in the neighbourhood, would be something like £1,500. He has been trying to get £8,000 on the basis that he was selling the house with vacant possession, I suppose. So what this case is about is whether Mr Cohen should get the ordinary site value compensation for this property or some much higher sum.

I entirely agree, for the reasons which my Lord has given, that the notice to treat which was served in fact on June 24, although dated April 14, was a perfectly good notice. There seems to be nothing whatever in the section to support the learned judge's conclusion
E that the notice to treat must be served before entry into possession. I see nothing in the section to that effect. If it were otherwise, it would lead to some very strange conclusions; but the section is quite clear. Once the compulsory purchase order has been made, in order to start the procedure rolling notice to treat is the first step. I can see no reason why that notice to treat should not be given at any stage so long as it is not outside the time limit fixed by the Act — which was not so in this case. Consequently, it is a perfectly good notice to treat.

All that remains, therefore, to be either agreed or determined by the Lands Tribunal in the event of dispute is the problem of compensation. As soon as that has been assessed, under the statute the
F council can execute the necessary deed to transfer the title to themselves.

The only effect of not serving a notice of entry, in my judgment, is to place the defendant acquiring authority here in the position of trespassers and subject, no doubt, to a claim for damages for any loss which the plaintiff might have suffered. The only function of the notice to enter is to prevent the acquiring authority being trespassers from the moment that they enter upon the land. I am unable to see any reason why, having wrongly entered by accident, as it turned out in this case, they should not be able to put themselves right by serving a notice under section 11(1) without going through the farce of moving out of possession momentarily, whatever that might involve. It has been suggested that it might involve G
taking down the hoarding around the property for 14 days and then issuing the notice. For my part, I should be very reluctant to have to come to a conclusion in this day and age that the law imposes quite such a farcical formality on people who are trying to do their best in the circumstances. It is eminently desirable that this legislation should be construed reasonably and not formalistically and, for my part, I see no reason why the notice to enter was not a perfectly good notice as at the date when it was served, which was June 24.

I do not find it necessary to say anything on the equitable side of this case; although, had it been necessary to do so, I, like my Lord, would have been disposed to agree with the way in which the learned judge put it in his judgment below. H

I agree that this appeal should be dismissed.

O'CONNOR LJ also agreed.

The appeal was dismissed with costs.

QUEENS'S BENCH DIVISION

February 13 1981

(Before Sir Douglas FRANK QC, sitting as a deputy judge of the division) J

E O N MOTORS LTD v SECRETARY OF STATE FOR THE ENVIRONMENT AND ANOTHER

Estates Gazette June 27 1981
(1981) 258 EG 1300

Compulsory acquisition — Acquisition of Land (Authorisation Procedure) Act 1946, Schedule 1 — Notice of compulsory purchase order — Complaint by tenants of part of subject land that they should have been served with the notice — Exception for "tenants for a month or any period less than a month" — Rent paid by reference to a week, but tenants had been in occupation paying rent for some years — Argument by tenants that they were to be regarded as tenants for more than a month as their tenancy had lasted for months — Held, rejecting this submission, that they were weekly tenants and that the length of their occupation did not covert their tenancy into a great interest — Hence tenants not entitled to a notice of the making of the order — Application dismissed K

This was an application by E O N Motors Ltd to quash a compulsory purchase order made by Newbury District Council and confirmed by the Secretary of State for the Environment in respect of land at Speenhamland Farm, Newbury, of 0.56 hectare L
in extent. The applicant complained that he had not received notice of the compulsory purchase order, contrary to Schedule 1 to the Acquisition of Land (Authorisation Procedure) Act 1946.

G Nurse (instructed by Charles Lucas & Marshall, of Newbury) appeared on behalf of the applicant; J Sullivan (instructed by the Treasury Solicitor) represented the first respondent, the Secretary of State; the second respondents, Newbury District Council, were not represented and took no part in the proceedings.

Giving judgment, SIR DOUGLAS FRANK QC said: On December 7 1978 the second respondents, exercising their power under the Housing Act 1957, made a compulsory purchase order on M
0.56 hectare of land known as Speenhamland Farm, Newbury. The order was submitted to the Secretary of State, who confirmed it on November 19 1979, there having been no objections. Before making the order, the council sent the freeholder a requisition for information, requiring him to state the nature of his interest in the land and that of any other occupier. In response to that, the owner sent by way of his solicitors the completed form, in which the only occupier stated was the lessee under a seven-year term. In fact, in about May 1975 the applicants in this court were granted a tenancy of part of the land. The agreement as stated by one of the directors was

"that the company would pay every quarter a rent calculated on the basis of £20 per week, and it was further agreed that the company would pay half the cost of repairs to the premises". Copies of invoices produced show that rent was paid at intervals of not less than four months, but calculated at the rate of £20 per week.

The Acquisition of Land (Authorisation Procedure) Act 1946 applied to this acquisition and in particular Schedule 1 paragraph 3(1)(b), which so far as relevant provides as follows:

> 3. (1) Before submitting the order to the confirming authority the acquiring authority shall:
>
> (a) . . . .
>
> (b) except in so far as the confirming authority directs that this provision shall not have effect in any particular case, serve on every owner, lessee and occupier (except tenants for a month or any period less than a month) of any land comprised in the order a notice in the prescribed form stating the effect of the order and that it is about to be submitted for confirmation, and specifying the time (not being less than twenty-one days from the service of the notice) within which and the manner in which objections thereto can be made;

It is the applicants' case that the compulsory purchase order should be quashed on the ground that they were entitled to have been served with a notice of it and that they were not (as is conceded) so served. The argument advanced for the applicants, put shortly, is that they did not fall within the exception in paragraph 3(1)(b) of Schedule 1 because their tenancy lasted for more than a month, so that they were a tenant for more than a month. In other words, a tenancy for a month is not the same as a monthly tenancy and once occupation goes beyond a month the tenancy is not for a month.

In my judgment, the tenancy granted here was not for a fixed term and therefore the term is determined by reference to the period for which the rent was payable and therefore it is a weekly tenancy. That much is not in contention. However, in my judgment a tenancy which commences as a weekly tenancy is a tenancy for one week and continues to be a tenancy for one week, however long it may last. The fact that a tenant stays for longer than a week does not convert the tenancy into one of a longer period. That view seems to be reinforced by the purpose of the paragraph, which cannot be for the benefit of those who have been tenants for a long time, but for the protection of those who might be entitled to remain for a long period.

The conclusion I have reached seems to be in line with the decision of the Lands Tribunal in *Selborne (Gowns) Ltd v Ilford Corporation* (1962) 13 P&CR 350 and of the Court of Appeal in *Benjamin v Newham London Borough* (1968) 19 P&CR 365. Accordingly, I hold that the applicants were not entitled to the service of a notice of the making of the order and therefore that this application fails.

I should add for completeness that Mr Sullivan, for the Secretary of State, argued that because the applicants became aware of the making of the compulsory purchase order before it was confirmed and made no inquiries as to their right to object, that they did not suffer substantial prejudice or that, in any event, the court should not exercise its discretion in favour of the applicants. However, had the applicants made inquiries as to their right of objection they would have discovered that they would have been out of time to object. Be that as it may, I do not think that it was incumbent upon the applicants to make the inquiries which Mr Sullivan said they should have made. Accordingly, had I been satisfied that the notice should have been served, I would have quashed the order.

The application was dismissed with costs, payable to the first respondent.

FEES AND COMMISSIONS

A **QUEEN'S BENCH DIVISION**

October 24 1980

(Before Patrick BENNETT QC, sitting as a deputy judge of the division)

RICHARD ELLIS v PIPE-CHEM (HOLDINGS) LTD

Estates Gazette April 25 1981

(1980) 258 EG 329

Estate agents' commission — Claim by estate agents for commission for finding suitable premises for the defendants — Commission to be payable by the defendants "subject to our offer . . . being accepted"
B **— The merchant bankers advising the lessors were not in favour of the grant of a lease to the defendants — Defendants suggested that the lease should be granted to another company who would grant an underlease to the defendants — This suggestion was adopted — Whether in these circumstances the plaintiffs were entitled to their commission — Plaintiffs argued that the interposition of an intermediate interest made no difference, whereas the defendants submitted that "our offer" meant an offer by the defendants directly and no one else — Held that no term could be implied which would entitle the plaintiffs to commission in these circumstances and their claim accordingly failed**

In this action the firm of Richard Ellis claimed commission from
C the defendants, Pipe-Chem (Holdings) Ltd, for finding suitable premises required by the latter for their business of making petrochemical equipment. The premises in question were Cray Valley House, Orpington, Kent, owned by Brigade Investments Ltd.

Alec Grant (instructed by Lawrance Messer & Co) appeared on behalf of the plaintiffs; Allan Levy (instructed by Lee & Pembertons) represented the defendants.

Giving judgment, PATRICK BENNETT QC said: In the early part of 1978 the defendants, a company specialising in the making of petro-chemical equipment whose main works are in the north of England, wanted to open premises in the south of England, and they approached the plaintiffs in this action (Richard Ellis, a well-
D known firm of chartered surveyors and estate agents) for their assistance in finding suitable premises. Mr A Whicheloe, an employee of the plaintiff firm, made a number of inquiries for suitable premises and communicated the results of those inquiries in correspondence to the defendants. The choice fell upon Cray Valley House, Orpington, Kent, owned by a company called Brigade Investments Ltd, the freeholders of the property, who had close links with a merchant bank (Keyser Ullman). Mr Whicheloe, on behalf of his principals, negotiated with the representatives of Brigade Investments Ltd, Leighton Goldhill & Partners, consultant surveyors and estate agents, and, in particular, with a Mr I A French [ARICS], who was also a director of Brigade Investments Ltd. As a
E result of those negotiations, on March 7 1978 Mr Whicheloe submitted an offer on behalf of the defendants to Brigade's representatives in respect of the property.

On the same day Mr Whicheloe, who had not hitherto negotiated the terms of remuneration with the defendants, had a telephone conversation with Mr Simon, a director of the defendant company, in which authority to act on behalf of the defendants was given to the plaintiffs, and the remuneration was discussed. That agreement was recorded in a letter dated March 7 1978 from the defendants to the plaintiffs as follows:

> Following our telephone conversation of today, we confirm we wish you to F
> act on our behalf in negotiating for the above-mentioned property and that we will accept a fee —

originally expressed at 5 per cent

> subject to our offer of £45,000 per annum for the first five years and a rent free period of six months being accepted.

The figure for commission was altered, by agreement, to one of 7.5 per cent, but that letter represents the gist of the telephone conversation and the agreement between Mr Whicheloe and Mr Simon; it was consequent upon that agreement that Mr Whicheloe wrote the letter of March 7 to which I have already referred. There is a second
letter dated March 15 1978 from the secretary of the defendant G
company to Mr Whicheloe where he also confirmed that:

> a commission of 7.5% will be payable to you when the lease has been completed.

However, both counsel agree that that second letter does not otherwise modify the terms of the agreement of March 7 1978.

The matter continued with the good offices of the plaintiffs. Brigade Investments Ltd required among other things a sight of the defendants' accounts for the last three years and also satisfactory references. One of the referees whose name was supplied by the defendants to Mr Whicheloe for onward transmission to Brigade
Investments Ltd was Amari Ltd, as appears from a letter dated H
March 28 1978. Amari Ltd were part of a group of companies in which Amari World Steel Ltd was the major creditor and a debenture holder of the defendant company.

Unfortunately, on receipt of the accounts of the defendant company, the surveyors of Brigade Investments Ltd sought the advice of their merchant bankers (Keyser Ullman) and it is clear that that advice was adverse to granting the defendants the lease they sought. The matter was then discussed directly between representatives of the defendants and Keyser Ullman. I say "directly" because it appears that from that stage onwards the negotiations which concluded the transaction took place between the defendants and
Brigade's merchant bank. J

It is common ground that the defendants put forward the suggestion that Brigade Investments Ltd grant a lease to Amari Ltd in terms identical to those sought by the defendants, that Brigade Investments Ltd would grant Amari Ltd a licence to grant an underlease of the property to the defendants on similar terms, but that the underlease would be one day shorter in duration than the head lease. This arrangement proved acceptable to Brigade Investments Ltd and their advisers, and they granted a lease to Amari Ltd on June 1 1978 in the terms sought by the defendants in their original offer to Brigade Investments Ltd. Brigade Investments Ltd granted Amari Ltd a licence to sublet the demised
premises to the defendants, and on July 16 1978 Amari Ltd made K
use of that licence and granted an underlease to the defendants in similar terms to those contained in the head lease; the term of the underlease was one day less than that granted to Amari Ltd by Brigade Investments Ltd.

On April 24 1978 the secretary of the defendant company wrote to Mr Whicheloe saying that he was instructed by his directors to advise him:

> that we no longer consider ourselves liable to you for commission as set out in my letter of March 15 1978

A that being the letter to which I have previously referred, which followed the letter of March 7 but which the parties accepted does not modify or alter the agreement of March 7 1978. The letter continued:

> as we were unable to obtain a lease on the above premises from the landlord. Although we are now hopeful of obtaining a lease on the premises via a third party, this has resulted from our own endeavours: an ability to introduce to the landlords a company acceptable to them as lessees.

The plaintiffs replied to that letter on April 28 1978 to this effect. The writer said that he had:

> B taken the opportunity to look into the matter, and bearing in mind the content of the letter dated March 7 of Mr Simon together with the letter of March 15, there is a clear obligation on you to pay my firm's fees if you have taken up a lease of the property. I understand that there is now an intermediate interest involved in the transaction, but notwithstanding this the property was introduced to you by this firm.

The essence of the conflict between the parties in this case is whether in those circumstances the plaintiffs are entitled to their commission on the terms agreed between them and the defendants.

One witness to whom I must make reference, Mr French, a director of Brigade Investments Ltd and a partner in their surveyors and estate agents, Leighton Goldhill & Partners, expressed a cen-
C sorious view of the defendants and said that in his view this was a devious means of avoiding a fee, due in law, to the firm of surveyors and estate agents. I acquit the defendants of any such deliberate act. I find that there is no evidence on which I could believe that this was anything other than a straightforward business view of the situation which existed. If this had been a deliberate attempt by the defendants to avoid payment by a cleverly contrived scheme, first of all by presenting unacceptable accounts to Brigade Investments Ltd and their merchant bank and then to put forward an acceptable proposed lessee, in spite of what Mr Levy has said I would be very surprised if this court could not grant relief to the plaintiffs in those circumstances. I am quite satisfied that those circumstances did not
D occur in this case.

What happened, as I find, was that the possibility that the defendants would not be acceptable to Brigade Investments Ltd was not considered, nor was the solution of the problem to which that unacceptability gave rise (that is, the solution at which Keyser Ullman and the defendants appear to have arrived, namely, by finding an acceptable lessee prepared to grant an underlease to the defendants) envisaged by either the plaintiffs or the defendants at the time when the agreement of March 7 1978 was entered into.

I have been referred by counsel to a number of authorities. The plaintiffs say, first, that the true and ordinary meaning to be attached to the words "subject to our offer being accepted" is wide
E enough to include the offer being accepted by Amari Ltd. That is, that in the phrase "our offer", to quote Mr Grant's words: " 'our' can mean by our side or our camp". With the very greatest respect to him, I do not think that without an unjustifiable straining of language it would be right to construe those words in that way. In my view, the normal and natural meaning of the words "subject to our offer being accepted" means "subject to an offer by us being accepted"; "us" being Pipe-Chem (Holdings) Ltd and not someone else. Mr Grant says, however, that if such a construction is not permissible I should imply a term into the agreement between the parties, to the effect that if the offer was made in the circumstances of this case (that is, by someone who thereafter would grant a
F sublease to the defendants; he uses the phrase "the defendants' nominees" to include those persons) then a term to that effect was impliedly necessary to give the agreement business efficacy.

As I said earlier, I have been referred by counsel to a number of authorities, the first of which was the case of *Luxor (Eastbourne) Ltd v Cooper* [1941] AC 108 at p 125, where Lord Russell of Killowen when discussing implied terms said:

> Implied terms, as we all know, can only be justified under the compulsion of some necessity. No such compulsion or necessity exists in the case under consideration. The agent is promised a commission if he introduces a purchaser at a specified or minimum price. The owner is desirous of selling. The chances are largely in favour of the deal going through, if a purchaser is introduced. The agent takes the risk in the hope of a substantial remunera- G tion for comparatively small exertion.

I do not wish to denigrate the skilled and persistent efforts of Mr Whicheloe, but that description by Lord Russell of Killowen is not entirely inappropriate to this particular case.

I was referred, also, to the case of *Trollope & Colls Ltd v North West Metropolitan Regional Hospital Board* [1973] 1 WLR 601 at p 609, where Lord Pearson said:

> An unexpressed term can be implied if and only if the court finds that the parties must have intended that term to form part of their contract: it is not enough for the court to find that such a term would have been adopted by the parties as reasonable men if it had been suggested to them: it must have been a term that went without saying, a term *necessary* to give business H efficacy to the contract, a term which, though tacit, formed part of the contract which the parties made for themselves.

The third case to which I was referred was *Liverpool City Council v Irwin* [1977] AC 239. In that case Lord Wilberforce analysed the various categories or the spectrum of implied terms. He does not accept the shade of colour (if that is the right expression) which the Master of the Rolls had sought to add to the spectrum. At p 253 Lord Wilberforce said:

> To say that the construction of a complete contract out of these elements involves a process of "implication" may be correct: it would be so if implication means the supplying of what is not expressed. But there are J varieties of implications which the courts think fit to make and they do not necessarily involve the same process. Where there is, on the face of it, a complete, bilateral contract, the courts are sometimes willing to add terms to it, as implied terms: this is very common in mercantile contracts where there is an established usage:

It was neither pleaded nor proved in this case that in these circumstances which Mr Whicheloe and Mr French described as "unique" there was any established mercantile usage:

> in that case the courts are spelling out what both parties know and would, if asked, unhesitatingly agree to be part of the bargain. In other cases, where there is an apparently complete bargain, the courts are willing to add a term on the ground that without it the contract will not work — this is the case, if K not of *The Moorcock* . . . itself on its facts, at least of the doctrine of *The Moorcock*. . . . This is, as was pointed out by the majority in the Court of Appeal, a strict test — though the degree of strictness seems to vary with the current legal trend — and I think that they were right not to accept it as applicable here. There is a third variety of implication, that which I think Lord Denning MR favours, or at least did favour in this case, and that is the implication of reasonable terms. But though I agree with many of his instances, which in fact fall under one or other of the preceding heads, I cannot go so far as to endorse his principle; indeed, it seems to me, with respect, to extend a long, and undesirable, way beyond sound authority.
>
> The present case, in my opinion, represents a fourth category, or I would rather say a fourth shade on a continuous spectrum. The court here is simply concerned to establish what the contract is, the parties not having them-selves L fully stated the terms. In this sense the court is searching for what must be implied.

At p 263, in support of this view that such a term as was required to be inserted in the contract of the *Irwin* case ought to be so inserted, Lord Salmon said:

> I find it difficult to think of any term which could be more necessary to imply than one without which the whole transaction would become futile, inefficacious and absurd as it would do if in a 15 storey block of flats or maisonettes, such as the present, the landlords were under no legal duty to take reasonable care to keep the lifts in working order and the staircases lit.

Applying those tests to the question: is this term necessary to give M the agreement business efficacy, I cannot find that it is. There was a bargain made at arm's length by the respective agents of the plaintiff firm and the defendant company; namely, Mr Whicheloe, who is a qualified professional man, and Mr Simon, a director of the defendant company. I cannot see that it is necessary to imply such a term into this agreement as that for which Mr Grant asks in order to give it business efficacy. It is clear that neither party considered the possibility of what did in fact happen, but I am not satisfied that had that matter been drawn to their attention by an officious bystander or an obtruding listener on the telephone, that both unhesitatingly would have said, "In those circumstances commission is nevertheless payable". It may well be that one side (that is, the beneficiary of the agreement) might have said, "Yes, I would agree to that"; but I am not satisfied, and I do not find, that both would have agreed to this modification had the possibility been pointed out to them. In the event I find that the plaintiffs fail in their claim and are not entitled to the commission claimed on the transaction.

QUEEN'S BENCH DIVISION

January 29 1981

(Before His Honour Judge STABB QC, sitting as a judge of the High Court)

IAN SCOTT & CO v MEDICAL INSTALLATIONS CO LTD

Estates Gazette May 9 1981

(1981) 258 EG 556

Estate agents' claim for commission fails — Instructions to agents to find an assignee of a lease of business premises at a premium in excess of £20,000 and at a rental of £10,000 a year — Potential purchaser found at £24,000 — Agents' employee then alleged that a third party had intimated that there was another purchaser who might be prepared to offer more — Vendors agreed that they would pay the third party, if his purchaser completed, a commission equal to half of any excess over £24,000 or the agents' scale fee, whichever was the smaller — Lease purchased for £29,000 by a firm who had in fact been one to whom the agents had sent the initial particulars — Agents claimed commission for themselves of £2,450 and also claimed £2,450 on behalf of the third party (the scale fee being less than one half of £5,000) — Suspicions were aroused in the minds of the vendors by various circumstances and they refused to pay the commission claimed on the ground of breach of fiduciary duty by the agents — During the proceedings the claim on behalf of the third party, who had not been identified, was dropped — Judge found that the agents were in serious breach of their fiduciary duty to their principals by seeking to extract a large sum of extra commission to which they were not entitled — Agents accordingly were not entitled to any commission at all on the transaction — Judgment for vendors with costs on common fund basis

This was an action by Ian Scott & Co, estate agents, of 20 Berkeley Street, London W1, plaintiffs, against Medical Installations Co Ltd, defendants, for commission in respect of the disposal of a lease of one floor of premises at 25 Buckingham Palace Road, London SW1, on the instructions of the defendants.

Miss J Beale (instructed by Manches & Co) appeared on behalf of the plaintiffs; Miss P Baxendale (instructed by Herbert Oppenheimer, Nathan & Vandyk) represented the defendants.

Giving judgment, JUDGE STABB said: In this action, the plaintiffs, Ian Scott & Co, a firm of estate agents carrying on business at 20 Berkeley Street, London W1, sought by their writ issued in January 1980 to recover from the defendants the sum of £5,267.50, made up of their own commission of £2,450 and an equal sum representing the commission of a third party, together with value added tax. That sum of £5,267.50 was subsequently reduced to £4,867.50 owing to what was said to be an error of £700 in the computation of the third party's commission. At the hearing, Miss Beale, on behalf of the plaintiffs, abandoned the claim for the commission alleged to be due to the third party, because in September 1980 the third party was said to have indicated that he was no longer asking for the money which was said to be due to him.

The story, as related by Mr Rodman, an employee of the plaintiffs, who had the conduct of the matter on their behalf which gives rise to this claim, can be briefly summarised as follows. The defendants wished to dispose of their lease of one floor of premises at 25 Buckingham Palace Road, SW1. They instructed the plaintiffs to find an assignee of the lease on payment of a premium in excess of £20,000, and at a rental of £10,000 a year. The plaintiffs found a buyer at a figure of £24,000, which was acceptable to the defendants, but at that moment a third party contacted them indicating that he knew of a buyer who might be prepared to purchase the lease for more and asking whether, if he revealed the name and the purchase went through, he could expect a commission. Mr Rodman reported this to the defendants, who agreed to pay this third party a commission equal to half the amount of the excess over and above the £24,000 for which the lease was sold, or the estate agents' scale fee, whichever was the lower. This was reported to the third party, who thereupon revealed the name of the interested buyer, who eventually purchased the lease for £29,000. This was £5,000 in excess of the original offer of £24,000. In the result, the plaintiffs presented the defendants with an account of their commission amounting to £2,450 based upon the RICS scale of fees, and £2,450 payable to the third party, since half the excess of £5,000 would have been more than the scale fee.

On that bare statement of the facts there would seem to be nothing untoward, and indeed the defendants were happy with the outcome, since they were to receive £29,000 less the double commission of £4,900, that is some £24,100, as against £24,000 less the one commission of the plaintiffs of £2,450, which would only have been some £21,550, so that they would be £2,550 better off. But when they received a copy of the third party's account, their suspicions were aroused, mostly because the account was on the printed notepaper of Misty Days Management Ltd, who turned out to have been business management agents for musicians, who knew nothing of the transaction and who had never heard of the purchasers, who equally had never heard of them. In the light of this, the defendants had reason to doubt the integrity of the plaintiffs, whom they had instructed to act as estate agents on their behalf, and who accordingly owed them fiduciary duties of various kinds. When in January 1980 the plaintiffs commenced this action to recover the double commission, the defendants refused to pay on the ground that the plaintiffs were in breach of those fiduciary duties, and, indeed, by a late amendment, alleged that the plaintiffs had acted fraudulently.

Having now heard what I regard as the quite unacceptable account of the events as given by Mr Rodman, the young employee who had the conduct of this transaction on behalf of the plaintiffs, I am driven to the conclusion that not only were the plaintiffs, as agents, in breach of their fiduciary duties towards their principals, the defendants, but that his whole account of the alleged events smacks of fraud.

This is his account. At the beginning of August his firm were instructed to find a buyer of the lease, and one finds, in a letter dated August 2 from the defendants to Mr Rodman at the plaintiff firm's address, a letter which reads as follows:

> I refer to our previous correspondence on the lease of the above property and your valuation last year of some £10,000. It now seems unlikely that our two direct contacts are interested in taking over this property, and we therefore wish to put the offices and suite on the market. I am enclosing a copy of the lease.
>
> Before I can give you instructions to proceed with the marketing of these offices I would be grateful if you could telephone me advising me of your terms of business.

On August 8 one finds a letter addressed to the defendants over the signature of Mr Rodman, which reads as follows:

> Following our telephone conversation, I write to confirm and thank you for your kind instructions for my company to act on your behalf in the disposal of your offices held in the above. I confirm we are to invite premium offers in excess of £20,000 for the benefit of your lease, fixtures and fittings and all the furniture, for your lease which expires on August 15 2002, which carries a rental of £10,000 a year.
>
> I confirm that our fees will be in accordance with the scales as laid down by the Royal Institution of Chartered Surveyors.

He enclosed a set of letting particulars for the defendants' attention.

Having received those instructions, Mr Rodman then set about to draw up particulars of the premises under the lease. He circularised various prospective customers and in particular circularised all the other tenants who were in that same building, sending each of them particulars of these premises, and among those tenants, be it noted, was the ultimate purchaser, Messrs Watermeyer Legge, a firm of

A consulting engineers. Mr Rodman also circularised and sent particulars to other estate agents who might be interested.

Mr Rodman says the chronology of the events that followed was this. First, an offer from a concern known as Wedge International through some estate agents known as Clive Lewis & Partners, and that offer was in the sum of £24,000. He says he then informed Mr Hatfield of the defendants, who said that the offer was acceptable. But within 24 hours of that, he said he had a phone call from a man who called himself Peters, who said that he knew someone who would be prepared to pay more than £24,000. It apparently seems that Mr Rodman had disclosed to this gentleman on the telephone that that was the offer that had been received up to that time. Mr
B Peters asked whether, if the person that he knew of subsequently purchased, he, Mr Peters, could expect some commission. Mr Rodman says that he told him that he would enquire of his clients and see what they were prepared to do. Accordingly, he says that he then phoned Mr Hatfield again, told him that a person or a party had made this proposition, and Mr Hatfield agreed to pay half the excess or the scale fees, whichever were the lower, so that they, the defendants, would in any event derive some benefit.

I pause at this juncture to say that Mr Rodman could not have let Mr Peters know the result, because he did not then or subsequently ascertain, nor has he to this day ascertained, who Mr Peters was, where he lived or worked, or what was his phone number. He just
C had to wait for Mr Peters to ring back.

Mr Rodman said that he then told Wedge International that they would have to make a higher offer, or at least make their highest offer, and in the result, by letter dated August 23, they did send in another offer which was higher, namely in the sum of £25,500. Then Mr Rodman said Mr Peters phoned back and he was told about the commission agreement that the defendants had reached with the plaintiffs, and Mr Peters then revealed his prospective purchaser as being Watermeyer Legge. One would have thought that Mr Rodman's immediate reaction would have been to say that he had already introduced Watermeyer Legge to this proposal to acquire the lease of these premises, but not a bit. He merely sent
D Messrs Watermeyer Legge another set of particulars, which he had already sent them previously, and made no enquiries of them about Mr Peters. It is quite in keeping with this wholly unacceptable account given by Mr Rodman that Watermeyer Legge had never heard of Mr Peters. Indeed, to this day, no one, not even Mr Rodman, knows who Peters is, and in the light of what followed, one is tempted to wonder whether he ever existed.

Watermeyer Legge eventually purchased the lease for £29,000, and on the wholly unsupported assumption that it was Mr Peters' information which led to the plaintiff sending a second set of particulars that was instrumental in bringing about the sale, the plaintiffs sought to extract for their principals a second commission
E of £2,450. This was supported — if "supported" is the right word — by an invoice which Mr Rodman says was subsequently received from Misty Days Management Ltd. Mr Rodman confessed that he had never heard of that company, but he blithely assumed that the account must have emanated from Mr Peters because the details in the account were right. How that account on that notepaper came to the plaintiffs it is difficult to understand. Mr Wright, a director of Misty Days Management Ltd, said that he had never heard of anybody called Peters, had never heard of Watermeyer Legge, and simply had no explanation for the fact that his company's notepaper had been used for the purpose of an invoice. So far as Mr Rodman is concerned, if and when he received this
F account from Misty Days Management Ltd, here at least was the golden opportunity to track down the shadowy figure of Mr Peters. From an honest agent it demanded an immediate enquiry as to what Misty Days Management Ltd was, how they had come into the affair, and what their connection with Mr Peters was. But Mr Rodman did nothing, except to ask the defendants for the money to pay the account, and as soon as the defendants saw a copy of the account which had been omitted from the first communication (Mr Rodman says in error) the defendants' suspicions were rightly aroused, and enquiries which they made of Misty Days Management Ltd only served to confirm those suspicions.

I have heard Mr Hatfield give evidence on behalf of the defendants, and his recollection is usefully supported by contemporane- G
ous notes. From what he says — and which is confirmed by those notes — it is perfectly plain that when Mr Rodman telephoned him on August 24 and put the "Peters' " proposition to him, he falsely told Mr Hatfield that the highest offer to date was £24,000. In point of fact, a letter dated August 23 addressed from Clive Lewis & Partners to the plaintiff company and to Mr Rodman in particular, which is stamped as having been received on August 23, showed that an offer was being made by Wedge International Holdings amounting to £25,500. By telling Mr Hatfield that £24,000 was the highest offer they had had to date he thereby obviously increased the excess of the final purchase price from £3,500, which it would have been if the offer of £25,500 had been disclosed, to £5,000, H
which was the excess over the figure of £24,000, and, therefore, by doing that, the second commission was increased from £1,750 to £2,450.

I really need add little more. As I have said, I find the evidence of Mr Rodman wholly unacceptable. He had been put forward by Miss Beale on behalf of the plaintiffs as a person who might well have been gullible and mistaken but not dishonest. I did not find him to display the characteristics of gullibility. He seemed to me to demonstrate a measure of self-assurance, and so far as his truthfulness is concerned, all I can say is that in one respect at least he was demonstrated to be telling what was not the truth with regard to the telephone conversation with Mr Hatfield, which he said took place J
well before August 23. By his conduct, the plaintiffs were in breach of their fiduciary duties to their principal by seeking to extract from them a large sum by way of extra commission to which I do not think anyone was entitled, and it was not surprising that in September of 1980, when these proceedings were well advanced, Mr Rodman reported that the still unknown and unseen Mr Peters had telephoned to say that he was withdrawing his claim. This at least enabled the plaintiffs to say that they were now only advancing their own entitlement to commission; but such a retraction is, in my judgment, too late. It does not purge the plaintiffs of their previous breaches of their fiduciary duties towards the defendants, and the law is clear on this point. I was helpfully referred to a number of K
authorities. I need refer to only one, the case of *Andrews v Ramsay & Co* [1903] 2 KB 635, where Lord Alverstone CJ in the course of his judgment had this to say, after referring to the case of *Salomons v Pender* (1865) 3 H&C 639:

> The principle there laid down is that, when a person who purports to act as an agent is not in a position to say to his principal, "I have been acting as your agent, and I have done my duty by you", he is not entitled to recover any commission from that principal. In *Salomons v Pender*, Bramwell B said, "It is true that . . . the defendant has had the benefit (if it be one) of the plaintiff's services. But the defendant is in a position to say, 'What you have done has been done as a volunteer, and does not come within the line of your duties as agent' ".

And then later in his judgment he said: L

> It seems to me that this case is only an instance of an agent who has acted improperly being unable to recover his commission from his principal. It is impossible to say what the result might have been if the agent in this case had acted honestly. It is clear that the purchaser was willing to give £20 more than the price which the plaintiff received, and it may well be that he would have given more than that. It is impossible to gauge in any way what the plaintiff has lost by the improper conduct of the defendants. I think, therefore, that the interest of the agents here was adverse to that of the principal.
>
> A principal, is entitled to have an honest agent, and it is only the honest agent who is entitled to any commission. In my opinion, if any agent directly or indirectly colludes with the other side, and so acts in opposition to the interest of his principal, he is not entitled to any commission. M

In my view, the plaintiffs here have demonstrated themselves as having been in breach, and serious breach, of their fiduciary duties towards the defendants and, that being the result, even though the defendants may have benefited by services in the sense that the lease of the premises was sold for £29,000, quite plainly the plaintiffs have forfeited any right to commission by reason of those breaches and by reason of their conduct generally. Accordingly, in my view, this claim must fail and there should be judgment for the defendants, with costs.

Judgment was given for the defendants with costs, the costs to be taxed on the common fund basis.

HOUSING ACTS

A **COURT OF APPEAL**

December 16 1980

(Before Lord Justice OLIVER and Lord Justice WATKINS)

ELLIOTT AND ANOTHER v BRIGHTON BOROUGH COUNCIL

Estates Gazette May 2 1981

(1980) 258 EG 441

Housing Act 1957, section 10 — Appeal from decision of county court judge in favour of local authority's claim for the expenses of
B **repairs carried out by the authority — Meaning of word "may" in the phrase "the local authority may themselves do the work" — Local authority's view that, in default of compliance by an owner with a notice under section 9(1), the authority had an enabling power which was coupled with a duty to carry out the necessary works — The authority operated an automatic procedure by which, on such default, steps were taken to carry out the works without any further review of the circumstances — Opposing submission by property owners that section 10 gave the authority a discretion to decide to carry out the works after giving consideration to the relevant circumstances — Held that section 10 conferred a discretion and that in the present case the authority acted wrongly in**
C **adopting a procedure by which works were carried out in reliance on the ministerial act of an official in pursuance of a standing instruction issued years before the event — Appeal allowed and authority's demand for expenses quashed — Incidental ruling that it was competent to challenge the validity of the section 10 notice by appeal to the county court and that it was not a matter which had to be raised by means of an application for judicial review under Order 53**

This was an appeal by Grant Elliott and his wife, Marilynne Therese Elliott, from a decision of Judge Grant at Brighton County Court dismissing an appeal under section 11 of the Housing Act 1957 against a demand by Brighton Borough Council, the present
D respondents, for the recovery of expenses incurred in carrying out remedial works under section 10 of the Act. The property in respect of which the works were carried out was a front basement flat at 1 Chichester Terrace, Brighton, Sussex.

Graham Platford (instructed by Kipling & Co, Cuckfield) appeared on behalf of the appellants; Nicholas Wood (instructed by the borough secretary) represented the respondent borough council.

Giving judgment, OLIVER LJ said: Section 9 of the Housing Act of 1957 places upon a local authority which is satisfied that a house in its area is unfit for human habitation a duty to serve on the person having control of the house a notice requiring him to carry out
E specified works which will, in the opinion of the local authority, make it fit for human habitation. That is something which, under the terms of the section, the local authority is obliged by law to do unless it is also satisfied that the house is not capable of being rendered so fit at reasonable expense.

Section 11 of the Act gives to any person who is aggrieved by the notice, or by a subsequent demand in pursuance of it, a right to appeal, within 21 days of the notice or demand, to the local county court. If he does that then no proceedings are to be taken under the notice or demand until the appeal has been determined.

The present appeal is from an order of His Honour Judge Grant F
made in the Brighton County Court on July 16 1979 dismissing the appellants' appeal to the county court under section 11 of the Act. That appeal was not against a notice served under section 9 but was against a demand for the expenses of repairs carried out by the local authority under section 10. I must therefore refer to the latter section so far as it is relevant. In the context of the instant case there are two relevant subsections, namely subsections (1) and (3). Subsection (1) provides:

If a notice under the last foregoing section — that is section 9 — requiring the person having control of a house to execute works is not complied with, then, after the expiration of the time specified in the notice or, if an appeal has been made against the notice and upon that appeal the notice has been G
confirmed with or without variation, after the expiration of twenty-one days from the final determination of the appeal, or of such longer period as the court in determining the appeal may fix, the local authority may themselves do the work required to be done by the notice, or by the notice as varied by the court, as the case may be.

Subsection (3) provides:

Any expenses incurred by the local authority under this section, together with interest from the date when a demand for the expenses is served until payment, may, subject as hereinafter provided, be recovered by them, by action or summarily as a civil debt, from the person having control of the house or, if he receives the rent of the house as agent or trustee for some H
other person, then either from him or from that other person, or in part from him and as to the remainder from that other person.

Then there is a proviso which I need not read.

A demand made under section 10(3), as I have mentioned already, is equally the subject-matter of an appeal to the county court under section 11. That is provided in subsection (1)(b); section 11 is in these terms:

(1) Any person aggrieved by — (a) a notice under the foregoing provisions of this Part of this Act requiring the execution of works, (b) a demand for the recovery of expenses incurred by a local authority in executing works specified in any such notice, (c) an order made by a local authority with respect to any such expenses may, within twenty-one days of the service of J
the notice, demand or order, appeal to the county court within the jurisdiction of which the premises to which the notice, demand or order relates are situate, and no proceedings shall be taken by the local authority to enforce any notice, demand or order in relation to which an appeal is brought before the appeal has been finally determined.

There is then an important specific provision relating to an appeal under subsection (1)(b) and (c) which is in these terms:

(2) On an appeal under paragraph (b) or paragraph (c) of the foregoing subsection no question shall be raised which might have been raised on an appeal against the original notice requiring the execution of works.

Subsection (3) shows the wide discretion which is conferred upon K
the county court. It provides:

On an appeal to the county court under this section the judge may make such order either confirming or quashing or varying the notice, demand or order as he thinks fit and where the judge allows an appeal against a notice requiring the execution of works to a house, he shall, if requested by the local authority so to do, include in his judgment a finding whether the house can or cannot be rendered fit for human habitation at a reasonable expense.

What occurred in the instant case was this. On February 3 1978 the respondents, the Brighton Borough Council, served a notice on the appellants under section 9(1) in respect of premises which consisted of a front basement flat at 1 Chichester Terrace, Brighton. The premises were undoubtedly damp and were not fit for human occupation. The appellants — and this is not disputed — were the persons who were in control of the premises which were let to a tenant. That notice was in the proper form and it called for the carrying out of a schedule of remedial works which does not matter for present purposes. It was served as the result of an inspection which had been made in November 1977 and which had been followed by a prolonged and inconclusive correspondence in the course of which the appellants contended that the works called for were more than merely remedial and they indicated that they thought the case a suitable one for an improvement grant.

The notice was not complied with, nor did the appellants exercise their right to appeal under section 11 in respect of the notice. Consequently, the respondents' environmental health officer, without any further instruction from the council or its housing committee, obtained an estimate for the necessary works. On May 16 1978 he wrote to the appellants — or, to be precise, to Mr Elliott — saying that the council would proceed with the work unless the appellants themselves did it. The appellants still did not do the work and it was in fact carried out by a contractor engaged by the environmental health officer.

On December 5 1978 the respondents made a demand on Mr Elliott for the sum of £1,283.70, the expenses incurred. That produced a dual response. There was first a letter of complaint about the quality of the work and, secondly, there was an appeal to the county court. It was that appeal which came before His Honour Judge Grant, the first hearing being on June 13 1979. At that hearing it emerged that the appellants were taking the point (which, it is fair to say, they had ventilated in correspondence before the work was done) that the word "may" in section 10(1) conferred on the local authority not just a power but a discretion which they were bound to exercise as a council only after a consideration of the circumstances which, it was suggested, would include any representations which the appellants had sought to make. That, it was said, had not been done and, therefore, the works carried out were *ultra vires*. Of course, since what was challenged was whether the works had been properly done under section 10, that was a point which could not have been taken on an appeal against the notice under section 9, so it was not excluded by the proviso.

The learned judge rejected the appellants' submission and a further hearing took place on July 16 1979 which was concerned with the adequacy of the works, an issue which the learned judge also decided against the appellants on the facts. He accordingly dismissed the appeal. There is no appeal to this court on the question of fact, but the appellants appeal against the rejection by the learned judge of the preliminary point of law and it is on that point alone that the argument before this court has centred.

There is no dispute that the works which were carried out were the result, not of a conscious decision of the council or even, in any real sense, of the environmental health officer, but of an automatic procedure flowing from a standing instruction which applied indifferently to every case of failure to comply with a notice under section 9(1) of the Act within the time-limit specified in the notice. The council never addressed its collective mind, either by itself or by its delegates, to the question whether it should or not carry out the particular work, for it had, some eight years previously, issued what was, in effect, a standing instruction in the form of a resolution dated January 28 1970. That resolution was in the following terms:

> RESOLVED: That in respect of notices served under section 9 of the Housing Act 1957 the chief public health inspector be authorised: (a) to obtain estimates immediately the time for compliance with a notice has expired and to begin default works promptly thereafter; (b) not to delay work done in default because of applications for improvement grants; (c) to inform owners of this intention when the notice is served and to give no further information of the council's intentions.

The reason for a resolution in this form being applicable to all cases was that the respondents' interpretation of the Act led them to the conclusion that, on a default in compliance with a notice under section 9(1), they had no option but to undertake the work themselves because the statute cast a positive duty on them in those circumstances to do so. It is quite clear, from the correspondence, that the environmental health officer took the view, which coincided with the council's view, that once a notice had been served and not complied with, for whatever reason, the council came under a statutory duty to undertake the work itself without regard to any considerations which might then exist pointing to the desirability or undesirability of doing so.

Mr Elliott had written to him (that is the environmental health officer) at the end of April, indicating that he might be applying for a possession order against the tenant and was contemplating major works of renovation which would render the proposed works of minimal benefit. Whether he genuinely had such an intention is at least questionable, but the environmental health officer, in his reply, made it quite clear that he did not consider himself free, whatever the circumstances, to do anything but carry out the works. It is, in fact, not necessary to trace this determination through the correspondence because it was, as I understand it, conceded before the learned judge. The case was argued before him on the basis that once there had been a default in compliance with the notice, there was an immediate statutory obligation upon the local authority to do the work which absolved it from making any decision in the matter or even considering the merits or demerits of so doing.

The learned judge accepted the respondents' view of the matter. He said, in his judgment:

> I accept the argument of the local authority. There is nothing in the authorities, as I understand it, which compels the court to construe the word "may" as necessarily conferring a discretion upon them which they must exercise. The word "may" in a statute, it seems plain from the authorities, can be construed as a pure enabling provision or as an enabling provision giving an absolute discretion and it seems to me that the framework of this Act and the intent of the Act plainly points to the word "may" in section 10 being used as enabling the local authority to do the works themselves in their absolute discretion, or may be as a matter of duty. But certainly I reject as mistaken the contention that a notice having been served and not complied with, no appeal having been brought against it, that at that stage when the opportunity for everything relevant being considered had passed, they should have begun an investigation whether or not to carry on with the works when plainly they were under a duty, namely, to see to it that the house unfit for human habitation should be rendered fit.

Mr Platford has presented to us a number of cogent arguments for saying that that was wrong; wrong at any rate to this extent, that the learned judge took the view that the word "may" in the section imposed a positive obligation. But after the conclusion of the arguments, indeed in the course of preparing our judgments in a case which is of considerable importance to local authorities generally, there arose a somewhat fundamental point which was not argued before the learned judge and was certainly not argued before us on the substantive hearing of the appeal. It is this. Section 11(1) of the Act confers the right of appeal to the county court and in relation to a demand for expenses incurred in carrying out work under section 10 it gives the right of appeal which expressly excludes any matters which might have been taken on an appeal against the section 9 notice. The court has power, on such an appeal, to confirm, quash or vary the demand. An appeal under this section is clearly appropriate when what is in issue is whether, for instance, the work has been properly carried out or is in accordance with the notice or whether the charges raised are excessive or unreasonable — all matters which, in the instant case, have been determined against the appellants. But is it appropriate where the point which the appellants wish to raise has nothing to do with the contents of the demand itself or the value or nature of the work in respect of which it is made, but consists of a challenge to the discretion of the local authority in carrying out the work at all? On the face of it this is not a matter which is within the contemplation of the section; it is a matter more appropriate for judicial review under Order 53 of the Rules of the Supreme Court.

What is sought to be challenged is the discretion of an administrative body charged with a public duty and the proper remedy in such a case, *prima facie*, would appear to be an application to the Divisional Court for a judicial review.

In the light of this we have sought the assistance of counsel by restoring the case for further argument before giving judgment.

A That assistance has been willingly given and we are grateful to counsel for the helpful arguments. Mr Wood has helpfully drawn our attention to two cases, *Metropolitan Properties Co (FGC) Ltd v Lannon* [1968] 1 WLR 815 and *Chapman v Earl* [1968] 1 WLR 1315, both of which raised a serious doubt whether, on an appeal at any rate under section 9 of the Tribunals and Inquiries Act 1958, it is open to an appellant to challenge the very proceedings in which the decision appealed from was made. They suggest that the only remedy in such a case may be by what is now an application for judicial review.

Mr Wood himself, however, felt bound to draw our attention, most properly, to other cases which suggest that in such cases there B may be concurrent remedies and, as Mr Platford has pointed out, the only right under the statute under consideration in the two cases to which I have referred above is a right to appeal against the decision, which may be said to postulate that a valid, if erroneous, decision has been made. The wording of section 11 of the Housing Act 1957, however, is very general and it confers a very wide discretion upon the county court judge. It gives a right of appeal against a demand for payment of expenses incurred and there is no doubt that such a demand has been made in the instant case. There is nothing in the section itself which would indicate that the court, on appeal, would be precluded from considering whether the demand was properly made and indeed subsection (3) gives the C judge jurisdiction to quash it.

Furthermore, there is authority of this court which seems to show that on an appeal to the county court under section 11(1)(a) of the Act the county court is the proper tribunal for considering the question of the validity of exercise of the local authority's discretion in issuing the notice under section 9 and there is no logical reason why the same should not apply to a demand made under section 10.

The point in relation to what is now section 9 was considered by the court in *Cohen v West Ham Corporation* [1933] Ch 814 which concerned the predecessor section to the present section 11, namely, section 22 of the Housing Act 1930, which was substantially in the same terms.

D Lord Hanworth MR at p 832 said this:

> The simple point that is before this court is whether those notices were valid or invalid. We have come to the conclusion that they were valid. Consideration must be given to sections 17, 18 and 19 of the Housing Act, also to section 22, which enables an appeal to be made to the county court, but makes it impossible to appeal after the lapse of twenty-one days from the date of the service of the notice. In the present case there was no appeal to the county court within the twenty-one days of the service of the notices, and the result is that they stand good in their terms if they are valid notices under section 17.

At p 834 Lawrence LJ says:

> In my judgment, the learned judge was right in holding that the notices E served by the council were notices given pursuant to section 17 of the Act of 1930, and therefore came within the purview of sections 18 and 22 of the Act, with the result that not having been appealed against they are valid notices, whether or not the council complied strictly with requirements of subsection (4) of section 17.

So that their Lordships were clearly of the view that the right way of challenging the validity of the notice was by an appeal to the county court.

A somewhat similar point, although this time in relation to a possession order sought against a council tenant, came before the court in *Cannock Chase District Council v Kelly* [1978] 1 WLR 1, where it seems to have been suggested that the exercise of the F council's discretion to seek possession could be challenged only on a judicial review. Megaw LJ at p 8, said this:

> The passage which I have already quoted from the note of Judge Allardice's judgment is, I think, consistent, in its reason for rejecting the defence, with the views which I have expressed. But earlier in his judgment the judge had expressed the view that if a council house tenant, having received from a local authority a notice to quit, desired to challenge it on the ground of bad faith or abuse of power, his correct course was to apply to the Queen's Bench Divisional Court for a prerogative order, or to the High Court for a declaration; and that, if he failed to do so before the notice had expired, the county court could offer no remedy. Counsel for the local authority, in this court, submitted as one of his arguments that the judge in the county court had a discretion to refuse to deal with such a matter himself and to leave it to the tenant to apply to the High Court for a declaration or a prerogative G order. As at present advised, I do not agree. It may well be that the county court would be overwhelmed, and the local authorities would be gravely hampered in carrying out their functions, if, as was suggested in the present case, a tenant could hold up the obtaining of an order for possession and could bring about delay and discovery of documents and ultimately a trial with much oral evidence, merely by asserting: "I am a good tenant: therefore you must have acted in abuse of your powers." But I hope I have made it clear that, if my judgment is correct, that will not do.

Quite clearly, his Lordship was rejecting there the contention that the only remedy was a resort to prerogative writ.

These cases satisfy me that the point taken in this case is one which can be taken on an appeal to the county court under section H 11 and I am reinforced in that by the fact that the section is expressed — and presumably deliberately expressed — in terms which confer the widest possible discretion upon the county court judge (see *Cochrane v Chanctonbury Rural District Council* [1950] 2 All ER 1134).

Turning to the present appeal, Mr Platford on behalf of the appellants argues that the learned judge's decision was wrong. The demand made upon the appellants can be sustained, he submits, only if the works which were done were carried out under a proper exercise of a statutory power; and a purported exercise of such a power, otherwise than as a result of a conscious decision to exercise it is, he contends, *ultra vires*. To treat section 10(1) as imposing a J mandatory obligation on a local authority to carry out the work regardless of the circumstances involves, he submits, effectively reading "may" in the subsection as "shall" or "must" and there is no context for that in the Act.

When one looks at the Act as a whole, there is undoubtedly a striking contrast between those sections where the legislature in terms imposes a mandatory obligation and those where it confers a power or discretion. Section 9(1) is, in terms, mandatory once the pre-condition is fulfilled of satisfying the local authority of the unfitness of the premises and that they are not incapable of being rendered fit at reasonable expense. Equally, the next section (which I have not yet read) is clearly discretionary, that is section 9(1)(A) K which was added subsequently and which is in these terms:

> Where a local authority, upon consideration of an official representation, or a report from any of their officers, or other information in their possession, are satisfied that the house is in such a state of disrepair that, although it is not unfit for human habitation, substantial repairs are required to bring it up to a reasonable standard, having regard to its age, character and locality, they may serve upon the person having control of the house a notice requiring him, within such reasonable time, not being less than twenty-one days, as may be specified in the notice, to execute the works specified in the notice, not being works of internal decorative repair.

It must be remembered that the provisions of section 10(1) apply L indifferently both to notices under section 9(1) and to notices under section 9(1)(A). Again, section 12(1), which authorises the purchase of property found on appeal not to be capable of being rendered fit for occupation at reasonable expense, clearly confers a discretion. There are many, similar provisions where the authority "may" do things and where it is clearly a matter for their discretion whether or not they do or not. For instance, section 10(5) which enables the authority to make provision for the recovery of expenses by instalments, section 25(1) which relates to the cleansing of premises subject to demolition orders and section 29(1) which relates to the purchase of condemned houses.

In striking contrast with these provisions are the provisions of M section 23(1) which provides that where a demolition order has become operative and the owner fails to demolish within the time specified the local authority "shall" itself enter and demolish.

I have referred to these provisions because they do, as it seems to me, support Mr Platford's submission that where the statute intends to impose a mandatory duty it does so in express terms. Moreover, Mr Platford submits that, quite apart from these contrasts and from the natural meaning of the word "may", section 10(1) really only makes practical sense if that word is construed in a permissive or discretionary sense rather than the obligatory and unnatural sense contended for by the council.

In the first place, the statutory provisions themselves envisage a lapse of time between the service of the notice and a resort to section 10(1). During that time circumstances may change entirely — for example the premises may be sold to someone anxious to redevelop them — and it requires little imagination to produce instances of matters which could arise which would radically affect the desirability or usefulness of having the work done either at all or by the local authority itself. To take only one instance, the notice requires that the specified work be "executed" and that must, I think, mean "completed", not merely started. Suppose that the owner had set about them but had not finished them when the time-limit expired. Is it to be said that the local authority is obliged — not merely empowered — to go in and turn the owner's contractors out so that they themselves can complete the work under section 10(1)? This seems an absurd suggestion but it must, I think, be the logical consequence of the construction for which the respondents contend, for they say that once the statutory machinery and time-limits are set in operation, the steps envisaged by section 10 follow inexorably without any further conscious act by the authority beyond, I suppose, the purely ministerial one of ordering their workmen on to the site, an act which the law obliges them to do.

Secondly, Mr Platford submits that, quite apart from any circumstances occurring in the time which necessarily must elapse between service of notice and default, there may be considerations which, though entirely irrelevant to the question whether it is appropriate to serve a notice under section 9(1), are nevertheless relevant to the question whether it is desirable for the work to be carried out by the local authority at that juncture. Thus, he submits, there is certainly nothing compulsively leading to a construction of the word "may" in section 10(1) in a sense which, in any event, is not its natural one.

For the respondents, Mr Wood argues that the purpose of the Act is to make owners of houses keep them in proper repair and that it casts on the local authority the duty of seeing that they do so. It is clear that where the conditions mentioned in section 9(1) apply, the local authority is under a duty to serve notice and once that has been done, and all rights of appeal by the person having control of the premises have either lapsed or been exhausted, their duty to secure that the premises are put into a proper state compulsively dictates, it is submitted, that, in default of compliance with the notice, the power in section 10(1) shall, without more, be exercised. It is, therefore, he argues, quite unnecessary for the local authority to apply its mind to the question whether it should go in and do the work because, if it did, it could come up with only one answer, namely that it must do it. So if, for instance, the building owner has been prevented by illness or other cause from doing the works within the time prescribed, but is willing and able to undertake them at his own expense, they must nevertheless be carried out by the local authority whether they wish to do them or not. He draws our attention to cases such as *MacDougall v Paterson* (1851) 11 CB 755 and *Sheffield Corporation v Luxford* [1929] 2 KB 180 in support of the proposition that a statute conferring a power may be construed as imposing a positive obligation to exercise it where the conditions for its exercise exist.

Both those cases were cases of judicial powers, but he submits that there is no essential distinction between a judicial and an administrative body. The discretion here, he suggests, is at the notice stage and it is here that the person affected has his right of appeal under section 11. His only other right of appeal under that section is in relation to the demand made for the expenses of an exercise of the power conferred by section 10(1) and it cannot, he submits, have been contemplated by the legislature that on an appeal under section 11(1)(b) he would be entitled to raise questions as to the propriety of the exercise of the power under which the expenses arise.

There are, Mr Wood submits, three possible meanings comprehended in the word "may". It may mean simply that the propositus is at liberty to do something which would otherwise be unlawful — and he would style this a pure "enabling" power which does not involve any notion of any exercise of will. It may mean a discretionary power and, finally, it may mean a power coupled with a duty. The power under section 10(1), he submits, must come within either the first or the third of these categories, because where the legislature intends to confer a discretion in this Act it points out what has to be taken into account in exercising it as in section 9 and section 16 where the authority has to be satisfied "upon consideration" or in section 17(2) where certain matters "appear" to the authority.

Speaking for myself, I am not persuaded of the validity of these distinctions. As Mr Platford suggests, once one finds a power conferred by an Act there is necessarily an act of will in deciding whether it should be exercised or not. It involves a decision and that is a discretion. What facts will have to be taken into account in arriving at the decision may be a matter for argument but a body cannot, to take the present case, go in and carry out repairs unconsciously. Nor am I persuaded that any sensible distinction is to be drawn between those sections of the Act which do and those which do not point out or delimit matters which have to be taken into account in determining any particular course of action which the statute authorises. It seems to me, for instance, really unarguable that the power to prescribe payment by instalments in section 10(5) is not a discretionary power simply because there is no reference to the matters which the authority must consider.

One is left, therefore, with Mr Wood's central proposition based on the cases to which I have referred above and it is essentially a point of construction. I find myself unable to accept his construction of the statutory provisions. It involves the conclusion that an entry by the local authority to do the work is the compulsory and automatic consequence of a default in compliance with the terms of the notice and that is neither what the statute appears to me to say on any ordinary construction of the words used nor is it, as it seems to me, a sensible construction.

It has to be borne in mind that section 10(1) applies indifferently both to default in complying with a notice under section 9(1) and to default in complying with a notice under section 9(1)(A) where the giving of the notice itself is quite clearly a matter not of duty but of discretion. It would be a very strange result and, as I think at any rate, an absurd result if the exercise of an admitted discretion under section 9(1)(A) committed the authority to an irreversible and immutable process which, from then on, it was powerless to control. But if Mr Wood is right on the construction which he seeks to place on section 10(1) he must either face this absurdity or he must say that the self-same word in the self-same section has alternative meanings according to the provision of the Act under which the notice was given; whereas the natural meaning which Mr Platford urges produces a perfectly reasonable and intelligible result in both sets of circumstances.

Of course, it may very well be in many cases, and indeed I strongly suspect it may be so in the present case, that if the matter were considered, it would appear that the only way of getting the work done and thus accomplishing the statutory purpose is by the local authority undertaking it itself so that the room for the exercise of a choice in the matter may be very narrow. But that does not mean that the power is not a discretionary power. It is not, I think, open to the respondents in this case to rely upon such considerations because, in the light of the concession made before the learned judge, that aspect of the matter was simply never investigated, nor indeed is there any respondents' notice in respect of it. They have, necessarily, argued their case (which necessarily makes this, as I think and as I am glad to say, an unusual case) on the footing that no consideration was given to exercising the power because none was necessary, since it followed as a matter of obligatory and inevitable statutory machinery. As to that I have reached the opposite conclusion to that reached by the learned judge.

I am, of course, very, very far from saying either that the necessity to exercise the discretion and come to a decision on the exercise of the statutory power obliges a local authority to engage in the sort of lengthy arguments and consultations which Mr Elliott obviously regards as appropriate or that, if the respondents in the instant case had applied their mind to the question at all, they would not and could not, quite properly, have reached a decision to exercise the power. But that is not the question. What they cannot, in my

A judgment, properly do is to make no decision at all because they never even considered the matter and then rely upon the ministerial act of an official pursuant to a standing direction issued years before the event as constituting an exercise of the election which the statute gave them. I am fortified in this conclusion by the following passage from the judgment of Maugham J, as he then was, in the *Cohen v West Ham* case to which I have referred, [1933] Ch 814 at p 827, where he says this:

> I do not wish to be misunderstood by my use of the word "discretion". I have come to the conclusion that this is not a case, as was *Hall v Manchester Corporation*, where the owner of the house or the person in charge of it is entitled to be summoned to attend the local council or committee before the B order is made: nor do I consider it necessary for the whole of the council to consider the matter. It seems to me quite sufficient if some delegate body, whether the housing committee or someone entrusted by it on behalf of the council, looks into the matter and studies the official representation or the official report, or the other information, and the notice which is proposed to be served. Upon that I think the local authority can properly act, but I do express the opinion that something of the sort is necessary.

With every word of that I agree and it seems to me it applies to the instant case.

It follows, therefore, from what I have said that the appellants were entitled, in my judgment, to succeed on the preliminary point that the power conferred by the statute, in so far as it can be said to C have been exercised at all by the instructions issued by the environmental health officer under the 1970 resolution, was exercised under a misapprehension that the statute compulsively dictated its exercise and was, therefore, not properly exercised and accordingly no demand for the recovery of the expenses could properly be made.

The appellants' merits, I must say, are not impressive. I have formed the very strong impression from the correspondence that the objections raised were insubstantial and a prevarication. Therefore I have reached the conclusion that I have with considerable reluctance.

The case is an unusual one and not, I think, one from which those D in a similar position to the appellants should take any encouragement to proceed with frivolous or time-wasting appeals. It is unusual because it is based on a concession made below which Mr Platford acknowledges to be his sheet anchor. I and my brother think it was a wrong concession, but there it is and with regret, therefore, I would allow the appeal. That leaves, I think, no alternative but to order that the demand be quashed.

Agreeing, WATKINS LJ said: I sympathise with the Borough Council of Brighton which is forced to be the respondent to an appeal which has no merit whatsoever.

The appellants own a number of houses in various places. In one of them, namely, 1 Chichester Terrace, Brighton, the tenanted E basement flat became so affected by damp as to render it, in the respondent's opinion, unfit for human habitation. The respondent has a duty under the provisions of Part II of the Housing Act 1957 to provide for the repair, maintenance and sanitary conditions of houses within the borough. In circumstances for which the Act provides it may itself, if the owner chooses not to do so, restore a house to a fit state for living in. It requested the appellants to repair the basement flat. They refused to comply, so the respondent served a notice to comply under section 9(1) of the Act. They neither obeyed the notice nor, as they were entitled to and clearly, having regard to their objections to repair, should have, appealed against it. If they had appealed their objection to what the notice F ordered them to do would have been considered by a county court judge who has a very wide discretionary power to set aside, vary, or leave untouched a statutory notice.

Having neglected the opportunity to have the notice considered judicially, they bombarded some of the respondent's officials and councillors with further objections to repairs being done to the flat. The respondent, using its power under section 10(1) of the Act, engaged a contractor to do this work. After the work was done, the respondent served a demand upon the appellants for the recovery of the expense of doing it. The appellants refused to pay and appealed against the demand. The appeal failed in detail. At the same time they also appealed under section 11(1)(b) to His Honour Judge Grant at the Brighton County Court against the respondent's G decision to use its powers to do the work under section 10(1) saying that this suggestion conferred a discretion upon the respondent which it had not exercised. Accordingly, it acted without authority. So the demand was unenforceable.

This point had been taken by the appellants in correspondence with the respondent; thus it was not surprising that it became the point of law, and the only one, relied upon in this part of the appeal to the county court. It also failed. The learned judge held that the word "may" in section 10(1) does not confer a discretion which must be exercised; it is an enabling provision giving an absolute discretion. The respondent was not called upon to investigate further or to consult with the recipient of a statutory notice. It was H under a duty only to ensure that a house unfit for human habitation was made fit.

It is now said that the learned judge was mistaken upon this point. The point in issue may be simply stated as, does "may" mean "must"? As a matter of common understanding of these two words, the answer surely is that it most certainly does not. The word "may" is indicative of a choice or of the exercise of a discretion; the word "must" is imperative. It is a word of command or of obligation to act in some way or another. "May" is undoubtedly used in this sense in some other sections of the Act — see sections 9(1)(A), 16(4) and 17(2). Can it possibly be said, therefore, to have a different meaning as it is used in section 10(1)? J

It is argued on behalf of the respondent that as used there it does bear a different meaning. It must be read in conjunction with the detailed provisions of section 9(1) and the overall duties of a council under Part II to ensure that houses within a borough are kept fit for human habitation. The rights of an owner of property are amply provided for, especially by the right of appeal against the statutory notice and the discretion given to a county court judge to set aside or to vary the terms of the notice so that it may justly reflect the situation by, among other orders, extending time for compliance. The introduction of yet another period for possibly reviewing the situation and consulting with the owner cannot possibly have been contemplated by or have been the intention of K Parliament. That is how the respondent has, for many years, regarded this matter. It has assumed that the word "may" means that a council is thereby given an enabling power to repair a house unfit for human habitation which is coupled with a duty so to do upon the failure of the owner to comply with the statutory notice in its final form. There is no lack of support from decided cases for the contention that the word can, in some legislation, bear the meaning which is attributed to it by the respondent.

The appellants contend that section 10(1) is a purely enabling provision. There are powerful indications that the use of the word "may" in the Act can only lead to the conclusion that it confers a discretion. The council may repair, upon failure by the owner to do L so after notice or, in its discretion, it may not. Circumstances may change during the period of the notice or after its termination. Council policy with regard to that part of the borough where the house is situated may be altered; an owner, having at some time felt unable to, may find himself able to, after all, effect the repairs; the terms of the notice as varied by a county court judge may change the situation so as to make it more sensible and practical that an owner and not the council should effect repairs.

However, I am in no doubt that if the respondent had exercised its discretion in the present case it would inevitably have concluded that it should carry out the repairs to 1 Chichester Terrace. The appellants were undoubtedly bent upon thwarting the respondent M at every turn. It had made all its decisions up to the time of effecting the repairs in a regular way. They are not complained about. It occurs to me that if the discretion contended for does exist, the occasions when it would be exercised in favour of an owner following failure to repair after notice to do so, will be rare. Nevertheless, the use of the word "may" in the context in which it is used in section 10(1) allied to the use made of that word in many other parts of the Act, lead me to believe that it bears its ordinary, commonly understood meaning.

For this reason and for those provided for in detail by my Lord, I have to disagree with the opinion of the learned county court judge.

A **When considering these reasons I began seriously to doubt whether the learned judge acted within his jurisdiction in entertaining this appeal at all. However, the further very helpful arguments of both learned counsel have removed that doubt entirely. Accordingly, I am also able to agree with my Lord on that point. I would, though not without regret, allow the appeal and quash the demand.**

The respondents were ordered to pay the appellants' costs here and below, the appellants' costs of application for leave to appeal out of time to be set off against other costs.

B

COURT OF APPEAL

March 5 1981

(Before Lord Justice ORMROD and Lord Justice GRIFFITHS)

BRIKOM INVESTMENTS LTD v SEAFORD

Estates Gazette May 23 1981

(1981) 258 EG 750

Section 32 of Housing Act 1961 — Liability to keep installations in dwelling-house in repair and proper working order — Construction of section 33(5) — Agreement for lease preceding execution of C lease — Whether a lease for a term of less than seven years or a lease for seven years — Definition of "lease" as including "an agreement for a lease" — If the length of the term was properly calculated from the date of the agreement for the lease, it was a lease for seven years; if from the date of execution of the formal lease, it was a lease for less than seven years — Held that in the present case, where the terms had been agreed and there had been part performance by entry into possession, the length should be taken from the date of the agreement — An agreement for a lease for a term of seven years or more is not caught by section 32, provided that the term begins on or after the date of the agreement, whether or not it is followed by a formal lease — Although the result of this, if it had stood alone, D would have relieved the landlords from liability under section 32 of the 1961 Act, they were estopped from denying liability under that section through having accepted fair rent registrations and demanded rents on the basis of assuming such liability — Landlords could take steps to have the fair rent corrected to reflect the true position in regard to repairs — Tenant's appeal allowed on estoppel point

This was an appeal by David Seaford, tenant of a flat, 72 Herga Court, Sudbury Hill, Harrow, from a decision of Judge Honig at Willesden County Court in an action by the landlords, Brikom Investments Ltd, for possession and arrears of rent. The landlords E claimed arrears of rent calculated on the basis that they were subject to the repairing liabilities under section 32 of the Housing Act 1961, but at the same time claimed that they were not liable under that section. The judge decided in favour of the landlords on the section 32 point, but did not deal with the estoppel point.

I E Jacob (instructed by J E Kennedy & Co, of Harrow) appeared on behalf of the appellant; N B Primost (instructed by A E Hamlin & Co) represented the respondents.

Giving the judgment of the court, ORMROD LJ said: This appeal concerns the liability, as between the appellant tenant and the respondent landlords, for certain internal repairs, as defined in section 32(1)(b) of the Housing Act 1961 in respect of Flat 72, F Herga Court, Sudbury Hill, a block of flats owned by the respondents, Brikom Investments Ltd.

The appellant puts his case in two alternative ways. First, he relies upon section 32 of the Housing Act 1961 under which there is to be implied into any lease to which the section applies a covenant by the landlord, *inter alia*, to keep certain installations in the dwelling-house for the supply of water, gas, electricity, etc in repair and working order, notwithstanding a covenant by the tenant in the lease to the contrary. Alternatively he says that the landlord, in the events which have happened, is estopped from disputing his liability for the repairs referred to in the defence and counterclaim.

The first point, which raises a difficult question of constructions of sections 32 and 33 of the 1961 Act, was decided against him by G his Honour Judge Honig, sitting at Willesden County Court, on June 27 1980. The second point was not specifically dealt with by the learned judge.

By section 33(1) of the 1961 Act section 32 applies, subject to the provisions of the section, to

> any lease of a dwelling-house granted after the passing of this Act, being a lease for a term of less than seven years.

So the question is whether the appellant had a lease for less then seven years or for seven years. He contends that his lease was for less than seven years; the landlords contend that the lease was for seven years. H

The dates in the case are, therefore, important. It is common ground that this flat was vacant in October 1969 and that the landlords' letting agents, Stackpole & Co, introduced the appellant to the flat in that month. He was shown a specimen lease for a term of seven years, which was the standard form used by the respondents. He decided to take the flat on the proposed terms. His references had to be approved by the managing agents, Waite & Sons, on behalf of the landlords, who approved them in due course, and, on November 1 1969, he was allowed into possession and paid in advance a proportion of the quarterly rent in respect of the period November 1 1969 to December 25 1969. He also signed the counterpart of the lease on that day. The landlords did not immedi- J ately execute the lease and it is not known precisely when they did so, but the lease itself bears the date November 12 1969. It was sent by post to the tenant, who received it on November 15 1969. The evidence was that the date on the lease was filled in by the landlords' solicitors but that it was "somewhat fortuitous", depending upon the conveyancing clerk's arrangements for stamping.

The habendum of the lease provides that the tenant is to hold the premises from November 1 1969 for a term of seven years at an annual rent of £385 payable by equal quarterly instalments, but it is clear that the lease was not executed by the landlords until some days after November 1 1969, and not delivered until November 15 1969. So, once again, the problem arises of determining, for the K purposes of a statutory definition, the length of the term of a lease.

It has been held in many cases, of which *Earl of Cadogan v Guinness* [1936] 1 Ch 515, *Roberts v Church Commissioners for England* [1972] 1 QB 278 and *Bradshaw v Pawley* [1980] 1 WLR 10, are examples (arising, however, in connection with different statutes) that a term, defined in a deed as beginning from a date prior to the delivery of the deed, say for 10 years from such a date, is not a term of 10 years. It is a shorter term beginning from the date of delivery of the deed and ending 10 years from the earlier date specified in the lease (see per Clauson J, as he then was, in *Earl of Cadogan v Guinness* at p 518), or, as Stamp LJ put it in *Roberts v Church Commissioners* at p 285: L

> It is well settled that the habendum in a lease only marks the duration of the tenant's interest, and that the operation of the lease as a grant takes effect only from the time of its delivery.

Mr Jacob, for the tenant in this case, accordingly submits that, although the habendum refers to a term of seven years from November 1 1969, the actual term created by the lease is two weeks short of seven years and is, accordingly, a lease for a term of less than seven years and, therefore, falls within section 32.

The question to be decided, however, is what does this phrase "being a lease for a term of less than seven years" mean in the context of section 33. Mr Primost, for the landlords, drew attention M to the subsection (5), which reads:

> (5) In the application of this section to a lease granted for a term part of which falls before the grant, that part shall be left out of account and the lease shall be treated as a lease for a term commencing with the grant.

He submitted that if subsection (1) is construed in accordance with the principle laid down in the cases cited subsection (5) is wholly unnecessary because in any event the term cannot start before the grant. So he says section 33(1) must refer to the term as described in the habendum, namely, seven years from November 1 1969; the term in the present case is therefore not a term of less than seven years. This submission, however, does not help him because if he is right such a lease is caught later by the same subsection (5); the part falling before the grant must be left out of account, and the term computed from the date of the grant. So, he is back to square one!

This is obviously an unsatisfactory conclusion. It is difficult to believe that Parliament intended that the application of section 32, which seriously affects the rights of landlords and tenants, should depend on something so essentially fortuitous as the date of the delivery of the lease. Fortunately, there is another way of approaching the problem which the learned judge in the court below in substance adopted.

This Act (unlike the Acts with which this court was concerned in other cases) contains a definition section which defines the word "lease" in relatively broad terms. Section 32(5) provides that "lease" includes, *inter alia*, "an agreement for a lease . . . and any other tenancy", and the word "term" is to be construed accordingly.

In the present case there was, undoubtedly, an agreement for a lease of seven years beginning on November 1 1969, made by the parties on or before that date, because by that time the terms of the lease as set out in the specimen lease or in the counterpart had been agreed and there had been part-performance by entry into possession and payment and acceptance of rent.

In *Roberts v Church Commissioners* there was, as Russell LJ emphasised, no agreement for a lease of the length required to satisfy the terms of section 3(1) of the Leasehold Reform Act 1967, namely a tenancy for a term of years certain exceeding 21 years. "Tenancy" in that Act means a "tenancy in law or equity" (section 37(1)). At p 284 of his judgment Russell LJ suggested a test which the tenant must pass to fulfil that definition, namely, that he

> must at some point of time be, or have been, in a position to say that, subject to options to determine, rights of re-entry and so forth, he is entitled to remain tenant for the next 21 years, whether in law or equity.

The tenant in the present case is in a position to fulfil that test, substituting seven years, which is the relevant period under the 1961 Act. So if this is the right approach to the 1961 Act, as we think it is, we are entitled to hold that for the purposes of section 33(1) there was an agreement for a lease, and therefore a lease as defined, for a term which was not less than seven years.

But the landlords have still to get over section 33(5), the language of which is not very apt to agreements for a lease unless the words "granted" and "grant" are to be read as equivalent to "made". If this is permissible the subsection will still be effective to prevent the mischief at which it was presumably directed, that is, to prevent a landlord granting or agreeing to grant a lease for less than seven years and back-dating the term so as to make it seven years from some anterior date. We do not think that such a construction does undue violence to the language of these sections read as a whole.

If this goes beyond the limits of construction, we think the same result follows from the application of section 33(3), which deals with consecutive leases. The tenant in this case was a person who immediately before the lease was granted — that is, delivered — was the lessee under another lease, ie under the agreement for the lease which for the purpose of these provisions is to be regarded as a lease. So he is within subsection (3)(a). The other lease — that is, the agreement for the lease — was not a lease to which section 32 applies because it was for not less than seven years; so subsection (3)(ii) is satisfied and, accordingly, section 32 does not apply.

In our judgment, therefore, an agreement for a lease for a term of seven years is not caught by section 32, provided that the term begins on or after the date of the agreement, whether or not it is followed by a formal lease. Accordingly we would hold, in agreement with the learned judge in the court below, that section 32 does not apply to the lease in the present case, and that the liability for internal repairs is governed by the terms of the original lease. That, however, does not dispose of this appeal, because if the appellant is right on the estoppel point he will still be entitled to succeed in this action.

The alleged estoppel arises in a curious way. In 1975 the landlords applied, under the terms of the Rent Act 1968, to the rent officer to register the rent of this flat. The application itself has not survived but, on September 23 1975, the rent officer determined the rent at the sum of £555 exclusive of rates with effect from May 16 1975 and duly registered it. In the notification of registration of rent the rent officer set out very clearly the basis of his determination. Under the heading "Allocation of liability for repairs" he referred to the lease and added the words "and subject to the provisions of sections 32 and 33 of the Housing Act 1961", plainly indicating that his assessment of the rent was made on the basis that the landlords were liable for the repairs specified in section 32(1)(b) and that the tenant's covenant in the lease did not apply to such repairs. This, of course, resulted in the registration of a higher rent than would have been the case if the liability had been on the tenant.

The landlords did not object and thereafter demanded and received rent at the enhanced rate. In June 1977 the landlords applied again for the registration of a fair rent. After a full hearing, at which the landlords' representative and several tenants gave evidence, the rent assessment committee, on January 30 1978, determined the fair rent of this and other flats in Herga Court. In the case of Flat 72 it was accepted by both sides that sections 32 and 33 applied, and that is that the landlords were liable for the section 32 repairs. On that basis the fair rent for Flat 72 was assessed at £671.92 exclusive. Again the enhanced rent was demanded and paid.

Another application was made in 1979 resulting in the registration of a fair rent in the sum of £930, still on the same basis.

About the same time correspondence took place between the parties on the subject of repairs, and it is clear that throughout the landlords accepted that they were liable for the section 32 repairs; but in the event the tenant did the repairs, paid for them, and withheld a proportion of the rent. Eventually the landlords brought the present action for possession and arrears of rent at the enhanced rate, and the tenant filed a defence claiming to set off the cost of the repairs against the rent. Then, for the first time, the landlords disputed their liability and claimed that section 32 did not apply to this lease. Thus, in the same proceedings, they are seeking to recover arrears of the rent fixed on the basis that they were liable for those repairs and claiming that they were not liable for them.

In our judgment it would clearly be inequitable to hold that the tenant was liable for the full amount of the arrears of a rent which reflects, in part, that the landlords were liable for the repairs, and at the same time that the tenant was liable for the cost of such repairs.

This is the classic situation which the doctrine of estoppel was designed to meet. Mr Jacob put his case in alternative ways. Either the landlords, by demanding a rent fixed on the basis of the rent officer's allocation of liability for repairs, represented that they accepted liability accordingly, or the landlords, by not taking steps to have the registered rent changed so as to reflect the true position and suing for the enhanced rent, had made their election and could not be heard, in these proceedings, to assert a claim inconsistent with the position they had adopted.

Mr Primost, however, contended that the representation was a representation of law and not of fact and, therefore, could not give rise to an estoppel, and that the tenant was seeking to use the estoppel as a sword, that is, to recover the cost of the repairs, and not, in the classic phrase, as a shield. He relied upon two cases, *London County Territorial and Auxiliary Forces Association v Nichols* [1949] I KB 35 and *Kai Nam v Ma Kam Chan* [1956] AC 358, in neither of which had the party alleging estoppel acted to his detriment, nor had the other party gained any advantage from the representation.

These dichotomies are dangerously neat and apt to mislead. Representations of fact shade into representations of law, and swords, with a little ingenuity, can be beaten into shields, or shields into swords. In this case the shield may have quite a sharp edge but it is none the less a shield and the representation was essentially one of fact, ie that the landlords accepted liability for the section 32(1)(b) repairs to the tenant's flat in return for the enhanced rent. We would hold that so long as the enhanced rent is claimed the landlords cannot put the burden of the section 32 repairs on the

A tenant. But they can take immediate action to have the fair rent corrected so as to reflect the true position in regard to repairs, and will then be entitled to the benefit of the tenant's covenant. The appellant, therefore, succeeds on this point.

The learned judge in the court below attempted to deal with the matter on broad commonsense lines by assessing the amount of excess rent paid by the tenant under the rent officer's assessments and allowing credit accordingly. But he had insufficient material on which to estimate the amount of the over-payment and, in our opinion, no jurisdiction to make such an adjustment, although he may have been under the impression that the parties were consent- B
ing to his taking this course. In fact, it seems clear from a letter from the tenant's solicitors written shortly after judgment was given that he had not agreed to it.

This appeal must, therefore, be allowed.

The appeal was allowed with costs. The case was remitted to the county court, to be heard by Judge Honig if possible, and to be decided in the light of the above judgment; the costs of the earlier hearing in the county court to be reserved to the judge dealing with the further hearing. Leave to appeal to the House of Lords refused.

LANDLORD AND TENANT
GENERAL

CHANCERY DIVISION

October 23 and 25 1979

(Before Mr Justice OLIVER)

REGIONAL PROPERTIES LTD v CITY OF LONDON REAL PROPERTY CO LTD

SEDGWICK FORBES BLAND PAYNE GROUP LTD v REGIONAL PROPERTIES LTD

Estates Gazette January 10 1981

(1979) 257 EG 64

Landlord and tenant —Problem of repairing a leaking roof —Interesting questions of landlord and tenant law canvassed in the course of interlocutory proceedings in two interrelated actions —Motions for injunction in first action by tenants of parts of an office block, including the roof, against the freeholders —Injunction sought to restrain freeholders, who had already obtained a judgment for damages against these tenants, from entering to repair the roof —These tenants were in breach of their covenant to repair roof but freeholders had reserved no right of entry to repair it —Motion for injunction in second action by tenants of other parts of the block against the tenants whose demise included the roof —The tenants of these other parts, who were not themselves in occupation, wished to protect their subtenants against damage caused by the penetration of water from the leaking roof —Hence they wished to carry out repairs themselves to the roof and sought an injunction to prevent the "tenants of the roof" from obstructing access for this purpose —Held on the first motion that the tenants of the roof were entitled to an injunction, but in view of mutual undertakings an order was unnecessary —The unusual relief sought in the second motion by the tenants of the other parts was refused after consideration of various submissions and taking account of the undertaking given by the tenants of the roof which satisfied the freeholders in the first action.

These two interlocutory motions for injunctions related to an office block known as Regis House in King William Street in the City of London. The first action was between Regional Properties Ltd, the tenants whose demise included the roof of the building, plaintiffs, and City of London Real Property Co Ltd, the freeholders, defendants. In the second action the plaintiffs were Sedgwick Forbes Bland Payne Group Ltd, tenants of parts of the building which did not include the roof, and the defendants were Regional Properties Ltd. Sedgwick Forbes Bland Payne Group Ltd had sublet the parts demised to them to various subtenants of which one was John Fryer & Co Ltd. This sublease contained a covenant by which Sedgwick covenanted to use their best endeavours to require the freeholders to keep the exterior in good repair.

REGIONAL PROPERTIES LTD v CITY OF LONDON REAL PROPERTY CO LTD

Peter Millett QC and Anthony Mann (instructed by Brecher & Co) appeared on behalf of the plaintiffs; Ronald Bernstein QC and Christopher Priday (instructed by Nabarro Nathanson) represented the defendants.

Giving judgment OLIVER J said: I have before me two motions in two separate actions. The parties to these proceedings are all substantial property companies and they are all interested one way or another in an office block in the City, known as Regis House, King William Street. The freeholders of the building are the City of London Real Property Co Ltd (to whom I will refer as "City") and they are the defendants in the first action. The building consists of a basement, a lower ground floor, a ground floor and seven upper floors, and the top floor, that is the seventh, is surmounted by a Mansard roof rising from a flat roof which is over the sixth floor.

The plaintiffs in the first action are Regional Properties Ltd (to whom I will refer as "Regional") and they are the tenants, as I understand it, of the lower ground floor and the first floor and all the common parts of the building, including the roof, which are not included in a lease of the basement and second to seventh floors granted in 1951 to the plaintiffs in the second action, Sedgwick Forbes Bland Payne Group Ltd (to whom I will refer as "Sedgwick Forbes"). The Sedgwick Forbes lease was granted in 1951 and it expires on June 23 1993. Under it City assumes the obligation of repairing the exterior and structure of the building including the roof. Sedgwick Forbes are not in occupation of any part of the building which is comprised in its lease, which is let to various subtenants. The important sublease for present purposes is that of the top three floors which are let until June 21 1993 to a company called John Fryer & Co Ltd, and that lease contains a covenant by Sedgwick Forbes to use its best endeavours to require the landlords to keep the exterior in good repair.

As I have mentioned, Regional are the tenants of the balance of the building including the roof. Their lease was granted in 1956 and was a demise of everything which was not let by the 1951 lease, the premises comprised in that lease being described as "the excluded premises". The relevant covenants in that lease are these: Clause 1 (6) of the lease is in these terms:

> The Lessee will at its own cost and charges from time to time and at all times during the said term well and substantially repair and maintain whiten colour paper pave cleanse amend and renew and keep so repaired maintained whitened coloured papered paved cleansed amended and renewed in every respect the said building and the Lessors' fixtures therein and the foundations and the roof thereof including the structure and the main walls and timbers thereof and the drains thereof and all buildings now or hereafter to be erected on the said piece or parcel of land and to cause the lift of the said building to be inspected and maintained at regular intervals by maintenance contractors.

Apart from a right of re-entry for breach of covenant the only relevant right of entry reserved for City was in clause 1 (10) of the lease and that was in these terms:

> The Lessees will permit the surveyor of the Lessors and the Superior Lessors or their respective agents with or without workmen and others at all reasonable times to enter upon the said building for the purpose of examining the state and condition thereof and in case any defect or want of reparation shall appear the Lessees will upon notice thereof in writing being given to it cause the same to be repaired in compliance with the aforesaid covenants in that behalf within three calendar months next after the date of such notice and also that it shall be lawful for the agents or workmen engaged or authorised by the Lessors or the Superior Lessors to enter and remain upon the said building at all reasonable times so far as may be necessary or useful in order to repair or rebuild any adjoining or contiguous premises belonging to the Lessors or the Superior Lessors or to cleanse empty or repair any of the sewers drains or gutters belonging to the same.

So that there was no right of entry reserved to the lessors for breach of the repairing covenant to the building itself, and the only remedies for breach of the covenant to repair, therefore, were, first of all, forfeiture or, secondly, damages and it is common ground

A that the roof has had leaks in it occurring from time to time since 1976, although there is a dispute as to the cause of such leaks. The correspondence which has been exhibited in evidence reflects the attempts made by the landlords to get the work done. Pressure was applied from Fryers on Sedgwick Forbes, and Sedgwick Forbes in turn applied pressure on City, and City applied pressure on Regional. Some work was done, but the allegation in the proceedings is that it has not been effective and that there is a substantial amount of water still penetrating in Fryers premises, particularly on the seventh floor. City does not (for reasons of its own) want to forfeit the lease and I surmise that Regional would not in fact be broken-hearted if its landlord did in fact re-enter and terminate its lease. A
B schedule of dilapidations was served on May 6 1976. Two writs were issued in the Queen's Bench Division for damages in June and October 1977, and in October of 1978 they were consolidated. Judgment by consent has been given for damages to be assessed by an Official Referee, and that inquiry is in train. But there is no dispute at all that Regional are in fact in breach of covenant and there is no question but that they are good for whatever damages may ultimately be awarded when damages have been assessed.

What has prompted the first action is a demand by City that its workmen should be admitted to the roof to carry out repairs, and Regional (for reasons of its own) have been unwilling to allow that, and they started this action and moved for an injunction. An ex
C parte injunction was granted during the vacation and the inter partes motion now before me was ordered to be heard at a convenient date. For reasons which I will come to in a moment, the matter has largely turned out to be academic, but I have been asked to give my judgment on matters which were in dispute when the motion was in contention, because it will have a relevance on the question of costs.

Since the matter came before the court in vacation Sedgwick Forbes has started its own action in which it claims an injunction restraining Regional from obstructing Sedgwick Forbes' entry to the roof for the purposes of abating a nuisance to its subtenants, and for other relief which is based upon an agreement relating to the
D common parts of the premises with which the present motion is not concerned. I will deal, then, first, only with the first action of Regional against City.

Mr Millett, who appears for Regional, says that their claim is a very simple one for an injunction in aid of a legal right. His clients, he says, are tenants of the roof. The landlord has reserved no right of entry to do repairs and if (for reasons of its own) it is unwilling to forfeit the lease, it has already recovered damages which are in the course of being assessed and the landlord has no right whatever to go on to the plaintiffs' property for any purpose other than the limited purposes which are specified in the lease. He submits that they cannot, on the one hand, claim to keep the lease on foot for the
E purposes of enforcing the covenant and receiving rent and, at the same time, ignore the exclusive possession which the lease confers upon the lessees. In aid of his submissions Mr Millett relies upon a series of authorities going back to *Barker* v *Barker* [1829] 3 C & P 557, *Doe* v *Rowlands* 9 C & P 734, and a decision of the Court of Appeal in *Stocker* v *Planet Building Society* [1879] 27 WR 877 where James L J says this:

> Where a reversioner has granted a lease with no power of reentry reserved on breach of a covenant to repair, can he give himself the right to enter and do the repairs? It is a plain invasion of the rights of property. He has no more right than any stranger has. There is no suit in point of law for what has been done. As a matter of law according to the present legal rights
> F in this country there is no right in a reversioner to go in and do necessary repairs.

and Brett L J was of the same opinion. He said he thought it was one of the clearest cases he had ever heard, and there was a wrongful act for which an action in trespass would lie; Cotton L J also in fact concurred. In the light of that decision, Mr Millett submits that there is no remotely arguable defence against a formal injunction and accordingly no question of any balance of convenience arises on this motion, for the court is not entitled, as he would submit, just because it may be convenient, in effect to authorise an entry on to somebody else's property in the absence of even a colour of a right to do so.

Against this, Mr Bernstein would argue, first of all as a matter of G
construction of the lease, that his client reserves a right of entry. That argument—which he accepts is his weakest argument—rests upon effectively rewriting what is erroneously described in the lease as a reservation but is in fact clearly a grant. I need not, I think, rehearse the argument in detail because it does not really appear to me to be a tenable one. Then Mr Bernstein argues that in any event the injunction should not be granted for a number of reasons. First of all, he says the plaintiffs do not come to the court with clean hands. The basis of that submission is that, although a plaintiff may if he wishes exert his legal rights so as to compel a defendant to provide him with some financial advantage, he ought not to receive the aid of the court in supporting his legal right if that is done to H
enable him—or if in fact it does enable him—to continue to be in breach of covenant. But, although it is accepted that the plaintiffs are in breach of covenant, in respect of which, of course, the defendant has elected to rely upon its remedy in damages, it is not accepted and I do not think that I can infer it in the light of the disputed evidence that there is, or will necessarily be, a breach of covenant beyond those covenants for which damages are already being assessed in the Queen's Bench Division and which will not be remedied. I find the case substantially indistinguishable from the *Stocker* case to which I have referred and, despite a note in *Woodfall* to the effect that the result of that case might be different today, I am not at all sure that I see why. No reason is given, and the mere J
fact that a hundred years has passed and that property rights are perhaps less carefully protected by the law than heretofore, does not, I think, justify me in declining to follow a decision of a strong and unanimous Court of Appeal. Secondly, it is submitted that damages are an adequate remedy, but, in saying that, it is also asserted that the damages are going to be nominal only. I do not think there is anything in this point and it seems to me that the case of *Shelfer* v *City of London Electric Lighting Co* [1895] 2 Ch 388, to which I have been referred, and particularly the judgment of A L Smith L J, is really a clear authority which runs directly counter to the proposition that Lord Cairns' Act authorises the courts, as it were, to grant a compulsory licence to execute threatened K
trespasses.

Finally, it is said that, although it has not yet been claimed, the defendants are entitled to, and will, claim specific performance of the covenant to repair. In the first place there is, as I see it, grave doubt whether such a covenant in the case of a tenant's covenant is capable of specific performance. That is based on a case of *Hill* v *Barclay* [1810] 16 Ves 402, and, although that case may logically be much weakened as an authority, if indeed it ever was more than a mere dictum, by the decision of Vice-Chancellor Pennycuick in *Jeune* v *Queens Cross Properties* [1974] Ch 97, the textbooks are unanimous in rejecting such a remedy. But allowing that that doubt might be resolved in the defendants' favour I really cannot see how L
it helps them on this motion. In the first place, this is a case where the defendants could quite easily do the repairs themselves if they were prepared to forfeit the lease. They declined to do that. They distinctly admitted the claim and they had actually got judgment for damages, and the whole basis of the way their case has been put in the Scott Schedule submitted to the Official Referee runs entirely counter to the notion that the covenant was to be performed by the plaintiffs. But, leaving that little difficulty aside, the concept of a decree of specific performance seems to me to be in direct antithesis of the relief which is claimed on this motion. A decree of specific performance would require Regional to carry out the work. What the defendants seek on the motion is to resist an injunction which M
would prevent them carrying out the work for themselves, and that is not specific performance.

I think here that the real key lies, as Mr Millett suggests, in the question: "Are the defendants, City, themselves entitled to an injunction in order to enable them to enter?" None has been actually sought. I am wholly unable to see how one could be sought in these circumstances. If, then, the injunction claimed by Regional is refused, the court is, in effect, giving its blessing to a position in which, in order to protect its own property in exercising its legal rights, the plaintiffs would have to resort to self-protection; while the defendants, if they seek to carry out their threat, can only do so

A not only by a deliberate trespass but possibly, and indeed probably, by violence to the property and even to persons. I do not think that this court can permit that. I am far from saying that the attitude taken by Regional is either reasonable or commendable, but it is, I think, one which the law permits them to take and for which they are entitled to seek the law's protection. So far, therefore, as the motion in the first action is concerned, the plaintiffs clearly would, I think, succeed were it not for a circumstance to which I will come in a moment, and I would have been prepared to grant an injunction in accordance with the terms of the notice of motion, and, of course, against the usual cross-undertaking in damages.

What in fact has now occurred is that, in connection with the
B second motion, the hearing of which has not yet been concluded, some evidence has been filed by Mr Millett indicating the sort of work which his clients say is necessary to put the roof into, at any rate, temporary shape, and he has stated in open court that his clients were proposing to carry that work out. Mr Bernstein then indicated that, if that could be framed in the form of an undertaking, he, indeed, would be prepared to give an undertaking in the terms of the notice of motion, and, since Mr Millett is, as I understand it, prepared to offer an undertaking to the court, the terms of which I think have now been settled (and I will hear counsel about those) it seems that there is really nothing left in this motion except the question of costs.

C Of course I imagine—but I will hear what Mr Bernstein has to say about this in a moment—that Mr Bernstein will not be desirous of pressing for the undertaking if in fact Mr Colyer succeeds on his motion and gets a liberty to enter to do the repairs himself.

Sedgwick Forbes Bland Payne Group Ltd v Regional Properties Ltd

J S Colyer QC and J K Reynolds (instructed by Walton & Morse) appeared on behalf of the plaintiffs; Peter Millett QC and Anthony Mann (instructed by Brecher & Co) represented the defendants.

Giving judgment, Oliver J said: I delivered judgment in the first motion in the first action the day before yesterday and I do not
D intend to rehearse the facts again. They are common to both cases. The only thing I wish to say about that is that there is a correction to that judgment that I ought perhaps to make which Mr Colyer has pointed out to me. I therein described all the parties as being substantial property companies when in fact the plaintiffs in the second action, Mr Colyer's clients, Sedgwick Forbes, are, as I understand it, insurance brokers. The distinction is not I think a very material one for present purposes.

I turn then to the motion in this action. The notice of motion asks for an order "that the defendants be restrained until trial of this action or until further order whether by their directors, officers, servants or agents or howsoever otherwise from preventing the
E plaintiffs and their contractors and workmen from entering the defendants' premises being the roof of Regis House, King William Street, London EC4, and repairing the said roof for the purposes of abating the nuisance being caused to the plaintiffs' subjacent premises at Regis House aforesaid". That claim is in fact a claim, as has now been made clear by Mr Colyer in the opening of the motion, to enter and carry out on the roof (which, for the purposes of this motion at any rate, is conceded to be the property of the defendants) a full schedule of dilapidations, which has been put in evidence, a full programme of works for putting the roof into a state of repair and it is made on the basis of the existence of a right in the plaintiffs to enter and do these repairs which Mr Colyer claims that
F he will be able to establish at the trial. Of course it is axiomatic to the claim at this interlocutory stage that there should be at least a serious case for arguing that he will be able to establish at the trial the right upon which he now claims to rely. That is a point which I shall have to consider in a moment.

At the outset it seems to me there are a number of points (I think six) to which I ought to draw attention in the context of this very unusual motion. First of all the claim here is for immediate relief before the rights of the parties have been exhaustively determined at the trial. That of course is a feature of virtually every interlocutory motion of this sort. Secondly, if granted, it is relief which pre-empts for all practical purposes the trial of this issue, which is after all the main issue in the proceedings, because if Mr Colyer G
succeeds in his claim on this interlocutory motion, there is nothing further left to be done. The repairs which are in issue will in fact have been carried out. Even if he loses at the trial and fails to establish that he has got any right whatever to do the repairs they will in fact have been done and there will be no issue left to be tried, except perhaps a claim, and possibly a shadowy one, under the undertaking in damages. Thirdly, what Mr Colyer is seeking on behalf of his clients is far more than mere abatement as the notice of motion suggests, the mere temporary packing up of the roof to preserve the status quo between now and the trial. The claim, and indeed the only claim, is to carry out a full schedule of works designed to bring the roof into a state in which, so it is alleged, it H
ought to be under the repairing covenant. Fourthly, this process which Mr Colyer seeks involves the court authorising, in effect, the carrying out by one party on another party's property of some permanent and irreversible work of alteration to the existing condition of the property, in effect taking the control of the property out of the hands of the defendants altogether *pro tanto* for the purpose of carrying out these repairs. Fifthly, it is a claim which is made in the face of an undertaking now offered by the defendants to carry out such immediate and urgent work as they say will effect at least a temporary cure or amelioration during the interim period between now and the trial. There is an issue between the parties as to how effective those works will be. It is not an issue which I can J
possibly decide on this motion and there is a conflict of evidence about it. Sixthly, it is a claim which is made by a party who has no occupation of the premises and whose only interest in the premises is a purely financial one.

Having regard to those features and to the unusual nature of the claim it is a bold claim to make on an interlocutory motion. I think that before any "balance of convenience" question arises, the court must require to be satisfied as to the basis for what on any analysis is a very novel claim, and indeed that there is a seriously arguable case for saying that Mr Colyer will be able, as he claims he will be, to establish at the trial the right and—I stress this—the legal right which he says exists to go on and do this work. I pause only to K
mention that in at least two of the ways in which he puts his claim, if he is right, the right not only exists and inheres in him but inheres also in his tenant, who is not a party to these proceedings. It is primarily I think for the protection of Mr Colyer's tenant that his clients are proceeding with this claim.

First it is said, on the authority of a number of cases, and in particular the case of *Lee-Parker* v *Izzet* [1971] 1 WLR 1688 and *British Anzani (Felixstowe) Ltd* v *International Marine Management (UK) Ltd* [1979] 2 All ER 1063, that a tenant who, after notice, carries out work which his landlord has covenanted to do but has failed to do is entitled to set off the cost of doing the work against the rent and it is said that that is so on the basis of these L
authorities even though the work has been done by what may be a technical trespass. Then it is said that that leads to an inference to this effect, that every lease containing a landlord's repairing covenant contains also by implication a grant of any necessary easements or rights of entry which are needed to enable the tenant to do the work if the landlord defaults. That grant, it is said, is analogous to an easement and binds the landlord's estate in the hands of all those who claim under him. In the first place, it seems to me, if that is right, it is really very difficult to reconcile the principles upon which Mr Colyer relies with the *Stocker* case, to which I referred in the judgment on the first motion; but I am bound to say that I cannot see any logical reason why the cases to which Mr Colyer has M
referred me should be thought to give rise to any such right at all. It is, as it seems to me, and as Mr Millett submits, a total *non sequitur*. No authority for such proposition has been cited and it does not seem to me that as a proposition it really bears examination at all.

The second way in which the claim is put is this. It is submitted that Sedgwick Forbes demonstrate a title which will enable them to enter Regional's land by the exercise of a remedy of self-help, that remedy of self-help which is open to an injured party who is damaged by a nuisance committed from his neighbour's land and such a person can, it is submitted, enter and abate if his neighbour does not remedy the nuisance. If his neighbour seeks to prevent him, the

A court, it is submitted, will enjoin him, that is the neighbour, from so preventing entry. This is a question upon which there has been considerable argument and I have been referred to a number of authorities, but in particular Mr Colyer relies upon the Privy Council decision in *Goldman v Hargrave* [1967] 1 AC 645 and the more recent decision of O'Connor J at first instance in *Leakey v National Trust* [1978] QB 849. That decision has been affirmed in the Court of Appeal and I have been given the opportunity of reading the transcript, although I think it has not yet got into the Law Reports. The effect of those cases, as Mr Colyer submits, is that where a danger to the plaintiff's land arises from the defendant's land due to a hazard on the defendant's land, even though that hazard may be
B due to natural causation, the defendant, once he knows about it and knows of the danger, is under a duty, in effect, to take steps to ensure that the hazard does not spread and affect his neighbour's land and if he fails to do so and abate the nuisance himself then he may be liable in damages.

I do not propose on this interlocutory motion to engage in a detailed analysis of the cases, but I have found the very greatest difficulty in seeing how the mere percolation of water through the non-repair of the defendants' premises can be said to be a hazard, except of course in the sense that all life is a hazard. Mr Millett has referred me to a number of cases establishing the proposition—Mr Colyer I think would submit only in relation to mining operations; I
C think the principle goes wider than that—that there is no liability on a landowner for the mere natural percolation of water through the soil or surface of his land. Of course, if he collects it, or something of that sort, then he does create a hazard. But mere presence of water in the form of falling rain and its percolation does not, as I see it, constitute a hazard of the type that is referred to in the *Leakey* case. I need, I think, say no more about this at this stage than this—because it is a matter which will no doubt have to be investigated fully with all the authorities relevant to this difficult branch of the law at the trial—that I remain wholly unpersuaded that this is a case where there is a common law liability in nuisance which would give rise to any right to abate of the type upon which
D Mr Colyer relies.

The third way in which the matter is put is this. It is said that the existence of the roof at the time of the demise of the interior of the premises to Mr Colyer's clients was an advantage enjoyed with the premises which were demised which passed as an easement under the doctrine of *Wheeldon v Burrows* [1879] 12 ChD 31, or I suppose under section 62 of the Law of Property Act, and that as an easement—the easement being the right to be protected against the fall of natural water from the skies—there is a correlative right implicit to do that which is necessary to maintain the easement. The existence of such a correlative right would seem to derive some support from cases such as *Jones v Pritchard* [1908] 1 Ch 630. I
E confess that I find this analysis of the position a much more attractive one than the two alternative ways in which Mr Colyer has put his case. But it is submitted by Mr Millett that that approach is precluded by the decision of the Court of Appeal, and it was a unanimous Court of Appeal, in the case of *Phipps v Pears* [1965] 1 QB 76. That was a case where the right of protection from the elements sought was one which was sought in relation to premises which existed side by side and which were not joined horizontally, as they are in this case. Lord Denning MR, who delivered the only reasoned judgment (the other two members of the court simply agreed with him), says this at p 83:

> F There is no such easement known to the law as an easement to be protected from the weather. The only way for an owner to protect himself is by getting a covenant from his neighbour that he will not pull down his house or cut down his trees. Such a covenant would be binding on him in contract: and it would be enforceable on any successor who took with notice of it. But it would not be binding on one who took without notice.

That is widely expressed and Mr Colyer submits that it is really restricted to the case of adjoining tenements in a vertical plan and that it can have no application to a case such as this where the only object of the superior premises is the protection of the inferior premises, or the subjacent premises, from the onslaught of the elements. I do not propose to decide that question on this motion. I think that it would be quite wrong that I should do so. There is this authority of the Court of Appeal which is of course binding upon G me. I have the strongest suspicion that the somewhat wide expressions used by the Master of the Rolls might not have been formulated having in mind the sort of position that one has here. I am not prepared to say, at any rate so far as this ground of attack by Mr Colyer is concerned, that it is unarguable. It seems to me that there are arguments which can be advanced in favour of the submission which he makes, serious arguments, which will have to be dealt with on total review of all the authorities when it comes to the trial, because a motion of this sort is not of course an appropriate proceeding for deciding what is on any analysis a very difficult question of law.

So I am prepared to approach the motion on the basis that there is H an arguable point here and that brings me to the question of the balance of convenience. Much has been said here about the necessity for making the premises wind and water-tight over the coming winter. Mr Colyer has pointed out that if he is allowed to go in with his workmen and do all such work as is necessary any damage which may be sustained by Mr Millett's clients is minimal and indeed Mr Millett's clients may in fact get a windfall out of it; therefore he says the convenience lies all on one side. I am bound to say that I regard this motion as a most unusual one. It was of course envisaged in the *American Cyanamid* case [*American Cyanamid Co v Ethicon Ltd* [1975] AC 396] that there might be situations of a very unusual nature where the general principles which were applied in that case J might require some modification. But in this case it seems to me that Mr Millett is right in saying that in the ultimate analysis the question between his clients and Mr Colyer's clients is one of damages and Mr Colyer's clients have no interest in these premises except a financial one and if they suffer damage as a result of continued lack of repair then damages will be an adequate remedy for them. Mr Colyer says the damages will be difficult to assess. I am bound to say I think that is greatly exaggerated. I do not see any real difficulty in assessing the damages if, as Mr Colyer's clients fear, there is some adverse consequence as a result of disrepair on the rent reviews which are due to take place. Secondly, of course, this is a case which I think I must approach with a good deal of caution in K view of the fact that what is really sought on this motion is to pre-empt the trial. That is not a fatal objection, of course, to relief on motion, but I think it is a circumstance which the court has to take into account in assessing whether it is right at this stage to grant all the relief which the plaintiff is claiming and that particularly is so where an undertaking is offered which, at least on one view of the facts, would be sufficient to preserve the position pending the hearing of the action. Mr Millett also suggests that to concede the unusual relief which is claimed in this case would, or might, expose his clients to yet an additional claim for damages in that Mr Colyer's clients, in addition to the damages which of course are being assessed between Mr Millett's clients and Mr Bernstein's clients, in the L Queen's Bench Division, might seek to withhold any sums which they expend on repairs from the rent and thus expose Mr Millett's clients to yet a further claim at the hands of Mr Bernstein's clients.

In all the circumstances I do not think this is a case where it would be right that I should grant to the plaintiffs the very unusual and stringent relief that they claim. Had Mr Millett failed to offer any sort of undertakings which might remotely appear to be satisfactory I might have taken a different view. But in the light of the undertaking, which was sufficient at any rate to satisfy Mr Millett's landlords, who are the persons with whom he has a direct contractual relationship, I think that I, too, should be satisfied with that undertaking, and that undertaking having been given I should not make any M order on the plaintiffs' motion in this action.

A **CHANCERY DIVISION**

October 23 1980

(Before Mr Justice NOURSE)

GMS SYNDICATE LTD v GARY ELLIOTT LTD and OTHERS

Estates Gazette April 18 1981

(1980) 258 EG 251

Landlord and tenant — Forfeiture — Alleged immoral user of basement premises used as sauna bath, gymnasium and health B club — Ground floor and basement in building of six floors let to first defendants, a company running retail men's clothing shops — Basement sublet to second and third defendants, persons carrying on the business of sauna bath etc — Covenants in both lease of ground floor and basement and sublease of basement against annoyance, inconvenience and nuisance and against illegal or immoral user — Direct covenant also between head lessors and sublessees to observe these covenants — Evidence of breaches of covenants by participation of some masseuses in immoral practices and by annoyance and nuisance to residential tenants by men calling at property — Evidence also that the second defendant knew that such practices were going on and that first defendants, after becoming C aware of them, took no reasonable steps to prevent continuance — Head lessor entitled to judgment for possession against first defendants, subject to the question of relief — No relief possible for sublessees of basement against forfeiture — Primary object of head lessors to get sublessees out of basement — Whether to forfeit head lease in toto or to grant relief against forfeiture in respect of part only of the subject premises, namely, the ground floor — A "novel and interesting question" — Dumpor's case in "Coke's Reports" relied on — Relief granted to head lessors by restricting order for possession to basement — Award of damages against second and third defendants under direct covenant with head lessors

This was an action for forfeiture brought by GMS Syndicate Ltd, D the freehold owners of premises at 125 Queensway, London W2, against the first defendants, Gary Elliott Ltd (who run the "Lord John" chain of men's clothing shops), lessees of the ground floor and basement of the premises, and the second and third defendants, Napleandra Ruparel and his wife, Yvonne Ruparel, the sublessees of the basement of the premises. The premises consisted of six floors, including the basement, and there were residential tenants in occupation of the upper floors. The second and third defendants were the proprietors, through a private company, of the Queensway Health Club, which carried on the business of a sauna bath, gymnasium and health club in the basement.

A G Bompas (instructed by Collyer-Bristow) appeared on behalf E of the plaintiffs (head lessors); R C Pryor (instructed by Bennetts) represented the first defendants (the lessees of the ground floor and basement); R J Moshi (instructed by Portner & Jaskel) represented the second and third defendants (the sublessees of the basement).

Giving judgment, Nourse J said: This is a forfeiture action based primarily on the alleged immoral user of basement premises known as the Queensway Health Club at 125 Queensway, London W2. Those premises are used as a sauna bath, gymnasium and health club. The plaintiff, GMS Syndicate Ltd, is the freehold owner of 125 Queensway. The first defendant, Gary Elliott Ltd, is the lessee of the ground floor and basement. The second and third defendants, Napleandra Ruparel and his wife, Yvonne Ruparel, are F the sublessees of the basement and the proprietors, through the medium of a private company, of the Queensway Health Club.

125 Queensway is built on six floors including the basement. The upper floors are let to residential tenants of the plaintiff. The tenants of the third and fourth floors are Mr and Mrs Joseph Ralph, who have been there since 1946. Their daughter, Miss Kathleen Ralph, was the tenant of the second-floor flat from June 1968 until September 1977. The tenants of the first-floor flat are Mr and Mrs Eric McGregor, who have been there since 1947.

By a lease dated April 2 1969 the plaintiff demised to the first defendant the ground floor and the basement for a term of 12¼ years from December 25 1968. That term will expire on March 25 1981.

Clause 3 of the lease contains covenants by the first defendant in G the following terms:

> 14. Not to do or suffer or permit to be done or suffered on the premises anything which may be or become an annoyance inconvenience or nuisance to the lessors or the other owners or occupiers of any adjoining or neighbouring property or to the neighbourhood or which may infringe any legislation for the time being in force.
>
> 15. Not to carry on or permit upon the premises or any part thereof any noisy or dangerous trade business manufacture or occupation or any nuisance nor use the same or allow the same to be used for any illegal or immoral purpose. . . .

The lease contains a common form proviso for re-entry on non- H payment of rent or on failure or neglect by the first defendant to perform or observe any of the covenants, conditions or agreements on its part therein contained.

The first defendant runs a chain of retail men's clothing shops under the name of "Lord John". By the end of 1969 the first defendant had made arrangements, with the consent of the plaintiff, to sublet the basement to a Mr Silvalingham and a Mr Roberts for use as a sauna bath, gymnasium and health club until September 1972, that being the period covered by a planning permission which had been obtained for that use. Those arrangements were put into effect by an agreement dated December 29 1969 and were continued after an extension of the planning permission until June 30 J 1977 had been granted in August 1972. The continuation was effected by a supplemental agreement dated January 30 1974 by which the first defendant agreed to let, and Mr Silvalingham and Mr Roberts agreed to take, the basement premises for a further term commencing on September 1 1972 and expiring on June 30 1977. That agreement incorporated (by reference to the earlier agreement of December 29 1969) covenants on the part of Mr Silvalingham and Mr Roberts in all respects identical to the covenants on the part of the first defendant contained in clause 3(14) and (15) of the head lease to which I have already referred.

By an assignment dated December 23 1974 Mr Silvalingham and Mr Roberts assigned to the second and third defendants, Mr and K Mrs Ruparel, the benefit of the agreement of January 30 1974. On the same day the plaintiff, the first defendant, Mr Silvalingham and Mr Roberts and Mr and Mrs Ruparel had executed a deed whereby, first, the plaintiff and the first defendant had granted their consents to the assignment to Mr and Mrs Ruparel and, secondly, Mr and Mrs Ruparel had covenanted with the first defendant, and as a separate covenant with the plaintiff, that as from the date of the assignment "and thenceforth during the residue of the term granted by the underlease as extended by the supplemental tenancy agreement" they would pay the rent thereby reserved and observe and perform the covenants and conditions on the part of the sublessees therein contained. There was therefore a direct covenant by Mr and L Mrs Ruparel with the plaintiff to observe and perform (among others), first, the covenant against annoyance, inconvenience and so forth and, secondly, the covenant against illegal or immoral user incorporated in the agreement of January 30 1974. One of the questions in these proceedings is whether that direct covenant expired with the contractual term on June 30 1977 or whether it continues while Mr and Mrs Ruparel hold over (as they now do) under Part II of the Landlord and Tenant Act 1954. They have been able to do that because in September 1978 an appeal by them against the Westminster City Council's refusal to extend the planning permission for the continued use of the basement as a sauna bath, gymnasium and health club was allowed. M

I have already said that the plaintiff's action is based primarily on the alleged immoral user of the Queensway Health Club. I must now state that and the other alleged causes of complaint in greater detail. First, the plaintiff alleges that since in or about July 1978, Mr and Mrs Ruparel have allowed the basement to be used for illegal and immoral purposes and in particular that they have employed as masseuses on the premises women who for reward commit lewd and immoral practices at the behest of customers of the business. Secondly, the plaintiff alleges that the business carried on by Mr and Mrs Ruparel in the basement has become an annoyance, inconvenience and nuisance to the plaintiff and the occupiers of adjoin-

A ing and neighbouring property (in particular the plaintiff's residential tenants at 125 Queensway), and the plaintiff says that the business has been so conducted as to encourage persons to attend at the premises seeking women with whom to perform lewd or immoral practices. The plaintiff says that Mr and Mrs Ruparel have thereby brought the basement and the building as a whole into disrepute and, further or alternatively, have caused persons seeking disreputable services to attend at and loiter near the building so causing nuisance and giving offence to the plaintiff and its residential tenants. The plaintiff alleges that in these two respects Mr and Mrs Ruparel have been in breach of the direct covenant with the plaintiff contained in the deed of December 23 1974 and for that
B the plaintiff claims damages. However, its principal claim is against the first defendant for forfeiture of the headlease on the grounds that the first defendant also has allowed the basement to be used for illegal and immoral purposes and that it has suffered or permitted a business to be carried on in the basement which is an annoyance, inconvenience or nuisance to the plaintiff and to the occupiers of the adjoining and neighbouring property. In this connection, the plaintiff alleges that on or before August 1 1978 the first defendant had become aware of the matters of which the plaintiff complains in relation to the use of the basement but that it has nevertheless failed and neglected to take any sufficient steps against Mr and Mrs Ruparel. If the plaintiff succeeds in its claim for forfeiture of the
C headlease then, subject to any question of relief, it will follow that Mr and Mrs Ruparel's subtenancy will be forfeited as well. I should at this stage say that the plaintiff's primary objective is to get Mr and Mrs Ruparel out of the basement. If it succeeds in that, it will be less concerned with its claim against them for breach of the direct covenant in the deed of December 23 1974.

The defendants either deny or do not admit that the matters complained of have taken place. Further, they say that, if they have, they have not allowed or suffered or permitted them to happen. They each, so far as may be necessary, claim relief against forfeiture.

It is now necessary for me to deal with the material facts and the
D relevant evidence in some detail. First, I shall start with the allegation that since in or about July 1978 the basement has been used for immoral purposes. This allegation is based on the evidence of two private investigators. First, Arthur Price gave evidence of the two visits he paid to the premises. The first was on July 14 1978 when he said that he was given a massage by an Indian woman, aged about 35, who then offered to perform and did perform with him an indecent act falling short of sexual intercourse, for which he paid her an additional £15 over and above the fee of £12.50 which he had already paid the receptionist. Mr Price's second visit was on July 3 1980, shortly before this action came on for hearing, when he was given a massage by an English woman in her mid-twenties who did
E not offer to perform and did not perform any indecent or immoral act. The second private investigator who gave evidence was Mark Terence Langley, who visited the premises on June 25 1979, about one year after Mr Price's first visit. He said that he was given a massage by a woman of Chinese origin in her late twenties or early thirties, who then offered to perform and did perform with him an indecent act falling short of sexual intercourse, for which he paid her an additional £20 over and above the fee of £7.50 which he had already paid the receptionist. In spite of one problem which I am unable to resolve, I have no hesitation in accepting in its entirety the evidence of Mr Price and Mr Langley, who were both reliable witnesses, whose evidence could not be disputed and who were not
F shown to have any reason for giving evidence which was not true. The one problem is that the evidence adduced on behalf of Mr and Mrs Ruparel was universally to the effect that only one woman of Indian origin was employed at the premises from July 1978 onwards and that was a Mrs Singh. She was produced in court during Mr Price's evidence, but he said quite positively that she was not the woman who had given him massage on his first visit in July 1978. As I have said, I am unable to resolve this problem. A number of explanations are possible. I need not go into these. In the end I am satisfied that Mr Price was telling the truth and that the events he described took place. Having accepted his evidence and that of Mr Langley, I conclude that between July 1978 and June
1979 the premises were on two occasions used for immoral pur- G
poses. And I would, if it is necessary for me to do so, infer that they were so used on other occasions during the same period. It is unnecessary for me to decide whether the user was also illegal and I do not do so.

It not having been suggested that Mr and Mrs Ruparel themselves used the premises for immoral purposes, I must next consider whether they allowed them to be so used.

The substance of the plaintiff's allegation in this respect is that Mr Ruparel knew or ought to have known of the immoral user, that he abstained from taking reasonable steps to prevent it when it was within his power to do so, and that he thereby "allowed" the premises to be so used: compare the judgment of Atkin LJ in H
Berton v Alliance Economic Investment Co [1922] 1 KB 742 at p 759. I should point out that Mrs Ruparel was not concerned in any way with the day-to-day running of the business. She helped with the correspondence, did the books and so forth, but that was the extent of her involvement. It is not suggested that she personally allowed the premises to be used for immoral purposes. It is enough for the plaintiff's purposes if it establishes that Mr Ruparel allowed them to be so used.

The plaintiff says that Mr Ruparel did know what was going on and took certain steps which were only consistent with his having that knowledge or at least with his not caring whether the premises were being used for immoral purposes or not. In particular the J
plaintiff relies on the fact that Mr Ruparel placed advertisements for the Queensway Health Club in indecent and pornographic publications which could only have led those who read them to expect that they could receive immoral services if they attended the premises. There was a vast amount of evidence on this topic. The essence of it was that in 1975 and 1976 advertisements were placed in three publications with substantial circulations whose contents, both photographs and text, were frankly admitted by Mr Moshi, who appears for Mr and Mrs Ruparel, to be pure filth. Mr Ruparel was then advised not to advertise in those publications by the Westminster City Council, which has a supervisory jurisdiction over establishments of this kind within its area of responsibility. K
However, in 1978 and 1979 he placed a further series of advertisements in three other publications with even larger circulations, in which the photographs were admittedly of a less offensive nature, but in which certain sections of the text were no better than the first. As to the advertisements themselves, the main part of the text in both series was as follows: "Come to West London's most luxurious sauna for a super private massage — assisted shower — VIP and Two Girl massage — beautifully friendly masseuses". Taking the text of the advertisements and their placing in the publications to which I have referred together, and having seen and heard Mr Ruparel in the witness box, I am satisfied that Mr Ruparel well knew that they would be likely to lead those who read them to L
expect that they could receive immoral services if they came to the premises. It was said that similar advertisements were placed in respectable papers and magazines as well, but that seems to me to be beside the point. The question is, what did Mr Ruparel think would be the likely effect of the advertisements when they were placed not in respectable papers and magazines, but in publications of an indecent or pornographic nature? That question admits only of the answer I have given. That answer is confirmed by the unsatisfactory, and at times evasive, manner in which Mr Ruparel gave discovery for the purposes of these proceedings.

There are other matters which tend to show that Mr Ruparel knew or ought to have known that the premises were being used for M
immoral purposes. First, it is said by the plaintiff that the photographs on display outside the premises necessarily suggested to the casual observer that he could receive immoral services inside. I think that there is something, although not very much, in this point. Secondly, in July 1978, there was a public inquiry to hear Mr and Mrs Ruparel's appeal against the refusal of the Westminster City Council to continue the authorised use of the premises for planning purposes. Before then the plaintiff's advisers had obtained a report of Mr Price's visit to the premises on July 14 1978 and its managing director, Robin Gibbon, asked for leave to call Mr Price to give evidence to the inspector. Having consulted the department, the

inspector refused that request. But Mr Gibbon, in the presence of Mr Ruparel, had already outlined in broad terms to the inspector what Mr Price was going to say, that he would give evidence of having attended at the premises and would say what he saw and what services he had been offered. I am satisfied that Mr Ruparel then, if not before, knew perfectly well what was being alleged against him by the plaintiff. In my judgment, this is an important point and it tells against Mr Ruparel. Thirdly, and perhaps equally important, in December 1978 the first defendant served a section 146 notice on Mr and Mrs Ruparel based on the alleged user of the premises for illegal and immoral purposes. Mr Ruparel said that he knew what this meant, that he called the masseuses together and he threatened any of them who was caught committing an immoral act with immediate dismissal. I have some doubts whether Mr Ruparel did do this, but if he did, Mr Langley's evidence of what happened in June of the following year shows that the steps he took were inadequate. Fourthly, there was evidence as to how much the masseuses earned from their employment. Mr Ruparel said that a masseuse made £30 to £40 for a week of about 30 hours: £15 of this would be a basic wage of 50 pence an hour, and the balance would be made up by commissions at varying rates not exceeding £1 per client. He made no mention of tips. Mrs Singh said that she expected to earn about £65 per week, of which about £35 to £38 would be wages (including commission) and the balance tips. I do not regard Mrs Singh as a reliable witness. It seemed to me that her evidence either exaggerated or played down events as the occasion required. I therefore conclude that Mr Ruparel's estimate of the normal earnings of a masseuse is likely to be nearer the truth. And then it might be said that Mr Ruparel ought to have realised that the women would not be content with £35 to £40 per week and might well have been earning additional remuneration by providing services of an immoral nature. However, it was not put to Mr Ruparel quite in that way and I therefore do not think it would be right to attach any great weight to this particular point.

On the other side, evidence was given by two clients of the business to the effect that there had never been any indication of impropriety when they visited the premises. That is not conclusive evidence of the absence of impropriety, particularly when it is alleged to have taken place behind the closed doors of the massage rooms. Nor is it at all conclusive as to the state of Mr Ruparel's knowledge. And it appeared that one client had only had about 10 massages in all, between about 1970 and 1978, for the specific purpose of treating his bad back. The other had never had a massage and had used the premises mostly for the purpose of visiting the gymnasium. There was also evidence that checks were carried out by Mr Ruparel himself, by the manager, Mr Mehra, and by Westminster City Council, but having considered the whole of that evidence I am satisfied that the checks were neither thorough nor regular enough to prevent immoral services being offered and accepted behind the closed doors of the massage rooms, and even though there were no locks on the doors. In particular, I think that Mr Mehra's evidence on this point was exaggerated and unreliable.

Having considered all the evidence on this question, and in particular that of Mr Ruparel himself, I am satisfied that he either knew that indecent and immoral acts were being committed on the premises or, it matters not which, that he shut his eyes to whether they were being committed or not. For this conclusion I rely in particular on the advertisements, on Mr Ruparel's knowledge of what Mr Gibbon told the inspector in July 1978, and on the service of the section 146 notice on him in December of that year. It is also clear that although it was within Mr Ruparel's power to prevent the immoral user, he abstained from taking reasonable steps to do so. To a large extent I have covered this point already. I need mention only two further matters. First, it appears that shortly before this action came on for trial the doors of the massage rooms were taken off. If that could be done then, it could and should have been done after Mr Ruparel heard what Mr Gibbon told the inspector in July 1978. Secondly, after July 1978, Mr Ruparel still continued with the second series of advertisements which advertised, among other things, "private" massage by "beautiful friendly masseuses". He clearly ought then, if not before, to have stopped advertising in that form. In all the circumstances I conclude that Mr Ruparel did allow the premises to be used for immoral purposes between July 1978 and June 1979.

I deal next with the question whether the first defendant allowed the premises to be used for immoral purposes. Shortly after the planning inquiry in July 1978 the plaintiff's advisers sent the first defendant a copy of the written report of Mr Price's visit to the premises on July 14. In all material respects that report was to the same effect as the evidence given by Mr Price in these proceedings. I need not deal in detail with what the first defendant did or did not do after receiving that report, because it is clear that it did not take reasonable steps to prevent the immoral user from continuing. Indeed Mr Pryor, who appears for the first defendant, while making no concession on the point, found it difficult to argue the contrary. The facts are that it accepted rent from Mr and Mrs Ruparel after serving the section 146 notice in December 1978. It did not serve a fresh notice. It did not take forfeiture proceedings against Mr and Mrs Ruparel. In my judgment, the first defendant ought to have taken those steps. Since it did not do so it is clear that it, too, allowed the premises to be used for immoral purposes.

On that footing and, subject to questions of relief, the plaintiff is entitled to judgment for possession against the first defendant. That will necessarily involve the forfeiture of Mr and Mrs Ruparel's subtenancy of the basement. That means that the plaintiff's secondary claim based on annoyance, inconvenience and nuisance becomes of less importance, but I must nevertheless deal with it as shortly as I can in relation to the plaintiff's claim against Mr and Mr Ruparel under the direct covenant contained in the deed of December 23 1974.

Mr Bompas, who appears for the plaintiff, emphasised that as against Mr and Mrs Ruparel this claim is not based on their having allowed or permitted or suffered something to happen. Clause 2(14) of the agreement dated December 29 1969, when suitably abstracted for present purposes, requires the lessees not to do on the premises anything which may be or become an annoyance, inconvenience or nuisance to the plaintiff or the other owners or occupiers of any adjoining or neighbouring property and so forth. Mr Bompas, assuming at this stage that Mr and Mrs Ruparel are still liable under the direct covenant contained in the deed of December 23 1974, says that they are in breach of this covenant because they have carried on on the premises a business which has become an annoyance, inconvenience and a nuisance to the plaintiff and its residential tenants of the upper parts of the building. As to this, evidence was given by Mr McGregor, Mr and Mrs Ralph and also Miss Kathleen Ralph. Without going into great detail, I am entirely satisfied on the evidence that the presence of Mr and Mrs Ruparel's business in the basement has caused annoyance and inconvenience and has been a nuisance to the plaintiff's residential tenants above. Many incidents were described in evidence. In particular, evidence was given of men ringing the doorbells to the flats late into the night on frequent occasions, asking if the sauna was open or whether massage was available and even explicitly asking for sexual intercourse. Mrs Ralph gave evidence that this request was made to her on one occasion in respect of a named woman in terms which were so offensive to her that she asked to be allowed to write them down on a piece of paper. Then there was ample evidence that the tenants and their visitors had been subjected to various forms of importunity and insult from persons standing in or outside the common entrance passageway which leads from the front door both to the stairs down to the basement and the stairs up to the flats. On one occasion Mrs Ralph, who is an elderly lady, was prevented from going down the passageway by two youths, one of whom threatened her in a way which frightened her badly. In general I have every sympathy with the tenants in respect of their complaints on these matters and I am quite satisfied that the events which they described were the direct result of the presence of Mr and Mrs Ruparel's business in the basement.

The convenient course will be for me to deal next with the question whether the direct covenant by Mr and Mrs Ruparel with the plaintiff contained in the deed of December 23 1974 expired with the contractual term on June 30 1977 or whether it continues while Mr and Mrs Ruparel hold over under Part II of the Landlord and Tenant Act 1954. As I have said, the material words are "and thenceforth during the residue of the term granted by the underlease as extended by the supplemental tenancy agreement". On one reading of those words it certainly might be said that the covenant expired with the contractual term, on the ground that that and no more was the term granted by the agreement of December 29 1969 as extended by that of January 30 1974. But it seems to me that such a reading ignores the effect of section 24(1) of the 1954 Act, which provides that a tenancy to which Part II applies shall not come to an end unless terminated in accordance with the provisions of Part II. It is now well established that this means that the term granted by the tenancy continues by way of a statutory extension and with a statutory variation as to the mode of determination: see, for example, *Cornish v Brook Green Laundry Ltd* [1959] 1 QB 394 at p 409. In the circumstances, it seems to me that the term granted by the first agreement as extended by the second is still subsisting, albeit that it has been further extended by the 1954 Act. It is still the same term. On that footing, it seems to me that the better reading of the material words in the direct covenant is to construe the obligation as continuing so long as Mr and Mrs Ruparel hold over under Part II of the 1954 Act. And I am fortified in that conclusion by the knowledge that the rival construction would produce an anomalous state of affairs which cannot have been within the contemplation of the parties to the deed. It follows that the plaintiff's claim against Mr and Mrs Ruparel for damages for breach of the direct covenant in the two respects I have mentioned succeeds. I will deal with quantum at the end of this judgment.

I deal next with the question of relief against forfeiture. Taking it by stages, it is clear, first, that I cannot, or at least that I should not, grant relief to Mr and Mrs Ruparel. It is the established practice of the court not to grant relief in cases where the breach involves immoral user, save in very exceptional circumstances such as those which were considered in *Central Estates (Belgravia) Ltd v Woolgar [No 2]* [1972] 1 WLR 1048. There are no such circumstances in the present case so far as Mr and Mrs Ruparel are concerned.

The position of the first defendant is more difficult. As I have said, the plaintiff's primary objective is to get Mr and Mrs Ruparel out of the basement. If the only way it can achieve that is to forfeit the head lease in toto, then that is what the plaintiff seeks. On the other hand, provided that the plaintiff can be sure of obtaining possession of the basement, it has said that it will not object to the first defendant remaining in possession of the ground floor. This has raised a novel and interesting question. Has the court got jurisdiction to grant relief against forfeiture in respect of part only of the property comprised in a lease?

I should start by saying that if the only way in which the plaintiff can get possession of the basement is to forfeit the head lease in toto, I would think it clear that this is not a case in which I ought to grant the first defendant relief: compare *Borthwick-Norton v Dougherty* [1950] WN 481. But if there is jurisdiction to grant partial relief, I would think it equally clear that I ought to do so. Indeed, since the plaintiff has said that it will not object to the first defendant remaining in possession of the ground floor, this is a question which has effectively been removed from my decision. I assume, of course, that the first defendant will be able to satisfy the plaintiff and the court that the plaintiff will not suffer in costs as a result of this action and that it will not be prejudiced in any material way by the severance of the ground floor from the basement. Accordingly, I turn to consider whether there is jurisdiction to grant partial relief. On this question, Mr Bompas, being apprehensive that it might not be possible to grant such relief without at the same time procuring a surrender by the first defendant of the head lease so far as it concerns the basement, thus leaving Mr and Mrs Ruparel as direct tenants of the plaintiff, has argued against there being jurisdiction. Mr Pryor has argued in favour. Mr Moshi has argued against.

It is to be noted that section 146(2) of the Law of Property Act 1925, under which the first defendant's application is made, does not specify the nature of the relief which the court can grant. It says that "the court may grant or refuse relief, as the court, having regard to the proceedings and conduct of the parties under the foregoing provisions of this section, and to all the other circumstances, thinks fit". It appears from cases such as *Bowser v Colby* (1841) 1 Hare 109 and *Dendy v Evans* [1910] 1 KB 263 that regard will readily be had to the practice of the old Court of Chancery before the Landlord and Tenant Act 1730 when relief against forfeiture was exclusively a matter for that court. This is not of direct assistance in the present case. In those days relief was confined to cases of non-payment of rent, and since rent issues out of the property as a whole there could have been no question of relief in part. But it is, I think, important to notice that the Court of Chancery, acting *in personam*, either restrained the landlord from proceeding to take possession at law or, if he had already done so, required him to grant a new lease. Either result could have been achieved by the landlord of his own volition. The court did not seek to impose a result which could not have been arrived at by the landlord himself.

Relief against forfeiture was a process by which equity restricted a landlord from enforcing his rights at law. In my judgment it could not then, and cannot now, be granted so as to impose upon the parties a legal relationship which could not have been procured by the act of the landlord. If, therefore, I am to grant partial relief in the present case I must first be satisfied that the plaintiff could, had it wished, have forfeited the lease so far as concerns the basement alone.

I start with the proviso for re-entry which is in common form and allows the plaintiff "into and upon the premises or any part thereof in the name of the whole to re-enter". It is therefore possible for the plaintiff to forfeit as to the whole by entering on part, but I do not read those words as restricting any right to forfeit as to part which is otherwise available to the plaintiff. As to that, my instinct was against the notion that a landlord can forfeit in part. But Mr Pryor referred me to *Dumpor's Case* (1603) 4 Co Rep, 119b, in which the Court of King's Bench established the once well-known rule of common law that a licence to do an act in breach of a condition in a lease determined the condition. That rule was abolished by section 1 of the Law of Property Amendment Act 1859 (Lord St Leonard's Act), which (with amendments) is now section 143 of the Law of Property Act 1925, and it is possible that this has caused the "divers points" which, as the report of *Dumpor's Case* says, were there "debated and resolved" to be forgotten. One of them is treated at 4 Co Rep 120b, in this way: "But it was agreed, that a condition may be apportioned in two cases. 1. By act in law. 2. By act and wrong of the lessee". Apportionment by act in law is then dealt with and the report proceeds as follows:

> 2. By act and wrong of the lessee, as if the lessee makes a feoffment of part, or commits waste in part, and the lessor enters for the forfeiture, or recovers the place wasted, there, the rent and condition shall be apportioned, for none shall take advantage of his own wrong and the lessor shall not be prejudiced by the wrong of the lessee. . . .

Having considered this passage with very great care, I can only conclude that the Court of King's Bench were of the opinion that it was in certain circumstances possible for a landlord to forfeit in part. Although the words "and the lessor enters for the forfeiture" do not expressly refer to an entry on part, it appears clear from the context as a whole that that is what they mean. Once that point is overcome, the whole passage is seen to contemplate forfeiture in part and a consequential apportionment of the rent and the conditions of the lease between the part taken by the landlord and that retained by the tenant. It is true that what was under consideration in that case was re-entry for breach of condition without an express power to re-enter. But I can see no distinction in principle between that and re-entry under a common form proviso such as that found in the present case. There being no express restriction on the right to re-enter on part, I conceive that that right must exist as much in the latter case as in the former.

In the circumstances, the researches of counsel having revealed no other authority either way, I think that I am entitled to treat *Dumpor's Case* as establishing that a landlord can in certain circumstances, such as those found in the present case, forfeit in part. Had it not been for that case I might have found some difficulty in acceding to Mr Pryor's arguments on this point. But there it is, and it seems to me to be a perfectly respectable peg on which to hang a decision which I believe will produce a fair and workable result as between the plaintiff and the first defendant. I emphasise that I do

A not intend to go beyond the circumstances of the present case, where the two parts of the demised property are physically separated one from the other and are capable of being distinctly let and enjoyed, and where the breaches complained of were committed on one part of the property and on that part alone. I have not considered what would have happened if the circumstances had been different.

I therefore propose to grant the first defendant relief against forfeiture by restricting the order for possession to the basement. I will make that order against all three defendants. I have already indicated the matters on which I shall first wish to be satisfied in favour of the plaintiff and I will discuss those matters with counsel B at the end of this judgment. I agree with Mr Pryor that the order for possession of the basement alone will not cause any conveyancing difficulties, because it is clear on the authorities that the apportionment of rent and any other questions consequential on the severance of the two parts of the property can, if necessary, be settled by the court. No doubt an attempt will first be made to settle those matters by agreement. I should add that I cannot see how the order which I propose to make could operate as a surrender by the first defendant of the head lease so far as it concerns the basement, with the consequences of which Mr Bompas was apprehensive. A surrender could not be made without the concurrence of the first defendant. The order for possession of the basement is made *in* C *invitum* as against the first defendant, which has claimed relief in respect of the whole and has only obtained it in part.

Finally, I return to the quantum of the damages which I should award the plaintiff for Mr and Mrs Ruparel's breaches of the direct covenant contained in the deed of December 23 1974. Having achieved its primary objective of getting Mr and Mrs Ruparel out of the basement, the plaintiff is less concerned with its claim against them for damages. I propose, therefore, to award a sum which makes every allowance in favour of Mr and Mrs Ruparel. The sum I award is £500.

D

CHANCERY DIVISION

December 12 1980

(Before Mr Justice NOURSE)

MARJORIE BURNETT LTD v BARCLAY

Estates Gazette May 16 1981
(1980) 258 EG 642

Landlord and tenant — Whether lease created a "perpetually renewable leasehold" operating as a demise for 2,000 years — Lease of shop premises with residential accommoda- E **tion — Correct approach to question of construction — Covenant for renewal in the present lease construed to provide that the like covenant in the second lease would allow only one further renewal, the effect being that the present lease could under its terms be renewed twice only — Landlords entitled to declaration to that effect**

This was an originating summons for construction, the declaration sought being that a lease between the plaintiff company, Marjorie Burnett Ltd, and the defendant, William Barclay, dated August 3 1971, for seven years from June 24 1971, was not a perpetually renewable lease but could be renewed twice only. The lease comprised a shop, dwelling-house and outbuildings at 179 F Ashley Road, Boscombe, Bournemouth.

J M Henty (instructed by Lennon & Co, of Swindon) appeared on behalf of the plaintiff company; the defendant did not appear and was not represented.

Giving judgment, NOURSE J said: This is an originating summons in which the primary relief sought is a declaration that upon the true construction of a lease it is not a perpetually renewable lease but can be renewed twice only.

The matter arises in this way. The lease is dated August 3 1971. It was made between the plaintiff in these proceedings, a company called Marjorie Burnett Ltd, as landlord of the first part, another company as tenant of the second part and certain individuals as sureties of the third part. G At some time between the date of the lease and 1975 it appears that it was assigned to the defendant in these proceedings, William Barclay. The property comprised in the lease is a dwelling-house, shop and outbuildings known as 179 Ashley Road, Boscombe, Bournemouth. It appears that those are shop premises with residential accommodation above or beside them. The lease was expressed to be for a term of seven years from June 24 1971 at the yearly rent of £650. It was mainly in common form and I do not need to refer to any of its other provisions until I reach clause 6. That is the clause which has given rise to the question which I now have to decide. It reads as follows:

> If the tenant shall be desirous of taking a new lease of the demised premises H after the expiration of the term hereby granted . . .

There are then provisions which are in a not very elegant form, but the effect of them is clear enough. It is that the tenant can give the landlord notice of the desire to take a new lease. Then clause 6 goes on as follows:

> then the landlord will at or before the expiration of the term hereby granted, if there shall then be no subsisting breach of any of the tenant's obligations under this present lease — and now I come to some important words — grant to the tenant a new lease of the premises hereby demised for a further term of seven years, to commence from and after the expiration of the term hereby granted at a rent to be agreed between the parties.

J There are then provisions for the rent to be fixed in default of agreement. Then come the final words of the clause, which are also important. They read as follows:

> And such lease shall also contain a like covenant for renewal for a further term of seven years on the expiration of the term thereby granted.

What happened was that in June 1975, some three years before the term granted by the lease expired, a firm of solicitors acting for the defendant wrote to the solicitors acting for the plaintiff, suggesting that the lease was a perpetually renewable lease for the purposes of paragraph 5 of the 15th Schedule to the Law of Property Act 1922 and that it had therefore taken effect as a lease for a term K of 2,000 years. Mr Henty, who appears for the plaintiff, says, and this must be right, that that contention could only have been based on clause 6 of the lease.

That suggestion was refuted by the plaintiff's solicitors, who said that in their view clause 6 gave the tenant options to renew for two periods of seven years and no more. At that stage the area of the dispute which I have to resolve was defined.

There were two further letters from the defendant's solicitors, the second in October 1975, which indicated that the defendant did not resile from the suggestion which had previously been made on his behalf. It appears that on December 15 1977 the defendant exercised the option for a new lease, pursuant to the provisions of L clause 6.

On June 24 1978 the term of seven years granted by the lease expired. On November 13 1978 the plaintiff company issued the originating summons which is now before me. That led to a letter from another firm of solicitors acting on the defendant's behalf. The letter was dated January 26 1979. It made an offer to the plaintiff that the defendant would be content to have these proceedings heard without argument on his part, provided that the plaintiff was prepared to abandon its claim for costs. That offer was accepted by the plaintiff by a letter of March 2 1979, and that is how the matter has now come before the court. Mr Henty, as I have said, is here to represent the plaintiff company, but the defendant does not appear M and is not represented. The position nevertheless is that Mr Henty has to make out a case for the declaration which the plaintiff seeks.

The way in which he has sought to do that is this. First he has referred me to *Woodfall on Landlord and Tenant*, 28th ed, vol 1, para 2071, where I find this passage:

> The leaning of the courts has been against perpetual renewals, so that in order to establish this construction the intention has to be unequivocally expressed, and a proviso in general terms that the lease to be granted shall contain the same covenants and agreements as the lease containing the covenant or option has been repeatedly held not to extend to the covenant or option for renewal.

A There is then a reference to three old cases, the most recent of which, *Moore v Foley*, was decided in 1801 and is in 6 Ves 232. At the beginning of the next paragraph, 2072, there is this sentence:

> But although, prima facie, a lessor was taken not to have intended to give such a covenant the court will give effect thereto if there is in the lease evidence of such intention.

Then Mr Henty referred me to two more recent decisions of the Court of Appeal, in each of which it was held that the provisions of the lease in question were sufficient to make it a perpetually renewable lease and thus to secure its conversion into a 2,000-year term under the provisions of the 1922 Act. The first was *Parkus v*
B *Greenwood* [1950] Ch 33 reported at first instance before Harman J. Harman J was of the view that there was not in the case of the lease there before him a covenant or obligation for perpetual renewal. His decision was reversed by the Court of Appeal, but I would like to refer to a passage in his judgment at p 37. Before I do that I should state that the words of the material provision in that case were these. The landlord agreed that he would grant a tenancy of the demised premises "for a further term of three years from the expiration of the said term at the same rent and containing the like agreements and provisions as are herein contained, including the present covenant for renewal". At p 37 Harman J said this:

> C Mr Albery . . . pointed out that in the old conveyancing precedents these words were used to create a perpetual right to renew and that a careful conveyancer if he wished to avoid trouble and did not wish to have it said that there might be such a perpetual right, would use the opposite words, namely "excluding this present covenant". All that I accept; none the less, in my judgment, this part of the Act only operates to create a term of 2,000 years where the lease is on the face of it perpetually renewable. You have to find expressly in the lease or agreement a covenant or obligation for perpetual renewal. I do not find any such covenant here. All I find is a covenant for renewal once.

Although, as I have said, Harman J's decision was reversed by the Court of Appeal [1950] Ch 644, that reversal proceeded on the footing that there was in fact a covenant or obligation for perpetual
D renewal. There is nothing in the Court of Appeal's decision to displace the learned judge's statement that you have to find expressly in the lease or agreement a covenant or obligation for perpetual renewal.

The second case in the Court of Appeal was *Caerphilly Concrete Products Ltd v Owen* [1972] 1 WLR 372. In that case it was again held that there was a covenant or obligation for perpetual renewal. Again I will read the material words. There was an agreement by the landlord to grant a lease for a further term of five years "at the same rent and containing the like covenants and provisos as are herein contained (including an option to renew such lease for the term of five years at the expiration thereof)". I find in the leading
E judgment of Russell LJ in that case at p 375 this passage:

> In the present case the brackets make it abundantly plain that the parties are explaining that "containing the like covenants and provisos" is a phrase intended to embrace an option. That is to say that the covenants and provisos contained in the first lease, which the first lease requires the second lease to contain, are not to be construed as a reference to those covenants and provisos other than an option to renew, but as a reference to all those covenants including an option to renew.

And then a little later on he said this in reference to the material words:

> If the words . . . are repeated in the second lease without the words in
> F parenthesis the second lease will not be carrying out the requirement of the first lease: it will not be granting an option for a further lease containing "the like covenants" as defined.

In construing clause 6 of the lease in the present case I must therefore approach the matter in this way. I must bear in mind that the leaning of the court s has been against perpetual renewals. I have to find expressly in the lease a covenant or obligation for perpetual renewal. And I have to look ahead to see what the second lease will contain when the requirements of the covenant for renewal in the first have been duly observed.

I now return to clause 6 of the lease. What the landlord has to do, if the tenant gives it notice of his desire to take a new lease, is to grant to the tenant a new lease of the demised premises for a further G
term of seven years at a rent to be agreed, and if not agreed to be fixed in the manner specified. Then it is provided that such lease shall also contain a like covenant for renewal for a further term of seven years on the expiration of the term thereby granted.

Mr Henty really puts his case on this primary question in two ways. First, he takes the simple course of asking me to see what provisions the second lease would contain if it were to be granted pursuant to clause 6. He says that it would inevitably be at a different rent from the £650 reserved by the first lease. Then he says that the second lease would contain a like covenant for renewal for a further term of seven years as that contained in clause 6 of the first lease. But that convenant ends with the part of clause 6 which deals H
with the provisions for fixing the rent in default of agreement. It does not seem to me that the second lease could possibly contain the words "and such lease shall also contain a like covenant for renewal for a further term of seven years on the expiration of the term thereby granted", because those words are not part of the covenant for renewal and to include them would be to go further than clause 6 requires. And so I agree with Mr Henty that the second lease would be at a new rent and that it would contain the whole of clause 6, except for the last three lines or so which I have just quoted. On that footing it is clear that there is no express covenant or obligation for perpetual renewal. Indeed the contrary is the case. There is an express provision in the lease to the effect that it can be renewed J
twice only.

That would in itself be enough to dispose of the primary question in these proceedings. But Mr Henty goes on to take a second point, which appears to me to be one of equal force, and that is this. He says that even supposing his first argument were wrong I must bear in mind that what will happen if this is a perpetually renewable lease is that it will be converted by the 1922 Act into a lease for a term of 2,000 years. He says, and I can see no answer to this, that the notion of a 2,000-year term is completely inimical to a lease which contains provisions for rent review every seven years. And so again he says that as a matter of construction clause 6 could not possibly have the effect for which the defendant has contended. I agree with that K
contention also.

In the circumstances it seems clear to me that the plaintiff company is entitled to the primary declaration which it seeks. On that footing it is not necessary for me to consider the second possible alternative raised by the originating summons.

QUEEN'S BENCH DIVISION

October 22 1980

(Before Mr Justice GLIDEWELL) L

DRUMMOND v S & U STORES LTD

Estates Gazette June 27 1981
(1980) 258 EG 1293

Landlord and tenant — Main issue quantum of damages for which tenants of shop premises were liable to landlord for breach of the covenant to repair — Useful statement of the law as to the measure of the damages and analysis of items in the schedule of works required to repair the premises — Proper measure the diminution in the value of the reversion — Smiley v Townshend — In the absence of direct evidence the cost of repairs may be a M
useful guide — Consideration of schedule in detail — Damages included value added tax, the landlord not being registered for such tax — Conflicting submissions as to the extent to which disrepair or lack of decoration would influence the bid of a new tenant — A landlord's arrangements with new tenant "res inter alios acta" as regards old tenant — Judgment a practical lesson in how the court deals with a schedule of dilapidations

In these proceedings the landlord, the freehold owner of shop premises at 14 Brentgoval Street, Bury St Edmunds, brought a number of claims against the tenants, S & U Stores Ltd, but the issues became narrowed down to a claim for damages for breach of the covenant to repair in the lease and a claim for costs incurred by the landlord in relation to sections 146 and 147 of the Law of Property Act 1925.

D Lamming (instructed by Foskett, Marr, Gadsby & Head, of Epping) appeared on behalf of the plaintiff, Mrs R W Drummond; P Pelham (instructed by Edge & Ellison, Hatwell, Pritchett & Co, of Birmingham) represented the defendants.

Giving judgment, GLIDEWELL J said: In this case the plaintiff is the freehold owner of shop premises at 14 Brentgoval Street, Bury St Edmunds, in Suffolk. The premises, which are shown in photographs which have been agreed and put before me, were, by a lease dated July 2 1974, let to the defendants for a term of years which expired on June 24 1980 at a rent which, until June 24 1977, was £1,100 per annum. (From now I shall refer to the plaintiff and the defendants as the landlord and tenants.) There was a rent review clause in the lease, and as a result of the review, and the arbitration under that review, the rent was increased for the remainder of the term to £1,475 per annum.

The lease contained, as is commonly the case, covenants by the lessee to repair and decorate the premises. In this case the covenant (clause 2 of the lease) was to perform and observe the covenants and stipulations contained in the First Schedule, and the relevant clauses of the First Schedule are no 4, a general repair covenant, "from time to time and at all times during the said term well and substantially to repair, cleanse and put and keep in good and substantial repair and condition, the premises"; and then two further clauses, one of which, as a result of an agreement between the parties during the course of the hearing, or to be more precise, immediately before the hearing commenced, has not troubled the court. Those clauses, 5 and 6 in the schedule, are clauses dealing respectively with the decoration of the exterior and of the interior of the premises, and it is the interior decoration clause which is still germane to the matters that I have to decide. That required that during the first year of the term and in the last three months, that is to say in the three months ending June 24 1980, whether determined by effluxion of time or not (and, as I shall say, this lease in the event was determined by the effluxion of time) the tenants were required to paint and decorate the inside walls and ironwork. The nature of the decoration is specified in some detail in the remainder of the clause.

Paragraph 7 of the schedule required the tenants at the expiration of the term to yield up the premises in good and substantial repair and condition.

Finally, a different matter, which is really a separate issue of its own, paragraph 18(1) required the tenants to pay to the landlord all costs, charges and expenses, including solicitors', counsel's and surveyors' costs, charges and fees incurred by the lessor in, or in contemplation of, any proceedings in respect of this lease under sections 146 and 147 of the Law of Property Act 1925.

The lease contained a proviso for re-entry entitling the plaintiff to re-enter and take possession if the rent should be unpaid for 21 days after the date for payment, or if the tenant should fail to perform any of the other covenants. Reference is then made in the statement of claim to the covenant to pay costs, charges and expenses, to which I have just referred.

The proceedings were started by a writ issued on July 24 1978 and the statement of claim alleged that the defendants, firstly, had failed to pay the rent due and owing at June 24 1978 and, secondly, that they had failed to put and keep the premises in good and substantial repair and had not painted or decorated them in accordance with the particular covenants to which I have referred.

It is alleged, as was the fact, that on May 12 1978 the landlord's solicitors had served on the tenants a notice in accordance with section 146 of the Law of Property Act 1925, specifying the breaches of covenant and requiring the tenants to remedy them and to make compensation in money for them. The statement of claim alleged that the tenants had failed to remedy those breaches and claimed damages for the diminution of the value of the reversion. The cost of remedying the breaches at that time was put at the sum of £6,404.40 and there was also a claim for the costs and expenses referred to in paragraph 5 of the statement of claim, those dealt with in paragraph 18 of the schedule to the lease.

Thus, there was claimed: possession of the premises; the arrears of rent; mesne profits until possession was delivered up; damages for breach of covenant; and the amount of any costs under paragraph 5.

Since the proceedings were commenced, indeed since the pleadings were completed, a number of events have supervened. First of all, as is apparent, there has not been any order of this court for possession, but the lease has expired by effluxion of time, on June 24 1980. The tenants, I was informed by counsel, vacated the premises and thus there is now no longer any issue between the parties about the possession of the premises, and I have not been asked to make any order in respect of that matter.

Secondly, it was agreed between the parties that sums which the landlord has accepted as mesne profits, which equal the amount of the rent due up to the date when the lease terminated on June 24, have been paid and thus I am no longer concerned with any question of arrears of rent. What has been pursued is the claim for damages for breach of the covenant to repair and, secondly, the claim in respect of the costs.

On the last matter no evidence has been advanced on behalf of the landlord as to what the amount of those costs is, has been, or may be. Why that evidence was not advanced, I do not know. It seems to me that it would be perfectly simple for the landlord to put forward evidence as to the costs to which the clause referred. However, she chose not to do so, and this morning Mr Lamming, on behalf of the landlord, asked that rather than giving judgment for the amount, which I really cannot do because I do not know what the amount is, I should instead grant a declaration that the landlord is entitled to recover those costs. Having taken instructions, Mr Pelham, on behalf of the tenants, agreed to treat the matter in that way and, subject to an agreed form of amendment to the statement of claim being handed in, in due course, so that it may be attached to the file, I propose to deal with the matter in that way.

The more important issue concerns the damages for the breach of the covenant to repair, and there, too, there has been an alteration because, as I pointed out at the beginning of this morning's hearing, the claim on the statement of claim is of course for damages as at the date of the writ, July 1978, and it is a commonplace that the cost of all building works and decorating works, largely due, although not I think entirely due, to the fall in the value of money, has substantially increased in the two and a quarter years that have passed since that date. If I gave judgment for the landlord for an amount assessed as at July 1978, it would have, at least in theory, been open to the landlord to bring another action for any further damages accrued since that date up to the date when the lease expired. Moreover, I would find it extremely difficult to deal with the matter as at July 1978, because both parties advanced evidence to me based very largely on present-day costs and circumstances rather than those appropriate to 1978. So I have been asked by counsel, who have agreed that the matter should be dealt with in this way, to treat this action as being an action for damages for breach of the covenant to repair right up to the date of the termination of the tenancy, June 24 1980. I have accepted that submission, and I so propose to treat it.

I should have said that the tenants, who have been in occupation of the premises from 1974 (and indeed, for all I know they may have been in occupation of the premises before that — I have not been told that), ceased to occupy the premises in June 1976, but they did not surrender. They apparently hoped to arrange an assignment of the premises to another occupier, but after some little discussion nothing came of that and the premises have therefore now stood empty for something over four years.

The issues which I have to decide were helpfully summarised by Mr Lamming in his opening, as being the following: firstly, as to the sum required to be spent on the interior to put it in a condition which complies with the repairing covenants in the

A lease; secondly, as to the amount by which the reversion is damaged by reason of breaches of the repairing covenants; and thirdly, as to the application of the clause in the lease dealing with the costs and expenses of the landlord related to the section 146 notice.

Before I go on to consider the facts and to say something about the evidence called in relation to these matters, I think it right that I should set out my view of the law that I am seeking to apply. In my judgment, the proper measure of damages for breach of a covenant to repair is in the first place based upon the amount by which the reversion is depreciated. Authority for that proposition, which is well established, is to be found in the
B decision of the Court of Appeal in *Smiley v Townshend* [1950] 2 KB 311, in which, in the course of his judgment, Denning LJ (as he then was) said at the bottom of p 319:

> I take it from authorities which this court is not at liberty to disregard that the proper measure of damage is the difference in value of the reversion at the end of the lease between the premises in their then state of unrepair and in the state in which they would have been if the covenants had been fulfilled. That test was laid down by this court in *Hanson v Newman* . What has to be considered is the diminution in value of the reversion at the end of the lease; and I take the word "reversion" to mean the landlord's then interest in the premises.

He then went on to deal with a matter which was germane to that
C case, but which is totally irrelevant in the present case, and he then said:

> The question is, therefore: how much was the market value of the landlord's interest diminished at the end of the lease by reason of the disrepair for which the then assignee was responsible?

That then is the first principle. But it is apparent that evidence of the value of the premises in their state of disrepair may be difficult to come by, because of course, since most leases, particularly of commercial premises, envisage the premises being let in repair, or at any rate subject to the liability of a previous tenant to repair, the majority of comparable evidence relates to
D lettings of property in repair rather than in disrepair. Thus it may be very difficult indeed to obtain and put forward evidence of a comparable property in disrepair. Certainly no such evidence has been called before me. Thus it follows that evidence of the cost of putting the property into the state of repair required by the lease is relevant evidence and will very often be *prima facie* evidence, or at any rate the starting point, from which the amount of the diminution in the value of the reversion may be deduced. In *Smiley v Townshend*, Denning LJ dealt with that matter at the bottom of p 322 where he said this:

> In cases where it is plain that the repairs are not going to be done by the landlord, the cost of them is little or no guide to the diminution in value
> E of the reversion, which may be nominal: see *Espir v Basil Street Hotel Ltd* and *James v Hutton*. But in cases where the repairs have been, or are going to be, done by the landlord, the cost may be a very real guide. That is shown by the recent case of *Jones v Herxheimer* to which we were referred. In cases where it is open to question whether the repairs will be done by the landlord, as in the present case

which I should say was a case where the premises were under requisition, so contained facts peculiar to its own situation

> then the cost may afford a starting figure; but it should be scaled down according to the circumstances, remembering that the real question is: what is the injury to the reversion? That is what the judge did here, as I read his judgment: he used the cost merely as an aid in assessing the
> F diminution in value of the reversion.

Singleton LJ at p 328 in the same decision, said this:

> In the course of his judgment in *Landeau v Marchbank* the judge

and he means by that Lynskey J, who was the judge at first instance both in *Smiley v Townshend* and in *Landeau v Marchbank*

> appears to have said: "It seems to me that, having regard to those decisions, the fact that repairs are necessary is not in itself even *prima facie* evidence of damage to the value of the reversion. In certain circumstances, it may be an important factor, but of itself the fact that a G
> certain amount of repair needs to be done does not, in my view, necessarily mean that there is damage to the reversion." If I may respectfully say so, I agree with the judgment in that case except that I would substitute the word "conclusive" for the words "even *prima facie*". Evidence of lack of repair is not conclusive evidence of damage to the value of the reversion; but it may well be, and is generally held to be, *prima facie* evidence of it.

I note in passing that where Denning LJ spoke of the situation where repairs had been or are going to be done by the landlord, nobody in the circumstances of that case was considering a situation where it might in fact be a future tenant who actually arranged for the repairs to be done in agreement with the H
landlord, some financial satisfaction for his doing so being arranged between them. No doubt, if that matter had been in issue there, his Lordship would have considered that if the future tenant actually does the repairs, but receives some satisfaction from the landlord for doing them, either by direct payment or by reduction in the rent which he has to pay, then that is equivalent to the landlord doing the repairs. Certainly that is my view.

The third matter of which I remind myself is that superimposed on the common law, as I have sought to set it out, is section 18(1) of the Landlord and Tenant Act 1927, which provides:

> Damages for a breach of a covenant or agreement to keep or put J
> premises in repair during the currency of a lease, or to leave or put premises in repair at the termination of a lease, . . . shall in no case exceed the amount (if any) by which the value of the reversion . . . in the premises is diminished owing to the breach of such covenant or agreement as aforesaid. . . .

In other words, even in a case where it is right to treat the cost of doing repairs as a starting point or as *prima facie* evidence, nevertheless, if it is shown by other evidence clearly that the loss to the reversion is some lesser sum, then the damages cannot exceed that lesser sum.

I turn now to consider, on the facts and evidence in this case, what, in my judgment, is the cost of putting this property into K
repair. There was set before me, most helpfully, a schedule in the form of a Scott Schedule, a description of the works which, in the opinion of the landlord's surveyor, Mr Fullerton, require to be done in order to put the premises into the condition of repair required by the lease, and that contains not merely such a detailed description in relation to the various parts of the premises and various parts of the structure, but also contains, tabulated, the price which the landlord says it would cost, or it would have cost in June 1980, to carry out those repairs; and a column containing the figures which the tenants say should be expended. The two columns are not strictly comparable because in that exercise the tenants did not essay to decide what it would L
cost to put the premises into the condition required by the lease, but what it would cost to put it into some lesser condition. That is a matter to which I shall have to return. Then there is a column containing the tenants' comments.

The first eight main items in that schedule relate to various parts of the exterior of the premises. My task has been eased because, before the hearing commenced, the parties agreed that a reasonable figure for carrying out such works to the exterior of the premises as are required to put the property into the condition required by the lease is the sum of £1,976.50. That is really all I need to say about the exterior. That means that I can concentrate, so far as this part of the case is concerned, upon the M
interior.

As to the interior, there were a number of issues. The first was: what is the state of disrepair? As to that, which of course is a purely factual question, there is little, if any, disagreement, save on one item as to the floorboards, and I will have to come back to that. To a considerable extent the state of disrepair in the interior is disclosed in photographs which have been put before me.

The second point is: what needs to be done to put the premises in the condition required by the lease? On that Mr Pelham, on behalf of the tenants, again most helpfully, admitted that the items set out in the Scott Schedule were, and are, necessary in order to put the property in the condition of repair and decoration required by the lease, but with a number of specific reservations from that admission. The specific reservations related to the follow items, and now I am necessarily having to get down to detail:

Item 9(1); which related to cleaning out rubbish and so on from the premises;

Item 9(2); which in terms says "Obtain a report from the Eastern Electricity Board on the condition of the defective lighting", but which really meant not only get a report but also do the work which the report requires to be done;

Items 10(4), 11(4) and 12(4); all of which related to worn and defective floorboards; and

Items 11(5) and (6), and 12(5); which related to a firescreen which had been partially erected alongside the staircase leading up from the ground to the first floor and between the staircase and the passageway on the first floor. When I say "partially erected" I mean that the necessary framework for that screen has been erected and the fireproof panelling on one side has been put on; but the other side has never been panelled. Therefore there is a screen there which is not complete and which either requires to be completed or to be taken down.

I must deal with each of those in turn. First, as to the clearing away of rubbish, fortunately a relatively minor item, it transpires that since the drawing up of the schedule, the tenants have done a good deal of clearing up of the considerable amount of debris that is shown in the photographs, but there is still a certain amount to be tidied up, and the parties have agreed that that would cost some £30, so I need not trouble myself any more with that.

As to the floorboards, I heard conflicting evidence from Mr Fullerton, on behalf of the landlord, and Mr Edwards, an architect called to give evidence on behalf of the tenants, about this. Mr Edwards' evidence was that there was not really anything very much wrong with the floorboards at all. Certainly they were not so unsafe that any of them really required to be replaced. At most a few of them required some refixing. One or two of them, I think, required some remedial work to smooth them out or plane them down, or matters of that sort. On the other hand, Mr Fullerton said that there were a number of defective and worn floorboards. In his view, something like 10 per cent of the total area of the wooden floors throughout the building required to be replaced. On that issue it is after all for the plaintiff to satisfy me, on a balance of probabilities, that she makes out the case that she puts forward, and I must confess that I find that approach too broad and general to be satisfied by it. I find that there is not much that needs to be done to the floorboards to put them into the condition required by the lease in a building of this age and of this type. One must not forget that this is not a brand new building; it is a building which I am told is something between 100 and 120 or 130 years old. I take the view that there was nothing materially wrong with the floorboards and certainly only the most minor repairs of the sort indicated by Mr Edwards are required in that respect. Unfortunately, I do not have any specific figure for the floorboards themselves because in each case the total of the floorboards was assumed in a larger total in relation to the redecoration of each floor of the premises. So my approach is to make a deduction in respect of floorboards from the total sum at which I had arrived in respect of repairs to the interior.

Then there is the question of the firescreens. Fortunately, after a little examination this does not really present any problem because, as I have said, there are two ways of dealing with the problem. One is to complete the firescreening and fit appropriate doors. The other is to pull down the existing screening and refit some doors which were there before it was put up. The cost of doing the work suggested in the Scott Schedule totals £189. The cost of pulling the firescreen down and putting up the alternative doors has been put forward by the tenants, and I accept as being about £150, so there is not very much in this. Again I am going to adopt the tenants' figure in respect of that one.

The electricity is a much more substantial item. The evidence here came to me in two parts. Indeed, the evidence generally on behalf of the landlord came in two parts because what has happened was that Mr Fullerton, the surveyor, had surveyed the premises and had drawn up the Scott Schedule, and had decided what work needed to be done. He had then obtained the assistance of a Mr Shillingworth, who is a senior surveyor with a substantial firm of builders in that locality (this is W J Baker (Builders) Ltd), and has been with them for some 16 years. Mr Shillingworth gave an estimate, a quotation I should say, of the total cost of the repairs upon which the claim now put forward by the landlord is based. That originally came to some £8,515, but included in that figure was a sum of £2,900 odd, and since that has been reduced by agreement to £1,976.50, the total sum would now, on the landlord's claim, be reduced to, in round figures, £7,500.

As I have said, generally speaking, what Mr Shillingworth did was to give, from his knowledge and experience, prices for the doing of the work. But he is not an electrician and he is not an electrical contractor, and what happened in relation to the electrical works was that Mr Fullerton, who also is not an electrician and is not an electrical contractor, decided what he thought the electricity board would require, and Mr Shillingworth decided what he thought an electrical contractor would charge to do the work which Mr Fullerton thought the electricity board would require. So there are two matters being put before me and informed guesses involved in that particular exercise. The amount which Mr Shillingworth put in as a provisional sum and which, as he says, was based on his experience, which is very considerable of course, was some £1,700.

I did not have evidence from anybody on behalf of the electricity board, nor did I have evidence from an electrical contractor. The position about the electricity board, the evidence establishes, is this: that generally speaking if a property changes hands but the electricity supply is not disconnected, they do not concern themselves with the state of the electrical installation in the premises, unless it is drawn to their attention in some way or another, and so the work required would be only such work as was quite patently required to bring the premises up to the condition required by the lease. Of course, other work might well be required by the incoming tenant for specific purposes of his own. But where an installation is disconnected, then the electricity board do not in practice get interested because the new tenant will have to make an application for a reconnection of supply and the board will not agree to reconnect the supply until they are satisfied that work has been done to bring the installation in the premises up to the standard that they regard as proper in modern circumstances.

The electrical installation in these premises is quite patently, at least in part, an old one. Among other matters, apparently it has a number of brass light switches and lampholders included in it. But be that as it may, the fact that the electrical installation has been disconnected in this case is of course entirely the responsibility of the tenants, because it was disconnected as a result of their vacating the premises, no doubt at some time after they did vacate, and thus they cannot complain about the fact that the electricity board will now require more work to be done than might otherwise have been the case.

On the other hand, I do not find the evidence on this aspect of the matter, on behalf of the landlord, anything like as compelling as the rest of the landlord's evidence about the value of the work. I shall also say that Mr Fullerton and Mr Edwards, a few days ago, did meet at the premises and they had a discussion, the object of which was to attempt really to solve this whole dispute. As far as the whole dispute is concerned they failed, but as far as the exterior is concerned they succeeded, as I have already said; but they did consider the interior also, and although there is no doubt that their discussion was without prejudice, rather oddly I have been told all about it. What I have been told is that they did arrive at an agreement as to a figure which would put the premises into some state of repair, which I find difficult to define because they themselves were not agreed about what they had

A agreed. Mr Edwards thought that they had agreed that it would be put in a state of repair which an incoming tenant would regard as acceptable. Mr Fullerton thought that it was something less than that, which he described as "bodged up". Neither of them suggested that it was a state of repair which would satisfy the lease. But among other matters they did agree that for an expenditure of £500 the electrical installation could be put in a state which certainly would not satisfy the lease but, on the other hand, would probably mean that the electrical supply system would keep going satisfactorily for, at any rate, a few years.

That evidence, generally, in relation to this agreement, if there ever was an agreement between them, does not really help me B much at all, but the figure for the electricity supply does give me some modest assistance. On the one hand, I have got a partial agreement in relation to something less than the works required to put the electrical system into the state required by the lease. On the other hand, I have got evidence, which is far from satisfactory, from the landlord as to the £1,700 which she or her advisers say would be required to comply with the electricity board's requirements. In this particular matter I have been put very much in the position which one of the decisions I have been referred to calls the position of the jury. I have come to the conclusion that it would be proper to allow £1,000 for the electrical installation.

C As to the remainder of the works required on the interior of the premises, as I have said, the evidence before me from the landlord, the evidence of Mr Fullerton, tells me what is required to put the premises in the condition which accords with the lease. There is a concession that indeed those works are required for that purpose, once one has discounted the particular matters to which I have already referred.

The only evidence that I have as to the cost of carrying out such works is the evidence of Mr Shillingworth, on behalf of the landlord, because the tenants, taking the view on advice that this is not the proper measure at all, just have not sought to advance evidence as to what would be the proper figure, if that were the D standard to adopt. The only figures I have on behalf of the tenants are figures which would suffice to bring the premises up to some lesser standard. So I can only accept, and I do accept, the evidence on behalf of the landlord as to the cost of bringing the interior of the premises up to the standard required by the lease with the exception of those items to which I have already specifically referred and with which I have dealt, that is to say, the electrical installation, the small items for clearing rubbish, the firescreen, and the defective floorboards.

Those figures do not take into account value added tax, and it is the landlord's case that since the tenants have not done the work, the landlord may well wish to put the premises into the E condition required by the lease, and if she were going to do so, she would incur value added tax. She would have to pay value added tax, and since she is not registered for value added tax, she would not be able to reclaim it. On the other hand, the tenants say that what the landlord is most probably going to do, on the evidence, is to relet these premises to somebody else. Any tenant, they say, is almost bound to be registered for value added tax and thus if the landlord arranges for the new tenant actually to do the work, whether there is a financial arrangement between them matters not for this purpose, value added tax will not enter into it because the new tenant will be able to claim back the value added tax which the landlord would not be able to claim F back. Thus, says Mr Pelham, the landlord being under an obligation to mitigate her damages, is under an obligation to adopt that method of carrying out the work in order to save the tenant from being responsible for payment of value added tax as part of the damages. (If this is the proper method of approach at all!)

My mind has fluctuated about that question, but I have come to the conclusion in the end that the argument on behalf of the landlord is correct in relation to this matter. There are two reasons for that conclusion: the first is, as I suggested in argument, that if Mr Pelham's argument is correct it means that, albeit the landlord may very well in fact relet the premises, nevertheless the effect of depriving her of value added tax on the G damages, or as part of the damages, would effectively be to close her other options, that is to say, if she changed her mind and decided that she would like to sell the premises and put the money to some other use but that she wanted to sell them in lease condition, then she would have to incur value added tax and she would not have recovered it as part of the damages.

I was reminded by Mr Lamming that a decision to which he had referred me during the course of his opening, that in *Haviland v Long* [1952] 2 QB 80, another decision of the Court of Appeal, while not exactly on all fours with this, was helpful. That was a case in which the tenants were in breach of a covenant to repair and the landlords entered into a fresh new H lease with other tenants who agreed to carry out the repairs. The landlords agreed to reimburse them out of the sum they recovered from the old tenants by way of payment for the dilapidations. The old tenants said, "Well, you are going to get the new tenants to do the repairs. They only get anything from you if we pay you anything, so that there is no need for us to pay you anything. Thus, no obligation on you will arise to pay the new tenants anything." But the court held that that was a matter which was *res inter alios acta*. In other words, it was nothing to do with the respective obligations or rights as between the landlord and the old tenant, and that really this was merely a mechanism by which the work of repair was to be done. That J authority does seem to me to establish the principle for which Mr Lamming is contending, and I have come to the conclusion that the damages, if properly assessable on the basis of the cost of repair, should properly include also value added tax, or rather, should include an item equal to the amount of value added tax which the landlord would suffer on such costs.

As I have said, that is dealing with the matter on the basis of the landlord's approach; but the tenants' approach was quite different. The evidence of Mr Hazell, a Fellow of the Royal Institution of Chartered Surveyors, and a partner in the firm of Rutters, of Bury St Edmunds, called on behalf of the tenants, was effectively this: that while he conceded that there was a K general need for redecoration, internally and externally, and some need for repairs, he believed that most potential purchasers would anticipate, and a tenant of the lease for 14 years or longer would be prepared to accept, the state of disrepair and would discount the cost of decoration. His final conclusion was that the proper rental value of the property was some £2,500 per annum. He capitalised that to give a capital value of £25,000 in a good state of repair, and he deducted from it what he described as a spot figure of £2,500, which he believes was the kind of sum which a purchaser would seek to deduct for initial essential repairs. He then came back to the question of rental value and said that although a purchaser would, he thought, deduct his spot L figure of £2,500, he did not think a prospective tenant would deduct anything at all. I must confess I fail to understand the logic of that distinction, but nevertheless, that was his evidence.

What in effect was being said to me in this respect by Mr Hazell was that I should really completely disregard all the detailed evidence about the cost of repairs; that, albeit the premises did require repair and a good deal of decoration, except for possibly some element of structural repairs, a new tenant (and it was firmly argued that the probability is that these premises will be let) would want to put in a scheme of decoration and install fixtures and fittings which would suit his own business. Thus, if the landlord carried out the decorations beforehand, it M might well be that a great deal of what the landlord did would be totally wasted because the new tenant would simply scrap it and put in something different, or carry out a different scheme of decorative repair.

That argument was particularly advanced because there was evidence from Mr Fullerton that he was, and is, and has for some time been, in negotiation with a bus company to lease these premises to them. I should say that the premises are situated on a street corner, and it looks very much as though they were purpose built as a corner shop, with living accommodation alongside and above, at some time in the middle or later half of the 19th century, and they stand some little distance from the centre of Bury St Edmunds in a position that has been described as a secondary shopping position, but they are immediately across the crossroads on which they stand from a bus station. Thus it would no doubt be extremely convenient to the bus company to locate themselves in the premises, and if they do take a lease of the premises they apparently intend to use them for the purposes of a ticket office and a travel agency generally. I am not sure whether that includes international as well as national travel, but that is perhaps irrelevant.

In town and country planning terms, that does not involve of itself a material change in the use of the premises because a shop is defined in the relevant order as including a ticket office or agency, but it is, in ordinary terms of course, a somewhat different type of institution from a retail shop selling clothing or any other article. There is obviously a difference between the way in which a building is used if, on the one hand, the occupier is a retail trader with a shop and all the accoutrements of a sales shop and storage accommodation above, and the way in which the premises will be used if the front part were a ticket office and a travel agency and the remainder were used partly for storage of documents and the like, but very largely as office accommodation. What is urged upon me is, particularly if the bus company go there, they will want, to a large extent, the property to be fitted out and decorated to meet their own needs.

Mr Fullerton on the other hand, while I think conceding that to a modest extent a tenant, particularly the bus company, might have special needs, said that of course the bus company was not the only potential tenant and if they were not satisfied he envisaged that the tenant might well be a small shopkeeper. Small shopkeepers are notoriously short of capital and, in his view, it would be quite common for a small shopkeeper to welcome the fact that the landlord had had the premises redecorated, and simply to enter them, accepting the landlord's decorations virtually as they stood, and immediately start trading. So Mr Fullerton's evidence was that of course all, or certainly most, of the expenditure required to put the premises into the condition required by the lease would be to the advantage of a new tenant and would not be wasted.

Faced with these two conflicting views of gentlemen equally eminent in their own professions, I can only base myself, to an extent, on my own experience, such as it is on these matters, and obviously it is somewhat limited. I have come to the conclusion that Mr Hazell's point is only partially right. I do think that it is probably right in relation to the shoproom in particular, and possibly in relation to the second room on the ground floor, that any tenant would want to put in fittings and fixtures to suit his own trade, and would want his own decorative scheme. He probably would want his own shopfront on the exterior, though the ability to install anything very much in this particular situation must be limited.

On the other hand, I cannot see why, if the upper floors and the ancillary rooms on the ground floor were cleaned up and redecorated, they should not be equally useful for the purpose to which the bus company would put them, as they would be for storage purposes. It may be that the tenant may not like the exact shade that the landlord chooses, but he will get newly decorated premises. I do not accept that if he got them undecorated he would expect to pay the same rent. I think that any tenant coming into these premises, if asked to go into them in their present situation, would say, "Well, yes, but look at the state of repair and decoration. I am going to have to spend quite a lot of money". While I accept that he might not necessarily do the sort of detailed calculation that Mr Fullerton has done, I do think that he would expect in some way or another to be compensated for taking the premises in their present state of repair and decoration.

While I have said that the proper approach is to see whether in the first place there is evidence from which one can go straight to the diminution in the value of the reversion, that demands having clear valuation evidence, both of the value of the premises in a proper state of repair (that required by the lease) and in the present state of repair. As to the first, the value is disagreed to a very considerable extent, although I have got clear evidence from them about it. As to the second, I really have no clear evidence at all. All I have got is Mr Hazell's deduction of what he calls a spot figure of £2,500 which, with the greatest respect to him, he did not seek to base upon anything more than his experience, which is a very generalised way of saying, "Well, that seems about right". That is not sufficient, in my view, for me to enable myself to say that is the diminution in the value of the reversion, and it is not sufficient to displace the evidence of the cost of repairs as being a useful way, or at any rate a starting point, for deciding the diminution in the value of the reversion.

So I go back to what I have already said about the cost of the repairs. The exterior, as I have said, has already been agreed at £1,976.50. The rubbish has been agreed at £30 — for clearing the rubbish. The electrical works I have already assessed at £1,000, and the firescreen at £150. The balance in the Scott Schedule is some £3,584. My arithmetic can be checked, but it is certainly a figure in that order, after disregarding deducting the exterior and the other items to which I have just referred. From that, as I have said, some discount should properly be made because of my view that the tenant would not require or want a complete redecoration of the ground floor but would only want cleaning up and some partial preparation towards redecoration, and also because, as I have already said, I am not satisfied by the landlord's evidence about the floorboards.

All in all I have come to the conclusion that the repairs figure, which I should adopt as being a reasonable one, is a sum of £6,250. I have already indicated that it is my view that value added tax should be added to that. Subject to what counsel say, I have calculated that at £937.50.

There remains the question of anything over and above that. Mr Fullerton said that his total claim for the diminution in the value of the reversion was originally £10,000, of which some £8,500 was the cost of repairs. By definition, when he reduced the exterior repairs by about £1,000, the total figure must have come down by about £1,000, so that reduced it to £9,000, of which £7,500 was the cost of repairs. The other £1,500 broadly was made up as follows: that in Mr Fullerton's view the landlord is entitled to be compensated for the fact that whoever does the repairs, time will elapse before they can be completed, and thus the landlord will be receiving no rent during that time. Moreover, Mr Fullerton suggested that there should be added to that a further period of time to enable the landlord to enter into a new lease with the tenant. All in all, said Mr Fullerton, what the landlord should have is compensation for something like six months' loss of rent which, though he had given evidence that in his view the proper rent was £4,000, he said, "Well, let us put that at £1,500". I hope I have followed his thought process correctly. I have really summarised a series of propositions which he put forward.

The evidence, as it emerged, about the time that would be necessary to enable the work to be carried out, eventually was very near agreement. The respective experts take the view that from the time when the tenants are first under notice to carry out works to the time (I am now talking about notice on the old tenant) when the works could have been completed would be something between 13 and 16 weeks. I think it right to accede to Mr Fullerton's suggestion that there should be a quarter's rent, or compensation to the landlord for loss of a quarter's rent during that period of time.

On the other hand, I see no reason why there should be any more than that. After all it has been absolutely apparent to this landlord for a very long time that this tenant was not going to carry out the repairs. It must have been apparent to the landlord that the tenant was not going to reoccupy. The tenant, as I see it, had no rights to a new lease under the Landlord and Tenant Act 1954 and none has been suggested. I can see no reason why the landlord, had she wished to do so, should not have negotiated with a new tenant before the expiry of the old lease, and thus been ready the moment the old lease expired either to get a new tenant to do the repairs or to do them herself if she had sufficient

A funds to enable her to do so. The fact that she did not is a matter for her and her advisers. I suspect that it was because of these negotiations with the bus company that were likely to produce a more attractive proposition than any other tenant was able to offer, but that is perhaps an irrelevance. That, however, is my reason for not giving more than compensation for one quarter's loss of rent.

That, added to the matters to which I have already referred, adding £6,250; value added tax at 15 per cent on that sum, £937.50; and a quarter's rent, which is half of Mr Fullerton's figure of £1,500, that is to say £750, brings me to a total of £7,937.50. Subject to my arithmetic being checked, I give B judgment to the plaintiff for that amount.

There remains the question of the costs, that is to say, whether the landlord is entitled to anything by way of costs under paragraph 18 in the schedule to the lease. The argument here, on behalf of the tenants, is that the action was started unreasonably soon after the section 146 notice had been served. If this action had been heard immediately the writ was issued, there would have been an absolute answer to the claim for forfeiture in that not sufficient time had been given to enable the works to be done. On the evidence before me that submission is entirely justified on the facts because, as I have already said, it is agreed that at least 13 weeks would have been required, and the writ C was issued substantially less than 13 weeks after the service of the section 146 notice. Thus, submits Mr Pelham, since that, in effect, invalidates the notice, or at any rate renders the proceedings in respect of forfeiture abortive at the time when they were served, the landlord should not be entitled to recover any of the costs of that section 146 notice or associated with it. On the other D hand, it is in my view clear that this clause is intended to ensure that if the landlord, in order to get the repairs done, or to get damages for paying to have them done, as well as obtaining possession of the premises, does have to serve a section 146 notice, that the landlord should not have to bear any part of the cost occasioned by the service of that notice. I have come to the conclusion that if any part of that cost is not claimable as costs in these proceedings (a matter which happily I do not have to decide — it is a matter for argument on taxation) then the landlord is entitled to recover such additional or earlier costs and expenses.

Since the paragraph in the lease reads: "All costs, charges and E expenses incurred by the lessor in or in contemplation of any proceedings in respect of this lease", despite Mr Pelham's argument on this matter and the authority to which he referred me, namely, *Re Adelphi Hotel (Brighton) Ltd* [1953] 1 WLR 955, I have concluded that such costs should be on a solicitor and client basis rather than on a party and party basis.

Whether there is anything additional, I have no idea, but, subject to Mr Lamming's formulating the form of the declaration which he seeks, I am prepared to declare that if there is any cost falling within paragraph 18 of the schedule to the lease which is not recoverable as costs in this action, then the landlord is F entitled to recover such costs and expenses from the tenants on a solicitor and client basis.

Judgment was given in favour of the plaintiff for £7,937.50 with costs.

LANDLORD AND TENANT
BUSINESS TENANCIES

CHANCERY DIVISION

November 20 1980

(Before Sir Robert MEGARRY, Vice-Chancellor)

ALLNATT LONDON PROPERTIES LTD v NEWTON

Estates Gazette January 17 1981

(1980) 257 EG 174

Landlord and Tenant Act 1954, Part II, Section 38(1) — Agreement to surrender void as purposting to preclude tenant from making an application or request for a new tenancy — Elaborate clause in lease, an "ungainly bit of drafting", prohibiting any assignment of premises as a whole, provided that if the tenant wished to assign and first offered to surrender the lease to the landlords on certain terms, and the landlords either refused or failed to accept it, the tenant was free to assign within a stated time, landlords not to withhold consent unreasonably — Joseph v Joseph followed — Landlord's claims for declaration that agreement was enforceable, for decree of specific performance of for award of damages rejected — Tenant's claim, on the other hand, rejected for a declaration that so long as the tenancy was within Part II the tenant was entitled to assign or sublet the whole of the demised premises with the consent of the landlords (such consent not to be unreasonably withheld) without first offering to surrender the residue of the term.

This was an action by the landlords, Allnatt London Properties Ltd, against the tenant, Ronald James Newton, for specific performance and other relief in respect of an agreement to surrender a lease of business premises in Hounslow demised by the plaintiffs to the defendant for 21 years from March 25 1972. The dispute turned upon the effect of a clause in the lease dealing with assignment. The clause is set out at length in the judgment of the Vice-Chancellor.

J H G Sunnucks (instructed by Marshall, Shortland & Co) appeared on behalf of the plaintiffs; Paul de la Piquerie (instructed by Paul Gromett & Co) represented the defendant.

Giving judgment, THE VICE-CHANCELLOR said: This action raises two points on business tenancies under the Landlord and Tenant Act 1954, Part II. Under a lease dated April 20 1972 the plaintiffs, the landlords, demised business premises to the defendant, the tenant, for a term of 21 years from March 25 1972. The lease contained an elaborate clause, clause 3(21)(b), which I will set out in due course. This prohibited any assignment of the premises as a whole, though if the tenant wished to assign them and first offered to surrender the lease to the landlords on certain terms, and the landlords either refused the offer or failed to accept it within 21 days, the tenant was free to assign the premises within a stated time, and the landlords were not to withhold their consent unreasonably.

What happened was that after the lease had been running for nearly seven years the tenant wished to assign it, and he approached the landlord for consent. They refused this and required the tenant first to offer to surrender his lease. This he did by letter dated December 20 1979, which offered to surrender the lease "in accordance with the terms of the lease"; and on January 9 1980 the landlords accepted the offer. An exchange of letters in February and March showed that whereas the tenant expected to receive £45,000 for the surrender, the landlords were willing to pay only £10,000. The difference, I was told, was in the main due to the landlord's offer being based on the value of the unexpired term, with nothing for goodwill, whereas the tenant's value included what an assignee would be expected to pay for goodwill. When this emerged, the tenant withdrew his offer to surrender. This was on March 6 1980: and on April 11 the landlords issued the writ. This claims a declaration that the contract for surrender was enforceable, specific performance of the contract, and damages. The tenant's defence and counterclaim, as amended and re-served on July 8, is that section 38 of the Landlord and Tenant Act 1954 renders the agreement to surrender void; and the tenant seeks a declaration to this effect, and also a declaration that so long as the tenancy is subject to Part II of the Act of 1954 the tenant is entitled to assign or sublet the whole of the premises with the landlord's consent (such consent not to be unreasonably withheld) without first offering any surrender.

Section 38(1) of the Act of 1954, as amended by the Law of Property Act 1969, runs as follows:

> Any agreement relating to a tenancy to which this Part of this Act applies (whether contained in the instrument creating the tenancy or not) shall be void (except as provided by subsection (4) of this section) in so far as it purports to preclude the tenant from making an application or request under this Part of this Act or provides for the termination or the surrender of the tenancy in the event of his making such an application or request or for the imposition of any penalty or disability on the tenant in that event.

I pause to say that it became common ground that the second part of that provision, beginning with "or provides for the termination", did not arise in the present case, for there was nothing which provided for the termination or surrender of the tenancy "in the event of his making such an application or request". What is at the heart of the matter is the provision "in so far as it purports to preclude the tenant from making an application or request under this Part of this Act".

In subsection (4) of the section, para (a) admittedly does not apply, and so I read it without that paragraph.

> (4) The court may:
>
> . . .
>
> (b) on the joint application of the persons who are the landlord and the tenant in relation to a tenancy to which this Part of this Act applies, authorise an agreement for the surrender of the tenancy on such date or in such circumstances as may be specified in the agreement and on such terms (if any as may be so specified);
>
> if the agreement is contained in or endorsed on the instrument creating the tenancy or such other instrument as the court may specify; and an agreement contained in or endorsed on an instrument in pursuance of an authorisation given under this subsection shall be valid notwithstanding anything in the preceding provisions of this section.

It is common ground that the agreement for the surrender is not contained in or endorsed on the lease, or in "such other instrument as the court may specify"; no application to the court on this point has been made. Accordingly, subsection (4) does not apply, and the case turns on the first limb of subsection (1).

With that, I can come to the terms of the lease itself. Clause 3 contains the tenant's covenants. The lease refers to "tenants" in the plural because, I think, the tenant was contemplating the incorporation of his business, though in fact he never did this. Clause 3(21) begins by imposing an absolute prohibition on any assignment of part of the demised premises:

> (a) Not to assign underlet or part with the possession or occupation or share the occupation of any part (as a part) of the demised premises.

A Nothing directly arises under that. There is then subclause (b), which is at the heart of the matter. It is an ungainly bit of drafting, with three provisos, and it shuttles to and fro between the first stage, that of the offer to surrender, and the second, that of an assignment if the offer is not accepted. For ease of reference I have inserted numerals which divide the subclause up into separate limbs, though of course this subdivision cannot affect the construction of the subclause. So divided, it reads as follows:

(1) Not (subject to the proviso to this clause) to assign underlet part with the possession or occupation of the demised premises as a whole (2) Provided Always that if at any time and so often as the Tenants shall desire to assign or underlet the demised premises as a whole the Tenants shall make to the
B Landlords an offer in writing in the terms hereinafter mentioned (3) then if the Landlords shall refuse such offer or shall not within twenty-one days after the receipt by the Landlords of such offer accept the same (4) the Tenants shall be at liberty within one month after such refusal or otherwise within two months after the making of such offer to apply to the Landlords in writing for consent to assign or underlet the demised premises as a whole (5) and such consent shall not be unreasonably withheld (6) Any such offer shall be in the terms following namely the Tenants shall offer to surrender the Lease with vacant possession and otherwise free from encumbrances on a date three months from the date of the said written offer in consideration of the payment by the Landlords to the Tenants of a sum representing the net premium value (if any) of this Lease for the unexpired residue of the term such value to be agreed upon by the parties or in default of agreement
C as may be determined by a surveyor to be appointed upon the application of either party by the President for the time being of the Royal Institution of Chartered Surveyors (7) Provided further that any acceptance by the Landlords of any such offer so made to them shall be without prejudice to the rights and remedies of the Landlords in respect of any rent in arrear or any breach of any of the covenants herein contained and on the part of the Tenants to be observed (8) Provided further that if the Landlords shall agree to give consent for the Tenants to assign this Lease then it shall be a condition of any such consent that the Assignee shall enter into a direct covenant with the Landlords to pay the rents hereinbefore reserved and to observe and perform the covenants herein contained during the residue of the term hereby created.

D It will be seen that limb (1) imposes the prohibition, limbs (2), (3), (6) and (7) refer to stage 1, and limbs (4), (5) and (8) relate to stage 2.

With that in mind, I return to the accepted offer to surrender. This is plainly an agreement relating to a tenancy to which Part II of the Act of 1954 applies, within the opening words of section 38(1). That agreement does not fall within section 38(4), and so by subsection (1) it is void "in so far as it purports to preclude the tenant from making an application or request under this Part of this Act". Does it so purport?

Whatever might be thought about the meaning of the word "purport" in the abstract, that word has to be construed in its
E context: and this was done in *Joseph v Joseph* [1967] Ch 78. There, the Court of Appeal unanimously held that the phrase "purports to preclude", in its context, meant not "professes to preclude" but "has the effect of precluding". The question is not so much whether there is an agreement which in terms prevents the tenant from making an application or request, but whether, whatever the agreement says, it in fact operates to prevent the tenant from doing this. In *Joseph v Joseph* there was an agreement for a lease for 10 years, and after this had run for some two years there was an agreement to surrender the tenancy within the next two years. The inevitable effect of the agreement to surrender the tenancy, if carried out, would be to preclude the tenant from applying for a
F new tenancy when the tenancy agreement ran out some six years after the date of the agreement to surrender: for by section 24(2), section 24(1) would not operate to prolong a tenancy which had actually been surrendered. Section 38(1) invalidated that agreement to surrender and therefore the landlord could not obtain any damages for the tenant's breach of the agreement.

Joseph v Joseph, of course, binds me: and I have heard nothing to detract from its authority. It was followed by the Court of Appeal in *Jones v Wrotham Park Settled Estates* [1980] AC 74, and although it was challenged on appeal, the House of Lords found it unnecessary to decide anything on it and said nothing one way or the other to affect its authority: see at p 114. Certain other authorities were cited to me, but I do not think any of them affect what I have to G decide. Mr Sunnucks, who appeared for the landlords, had a point on the lease not being assignable by reason of the parties clause at the beginning referring to the parties only by name, without mentioning assigns. Mr de la Piquerie, for the tenant, met this by invoking the Law of Property Act 1925, section 79; but then the point blew up in Mr Sunnucks' hands when, very properly, he drew attention to clause 5(c) of the lease, which provided, *inter alia*, that "the Tenants", where the context so required or admitted, included their assigns. Mr Sunnucks also had a point on the contrast in section 38(1) between "tenant" and "tenancy", and he contended that the subsection gave no protection to a tenant who did not want his tenancy to continue but instead wished to terminate it. I did my H best to follow his submissions, but in the end I was quite unable to see anything in them which could suffice to exclude the operation of section 38(1) as construed in *Joseph v Joseph*, or to distinguish that case.

In the result, it appears to me that I must hold that the agreement to surrender is rendered void by section 38(1). It follows that the landlord's claim must fail. I cannot declare that the agreement is an enforceable contract, nor decree specific performance of it or award damages for its breach. On the counterclaim, the tenant has made out a sufficient case to support his claim to a declaration that the agreement to surrender is void by virtue of section 38; and in the absence of any reason why I should exercise my discretion to refuse J to grant declaratory relief, I propose to make a declaration to that effect.

That leaves the tenant's further claim to a declaration that so long as the tenancy is within Part II of the Act of 1954 the tenant is entitled to assign or sublet the whole of the demised premises with the consent of the landlords (such consent not to be unreasonably withheld) without first offering to surrender the residue of the term. That is the second point in the case.

Relatively little was said about this issue during the argument. It seems to me the most difficult part of the case. Mr Sunnucks contended that if (contrary to his main contention) section 38(1) harmed him, it did not merely avoid the contract to surrender, but K also rendered void or ineffective the machinery in clause 3(21)(b) for the tenant having to make an offer to surrender the lease before seeking to assign; and if it did that, it made the whole of clause 3(21)(b) void or ineffective. Mr de la Piquerie, on the other hand, argued that the provision for a surrender was merely a condition precedent to an assignment; and if the condition precedent was void, as he said it was, that left the provision for assignment in full operation.

I think the starting point must be to consider whether section 38(1) invalidates any of clause 3(21)(b), and if so, how much of it. What section 38(1) invalidates is an agreement in so far as it has the effect of precluding the tenant from making an application or L request under Part II of the Act of 1954. Is there anything in clause 3(21)(b) which does this? If one takes the subclause as it stands, without applying section 38(1), I cannot see that, without more, there is anything which has the effect of precluding the tenant from making an application or request. Limb (2) requires the tenant to make an offer to surrender: but an offer is not an "agreement", nor does an agreement to make an offer to surrender, by itself, preclude the tenant from making an application or request. If the landlords reject the offer, there is still no such precluding, and the tenant is free to assign under the conditions stated. If on the other hand the landlords accept the offer, then the resulting agreement to surrender (if enforceable) would have the effect of precluding the tenant M from making an application or request: and then at that point section 38(1) would make that agreement void. In short, until it is known whether the landlords have accepted or rejected the offer to surrender it cannot be known whether there is any agreement which will preclude an application or request, within the meaning of the subsection: there may or may not be.

It will be observed that all that clause 3(21)(b) does is to require an offer to be made, and it depends on what happens to that offer whether there ever comes into being an agreement which offends against the subsection. It is not as if the subclause gave the landlords an option or other right to require the tenant to surrender the lease.

In my view, the subclause does not fall within the subsection, but stands at one remove from anything that does. It seems to me that the subsection, as construed in *Joseph v Joseph*, is perfectly adequate to guard against the mischief which it envisages if it strikes down the actual agreement to surrender, and that there is no need to construe the subsection so as to make it extend to the mechanism for producing an offer which, if accepted, would be invalidated. Nor do I feel any more enthusiasm than was felt by the Court of Appeal in *Joseph v Joseph* for enabling either party to a lease to escape from his bargain further than is necessary to give effect to the subsection and its manifest purpose.

There is a further consideration. The subclause represents a bargain in which the rights of the parties depend on a balance of advantages and disadvantages to each. If statute declares void part of such a subclause which operates for the benefit of one party, I should hesitate to hold that the other party could thereupon obtain the benefit of the rest of the clause without suffering the burden of what has been declared void, unless the provisions were clearly severable. Limbs (2) to (8) of the subclause seem to me to be so interdependent that it would be making a different contract for the parties if the provisions for the offer to surrender were to be disregarded as being void and the rest of the limbs were then revised so as to put them into an intelligible and enforceable form. I certainly would not regard the provisions for the offer to surrender as being a mere condition precedent. This is not a case of independent provisions such as there were in *Gaskell v King* [1809] 11 East 165; and see *Pickering v Ilfracombe Railway Co* [1868] LR 3 CP 235 at p 250. Accordingly, if I am wrong in thinking that the subclause is not affected by the subsection, then I would hold that limbs (2) to (8) all fall together.

Linb (1), which prohibits the whole of the premises from being assigned, and so on, seems to me to be in a somewhat different position. It is, of course, in terms made "subject to the proviso to this clause", which presumably means all three provisos, and thus includes limbs (2) to (8). If all those limbs fall by virtue of the subsection, then limb (1) remains perfectly intelligible and capable of operation, though it then ceases to be subject to anything. Limb (1) seems to me to be severable from limbs (2) to (8) in a way that is not possible for the latter limbs *inter se*. However, none of this was argued before me, and I hesitate to reach a firm conclusion on it. For the reasons that I have given I base my decision on the subsection not having any application to the subclause.

In the result, it seems to me that the second declaration sought by the counterclaim ought to be refused. I do not think that the effect of the subsection is to permit assignments or sublettings of the whole with the consent of the landlords which must not unreasonably be withheld, without the prior offer of a surrender. In the result, the counterclaim fails save as to the declaration which merely reflects the failure of the claim.

COURT OF APPEAL

December 16 1980

(Before Lord Justice CUMMING-BRUCE, Lord Justice O'CONNOR and Lord Justice GRIFFITHS)

CHIPPERFIELD v SHELL UK LTD
WARWICK & WARWICK (Philately) LTD v SHELL UK LTD

Estates Gazette March 14 1981

(1980) 257 EG 1042

Landlord and Tenant Act 1954, Part II — Appeals by landlords against decisions granting new tenancies to stamp dealers in the Stamp Centre or "Bourse" in the Shell-Mex House complex, Strand — Landlords had originally opposed applications on the ground that they intended to occupy the premises for the purposes of their own business, but had to abandon this ground and only remaining issue was duration of new tenancies — County court judge ordered that "duration should be three years to commence with reference to the provisions of section 64" — Section 64 problem — Court of Appeal commended and adopted the "admirable formula", unfortunately overlooked by the profession, devised by Wynn-Parry J in Re No 88 High Road, Kilburn — Judge's order varied to provide that the new tenancies should end on July 31 1983 — Date of expiry of tenancy should be certain even if the beginning is uncertain by reason of the operation of section 64 — "The duration of the new tenancy should always be expressed in this way" — Various criticisms by landlords of judge's exercise of discretion in determining the length of the tenancies rejected — Judge had balanced the inconvenience to the landlords against the effect on the tenants — Appeals dismissed, with variation of orders to accord with the Wynn-Parry principle

These were two appeals raising the same points from decisions of Her Honour Judge Rowland sitting at Westminster County Court. The sole issue was the determination of the duration of tenancies granted pursuant to the provisions of the Landlord and Tenant Act 1954. The tenants were respectively Michael Chipperfield and Warwick & Warwick (Philately) Ltd and the landlords Shell UK Ltd in each case. The appeals were by Shell UK Ltd. The premises in each case were in the Stamp Centre or "Bourse" at 84 Strand, London WC2.

John Colyer QC and Matthew Horton (instructed by F D Duffield, of Shell UK Ltd) appeared on behalf of the appellants; Derek Wood QC and Stephen Furst (instructed by A E Hamlin & Co) represented the respondents.

Giving judgment O'CONNOR J said: The judgment that I am about to read is the judgment of the court. Unfortunately, Cumming-Bruce LJ cannot be here today. He has, of course, read the judgment. On October 10 1980 we dismissed these appeals but made a variation in the orders. We give our reasons.

In these cases the landlords, Shell UK Ltd, appeal against orders made in the Westminster County Court on February 26 1980 granting new tenancies of business premises to the tenants, Michael Chipperfield and Warwick & Warwick (Philately) Ltd, pursuant to the provisions of the Landlord and Tenant Act 1954. There is no distinction between the two cases and the sole issue for determination in these appeals is the duration of the new tenancies ordered.

The tenants are stamp dealers and each occupied premises in what is known as the Stamp Centre or Bourse at 84 Strand (hereafter I shall call it "the Bourse") under leases for seven years expiring on March 25 1978. The Bourse is located in an old news cinema which had been carved out of the ground and first floors of part of a large building, Cecil Chambers, itself part of the Shell-Mex House complex owned and occupied by the landlords. Seventeen stalls had been built around the walls of the auditorium of the cinema and leased off to stamp dealers. The tenants each had one unit. Access to the Bourse from the street was through what used to be the foyer of the cinema.

By the late 1970s this part of Strand had become the focal point of the stamp trade in London. Opposite the Bourse on the other side of the street are Stanley Gibbons and there are others in the immediate vicinity. There were eight or nine dealers in the Bourse some of whom occupied more than one unit. At about the same time, despite the size of Shell-Mex House (some 335,000 sq ft) the landlords were cramped and needed more space for their staff. They applied for planning permission to add a ninth floor to Shell-Mex House and decided to repossess, as and when they could, some business tenancies. A decision had been taken by the middle of 1977 to determine all the tenancies in the Bourse in March 1978 for that was when the head lease to the cinema company under which they had been created fell in. The area of the whole Bourse, less the foyer, is 2,100 sq ft; this is doubled by the insertion or reinsertion of the first floor.

Before we continue the narrative we find it useful to set out some propositions of law which are well established and were accepted by experienced counsel appearing in this case.

(1) A business tenancy within Part II of the Landlord and Tenant Act 1954 does not come to an end on its contractual date; it continues until a notice to determine it is given either by landlord or tenant: section 24.

(2) A landlord may determine the tenancy by not less than six

A months' nor more than 12 months' notice in accordance with the provisions of section 25.

(3) The tenant may ask for a new tenancy by notice in accordance with section 26.

(4) If landlord and tenant agree for a new tenancy the court is not involved and none of the provisions which have caused difficulty in this case apply: section 28.

(5) Where an application is made to the court for a new tenancy, the old tenancy does not come to an end until three months after the date on which the application is finally disposed of: section 64(1). It is made clear by section 64(2) that the three-month period is not to start until the date by which the proceedings on the application
B (including any proceedings on or in consequence of an appeal) have been determined and any time for appealing or further appealing has expired.

(6) A landlord may oppose the grant of a new tenancy on the grounds set out in section 30(1), and not otherwise. We shall have to refer to this section later.

Section 33 of the Act provides:

> Where on an application under this Part of this Act the court makes an order for the grant of a new tenancy, the new tenancy shall be such tenancy as may be agreed between the landlord and the tenant, or, in default of such agreement, shall be such a tenancy as may be determined by the court to be reasonable in all the circumstances, being, if it is a tenancy for a term of
> C years certain, a tenancy for a term not exceeding fourteen years, and shall begin on the coming to an end of the current tenancy.

The effect of this section when read with section 64 is that when the court of first instance makes an order the date when the new tenancy is to begin is uncertain; it may be postponed for a long time if there is an appeal to the Court of Appeal and for even longer if there is an appeal to the House of Lords.

We return to the narrative. In June 1977 the landlord served notices on the tenants determining the tenancies on March 25 1978. The tenants gave notices asking for new tenancies and the landlord gave notice of opposition on the ground that they required the premises for use in connection with their own business — that is,
D pursuant to paragraph (g) of section 30(1) which reads: "subject as hereinafter provided, that on the termination of the current tenancy the landlord intends to occupy the holding for the purposes, or partly for the purposes, of a business to be carried on by him therein, or as his residence." Section 30(2) provides that a landlord may not oppose the grant of a new tenancy on ground (g) unless he has been landlord for five years.

For the next year or so the parties were in negotiation, but when they found that they could not agree the matter was set down for hearing in the Westminster County Court on May 21 1979. A few days before the hearing a spanner was thrown into the works. There was another tenant in the Bourse, Cameo Stamps Ltd, who
E occupied two units. They had been served with a notice to determine on March 25 1978. They, too, had asked for a new tenancy and been given notice of opposition, but in their case the landlords relied not only on the grounds in paragraph (g) of section 30(1) but also on the grounds found in paragraph (f), which reads: "that on the termination of the current tenancy the landlord intends to demolish or reconstruct the premises comprised in the holding or a substantial part of those premises or to carry out substantial work of construction on the holding or part thereof and that he could not reasonably do so without obtaining possession of the holding."

It will be seen that this paragraph is not subject to section 30(2). We do not know why the landlords chose to break tacks in this way
F and it does not matter. On May 17 1979 Cameo gave notice that they proposed to challenge the validity of the notices served on them on various grounds which it is unnecessary to specify here. They followed this up on May 18 with a request for a new tenancy beginning on May 18 1980 pursuant to section 26(2) of the Act. On May 21 1979 there was a full day's hearing of the instant cases, which were adjourned part heard. On July 6 all the parties were before the court, including Cameo. The landlords conceded that the notices served on Cameo were invalid. They had to concede that as a result of the section 26 notice served by Cameo the earliest that they could gain possession of the Bourse was May 1980. That concession cut the ground from under their feet in the instant cases and they conceded that they could not maintain their opposition to G
the grant of new tenancies to the present tenants.

Thereafter the hearing of the present cases continued, spread over a number of adjourned hearings from September 20 to December 13 1979. The only issue relevant to the present appeals was the duration of the new tenancies. The learned judge gave judgment on February 26 1980. She decided this issue in these words: "Having considered all aspects including the present leases, I have come to the conclusion that the duration should be three years to commence with reference to the provisions of section 64".

The final orders dated February 26 1980 gave effect to that decision. We were told that orders in this form were not uncom- H
mon, but that the more usual order would have been to say that "the duration of the new tenancy should be three years". The effect is exactly the same, for the addition to the words "to commence with reference to the provisions of section 64" merely spells out the position that arises where the more usual order is made. The effect of such an order is that the actual duration of the new tenancy is uncertain. The present case is a good illustration. If there had been no appeal the new tenancies would have commenced in July 1980, that is, time for giving notice of appeal plus three months from February 26 1980. The bringing of the appeal deferred the date of commencement to a date at least four and a half months beyond the date of judgment in the Court of Appeal, that is, time for petition- J
ing the House of Lords for leave plus three months. If leave to appeal is given either by the Court of Appeal or the House of Lords the commencement is further deferred to a date three months beyond the date when judgment is given in the House of Lords.

The existence of this problem was recognised as long ago as 1959 by Wynn-Parry J in the case of *Re No 88 High Road, Kilburn* [1959] 1 WLR 279, and a solution was found. That case was heard in February 1959. The landlords did not oppose the grant of a new tenancy and the issues before the court were duration and rent. Having construed the effect of sections 33 and 64 of the Act the learned judge said at the top of p 284: K

> The result is, therefore, that the new tenancy must begin at the earliest three months from today, but its commencement will, of course, be further delayed if the tenants (as they have every right to do) decide to test this matter before the Court of Appeal.
>
> I consider that the result at which I am compelled to arrive is one which is unjust to the landlord. But I cannot mitigate that injustice as regards rent, for the new rent can only operate at the end of the period determined by section 64. I can, however, intervene so far as the term of years is concerned, and I do not propose to grant a term of 14 years, but to direct that, when the lease comes to be executed — that is, when the matter is finally disposed of — the term of the lease to be executed shall be for a term ending on June 24 1972.

L

It seems that this admirable formula has been overlooked by the profession. The case was not cited to the county court in the present cases. We think it is in the interest of both landlord and tenant that the end date of the new tenancy should be established when the order is made even though it is not possible to make sure when it will start. The shorter the time the more important this is. We think that the duration of the new tenancy should always be expressed in this way.

In the present cases we think it apparent by her reference to section 64 that the learned judge was intending to grant new tenancies of three years ending in July 1983. Had there been no appeal that is what would have happened. It is for this reason that when we M
dismissed the appeal on October 10 we varied the order as to duration by ordering that the new tenancies should end on July 31 1983.

The duration of a new tenancy determined by the court under section 33 of the Act is a matter of discretion; it is for the judge to decide what is "reasonable in all the circumstances". This court will not interfere unless it is shown that the learned judge in exercising her discretion has taken into consideration matters which she ought not to have done and/or failed to take into consideration matters which she ought to have done.

The landlords contend that the circumstances of the present cases

A are such that a proper exercise of her discretion required the learned judge to grant new tenancies of no more than 12 months' duration. They make a number of criticisms of the judgment, introducing them by a submission that the learned judge misapprehended the issues before her. They point to passages where, in her judgment, she says that the landlords contended that the tenants should be granted licences and not tenancies. It is common ground that this was never an issue, indeed it could not have been. We think that the error arose as a result of the evidence called on valuation. There was evidence that dealers such as the tenants often preferred to occupy under licence and to pay as much as for a tenancy. Be that as it may, the learned judge, having set up a
B contention that had never been made, rejected it. When we look at her judgment as a whole we do not find any grounds for saying that this mistake has influenced her thinking on the matters which she had to consider in the exercise of her discretion.

We turn to the substantive complaints which are found in the notice of appeal. In the first place, the landlords say that the learned judge failed to take into account or to give proper weight to the circumstances in which they consented to the grant of new tenancies. The gravamen of this complaint is that it was only a technical defect in their notices to Cameo which made it impossible for them to oppose the grant of new tenancies to the tenants and that but for that they would have opposed the grant successfully. They submit
C that this is a "near miss" case and rely on the decision of this court in *Wig Creations Ltd v Colour Film Services Ltd* [1969] 20 P&CR 870. In that case the tenants occupied two floors under separate leases each of 21 years' duration, in the one case from March 25 1947, in the other from March 25 1959. The landlords, who occupied another floor in the building, bought the reversions on March 1 1965, as they foresaw expansion of their own business. When the first lease fell in, notices were given for March 1968. Although they wanted the premises for their own business the landlords could not rely on paragraph (g) of section 30(1) by reason of subsection (2), the five-year rule. The tenants, who also wanted the floor for their business, asked for a new tenancy to expire in March 1980, the
D same date as their lease of the other floor. The judge granted a tenancy of three years; he took into account the five-year rule and said that but for that factor he would have acceded to the tenants' request. The tenants appealed. Dismissing the appeal, Lord Denning MR said at p 874:

> Section 33 is in very wide terms. It empowers the court to do what is reasonable in all the circumstances. Suppose a landlord bought five years ago, plus one day. He could resist a new tenancy altogether on the ground that he wanted the place for his own business. Suppose he buys it five years ago less one day. Should he be kept out of the place for several years simply by the two-day difference? I think not. The policy of the Act is to give a
> E landlord (who has purchased more than five years ago) an absolute right to get possession for his own business: leaving it to the court to do what is reasonable if he has purchased less than five years. In doing what is reasonable, the five-year period is a factor which it is permissible for the judge to take into account. The weight of it is for him.

In the present cases the landlords submit that, by a parity of reasoning, the duration of the new tenancies should have been such as to enable them to get possession of the Bourse by serving fresh notices to coincide more or less with the Cameo case. They submit that the grant to July 1983 produces a leap-frogging situation cutting the ground from under their feet in the Cameo case.

F I do not think that these complaints are well founded. The learned judge said at p 73:

> Although the landlords have failed in opposing a new tenancy they are entitled to have their intentions and the evidence given in respect of their opposition taken into consideration.

Then at p 74 she gave herself the following direction:

> Dealing with the issue of duration and other terms, where there is no agreement section 33 of the 1954 Act comes into operation. It provides discretion to grant terms reasonable in all the circumstances and entitles me to take into consideration all factors. It includes the grounds of opposition of landlords under section 30(1)(g), their intentions and plans and the hardships and needs of each party. I must give proper protection to the tenants G but not be unfair to the landlords, nor allow the grant of a new tenancy to defeat development or reduce the value of the building, and must weigh up and reasonably protect the interests of both parties.

We can find nothing wrong in this direction. It is true that there is no express reference in the judgment to the possible effect on the future Cameo case, but she was well aware of Cameo and in the end we do not think that it matters. If it was reasonable in all the circumstances to grant new tenancies to these tenants up to July 1983 then that exercise of discretion is not made unreasonable by the fact that the result may be a factor for consideration if and when the Cameo case comes before the court. Further, we do not think it right to conclude that the learned judge had forgotten and thus paid H no attention to the reason for the landlords abandoning their opposition to the grant of new tenancies: she recorded it accurately in her note of the proceedings before her on July 9 1979; she set it out in her judgment at pp 61-62. We think that the landlords are driven to asserting that the grant of three years demonstrates that the learned judge cannot have given any weight to this topic. To that assertion we can only say that we do not agree. We think she had it in mind and that the terms granted do not show that she must have disregarded it.

The landlords contended that the learned judge wrongly allowed questions relating to planning applications to influence her in determining the duration of the new tenancies. She had heard J evidence on these matters before the landlords abandoned their opposition. She recited some of this in her judgment as part of the general history but ended up by saying at p 68:

> I do not think it would be right to allow myself to be influenced by these factors. My task is to take an objective view and to reach my conclusions in the light of the requirements, needs and all the facts surrounding both parties. In the circumstances I am unable to take the representations in the correspondence (with the local planning authority) into account as affecting my judgment on the duration of the tenancies.

We do not see that she could have said fairer than that and we can find no substance in this contention. K

The landlords complain that the learned judge was wrong to take into account an offer made by the tenants to move up to the Strand end of the Bourse. This matter arose towards the end of the hearing: the tenants and Cameo made the offer which, if accepted, would have freed the greater part of the ground floor of the Bourse to the landlords. It must be remembered that the landlords wanted the ground floor for a hospitality area and proposed to insert a first floor for office use. The scheme was investigated and the landlords rejected it; they were quite entitled so to do. The learned judge said that she could take this matter into account in determining duration. We think she was right. So this is quite different from the cases which show that a tenant cannot attack a landlord's intention under L paragraphs (f) and/or (g) by saying that he could achieve his purpose in some other fashion. Here we are concerned with the exercise of discretion under section 33. We think that this was a matter which the judge could take into account in considering all the circumstances of these cases. She was doing no more than looking at the effect on the landlords of the grant of new tenancies. In the same way it seems to us that in comparing the relative area of Shell-Mex House and the Bourse she was engaged in the same exercise.

The landlords complain that, in considering the landlords' avowed intention of leaving Shell-Mex House because it had become too small for their requirements, she ought to have recog- M nised that the grant of three years would frustrate their intentions to use this space for their own purposes. The fact is that the grant of any term must necessarily prevent the landlords from implementing their scheme to that extent. It is said that, as the judge was contemplating that the landlords anticipated moving by 1985, the grant of new tenancies to July 1983 would completely frustrate their plans. We do not agree. All these dates were necessarily speculative. The judge was balancing the inconvenience to the landlords against the effect on the tenants. There was ample evidence that the landlords' plans were fluid. They themselves assert in the notice of appeal that the judge ought to have found that there was no realistic prospect of

A their moving out of Shell-Mex House before 1990. If that be the time-scale it is idle to contend that the three-year terms must frustrate their intentions.

In the course of her judgment, the learned judge said that there was no reason why the landlords should not put in and occupy the proposed first floor in the Bourse. This idea was not canvassed in the evidence and it appears for the first time in the judgment. It may or may not be feasible but the judge ought not to have speculated. The error is to our mind peripheral and in the context of the judgment as a whole cannot have had any material effect upon the length of the new terms which she decided to grant.

B In the result, we conclude that no good grounds have been established to show that the judge's discretion was not properly exercised.

The appeals were dismissed with costs. The orders below were varied by deleting the words "three years to commence with reference to the provisions of section 64 of the Landlord and Tenant Act 1927" and substituting the words "a term expiring on the 31st day of July 1983".

C

COURT OF APPEAL

November 25 1980

(Before Lord Justice LAWTON, Lord Justice BRANDON and Lord Justice TEMPLEMAN)

A J A SMITH TRANSPORT LTD v BRITISH RAILWAYS BOARD

Estates Gazette March 28 1981

(1980) 257 EG 1257

Landlord and Tenant Act 1954, section 30(1)(f) — Court's comments on procedure and tactics and criticism of merits of appeal — "Play-
D
ing the system" — British Railways Board wanted possession of part of a goods yard leased by them to the applicants in order to redevelop the yard in conjunction with a firm which would build a supermarket and car parks thereon — Evidence clearly in support of board's opposition under section 30(1)(f) to the applicants' claim for a new tenancy — Board had made a decision to demolish or reconstruct the yard and had only been prevented from taking further steps by delaying tactics of applicants — No misdirection by judge, who would indeed have acted perversely if he had decided otherwise than in favour of board — Appeal dismissed

This was an appeal by A J A Smith Transport Ltd against a
E
decision of Judge Blackett-Ord, Vice-Chancellor of the County Palatine of Lancaster, sitting as a county court judge at Blackburn County Court. The appeal was against the judge's dismissal of the claim by the applicants for a new tenancy of part of a goods yard at Clitheroe railway station in Lancashire. The claim was opposed by the respondents, British Railways Board, on the ground set out in section 30(1)(f) of the Landlord and Tenant Act 1954.

W H Goodhart QC and J Dowse (instructed by J L Lumley, of Clitheroe & Whalley) appeared on behalf of the appellants; Derek A Wood QC and J L Hand (instructed by Evan Harding, chief solicitor, British Railways Board) represented the respondent board.

F Giving judgment, LAWTON LJ said: This is an appeal by A J A Smith Transport Ltd against an order made by His Honour Judge Blackett-Ord, Vice-Chancellor, sitting as a county court judge on June 4 1980 whereby it was adjudged that their application for a renewed tenancy be dismissed on the grounds set out in section 30(1)(f) of the Landlord and Tenant Act 1954. The order was in these terms:

> It is ordered that the applicants' application for the grant of a new tenancy in respect of the premises, the warehouse and adjoining land in former goods yard at Clitheroe Lancashire be dismissed with costs.

That form of order is, in itself, interesting because it implies that the applicants were seeking to say that the respondents, the British Railways Board, had duly terminated their lease of premises and G
that, under the provisions of the Landlord and Tenant Act 1954, they were entitled to a new lease. The case was conducted before the learned Vice-Chancellor on that basis. No point was taken before the Vice-Chancellor to the effect that the respondents had not duly terminated the lease which the applicants at all material times had.

Having been dissatisfied with the Vice-Chancellor's order, the applicants appealed to this court. The notice of appeal is on the basis that the lease was duly terminated so as to give the county court jurisdiction to grant a new lease. Indeed, it says this:

> . . . that the applicants are entitled to a new tenancy of the above-mentioned H
> premises and that it may be ordered that these proceedings be remitted to the Blackburn County Court to determine the terms of such new tenancy in accordance with the above-mentioned Act. . . .

The court assembled expecting to have to deal solely, having regard to the terms of the notice of appeal, with the question as to whether, in all the circumstances of this case and as a matter of law, the Vice-Chancellor had been right in his determination that the applicants were not entitled to a new lease. The applicants today had the benefit of being represented by Mr Goodhart. He had not represented them in the Blackburn County Court; they had then been represented by another leading counsel. When Mr Goodhart rose to open the appeal he informed us that there were two points J
which could be taken on behalf of the applicants: one which was dealt with in the notice of appeal and one other which went to the validity of the notice terminating the lease, which he said he did not propose to deal with before us.

The court was somewhat surprised at this because it seemed to us that what Mr Goodhart on behalf of the applicants was doing was blowing hot and cold. He was asking the court to say that his clients were entitled to a new lease. They could only be entitled to a new lease if the existing lease had been duly terminated, but he went on to say that, on some other occasion if this court did say that his clients were not entitled to a new lease, his clients might say that the original lease had never been properly determined. This was an K
unusual situation for the court to be met with and it was one which we did not find attractive — the more so as there were indications that this appeal, which has little merit as I shall demonstrate shortly, was one step in a delaying action to avoid giving up possession of the premises in dispute. It seemed to us that the tactics which Mr Goodhart was adopting on behalf of his clients were tactics which were designed, if I may use a colloquial expression, to play the system so that there would be further delays.

We gave Mr Goodhart an opportunity of considering what he should do because, when we first considered the matter, it seemed to us that he should be put to his election as to whether he wanted to go on with this appeal or abandon it and take some other proceed- L
ings elsewhere to establish that his lease had never been properly determined. But, after argument, it occurred to us that there were difficulties in that approach. After further discussion with counsel the question arose as to whether this was not a case in which the unusual powers given by Order 20, rule 8, as applied to this court by Order 59, rule 10 and section 43 of the Supreme Court of Judicature (Consolidation) Act 1925, might be used. This court has jurisdiction to order an amendment of pleadings in order that the real matter in issue can be tried.

The real matter in issue in this case is as follows. The defendants want possession of the premises. The plaintiffs do not want them to have possession of the premises. They want to remain in possession. M
Nowadays, under an amendment to the Landlord and Tenant Act 1954, which has become section 43A of that Act, the applicants in the Blackburn County Court could have pleaded their case in the alternative. They could have said that the notice purporting to determine the tenancy issued pursuant to the Landlord and Tenant Act 1954 was a bad one. Alternatively, if it was a good notice, they could have asked for a new tenancy pursuant to the terms of that Act. They had done nothing of the kind; and it did occur to us that it might be a sensible course to order the applicants to amend their originating application in the Blackburn County Court so as to raise the issue as to whether the original notice served by the respondents

A was a valid notice, and also to amend their notice of appeal so that all the matters in dispute could be dealt with in this court.

This was a course which Mr Wood, on behalf of the defendants, was willing to consent to, but he went on to point out that he would not thereby be abandoning any defence of estoppel which he might have. Mr Goodhart, on behalf of the plaintiffs, submitted that this court had no jurisdiction to order any such amendment of the pleadings. We adjourned to consider the matter which had arisen and, without making any finding as to whether we had or had not jurisdiction to order an amendment of the type which we had discussed in argument with counsel, we came to the conclusion that we should give Mr Goodhart an opportunity of opening the appeal B as it stood on the notice of appeal. It was pointed out to him by Brandon LJ that he was, if he wished, at liberty not to go on with his appeal. He elected to go on with it as it was put in the notice of appeal and he advanced arguments in support of that appeal.

I, for my part, make no findings as to the legal consequences as to what has happened. It may be that hereafter counsel will have to argue what those consequences are but, for the purposes of my judgment, I am going to confine myself to the short facts in this case and the law applicable to them. The relevant statutory provision is contained in section 30(1)(f) of the Landlord and Tenant Act 1954. Subsection (1) reads as follows:

> (1) The grounds on which a landlord may oppose an application under C subsection (1) of section 24 of this Act are such of the following grounds . . .
> (f) that on the termination of the current tenancy the landlord intends to demolish or reconstruct the premises comprised in the holding or a substantial part of those premises or to carry out substantial work of construction on the holding or part thereof and that he could not reasonably do so without obtaining possession of the holding.

Before the Vice-Chancellor the sole issue was whether, on the evidence, the landlords — that is to say the British Railways Board — intended to demolish or reconstruct the premises comprised in the holding; it being accepted that, if the British Railways Board did so intend, they would require possession of the whole of the premises occupied by the applicants.

D The applicants are haulage contractors and, for some time now, they have held under lease from the British Railways Board and their predecessors a tenancy of part of the goods yard at Clitheroe Station in Lancashire. They had a tenancy which seems to have been, for most of the time at any rate, a periodic tenancy. In 1978 and 1979 the British Railways Board were minded to redevelop the old station yard. What they had in mind to do was to grant a building lease to a firm of grocers in Preston, E H Booth & Co Ltd, who undertook under that building lease to build a supermarket together with the necessary car-parks for such a business enterprise. In order to accomplish this end the British Railways Board had to obtain planning permission. They got planning permission and, as a E result, provided they had a binding agreement with the prospective lessees, they were in a position to go ahead.

There was one small difficulty in their way. The local authority, whose planning committee had granted planning permission, owned a small strip of land running alongside the old goods yard and, unless the British Railways Board had rights of way over this small strip of land, there would be no access to the planned car-parks. It followed that they had to be sure of being able to get control of that small strip of land before the building project could go ahead.

Now, the prospective building lessees and, presumably, the local authority too, knew that, until the British Railways Board could get F possession from those who were occupying the goods yard, it would be impossible to start work; and this was a very important matter because, as Mr Wood, on behalf of the British Railways Board, has told us, building costs were going up all the time. He informed us — and there is no reason to doubt the accuracy of what he told us — that, as a result of the delay which has come about by reason of this appeal, the building costs have gone up by something in the region of a quarter of a million pounds. It is clear that the fewer further delays there are the better it will be for all concerned, if not for the applicants.

It follows, therefore, that the actual signing of a building lease and building contracts and the like had to be delayed until such time as the British Railways Board could be sure of getting possession G and, until they had obtained an order against the applicants, they could not be sure of getting possession. The applicants, again playing the system, have used the lack of certainty about getting permission for various small matters as a reason for saying that the British Railways Board do not intend to demolish or reconstruct the premises.

The situation is as follows. E H Booth & Co Ltd have agreed terms, but they are not signed, with the British Railways Board for the development of the railway yard. They have agreed the terms but have not signed a building contract for the work which will be required to be done under the development plans. Attached to the development plans are approved plans. E H Booth & Co Ltd H retained an architect for the purposes of providing approved plans. The local authority have told British Railways Board that they will sell them the land which borders the old goods yard and which is necessary for access to the proposed car-parks for the sum of £1, which indicates to me that they do not attach much importance to that particular piece of land.

The only matter which is outstanding is this. E H Booth & Co Ltd at the trial before the Vice-Chancellor said that they had got a builder in mind. He was one who had done work of a similar kind for them elsewhere in Lancashire, and they had no doubt that he would be willing to undertake this work of development at Clitheroe Goods Yard. The applicants have fastened on to that one J matter, together with the problem of the little strip of land, in support of their submission that the British Railways Board's plans are still only tentative, exploratory and provisional and they have not yet moved into the realm of — to use Lord Asquith's words in another case — the valley of decision.

In my judgment, there is nothing whatsoever in that argument. Indeed, if the Vice-Chancellor had not found that the British Railways Board had made a decision that they would demolish or reconstruct the existing railway yard, he himself would have been acting perversely. The evidence in this case is overwhelming. The British Railways Board had got to the stage of decision and the only reason they have not taken a further step forward has been the K delaying tactics adopted by the applicants in this matter.

It is also said that the learned Vice-Chancellor misdirected himself in law at the trial. He reminded himself of what the law was by looking at and considering the case of *Reohorn v Barry Corporation* [1956] 1 WLR 845 at p 849, and he also considered the judgment of Denning LJ (as he then was) in the case of *Fisher v Taylors Furnishing Stores Ltd* [1956] 2 QB 78. It is not submitted by Mr Goodhart that either of those cases was wrongly decided. In any event, they are both binding on us and it is impossible to say that the learned Vice-Chancellor did not direct himself as to the legal meaning of the word "intends" as used in section 30(1)(f) of the Landlord and Tenant Act 1954. L

There is nothing whatsoever in this appeal except its use for the purposes of delay. It is a use of which I disapprove. I would dismiss the appeal.

I would add this. As a result of our decision to hear this appeal and not to deal with the matter by way of amendment of the pleadings, it may be that further proceedings will be necessary. It is not for me to make any comment about the form those proceedings should take or what defences should be available to anybody. All I do wish to say is this. They may be taken in a county court in Lancashire and it is to be hoped that the circuit administrator of that circuit and the county court judge who is asked to deal with the matter will do everything in his power to ensure that any further M proceedings are dealt with with the maximum possible speed and, in so far as it is possible to abridge time, time should be abridged. The same should apply if there are further proceedings in the High Court.

Agreeing, Brandon LJ said: I regard this appeal as coming as near to an abuse of the process of the court as it is possible to go without crossing the line. Under the relevant provisions of the County Courts Act 1959 an appeal in a case of this kind lies only on a question of law. There are only two questions of law that could be canvassed in this case.

The first was whether the learned judge had directed himself

A correctly as to the meaning of the word "intention" in the relevant statutory provision. As to that, it was as clear as it could be from his judgment that he had correctly directed himself, and it was not even suggested in argument that he had not done so.

The only other point of law that could be raised was whether there was any evidence upon which the judge could make the finding of fact about the intention of the British Railways Board which he did. As to that I agree with Lawton LJ that there was, in effect, no evidence upon which he could have come to any other conclusion, and that any other conclusion would have been perverse.

In these circumstances, for an appeal to be prosecuted to this B court comes, as I have indicated, as near to the line of abuse of the process of the court as can readily be imagined. I, too, would dismiss the appeal.

TEMPLEMAN LJ agreed with both judgments.

The appeal was dismissed with costs, to be taxed on the common fund basis.

CHANCERY DIVISION

C October 10 1980

(Before Mr Justice GOULDING)

PHILIPSON-STOW v TREVOR SQUARE LTD

Estates Gazette March 28 1981

(1980) 257 EG 1262

Landlord and Tenant Act 1954, Part II — Landlord's notice to terminate business tenancy — Whether notice valid — Requirement of section 25(6) of Act that the notice must, inter alia, state whether the landlord would oppose an application for the grant of a new D tenancy and, if so, on which of the grounds mentioned in section 30 he would do so — Notice stated that the landlord intended to carry out substantial "work of redecoration" instead of the words in section 30(1)(f) "work of construction" — After reviewing the authorities, which tended "to allow considerable latitude in satisfying the requirements of subsection (6) of section 25", the judge held that the notice was valid — It was not deceptive or misleading — It made clear which of the seven separate grounds the landlord was relying on, although the landlords might not be able to make its case good as there might be difficulty in showing that work of redecoration was work of construction

These were originating summonses seeking declarations in rela- E tion to applications for new tenancies pursuant to Part II of the Landlord and Tenant Act 1954. The premises were at 239 Knightsbridge, London, and the applications, made by the tenant, Mrs Philipson-Stow, were opposed by the landlords, Trevor Square Ltd. For reasons which are explained in the judgment, the landlords had served two sets of notices, but the issue discussed in the judgment related to the first set. As these were held to be valid, the second set became of no effect.

N Wood (instructed by Withers) appeared on behalf of the plaintiff; H M Morgan (instructed by Amery Parkes & Co) represented the defendants.

Giving judgment, GOULDING J said: I have before me two appli- F cations pursuant to Part II of the Landlord and Tenant Act 1954, the object of both being to obtain a new lease of certain property in Knightsbridge. The plaintiff on both summonses is Mrs Philipson-Stow, the tenant, and the defendant to both is Trevor Square Ltd, the landlord. It is my intention to refer to the plaintiff and the defendant as the "tenant" and the "landlord" respectively.

In fact, the property to which each summons refers is held under two separate leases relating to distinct parcels of land. They adjoin one another and both leases, so far as their contractual terms are concerned, were limited to expire at the same date, namely Christmas 1979. One lease was dated June 24 1968 and was limited for a term of 11½ years, and the other was dated June 24 1969 and was limited for a term of 10½ years from that later date. I shall speak G henceforth as though there were but one demised property and but one lease because precisely similar proceedings have been taken in respect of each of the two leases and what I say of either is equally applicable to the other.

Now as I have said, the contractual term was due to expire at Christmas 1979. On March 9 the landlord served a notice on the tenant, dated the previous day, and, after setting out the name and address of the tenant and the particulars of the premises, all under the heading "Landlord and Tenant Act 1954, Landlord's Notice to Terminate Business Tenancy", the document proceeded as follows:

(1) We Trevor Square Ltd, whose registered office is at Africa House, Kingsway, London WC2, landlord (note 7) of the above mentioned H premises, hereby give you notice terminating your tenancy on 25th day of December 1979 (note 1).
(2) You are required within two months after the giving of this notice to notify me in writing whether or not you will be willing to give up possession of the premises on that date.
(3) We would oppose an application to the court (note 3) under part II of the Act for the grant of a new tenancy on the ground (note 4) that on the termination of the tenancy we intend to carry out substantial work of re-decoration of the whole or part of the premises and we could not reasonably do so without obtaining possession of the premises.
(4) This notice is given under the provisions of section 25 of the Landlord and Tenant Act 1954. Your attention is called to the notes overleaf.

J Then follows the date, March 8 1979, and the signature of a director on behalf of the landlord with the address of the landlord's solicitors. On the back are the notes referred to both by the final sentence of the notice and by a number of parentheses (which I read) referring to particular notes. They are notes which are part of the statutory form of the landlord's notice under Part II of the Act. Paragraph 4 of the notes (note 4 as it is called in the notice) reads thus:

The court has no power to make an order for the grant of a new tenancy if the landlord, having stated in his notice that he will oppose an application to the court on one of the grounds specified in the Act, establishes that ground to the satisfaction of the court. The grounds specified in the Act are: K

and then follow seven paragraphs beginning with the respective letters (a) to (g).

The tenant was advised, for reasons which I shall come to in a moment, that that document was not effective as a notice to terminate a business tenancy under the Act of 1954. By her first application, the tenant sought a declaration that the notice was void and of no effect and, without prejudice to that contention, added a claim for the grant of a new tenancy or tenancies in case the notice should be held to be valid. That application was made by originating summons on July 6 1979.

The landlord, seeing the question to which it gave rise, then served a further notice, dated November 14 1979, in similar statut- L ory form but stating in paragraph 3:

We would oppose an application to the court under part 2 of the Act for the grant of a new tenancy on the ground that on the termination of the current tenancy we intend to demolish or reconstruct the premises comprised in the holding or a substantial part of those premises or to carry out substantial work of construction on the holding or part thereof and that we could not reasonably do so without obtaining possession of the holding.

The service of that notice was followed by the issue of the second originating summons by the tenant which is dated March 6 1980. Thereby the tenant seeks a declaration that if the March notice is held to be valid, then the further notice of November is void and of M no effect, on the ground that by the earlier notice the tenancy was (subject to the provisions of the Act) terminated at Christmas 1979, and in case the opposite result should be arrived at, so that the November notice turns out to be valid, then again the tenant seeks the grant of a new tenancy.

The matter comes before me to decide simply the question which, if either, of the landlord's two notices is valid. I must look briefly at the relevant provisions in the Landlord and Tenant Act 1954 Part II. Section 24(1) provides that:

A tenancy to which this Part of this Act applies shall not come to an end unless terminated in accordance with the provisions of this Part of this Act.

A And then section 25(1) provides that:

> The landlord may terminate a tenancy to which this Part of this Act applies by a notice given to the tenant in the prescribed form specifying the date at which the tenancy is to come to an end.

Then I think I need only read one other subsection of section 25, namely subsection (6), the all-important one in the present contest:

> A notice under this section shall not have effect unless it states whether the landlord would oppose an application to the court under this Part of this Act for the grant of a new tenancy and, if so, also states on which of the grounds mentioned in section 30 of this Act he would do so.

B That carries one on to section 30(1), which provides that:

> The grounds on which a landlord may oppose an application under subsection (1) of section 24 of this act are such of the following grounds as may be stated in the landlord's notice under section 25 of this Act or, as the case may be, under subsection (6) of section 26 thereof, that is to say . . .

And then follow seven subparagraphs (a) to (g). The first three relate to unsatisfactory conduct on the part of the tenant in that quality. (d) relates to the provision of suitable alternative accommodation justifying the landlord in terminating the tenancy of the existing premises. And (e) deals with a special case of subletting. (f) I must read:

C
> That on the termination of the current tenancy the landlord intends to demolish or reconstruct the premises comprised in the holding or a substantial part of those premises or to carry out substantial work of construction on the holding or part thereof and that he could not reasonably do so without obtaining possession of the holding.

That is the ground of which the precise wording is here material, but there is one more ground, namely (g), which deals with a case where the landlord intends to occupy the premises for his own purposes.

Those subparagraphs (a) to (g) correspond in wording *mutatis mutandis* with the subparagraphs (a) to (g) in note 4 on the back of the statutory form of notice which the landlord here has used. The tenant's case is quite a simple one. The tenant says the notice of
D March 8 1979, although saying in words that the landlord would oppose an application to the court, does not state any ground mentioned in section 30 of the Act. There is some resemblance, the tenant says, to the words of subparagraph (f) of section 30(1), but redecoration is not demolition, reconstruction or a work of construction and, accordingly, no ground provided by section 30 is mentioned. The landlord, therefore, has not stated, as required by section 25(6), on which of the grounds mentioned in section 30 he would oppose an application for a new tenancy. Accordingly, the notice is of no value to the landlord. I use that vague form of words deliberately because if the tenant's contention is so far correct, Mr Wood, on the tenant's behalf, has put forward two alternative
E propositions as to the consequences of the defects in the notice. However, before I consider them, it will be necessary to decide the main point on the true construction of the notice. Does it state on which of the grounds mentioned in section 30 of the Act the landlord would oppose an application for a new tenancy? The landlord, who is represented here by Mr Morgan, contends that if you look at the notice properly and construe it in accordance with reported authority, it does contain a sufficient statement of one of the statutory grounds, namely that in subparagraph (f) of section 30(1).

Counsel have cited a number of authorities. They constitute a broadly consistent line of judicial opinion tending to allow con-
F siderable latitude in satisfying the requirements of subsection (6) of section 25. I do not think it is necessary to lengthen this judgment by reading extracts from all those cases. I will refer to them by title and report and I will read one or two passages that seem to me to put the effect of the authorities particularly clearly. The first, I think, in order of date is the case of *Biles v Caesar* [1957] 1 WLR 156 and then comes *Bolton's (House Furnishers) v Oppenheim* [1959] 1 WLR 913. Both those were cases in the Court of Appeal. And then came one at first instance before Barry J, *Barclays Bank v Ascott* [1961] 1 WLR 717. That was a case in which the learned judge thought that such latitude as allowable, having regard particularly to the *Bolton's* case, did not assist the landlord, Barclays Bank, and he held that particular notice invalid. Then there was G
another case in the Court of Appeal, *Marks (Morris) v British Waterways Board* [1963] 1WLR 1008. I think it is helpful if I read some short extracts from the judgment in that case. It should be borne in mind that the court there was considering not a landlord's notice under subsection (6) of section 25 of the Act but a landlord's counternotice under subsection (6) of section 26 thereof, which has to be served if the landlord desires to oppose a tenant's request for a new tenancy under section 26. Subsection (6) in section 26, like that in section 25, requires the counternotice to state on which of the grounds mentioned in section 30 of the Act the landlord will oppose the application. Accordingly, in my view, *Mark*'s case is *in pari materia* with the cases on landlord's notices under section 25 H
although referring to the following section. Lord Denning MR at p 1015, having referred to an argument that the notice in that case was bad because it did not correctly represent the landlord's intention at the time, said:

> I think the correct answer is that the notice opposing the new tenancy — a landlord's notice — is to be regarded as in the nature of a pleading. It is a pleading which comes for proof at the date of the hearing. It is sufficient as long as it gives notice to the tenant of the case he has to meet. So long as it is not deceptive or misleading it avails the subsequent owner of the property who is the landlord at the date of the hearing.

Harman LJ at p 1018 said:

> The only object, as Lord Simonds said in the *Betty's Cafés* case, of a counternotice is to inform the tenant of the case which will be made against him when the hearing comes on and provided it does that it is, I think, a good counternotice. . . . J

And, omitting two or three sentences, he continued:

> You must not mislead the tenant. You must not say anything which is fraudulent but if your notice is given in good faith and the fact about reconstruction can be substantiated by the person who is the landlord when the hearing comes on, I think the counternotice really has served the purpose which the legislature can be said to have required of it.

And he finished that paragraph of his judgment by saying:

> It is true the notice is in a most unhappy form. It was not candid as it should have been. If I had thought it could have misled Marks I should be inclined to say it was a bad notice, but I do not think it did, having regard to the prior notice served and the knowledge of the facts which he had. Therefore, although I think we do perhaps rather stretch the section in doing so, I agree with the county court judge that this counternotice was a valid counternotice. K

And, at the very end of the report, at p 1021, Pearson LJ says of the counternotice:

> It gave him [that is the tenant] adequate warning of what was the contention on the other side which he would have to meet at the hearing and, therefore, it performed its function adequately.

It will be observed, I remark in passing, from Harman LJ's judgment in the passage which I have cited, that in the *Marks* case there was some extrinsic evidence of the tenant's knowledge of material facts at the time of service of the notice. Nothing of that sort is before the court here and I have to construe the notice simply as a self-contained document, without reference to prior knowledge of either party. L

Then there is another case in the Court of Appeal *Housleys Ltd v Bloomer-Holt Ltd* [1966] 1 WLR 1244 and I cite from that a very compendious and generally expressed observation by Diplock LJ (as he then was). He said at p 1251:

> First as regards the technical point that the notice by the landlords given under section 25(6) of the Act did not follow precisely the wording of paragraph (f) of section 30(1) of the Act. This is a highly technical point. It is not, in my view, in any event an attractive one nor do I find it well founded. It is sufficient if a notice under section 25(6) makes it clear to the tenant which of the seven separate grounds specified in section 30(1) the landlord is relying on. The purpose of the notice under section 25(6) is to let the tenant know on what grounds his request for a new tenancy, if made, will be resisted, so that he may make up his mind whether to exercise his rights and bring the matter eventually before a court for a decision. M

Other cases cited were *Lewis v MTC (Cars) Ltd* [1974] 1 WLR 1499, a decision of Templeman J, another decision at first instance of *Carradine Properties v Aslam* [1976] 1 WLR 442, a recent case, unreported, in the Court of Appeal of *Jomac Securities v Speigel* whereof I have seen a transcript of the court's judgment dated June 14 1978, and a still more recent case at first instance of *Safeway Food Stores Ltd v Morris* (1980) 254 EG 1091.

Having gone through those authorities, from which I have cited one or two passages, I ask myself these questions about the notice dated March 8 1979. Did it give notice to the tenant of the case she had to meet? Was it deceptive or misleading? Did it make clear which of the seven separate grounds specified in section 30 the landlord was relying on? It is all-important, I think, in looking at the notice, to observe that the notes on the back are part of the statutory form; that the body of the notice expressly calls the tenant's attention to those endorsed notes and that the statement of the landlord's ground of opposition is qualified by a parenthesis referring in particular to note 4 where the alternatives (a) to (g) are set out. If the tenant compared the language in paragraph 3 of the notice with that of the subparagraphs of note 4, she could, in my view, have no doubt that the landlord was aiming at subparagraph (f). There is a close resemblance between the language actually used and part of that in subparagraph (f) and there is not the slightest resemblance to or close relation with any of the other subparagraphs. A reasonable tenant, reading this notice, would I think conclude that the landlord was seeking to rely on subparagraph (f), but the same hypothetical reasonable tenant would also conclude that there was considerable doubt whether the landlord could make its case good, because it would be indeed strange if a substantial work of redecoration could be shown to be, in the statutory language, a substantial work of construction. Therefore, I answer my questions by saying: "Yes, on a fair and not too technical construction the notice does give notice to the tenant of the case she had to meet"; and I would say "No, it is not deceptive or misleading", although it is inaccurately expressed. And then I would say "Yes, it does make it clear which of the seven separate grounds the landlord is relying on". I am conscious, like Harman LJ, in the case of *Marks*, of a feeling that I am perhaps rather stretching the section, but I think the Court of Appeal has already done so in the laudable desire to extend to these every-day notices between landlord and tenant of business premises the principle whereby the court desires *ut res magis valeat quam pereat.*

Accordingly, after close examination of the notice and of the authorities, I conclude that, on its true construction, the notice does state on which of the grounds mentioned in section 30 of the Act the landlord will oppose the application for a new tenancy. Therefore it is unnecessary for me to examine the alternative arguments that Mr Wood put forward in case I concluded the notice was bad, and I must refuse the tenant's application in the first originating summons for a declaration that the notice — and here I return to the circumstance that in fact there were two notices — that the two notices were void and of no effect. That application is refused. The rest of that first originating summons dated July 6 1979 has not been adjourned to me and, of course, remains to be dealt with.

Then under the second originating summons of March 6 1980 it appears to me that, having held that the March notices were valid, I must logically declare that the November notices, dated November 14 1979, were of no effect. That deals, I think, with the first two paragraphs of the second originating summons, which again is all that is before me.

No order was made as to costs.

COURT OF APPEAL

November 6 1980

(before Lord Justice STEPHENSON, Lord Justice DUNN and Sir David CAIRNS)

R v SECRETARY OF STATE FOR THE ENVIRONMENT AND ANOTHER, EX PARTE POWIS

Estates Gazette April 4 1981
(1980) 258 EG 57

Application for judicial review — Secretary of State's certificate under section 57 (1) of the Landlord and Tenant Act 1954 challenged — Landlord's interest held by local authority — Certificate given by Secretary of State that it was requisite for the purposes of the authority that the use and occupation of a piece of land let to the applicant should be changed — Appeal to Court of Appeal from refusal of Divisional Court to grant applicant an order of certiorari to quash certificate — Appellant a car breaker and scrap-metal dealer carrying on business on the land, a site of about one acre — Authority wished to obtain and use land for other purposes, the one finally relied upon in application for the certificate being that the site was required as a store for highway materials — Certificate attacked by appellant on various grounds, bad faith, misdirection as to meaning of "requisite", lack of evidence to support certificate; and refusal of Divisional Court to admit fresh evidence (although considered *de bene esse*) criticised — Submissions unsuccessful — Allegation of bad faith rejected — Secretary of State correctly interpreted "requisite" as "reasonably necessary" — Sufficient material before him to support decision — Rules for admission of fresh evidence in judicial review cases discussed — Not admissible in present case — Appeal dismissed

This was an appeal from a decision of the Queen's Bench Divisional Court (Shaw LJ and Kilner Brown J) refusing an application by Arthur Glyn Powis by way of judicial review for an order of certiorari to quash a certificate given by the Secretary of State under section 57 (1) of the Landlord and Tenant Act 1954. The certificate was in respect of a piece of land in High Heavens Wood, Marlow Bottom, Buckinghamshire, which had been let to the appellant by High Wycombe Borough Council, predecessors of the second respondents, Buckinghamshire County Council.

R G Rougier QC and R Marshall-Andrews (instructed by A Banks & Co) appeared on behalf of the appellant, the applicant below; Simon Brown (instructed by the Treasury Solicitor) represented the Secretary of State; G N N Huskinson (instructed by the Solicitor, Buckinghamshire County Council) represented the council, the second respondents.

Giving the judgment of the court at the invitation of Stephenson LJ, DUNN LJ said: This is an appeal by leave from the refusal of the Divisional Court on judicial review to grant an order of certiorari to bring up and quash a certificate of the Secretary of State for the Environment given under section 57(1) of the Landlord and Tenant Act 1954.

The material statutory provisions are as follows:

57(1) Where the interest of the landlord or any superior landlord in the property comprised in any tenancy belongs to or is held . . . by a local authority, . . . the Minister or Board in charge of any Government department may certify that it is requisite for the purposes of the first-mentioned department, or, as the case may be, of the authority, . . . that the use or occupation of the property or a part thereof shall be changed by a specified date.

(2) A certificate under the last foregoing subsection shall not be given unless the owner of the interest belonging or held as mentioned in the last foregoing subsection has given to the tenant a notice stating—

(a) that the question of the giving of such a certificate is under consideration by the Minister or Board specified in the notice, and

(b) that if within twenty-one days of the giving of the notice the tenant makes to that Minister or Board representations in writing with respect to that question, they will be considered before the question is determined, and if the tenant makes any such representations within the said twenty-one days the Minister or Board shall consider them before determining whether to give the certificate.

Subsections (3) and (4) of section 57 provide in effect that the section has no immediate effect on the rights of a contractual tenant of business premises. But once the contractual tenancy is validly

A determined by a landlord's notice under section 25 of the Act, then where a certificate has been given the tenant is precluded from applying for a new tenancy and the court has no power to grant the tenant a new tenancy expiring later than the date specified in the certificate. Section 59 of the Act provides for compensation to a tenant in respect of whose property a certificate has been given. Such compensation is in accordance with section 37 of the Act.

The result of these provisions is that business tenants of local authorities and other public bodies may be deprived by certificate under section 57 of the rights which they would otherwise have had of applying to the county court to obtain a new tenancy. They are deprived of the opportunity of testing in court any objection by the
B landlord under section 30 of the Act to a grant of a new tenancy, with the advantages of a public hearing, discovery of documents and cross-examination of witnesses. The decision under section 57 is a ministerial decision which can only be challenged in the courts on the well-established principles laid down in *Associated Provincial Picture Houses Ltd v Wednesbury Corporation* [1948] 1 KB 223 and other similar cases. As Lord Evershed MR said in *X L Fisheries Ltd v Leeds Corporation* [1955] 2 QB 636 at p 648, the section confers certain particular privileges on public authorities.

The appellant is the tenant of about one acre of land at High Heavens Wood, Marlow Bottom, Buckinghamshire. He held originally under a lease from the High Wycombe Borough Council
C dated June 1 1964. The lease comprised, in addition to the land, a right of way to afford access. It was a yearly tenancy determinable by a quarter's notice given by either party and the yearly rent was £5. Among the tenant's covenants there was a covenant "not to use or permit the said piece of land to be used in connection with any trade or business other than that of breaking up derelict motor vehicles". The demised site was immediately adjacent to a large rubbish tip owned by the borough council at High Heavens Wood and extending to some 64 acres. On December 11 1972 a new lease in the same terms was entered into between the borough council and the appellant. The appellant also had an agreement with the borough council for the disposal of abandoned vehicles which had
D been removed by the borough council pursuant to their powers and duties under section 21(1) of the Civic Amenities Act 1967, and in addition he carried on a general business on the site of a car breaker and scrap-metal dealer with the concurrence of the borough council. At that time there were two other traders, a Mr Jackman and a Mr Bowan, carrying on similar businesses at different sites. Mr Jackman was a tenant of the borough council, Mr Bowan was not. As a result of the local government reorganisation in 1974 the Buckinghamshire County Council took over the relevant functions of the High Wycombe Borough Council. The agreement for the disposal of vehicles with the appellant was terminated, and on April 1 1974 the reversion of the lease vested in the county council and they then
E became the appellant's landlords.

On August 4 1975 outline planning permission for 2½ acres of land immediately north of the refuse tip was granted by the county council to itself for use as a gipsy caravan site. This permission provides:

> (4) a landscaping scheme shall be submitted to and approved by the County Council . . . including the planting of trees and shrubs where necessary . . . in order to preserve the amenities of the neighbourhood.

The southern part of this site encroached on to the northern part of the tenant's land. In fact the local authority did not use the northern part of the land because it was in other ownership and would have required a compulsory purchase order. So the limits of the caravan
F site were confined to the southern portion.

In March 1976 the county council, as they were entitled to do, also granted themselves planning permission for the erection of a refuse pulveriser and transfer station for the treatment of refuse and salvage at the High Heavens refuse tip. Final planning permission was granted on March 23 1976. The permission was subject to the following condition:

> (1) The screen bank along the south side of the transfer station shall be completed and landscaped within one year of the station being brought into operation, and shall be constructed so as to effectively screen the buildings, treatment area, highway depot and car breaking areas from the south, and the trees shall be planted in the first planting season following completion of any section of the bank.

The reasons for imposing the above conditions are: G

> To minimise the effect that the development is likely to have on the amenities of the locality.

It was not surprising that that condition was inserted, because High Heavens Wood was scheduled as an area of outstanding natural beauty.

There was no other condition attached to the planning permission although, according to the county council in representations which they subsequently made to the Secretary of State, the planning permission was granted on the understanding that the area to be used for waste-disposal purposes would be kept to the minimum, that the rest of the site would be restored to an agricultural use, and H
that the appearance of the waste-disposal facility would be as tidy as the nature of the business would allow. Two banks of earth were erected by the county council to mark the southern limit of the refuse treatment area and to screen it from the south.

Thus the appellant's site was sandwiched between the proposed gipsy caravan site to the north and the refuse treatment area to the south. Meanwhile on August 27 1975 the county council served notice on the appellant under section 25 of the Landlord and Tenant Act 1954. That notice stated that on the termination of the current tenancy the county council intended to demolish or reconstruct the whole or a substantial part of the premises or to carry out substantial work of construction on the whole or part of them, and J
that they could not reasonably do so without obtaining possession of the premises. This constituted reliance on section 30(1)(f) of the Act. The appellant did not serve a counternotice under section 26 but filed a defence alleging that the section 25 notice had been waived by subsequent acceptance of rent by the county council. The judge upheld that defence and gave judgment for the appellant on May 5 1977. So the appellant still holds under the tenancy of December 11 1972.

Certainly by 1976 the county council had decided to reduce the number of car breakers from three to one. They sent out invitations to tender for the lease of the site occupied by Mr Jackman, opposite the appellant's site to the west of the access road to the tip. The K
appellant duly tendered on July 23 1976 but his tender was not accepted. Mr Jackman also tendered but his tender was not accepted, and the successful tenderer was a Mr Powell. Mr Jackman had by then given up his site pursuant to a possession order against him in the county court on October 30 1975. The third breaker, Mr Bowan, remained in occupation of his site just to the south of the appellant's site but he has now also given up his site and left the area.

In May 1977 shortly after the county court judgment the county council caused the appellant's telephone to be cut off at the site. We mention this because it is relied on as evidence of bad faith on the part of the county council. The appellant at once issued a writ in the L
Queen's Bench Division and obtained an interim injunction on June 3 1977 which was discharged on June 16 on undertakings by the county council in effect not to take any steps to remove the telephone. The appellant was granted his costs in any event.

On July 15 1977 the county council applied to the Secretary of State for a certificate under section 57. The application was supported by reasons which were sent to the minister under a covering letter, although the county council at first refused to send their reasons to the appellant. However, these were later sent to him by the Secretary of State. On August 4 the appellant submitted written representations to the minister and there were further representations and reasons by both parties including a report from a sur- M
veyor instructed by the appellant. All this material was put before the Secretary of State in writing.

The reasons given by the county council in support of their application were that the site was required: (1) partly for a gipsy caravan site, (2) for a store for highways material and (3) for the erection of fencing round the pulveriser and transfer stations. The appellant having agreed to surrender the northern part of his site for incorporation into the gipsy caravan site, and the proposed fencing having been shown to be unnecessary, the only ground finally relied on by the county council was that the site was required for use as a store for highways materials.

On August 7 1978 the Secretary of State issued his decision by letter. Paragraph 2 of that letter reads:

> The Secretary of State, having considered the representations of the tenant, is satisfied that it is requisite for the purposes of the Council that the use and occupation of the property should be changed by February 23 1979 and hereby certifies accordingly.

On October 31 in reply to a request by the appellant, the Secretary of State gave his reasons for his decision although, as he said, he was not bound to do so. He set out the nature of the application, the submissions of both parties and his conclusion, which is in the following terms:

> 5. The Secretary of State carefully considered the submissions of both parties. It was noted that the pulveriser complex proposals had gone ahead, apparently without your client's land being required in connection with preparatory work for fencing, and account was taken of your client's willingness to give up part of his land for the proposed gypsy site. However, the evidence showed that only a very small area was required in connection with the fencing proposals and it was not accepted that the Council's decision not to proceed in accordance with the permission granted for a gypsy caravan site in 1975 indicated that they did not really require Mr Powis's land. The majority of your client's land was, in fact, required in connection with the Council's proposals for a highways materials store; and the Council's submission was accepted that the land was suitable for this purpose and that, whilst they did not necessarily have to show that this was the case, no alternative site was available, either within the area covered by planning permission or elsewhere. Your client's personal reasons for opposing the grant of a certificate were taken into account but, on the evidence as a whole, the Secretary of State was satisfied that it was requisite for the purposes of the Council that the use or occupation of the land should be changed by the date specified in the certificate. He was satisfied that the Council genuinely intended to carry out their proposals, to which there was no obstacle such as a refusal of planning permission; and that the public need to repossess the land in question justified taking your client's tenancy out of the normal operation of the provisions of Part II of the 1954 Act.

In summary form the case of the appellant may be put in this way. The county council had available a refuse tip extending to 64 acres. They had voluntarily reduced the available area of that tip to about 10 acres described as "refuse treatment area" by using the southern part of the tip for forestry, which was a limitation not imposed on them by the planning permission of March 23 1976. But even with the remaining 10 acres there was ample room for the county council to store the highways materials for which less than one acre was needed. As Mr Bowan had given up his site the existing highways material store could remain in its present position without interfering with the activities of the county council in the refuse treatment area, and it was not necessary to use the appellant's site for the storage of highways materials. It was said that no reasonable Secretary of State could have come to any other conclusion.

It was submitted on behalf of the appellant that the history of the matter showed that the application of the county council was not made bona fide but for the purpose of evicting the appellant from his site for a purpose unconnected with the application, namely so that the appellant could not compete with Mr Powell, the successful tenderer on the site opposite.

Secondly it was submitted that the Secretary of State misdirected himself in law as to the meaning of the words "requisite for the purposes of the local authority" in section 57(1). It was said that those words meant more than "required" or "reasonably required" and that the word "requisite" was synonymous with "necessary" or "reasonably necessary for the purposes of the county council". The Divisional Court held that the word "requisite" meant "required by circumstances" and it was submitted on behalf of the appellant that the Divisional Court having given that meaning to the word "requisite" should have held that the minister had misdirected himself in the phraseology he used in paragraph 5 of his letter of October 31 1978.

Thirdly it was submitted that whether or not the use of the appellant's site was requisite or necessary for the purposes of the county council was a question of fact not planning policy, and that there was no evidence apart from the bald assertion of the county council to establish that fact. It was said that the Secretary of State should not have based his decision on that assertion. Reliance was placed on the decision of this court in *Coleen Properties Ltd v Minister of Housing and Local Government* [1971] 1 WLR 433. It was submitted that the Divisional Court in upholding the minister's decision on the evidence misdirected itself as to certain important issues of fact, and made findings of fact which were unsupported by the evidence.

It was finally submitted that the court should consider fresh evidence considered *de bene esse* by the Divisional Court and that the Divisional Court were wrong to refuse to allow such evidence to be admitted on judicial review.

We were told that this was the first case in which the decision of a minister to give a certificate under section 57(1) has been the subject of judicial review, and we consider first the meaning of the word "requisite" in the subsection. Taken by itself the natural and ordinary meaning of the word "requisite" is "required by circumstances" or "necessary" as used in the Book of Common Prayer in the phrase "to ask those things which are requisite and necessary as well for the body as the soul". But it is said on behalf of the Secretary of State and the county council that a looser and more subjective meaning should be put on the word in the context of this particular Act of Parliament. It is said that in that context the word means "reasonably wanted for the purpose which the local authority genuinely and bona fide intends to carry out".

Section 57 appears in Part IV of the Act. Part II of the Act enables certain tenants of business premises to obtain new tenancies in certain cases. Some business tenancies, eg agricultural tenancies which are covered by other legislation, are excluded from Part II. The material sections of Part II are expressed to be subject to Part IV, which contains a group of sections giving any minister power to give a certificate in effect excluding the tenant from his rights under Part II. Different language is used in each of these sections as to the circumstances in which the minister may give a certificate of change of use. The words in subsection (1) of section 57 are, as I have already quoted, "requisite for the purposes of the first-mentioned department".

The words in subsection (5) are that the minister:

> may certify that it is necessary in the public interest . . .

In subsection (7) are the words:

> Where the interest of the landlord . . . belongs to the National Trust the Minister of Works may certify that it is requisite, for the purpose of securing that the property will as from a specified date be used or occupied in a manner better suited to the nature thereof, . . .

In section 58, which is concerned with "Termination on special grounds of tenancies to which Part II applies", subsection (1) provides that the minister may certify:

> that for reasons of national security it is necessary that the use or occupation of the property should be discontinued or changed.

Subsection 3(a) of that section provides that:

> Where the landlord's interest . . . is held by statutory undertakers, nothing in this Act shall invalidate an agreement to the effect —
> (a) that where the Minister . . . certifies that possession of the property comprised in the tenancy or a part thereof is urgently required for carrying out repairs (whether on that property or elsewhere) which are needed for the proper operation of the landlord's undertaking, the tenancy may be terminated . . .

Section 60, which deals with "Special provisions as to premises provided under the Distribution of Industry Acts" provides that the:

> Board of Trade certify that it is necessary or expedient for achieving the objects of the said Acts that the use or occupation of the property should be changed. . .

These different words were described by counsel in *X L Fisheries Ltd v Leeds Corporation* as "a rising scale of urgency", and in this case by counsel for the county council as a "hierarchy of terms", and it was submitted that the word "requisite" comes low in the hierarchy and that it must mean something less than necessary. We accept that in the context of Part IV of the Act the word "requisite" should be given a less exacting meaning than "necessary" in the sense of "indispensable" and that the word "necessary" must be qualified so as to give effect to that. We would say that "reasonably necessary" was compatible both with the ordinary and natural meaning of the word "requisite" and the context of the section in which it appears in this part of the Act.

Little assistance is to be derived from the construction given by courts to the same word used in other statutes. The only case which

A has been cited to us in which the word "requisite" has been construed in anything like the present context was *Robinson v Minister of Town and Country Planning* [1947] 1 KB 702, in which the word "requisite" in section 1 (1) of the Town and Country Planning Act 1944 fell to be construed. At p 713 Lord Greene MR said:

> The words "requisite" and "satisfactorily" clearly indicate that the question is one of opinion and policy, matters which are peculiarly for the Minister himself to decide. No objective test is possible.

It is important to observe the context in which the word "requisite" appears in secton 1(1) of the Town and Country Planning Act 1944. The subsection provides:

> B 1(1) Where the Minister of Town and Country Planning . . . is satisfied that it is requisite, for the purpose of dealing satisfactorily with extensive war damage in the area of a local planning authority, that a part or parts of their area, consisting of land shown to his satisfaction to have sustained war damage or of such land together with other land contiguous or adjacent thereto, should be laid out afresh and redeveloped as a whole, an order declaring . . . the land to be . . . land subject to compulsory purchase . . . may be made by the Minister.

The case affords little if any assistance as to the meaning of the word itself, though it is valuable as to the nature of the minister's decision.

In *Coleen Properties Ltd v Minister of Housing and Local Government, supra*, the court had to consider the words "reasonC ably necessary", which is the meaning we have put on the word "requisite". Lord Denning MR at p 438, described it as an inference of fact. Sachs LJ at p 439 C said this:

> The question before [the minister] was not, to my mind, one of policy: it was in essence a question of fact that had to be established as a condition precedent to the exercise of the powers to take away the subject's property. It was no less a question of fact because it involved forming a judgment on matters on which expert opinion can and indeed ought to be given. (I rather doubt whether there is much material difference between the view I have just expressed and that of Mr Slynn who has argued that the question was simply a matter of planning judgment which had to be based on evidence.)

We agree with the words of Sachs LJ which I have just read, which appear in brackets in the last sentence, as referable to this case. We D think it matters not whether the decision of the minister is described as one of opinion and planning policy or as forming a judgment on matters of fact and opinion. In either case the minister has to arrive at a value judgment based on facts. If in arriving at that decision the minister asks himself the wrong questions and so errs in law, or if he takes into account matters he should not have taken into account or fails to take into account matters he should have taken into account, or if there is no evidence to support his decision so that it is contrary to all the evidence before him, then the court can interfere with his conclusion.

Did the minister err in law by putting the wrong meaning on the word "requisite"? If he did his decision cannot stand. The answer to E this question depends on the terms of the minister's reasons of October 31 1978 and in particular paragraph 5. Although the minister used the word "required" in three different places in that paragraph in the sense of "wanted", which is different from "reasonably necessary", he ended the paragraph in this way:

> . . . and the Council's submission was accepted that the land was suitable for this purpose and that, whilst they did not necessarily have to show that this was the case, no alternative site was available, either within the area covered by planning permission or elsewhere. Your client's personal reasons for opposing the grant of a certificate were taken into account but, on the evidence as a whole, the Secretary of State was satisfied that it was requisite for the purposes of the Council that the use or occupation of the land should be changed by the date specified in the certificate. He was satisfied that the F Council genuinely intended to carry out their proposals, to which there was no obstacle such as a refusal of planning permission; and that the public need to repossess the land in question justified taking your client's tenancy out of the normal operation of the provisions of Part II of the 1954 Act.

What the minister was there saying was:

> I think that the word requisite means genuinely required, but if I am wrong about that and if I have to consider whether alternative sites were available so that I can be satisfied that the appellant's site was reasonably necessary for the purposes of the county council, then I am satisfied that it was.

Despite the submissions of counsel for the appellant, there is nothing wrong in the minister expressing himself in this way. It is a type of formula commonly used by judges and we cannot find that the minister erred in law in using it.

But then it was said that the only evidence before the minister G was the bald assertion of the county council that no alternative sites were available and that accordingly the appellant's site was reasonably necessary for their purposes. It was said that the minister should not have acted on that bald assertion in the light of the evidence on behalf of the appellant and especially the statement of the surveyor which was before him. It was said that the county council made no attempt to put before the minister evidence of the acreage required for each of their activities within the refuse treatment area, so as to demonstrate that it was reasonably necessary for them to use the appellant's site for a highways material store.

Reliance was placed on the statement of Lord Denning MR in *Coleen Properties Ltd v Minister of Housing and Local Government, supra* at p 437 H: H

> At any rate, I am quite clear that the mere *ipse dixit* of the local council is not sufficient. There must be some evidence to support their assertion. And here there was none.

It is said that that statement exactly reflects the situation in this case in that the only evidence before the minister was the *ipse dixit* of the county council. It is important to see the facts of *Coleen Properties Ltd v Minister of Housing and Local Government*. There had been a public inquiry. Evidence had been called before the inspector. Witnesses had been examined and cross-examined. The inspector had made a report containing his findings of fact, conclusions and J recommendations. The minister reversed his recommendations. There was no evidence on which he could do so except the statement of the advocate for the local authority that the land was reasonably necessary for the proposed redevelopment. The situation here is quite different. Subsection (2) of section 57 does not envisage an inquiry or other hearing and no inquiry or hearing was held. The tenant was entitled to make representations in writing, which he did. There is not even a provision in the subsection for reasons to be given by the local authority, although reasons were in fact given. And there were counter-reasons and counter-representations by both parties. All this material was before the minister and it amounted to considerably more material than was K envisaged by the subsection.

The nature of the material on which a minister is entitled to rely in reaching a decision must depend on the statutory provisions and the circumstances of each case. It may well be that where there is a public hearing the minister should not rely on bare assertions unsupported by evidence. But where, as here, there was no public hearing, the minister must assess the submissions and reach his conclusion as best he can on the material put before him. It is a matter for him to weigh the statements submitted to him and reach his decision in accordance with them.

The minister summarised the submissions of the county council in paragraph 3 of his letter in this way: L

> The Council submitted that it was essential that the materials store be between the pulveriser complex and gypsy site for the pulveriser complex to remain secure and this would also facilitate use of the containers and the movement of vehicles in and around the pulveriser. An earth-bank screen was being constructed south of the transfer station and treatment area, in accordance with the planning consent, beyond which only forestry would be undertaken and the council argued that, for that reason, the treatment area could not be moved further south. Because a number of activities would be concentrated in the treatment area, it was stated to be impossible for highways materials to be stored there as well, and it was said to be impracticable to locate a highways depot on the far side of the waste disposal complex. The Council submitted that the depot at High Heavens Wood was used by Wycombe District Council, as agent authority for the County Council, to keep materials for use on roads within the boundaries of the M former borough of High Wycombe and had to be easily accessible in times of emergency.

Then the minister deals with alternative depots at different sites which were not relied on in this court by the appellant.

The minister accepted those submissions in paragraph 5. It was not necessary for the minister to go further and insist on evidence of the acreage of each activity proposed by the county council in order to check its submissions. There were other relevant considerations than mere acreage, eg the access to the various sites and the security of the pulveriser installation. It is impossible to say that the minister was wrong in taking into account the material put before him by the

A county council or that because he failed to take into account the exact acreage of the various sites his decision was thereby vitiated.

The Secretary of State and the Divisional Court both rejected the submission of bad faith on the part of the county council. Bad faith was, however, relied on but not pressed in this court and we think the Divisional Court were right to reject it. At the same time we cannot help thinking that it was foolish of the county council to threaten to cut off the appellant's telephone and unwise of them not to serve another notice under section 25 after the dismissal of the county court proceedings, and seek to obtain possession under Part II of the Act as they did in the case of Mr Jackman. But those matters were all before the Secretary of State and it is impossible on
B judicial review to disturb his finding that the county council genuinely required this site for their purposes.

Finally there was an application on behalf of the appellant to admit fresh evidence which the Divisional Court had refused to admit. Like the Divisional Court we considered the evidence *de bene esse*. What are the principles on which fresh evidence should be admitted on judicial review? They are (1) that the court can receive evidence to show what material was before the minister or inferior tribunal (*Ashbridge Investments Ltd v Minister of Housing and Local Government* [1965] 1 WLR 1320 per Lord Denning MR at p 1327): (2) where the jurisdiction of the minister or inferior tribunal depends on a question of fact, or where the question is
C whether essential procedural requirements were observed, the court may receive and consider additional evidence to determine the jurisdictional fact or procedural error (see de Smith's *Judicial Review of Administrative Action* 4th Ed at pp 140 and 141 and cases there cited): (3) where the proceedings are tainted by misconduct on the part of the minister or member of the inferior tribunal or the parties before it. Examples of such misconduct are bias by the decision-making body, or fraud or perjury by a party. In each case fresh evidence is admissible to prove the particular misconduct alleged (see *R v West Sussex Quarter Sessions* [1974] QB 24 per Orr LJ at p 39 and Lawton LJ at p 42).

D There was discussion at the Bar as to the situation where a party deliberately suppressed material facts with the intention of misleading the minister. If that were the situation then it would be for the court to consider whether the conduct of that party could be described as fraudulent so as to permit the admission of fresh evidence.

It is said that there is a distinction between cases such as the *West Sussex* case, where certiorari was sought to quash the decision of an inferior tribunal after a hearing, and cases such as the present where it is sought to quash the decision of the minister where there has been no hearing. We can find no such distinction and the remarks of the majority in the *West Sussex* case are quite general in their
E application to certiorari.

Reliance was placed by counsel for the appellant on the case of *Secretary of State for Education and Science v Tameside Metropolitan Borough Council* [1977] AC 1014. In that case fresh evidence was admitted to the Court of Appeal by consent. Certain passages in the judgments of Scarman LJ in the Court of Appeal at p 1030 E, and of the speeches of Lord Wilberforce at p 1047 D and Lord Diplock at p 1064, were particularly relied on by counsel for the appellant. It was said that those statements support the proposition that if the minister is misled by the failure of one party falling short of fraud to put before him evidence which should have been put before him, and on which he would probably have come to a
F different conclusion to that to which he came on the evidence which was put before him, then the court on judicial review should admit that evidence as fresh evidence. It was said that to fail to do so would be a breach of natural justice.

The *Tameside* case was a very special case. In effect the minister was himself reviewing the decision of the local authority and his decision was analogous to the position of the court on judicial review. The evidence was admitted in the Court of Appeal by consent. The case is not authority for the proposition relied on by counsel for the appellants. Indeed there are indications in the speeches in the House of Lords to the contrary (see Lord Wilberforce at p 1052 C):

> To rephrase the question: on June 11 1976 (this is the date of the direction, G
> and we are not entitled to see what happened thereafter), could it be said that the authority was acting unreasonably in proceeding with a selection procedure which was otherwise workable in face of the possibility of persistent opposition by teachers' unions and individual teachers, or would *the only* (not "the more") reasonable course have been for the authority to abandon its plans? This is, I think, the ultimate factual question in the case. And I think that it must be answered in the negative — ie that it could not be unreasonable, in June 1976, and assuming that the Secretary of State did not interfere, for the authority to put forward a plan to act on its approved procedure. The teachers, after all, are public servants, with responsibility for their pupils. They were under a duty to produce reports. These reports and the records in the primary schools are public property. I do not think that it could be unreasonable (not "was unreasonable") for the authority to take the view that if the Secretary of State did not intervene under his statutory powers the teachers would co-operate in working the authority's H
> procedure — a procedure which had, in similar form, been operated in part of this very area.

And Lord Russell at p 1076 E said:

> The question whether the Secretary of State was justified in his conclusion that the proposals of the local authority were unreasonable falls to be decided at the date of his conclusion, June 11: that is common ground. I would not however subscribe to the view that facts subsequently brought forward as then existing can properly be relied upon as showing that the proposals were not unreasonable unless those facts are of such a character that they can be taken to have been within the knowledge of the department.

If there is doubt as to the construction of those words of Lord J
Russell, we agree, save in one respect, with the construction put upon them by the reporter in the headnote at p 1016 C:

> Facts subsequently brought forward as existing on June 11 could not properly be relied on as showing that the authority's proposals were not unreasonable unless they were of such a chracter that they could be taken to have been within the knowledge of the Secretary of State.

The words "the department" used by Lord Russell at the end of his speech are distinguishable from the words "the Secretary of State" used by the reporter at the end of the headnote, unless those words are to be interpreted as meaning "the Secretary of State or his department". The converse of Lord Russell's proposition as stated by him is that fresh evidence may be admitted of facts within K
the knowledge of the department though unknown to the minister at the time of his decision. This is an example of one of the classic reasons for interfering with the ministerial decision within *Associated Provincial Picture Houses Ltd v Wednesbury Corporation* [1948] 1 KB 223, namely, that such facts are plainly facts which the minister should have taken into account. If he failed to do so, fresh evidence of the facts within the knowledge of the department could be admitted as a ground for quashing his decision.

This court is bound by the majority judgments in *R v West Sussex Quarter Sessions supra*, which formed the *ratio decidendi* of the case. *Secretary of State for Education and Science v Tameside Metropolitan Borough Council* is not authority for the proposition that L
the circumstances in which fresh evidence on judicial review may be admitted can be extended beyond the categories laid down in *R v West Sussex Quarter Sessions*.

There is nothing in the fresh evidence in this case to bring it within any of the categories of evidence which may be admitted on judicial review. We agree with the Divisional Court that the fresh evidence is not admissible. But we do not agree with the test proposed by the Divisional Court "so long as a minister or a tribunal has balanced the factors and weighed the evidence before him or it, no superior court should intervene unless it is plain that there has been such manifest failure to administer justice that the decision is perverse". We have already stated the principles upon M
which fresh evidence may be admitted on judicial review.

Although it is true there are certain mistakes in the statements of fact at page XI of the judgment of the Divisional Court, we do not think that those mistakes affected the conclusion at which the court arrived. We think that they were right to refuse the application for an order for certiorari and for the reasons which we have stated in this judgment the appeal is accordingly dismissed.

The appeal was dismissed, the order for costs being that the appellant was to pay half the costs of each respondent. The order for costs in the court below was not disturbed. Leave to appeal to the House of Lords was refused.

A

COURT OF APPEAL

February 17 1981

(Before Lord Justice CUMMING-BRUCE, Lord Justice TEMPLEMAN and Lord Justice OLIVER)

CAM GEARS LTD v CUNNINGHAM

Estates Gazette May 23 1981

(1981) 258 EG 749

Landlord and Tenant Act 1954, Part II – Appeal by tenants from county court judge's refusal to grant a new tenancy of a car park used in connection with tenant's business — New tenancy opposed B **by landlord who wished to occupy the land for the expansion of his business of testing vehicles for the MOT test — Submission by tenants that the landlord did not intend to occupy "the holding" for the purpose of his business within the meaning of section 30(1)(g) of the 1954 Act — Tenants argued that the landlord's proposal, which included the erection of a building on the site for a workshop, office, waiting rooms, vehicle inspection pits etc, was the creation of a new holding, not an occupation of "the holding" — Submission rejected — Difficult case of Nursey v P Currie (Dartford) Ltd considered and distinguished — In present case "the holding" consisted of the vacant site on which the landlord proposed to erect a building, so that he would in fact occupy "the holding" plus something** C **else — Appeal from decision of county court judge in favour of landlord dismissed**

This was an appeal by Cam Gears Ltd, tenants of premises consisting of a vacant site used as a car park on the corner of Selbourne Road and Covent Garden Close in Luton, from a decision of Judge Kingham at Luton County Court in favour of the landlord, David Lawrence Cunningham. The judge refused to grant the tenants a new tenancy of the site, the landlord having opposed the grant under section 30(1)(g) of the Landlord and Tenant Act 1954.

Nicholas Patten (instructed by Slaughter & May) appeared on behalf of the appellants; Allan Levy (instructed by John Photiades D & Co, of Luton) represented the respondent.

Giving the first judgment at the invitation of Cumming-Bruce LJ, OLIVER LJ said: This is an appeal by a tenant of business premises from a decision of His Honour Judge Kingham, which was given on January 11 1980, refusing the grant of a new tenancy under the provisions of the Landlord and Tenant Act 1954.

The premises with which the application is concerned consist of a vacant site — that is to say, a site unencumbered by buildings but with a concrete surface — constituting a car park and situate on the corner of Selbourne Road and Covent Garden Close in Luton. The appellants, the tenants, carry on business nearby, and the premises have been at all material times used as a car park in connection with E that business. The premises were originally let by the respondents to the appellants on a lease dated May 5 1972 for a term of seven years from January 1 1972; that term expired in 1979.

The proper notice under the Act was given by the respondent in July 1978, and there is no dispute that the correct sequence of events envisaged by the Act has taken place; I need not go through its various stages.

The present dispute arises out of the respondent landlord's opposition to the grant of a new tenancy which, in the notice which was served on the tenant, he put on the ground set out in section 30(1)(g) of the 1954 Act:

F "That on the termination of the current tenancy the landlord intends to occupy the holding for the purposes, or partly for the purposes, of a business to be carried on by him, or as his residence".

The respondent, Mr Cunningham, has been in business for some time, originally, I think, in partnership, but more recently through the medium of a limited company, Technical Brakes Ltd, of which he is a director and in which he holds 74 out of the 99 issued shares. That company runs a garage business and, as I understand it, the principal facet of that business is the provision of Ministry of Transport testing facilities for vehicles which are more than three years of age. In the course of its business, at present carried on in Wingate Road, Luton, it tests about 10,000 vehicles a year.

What the respondent wishes to do is to expand his business. The G lease of the present business premises expired in, I think, June of last year and he needs two new test lanes.

As I have mentioned, the premises the subject-matter of this appeal are used at present as a car park. They consist simply of a vacant site with no buildings on it, and the landlord's project, if he can get possession, is to have erected on the site what is known as a Banbury prefabricated commercial building, which will cover about one-third of the total area of the site and will furnish a workshop, an office, a waiting room, toilet facilities and inspection pits for vehicles. The total cost of that building is likely to be in the order of £16,000, the total cost of the operation, including erection, being estimated by the respondent before the learned judge at £20,000. H His evidence, which was accepted by the learned judge, was that facilities would be made available to him by his bankers. The learned judge accepted the respondent's evidence in toto and he found as a fact that he was bona fide in forming his intention and that the intention was a practicable one, albeit that he would have to get a detailed planning permission, satisfy the building regulations and obtain a Ministry of Transport licence. The learned judge was satisfied that these were not problems which would stand in the way of a successful outcome of the project.

The appellants' contention before the learned judge and in this court was that since the only ground of opposition to the grant of a new tenancy under the Act specified in the respondent landlord's J notice was that which appears in section 30(1)(g), the landlord had to show that his intention was not just to occupy the site, but that he was to occupy "the holding"; that the holding consisted of the present vacant site used as a car park and that the proposed erection on that site of a new building to be used for the purpose of a testing workshop constituted the erection of a new and different holding and that therefore the landlord was not proposing to occupy "the holding" for the purposes of the business.

The erection of the proposed building was, no doubt, a major work of construction, which could be carried out only if the landlord obtained possession, and it might indeed have justified opposition to the grant of a new tenancy under section 30(1)(f); but that was K not the ground that was specified, and since the only ground relied upon was the ground in subparagraph (g), that is the ground upon which the landlord must rely.

The learned judge rejected that contention, which was based upon the decision of this court in the case of *Nursey v P Currie (Dartford) Ltd* [1959] 1 WLR 273. That was a case which bears a superficial resemblance to the present case, in that the landlord's notice in that case was confined to the ground specified in section 30(1)(g), and what the landlord company proposed to do was to demolish the existing buildings, which were standing in a yard which was occupied by the landlord, and redevelop the site as part of a petrol station which it proposed to carry on there. The Court of L Appeal held that the ground of opposition was not made out; and it did so because of the limiting definition of the word "holding" in section 23(3) of the Act.

That subsection defines the word "holding" as meaning "the property comprised in the tenancy" — I close the quotation there; there is some more but it is not material for present purposes. The court held in that case that that definition included the existing buildings. The landlord, since it intended to remove the existing buildings and replace them with others, did not therefore intend to occupy "the holding" but intended to occupy the new buildings to be erected on the land forming part only of "the holding" viewed as a totality — at any rate, that is what I apprehend is the ratio of the M decision in the *Nursey* case, and that is the view of it which was adopted by this court in the case of *Metal Developments Ltd v Jones* [1971] 1 WLR 168; I refer particularly to the judgment of Salmon LJ, as he then was, in that case.

I confess that the *Nursey* decision is one which I find far from easy to understand. The only argument, so far as can be deduced from the report at p 275, had been that since the landlords intended to demolish the buildings, they could not be intending to occupy "the holding" which included the buildings. But Wynn-Parry J, who delivered the first judgment, seems to have taken the view, at any rate on one reading of his judgment, that it was fatal to the land-

A lord's claim that the holding was to be occupied as part of a larger complex and not as a separate holding on its own. He said, at p 277:

> It seems to me that that language circumscribes the use of the phrase "the holding" in that paragraph — that being subparagraph (g) — and makes it necessary to concentrate the whole of one's attention on the particular piece of land, whether it has buildings on it or not, which is the subject-matter of the tenancy in question. So viewed, it appears to me that the contention for the landlords in the present case is too wide, and that when one is looking at the material time at "the holding" under paragraph (g), it is not permissible to take into account the wider scheme which the landlords had in mind, and merely to treat the land comprised in the holding as land which, in one way or another, will be used for the purpose of the
B wider undertaking.

I cannot think that the learned judge can have intended to do more than to anwer the question, which is: Is the holding which the landlord intends to occupy the same holding as that comprised in the tenancy? Construed in the wider sense that I have indicated, it would follow that a landlord who carried on a business next door to the demised premises and who wanted to occupy those premises as one with his existing shop for an expanding business, would be unable to rely upon section 30(1)(g) and would be able to resist a new lease only if he intended to reconstruct. For my part, I cannot ascribe so eccentric an intention to the legislature. Certainly, Willmer LJ confined his judgment to the narrow ground that the defi-
C nition of "the holding" simply involved reading into the subsection, in place of the words "the holding", the parcels of the lease — a ratio which hardly helps the present appellants, since the lease in the present case merely refers to "all that piece or parcel of land delineated for the purposes of identification only on the plan annexed hereto and thereon edged red", and that is precisely what the landlord intends to occupy.

But even if I am wrong in the limits within which, as I think, the judgment of Wynn-Parry J must be read, and even assuming that the concurrence in that decision of Hodson LJ renders the wider construction binding upon us, it still does not seem to me to help the appellants in the instant case. There is no wider scheme here in
D which the holding is proposed to be incorporated. The landlord simply intends to place a building on the site and to use the whole site, together with the new building, for the purposes of his business.

Mr Patten, who has argued this appeal, if I may say so, with conspicuous frankness and acumen, seeks to steer a course midway between the construction of Wynn-Parry J's judgment to which I have referred and the very limited ratio adopted by Willmer LJ. He suggests that the ratio of the *Nursey* case is that you have to look at the holding as it is at the termination of the tenancy and to ask yourself the question: Does the landlord intend to occupy the holding for the purposes of *his* business in substantially the identical
E condition as it was at the date of termination? If he intends to occupy the whole of it, but to make any material alteration to its condition, then he is intending to occupy a different holding.

For my part, I find myself unable to follow Mr Patten through the gap which he thus seeks to make between the prongs of Morton's fork. Whatever may be the true ratio of the *Nursey* case, I am unable to extract that from it and indeed to do so would, I think, be to attribute a wholly irrational and capricious intention to the legislature. Accepting as I must that *Nursey* is binding upon this court, I certainly do not feel disposed to strain it beyond the narrowest limits within which it is capable of being confined.

I think that the determining feature of the *Nursey* decision was
F not the *purpose* for which the holdings were occupied by the tenant, or the particular condition at the time of the determination of the tenancy, but the fact that the holding consisted of the buildings which, under the landlord's proposals, were to be demolished. That may or may not have been a logical or reasonable construction of the section, and I bear in mind Salmon LJ's reservation in the *Metal Construction* case, as to whether it was correctly decided, although G
the combined industry of counsel has not succeeded in unearthing the inconsistent unreported case in the Court of Appeal to which Salmon LJ referred. But it cannot in any event, in my judgment, have any possible application to a case such as the present, where "the holding" consists solely of a vacant site upon which the landlord proposes to erect a building, so that what he will occupy is "the holding" plus something else. He proposes to occupy everything that is there at the moment, with the sole exception of two lengths of topsoil and subsoil which will be removed to sink the inspection pits. *Nursey v Currie* is, in my judgment, of no assistance to the tenant in such circumstances.

The learned judge rejected the tenant's submission. So do I, and I H
would dismiss the appeal.

Agreeing, TEMPLEMAN LJ said: By a lease dated May 5 1972 the landlord demised to the tenant "All that piece or parcel of land edged in red on the plan annexed hereto". The tenant covenanted to use the demised premises only for the purposes of a car park and not to erect on the demised land or any part thereof any buildings or structures other than those necessary for the parking of vehicles and, should the tenant so require, for providing shelter for vehicles. The landlord proposed to occupy the demised premises for the business of an MOT car testing centre and for that purpose needed and planned to erect a building on part of the land.

By section 30(1)(g) of the 1954 Act, the landlord having duly J
determined the lease by notice is entitled to resist the tenant's present application for a new lease if the landlord, on the determination of the current tenancy, intends to occupy the holding for the purpose of a business to be carried on by the landlord thereon. By section 23(3) "the holding" means the property comprised in the tenancy. The lease will determine at the expiration of three months after the final refusal of the tenant's application for a new lease. If there is no appeal from this court the lease will determine three months from today.

Unassisted by authority, it seems to me that the learned county court judge having found that the landlord in fact intends to occupy the whole of the demised land for the purposes I have mentioned, K
and has the financial resources and the ability and the will to do so, it follows that the landlord made good his opposition under the Landlord and Tenant Act 1954.

We were pressed by the decision of this court in *Nursey v Currie*, of which I make the melancholy observation that two bad reasons do not make one good reason although both may be binding on this court. It is plain, however, that the facts of the present case are distinguishable. Taking the test adumbrated by Wynn-Parry J: Is this a case where the holding which is to be developed is the same as the holding demised? the answer is Yes; it does not form part of any larger undertaking. Taking the test adumbrated by Willmer LJ, what was comprised in the demise in the present case is a piece of L
land, and the landlord's business will be carried on upon that piece of land. The fact that for the purposes of enhancing that business, or enabling it to be carried on, the landlord intends to put a workshop on part of the land is neither here nor there. The object of paragraph (g) is not to hand the land back to the landlord in a sterilised form, so that he has to put his hand on his heart, saying in effect: "I do not intend to make any alteration". The purpose of the subsection is to hand the land back to the landlord if he wants to carry on his own business there, and that indeed is what this landlord intends to do and that is what he is entitled to do.

Accordingly, I would dismiss the appeal. M

Agreeing, CUMMING-BRUCE LJ said: I agree with both judgments, hoping that such agreement does not give rise to the difficulties which have arisen as a result of Hodson LJ's similar agreement in the case of *Nursey v Currie*.

The appeal was dismissed with costs. Leave to appeal was refused.

LANDLORD AND TENANT
LEASEHOLD REFORM

A **CHANCERY DIVISION**

May 9 1980

(Before Mr Justice VINELOTT)

HARRIS v PLENTEX LTD

Estates Gazette May 2 1981

(1980) 258 EG 447

Leasehold Reform Act 1967 — Qualifications for enfranchisement — The plaintiff, who had been a regulated tenant of part of a leasehold house since 1967, acquired the residue of the 99-years' lease of the house in 1979 — Lease for 99 years from 1881 at a low B **rent — The plaintiff gave notice of his desire to acquire the freehold, but the defendant company, the freeholders, objected that the plaintiff had not satisfied the qualifying conditions in section 1 of the 1967 Act — Held that, although in 1979 he was a tenant holding under a long tenancy at a low rent, and although he was to be treated as having occupied the house as his residence since 1967, he had not been a tenant of the house under a long tenancy at a low rent "for the last five years or for periods amounting to five years in the last ten years" — He was not, therefore, qualified to acquire the freehold reversion — Declaration accordingly**

This was an originating summons taken out by the plaintiff, C Thomas Richard Harris, against the defendants, Plentex Ltd, for a declaration on the question whether on the true construction of section 1 of the Leasehold Reform Act 1967 and of a lease vested in the plaintiff, and in the events which had happened, he was entitled to acquire the freehold reversion of a house at 33 Beauchamp Road, Lavender Hill, Clapham Junction, London SW1.

J H G Sunnucks (instructed by G H Gelberg, Parkus & Co) appeared on behalf of the plaintiff; M K Zeidman (instructed by Cowan, Lipson Rumney) represented the defendants.

Giving judgment, VINELOTT J said: This application raises a short question of construction of the Leasehold Reform Act 1967. On July 12 1979 the plaintiff, Thomas Richard Harris, was the lessee of D a house at 33 Beauchamp Road, Lavender Hill, under a lease dated October 16 1884 whereby the owner of the freehold granted a building lease for a term of 99 years from March 25 1881 at a rent of £7 4s 6d a year. On July 12 1979 the rateable value of the house was £347. Mr Harris had acquired the lease by transfer dated May 4 1979. He had been in occupation of part of the house since July 24 1967 when he became a regulated tenant of a part. That part has been his main or only residence since that date.

On July 12 1979 Mr Harris gave notice under section 8 of the 1967 Act to the defendant Plentex Ltd, the successor in title of the original lessor and the owner of the freehold reversion immediately expectant on the term granted by the lease, of his desire to acquire E the freehold. It is common ground that on July 12 1979 Mr Harris was a tenant of a leasehold house in Greater London of a rateable value of less than £400 holding under a long tenancy at a low rent, that he had occupied part of the house for more than five years and that under section 1(2)(a) of the 1967 Act he is to be treated for the purposes of the Act as if his occupation of part had been occupation of the whole. By Schedule 3, paragraph 3(1), Mr Harris having made a claim to acquire the freehold, the lease, which would otherwise have expired on March 24 1980, is in effect extended during the currency of the claim and for three months thereafter.

Mr Harris has made an application in the county court which, F under section 20 of that Act, has exclusive jurisdiction, but that application is still pending. In the course of those proceedings a question of construction of section 1(1)(b) of the Act arose, and with a view to avoiding the expense and delay of a full hearing of Mr Harris's application and possibly a subsequent appeal, Mr Harris has issued an originating summons in this court asking for that question of construction to be decided. Mr Zeidman, who appeared for the defendant company, accepts, and I think rightly, that the originating summons is not one which invokes "any jurisdiction expressed to be conferred on the court by Part I of the 1967 Act" and that accordingly I have jurisdiction to entertain it.

Before explaining the question of construction which has arisen it G will be convenient to read section 1 of the Act.

> (1) This Part of this Act shall have effect to confer on a tenant of a leasehold house, occupying the house as his residence, a right to acquire on fair terms the freehold or an extended lease of the house and premises where —
> (a) his tenancy is a long tenancy at a low rent and subject to subsections (5) and (6) below the rateable value of the house and premises on the appropriate day is not (or was not) more than £200 or, if it is in Greater London, than £400; and
> (b) at the relevant time (that is to say, at the time when he gives notice in accordance with this Act of his desire to have the freehold or to have an extended lease, as the case may be) he has been tenant of the house under a long tenancy at a low rent, and occupying it as his residence, for the last five years or for periods amounting to five years in the last ten years; H
> and to confer the like right in the other cases for which provision is made in this Part of this Act.
> (2) In this Part of this Act references, in relation to any tenancy, to the tenant occupying a house as his residence shall be construed as applying where, but only where, the tenant is, in right of the tenancy, occupying it as his only or main residence (whether or not he uses it also for other purposes); but
> (a) references to a person occupying a house shall apply where he occupies it in part only . . .

As I have said, it is common ground that on July 12 1979 Mr J Harris was a tenant holding a long tenancy at a low rent of a house within the limits of rateable value within section 1(1)(a) and that he is to be treated as having occupied the house as his residence since July 24 1967. The question is whether on July 12 1979 he had been "tenant of the house under a long tenancy at a low rent and occupying it as his residence for the last five years or for periods amounting to five years in the last ten years".

Mr Sunnucks, who appeared for Mr Harris, argued that section 1(1)(b) requires an applicant for a long lease to satisfy two conditions. First, he must show that "he has been tenant of the house at a low rent"; secondly, he must show that he has been "occupying it as his residence" for the requisite period. That construction seems K to me to face two insuperable difficulties. The first difficulty is that the first of Mr Sunnucks' two conditions adds nothing to the opening words of section 1(1), which only confers a right to acquire the freehold of a house on "a tenant of a leasehold house" and therefore only confers a right to acquire the freehold of a house on someonc who at the relevant date was a tenant of a leasehold house. Any such person must have been a tenant for a period of time however short. Secondly, section 1(2) provides that references to a tenant occupying a house as his residence are to be construed as applying "where the tenant is, in right of the tenancy, occupying it".

A Mr Sunnucks said that the word "is" points to the moment of time when the notice is served. I think it is more natural to read it as applying throughout the period during which the tenant claims to have been occupying the house as his residence. But, apart from these difficulties, the words in section 1(1)(b) "for the last five years or for periods amounting to five years in the last ten years" are I think most naturally read as qualifying the words "has been tenant of the house under a long tenancy at a low rent". The words "and occupying it as his residence" then introduce a further condition which must also be satisfied during the period relied on by the tenant.

In my judgment, therefore, on the agreed facts, Mr Harris did not B at the relevant time, that is July 12 1979, satisfy the conditions in section 1(1)(b) and had not and has not under the Act any right to acquire the freehold reversion and I so declare.

COURT OF APPEAL

February 20 1981

(Before Lord Justice ORMROD, Lord Justice WATKINS and Lord Justice GRIFFITHS)

C

TANDON v TRUSTEES OF SPURGEON'S HOMES

Estates Gazette May 9 1981

(1981) 258 EG 553

Leasehold Reform Act 1967 — Whether building a "house", so as to qualify for enfranchisement under the Act, or a shop — Court of Appeal divided — Majority, reversing decision of county court judge, held that the building in question was a shop with living accommodation above it — Ormrod LJ, dissenting, found difficulty in distinguishing the case of Lake v Bennett and considered that it was a house, in agreement with county court judge — The premises in question were in a small parade of shops with living accommodation D above — About 75 per cent of the area of the premises was used as a shop, including a covered yard at the rear, and about 25 per cent consisted of modest living accommodation — Appeal allowed, but, in view of difference of opinion, leave given to appeal to House of Lords

This was an appeal from a decision of Judge Coplestone-Boughey at Wandsworth County Court, holding that the tenant, Ajeet Kumar Tandon, was entitled to acquire the freehold in certain premises at Mitcham Lane, Streatham, from the landlords, the Trustees of Spurgeon's Homes, under the Leasehold Reform Act 1967. The trustees appealed.

E Patrick Ground (instructed by Pothecary & Barratt) appeared on behalf of the appellants; David Parry (instructed by Radia & Co) represented the respondent.

Giving the first judgment at the invitation of Ormrod LJ, GRIFFITHS LJ said: This is an appeal from a decision of His Honour Judge Coplestone-Boughey, in which he held that a tenant was entitled to the enfranchisement of his lease pursuant to the provisions of the Leasehold Reform Act 1967. There is no dispute in this case that the tenant was entitled to enfranchisement provided that the building in question of which he was tenant could be said to be a house "reasonably so called" within the meaning of section 2 of the Act.

F I cannot improve upon the description of the property contained in the learned county court judge's judgment and I now read the description from that judgment:

> The property is clearly shown by the agreed plan, description and photographs which are before me.

I interpolate to say that the members of this court have had the advantage of seeing that plan and photographs.

> The site is a long narrow plot with a frontage of about 21 ft upon Mitcham Lane. The ground floor is a shop fronting the main road and is one of four identical shops with varying shop fronts. There is a first floor above containing three effective rooms in the main part of the building and a back addition. Behind this building there was originally a narrow open path G running beside the back addition and behind that an open space leading to a two-storey stable which in turn gave access to a rear service road. At some uncertain time but certainly before 1962 when Mr Kirkaldie inspected the premises a roof, treated for London Building consent as temporary, was erected over the whole of the yard except an outside water closet. This roof runs from and is attached to the rear of the main building and extends right back to the stable. The doors are arranged so that there is internal communication from the shop in front through the original open space to the stable and service road at the back. The interior layout is that at the rear of the main building there is an opening to the right which forms the access to the upper part of the building and also to the combined bathroom and water closet at ground level. The stairs rise by a right angle to the right and open upon the first floor. At present the whole of the ground floor including the H covered yard and stable are used as a shop and the first floor is residential. In 1962 the layout was the same but the first floor was used as offices.

We have seen the photographs of the premises and they consist of one of a small parade of shops. Each shop has living accommodation above it; each consists essentially of a flat above a shop. The premises now consist, as to their area, of about 75 per cent of the area being used as a shop, because the whole of the yard which is covered in is now used as storage for the shop, and about 25 per cent for the modest living accommodation, comprising living room and two bedrooms on the first floor.

I turn now to section 2, subsection (1) of the Act, which provides: J

> For purposes of this Part of this Act, "house" includes any building designed or adapted for living in and reasonably so called, notwithstanding that the building is not structurally detached, or was or is not solely designed or adapted for living in, or is divided horizontally into flats or maisonettes.

These premises were originally let by a lease in 1881 and were originally let not solely as a house but (I quote from the lease) as a "messuage or dwellinghouse and shop". So these are premises which started as a shop with living accommodation.

The learned judge was referred to the decision of this court in *Lake v Bennett*, reported in [1970] 1 QB 663. He concluded that that decision applied to the facts in this case and concluded it in favour of the tenant. The premises in *Lake v Bennett* were, in my K view, of a markedly different character to the present premises and we have had the advantage of seeing photographs of them. In *Lake v Bennett* the premises in question consisted of a house in a parade of houses, which, from looking at the photographs, were quite clearly built as a terrace of dwelling-houses. It was a house that consisted of a basement, a ground floor and first and second floors. In the course of time the ground floor was converted into a betting shop. The Court of Appeal held that, notwithstanding that the ground floor was now a betting shop, the building as a whole was "reasonably so called" a house, and that conclusion differed from that of the learned county court judge who tried the case at first instance. The Court of Appeal held that they were free to differ L from the learned judge because, he having found the primary facts, this court was free to say whether or not, on those primary facts, the house fulfilled the condition in section 2(1), namely, that it was "reasonably so called" a house.

The learned judge having found the primary facts in this case, in my view this court is free to apply its own views of whether these premises are reasonably called a house. In approaching this decision, I bear in mind that the right to enfranchisement is given only to householders. A person who lives in a flat has no right to enfranchise the flat. Furthermore, I bear in mind that Parliament does provide protection for those who carry on businesses and when their leases end they can apply, under the relevant statutory pro- M vision, for a new lease, and if they cannot agree with their landlords then there are provisions for the court to fix a fair rent for the parties.

In *Lake v Bennett*, although the court in that case, on those facts, was satisfied that that building was reasonably to be called a house, Lord Denning realised that in future other difficult questions might arise in regard to other buildings. I, for my part, think that this is such a case. I am bound to say that, looking at the photographs and taking into account that this was at all times a purpose-built shop with living accommodation above it and part of a parade of such shops, I do not think that it can reasonably be called a house. I share

A the view of the very experienced surveyor who gave evidence before the learned judge and who said it was not properly to be described as a house but as a shop with living accommodation above it.

I appreciate that, in the course of their judgments in *Lake v Bennett*, both Salmon LJ and Cross LJ pointed out that there might be a case in which one person would describe the building as a shop with living accommodation above it but equally another person could reasonably describe it as a house with the ground floor made into a shop. Of course, there will be such borderline cases, but in my view this case is not a borderline case and for my part I do not think that these premises could reasonably be described as a house. Although B the learned judge, I accept, applied the right test, and I take into account that he said that he would have come to the conclusion, quite apart from *Lake v Bennett*, that it could reasonably be called a house, I take the view that, on the facts that he has found, he came to the wrong conclusion. Accordingly, for my part, I would, for those reasons, allow this appeal.

I should perhaps just add that we were also referred to *Peck v Anicar Properties Ltd*, reported in [1971] 1 All ER 517, where, in respect of premises which appear to me to be almost identical to those in *Lake v Bennett*, it was assumed without argument that the premises were a house. I do not think that case carries the matter any further than the authority of *Lake v Bennett*.

C For those reasons, I would allow this appeal.

WATKINS LJ agreed.

ORMROD LJ (dissenting) said: The definition in section 2(1) of the Leasehold Reform Act 1967 is not at all easy to construe. It begins by saying that " 'house' includes any building designed or adapted for living in". It goes on to say "and reasonably so called" and then proceeds with the words "notwithstanding that the building is not structurally detached, or was or is not solely designed or adapted for living in". So Parliament plainly contemplated that a house within the meaning of this Act might well be a house with a D double user, partly for living in, partly for some other purpose. So the fact that this house was originally built in 1881 as a dwelling-house and shop does not take it outside, in my judgment, the definition in section 2(1).

It seems to me that one has to look at the building as a whole and answer the question, allowing for the fact it is used for two quite different purposes, whether it still can reasonably be called a house. That seems to me to introduce a degree of flexibility into the application of the definition, which is highly unusual in a statute but, of course, is designed to enable the court responsible for making the decision to arrive at what one might call a broadly sensible conclusion without being tied down by a technical definition.

E Consequently, I have some considerable difficulty in overruling the learned judge, who, having seen the property and having considered the matter and applied, as I think, the right tests, came to the independent conclusion, as he plainly did because he says in terms, "Even if the matter were free from authority I should conclude that the present building was reasonably called a house", I find great difficulty in differing from him on facts as narrow as these in the present case.

To my mind the approach which this court adopted in *Lake v Bennett* makes it still more difficult to hold that these particular F premises are not reasonably described as a "house". The approach which the court adopted in *Lake v Bennett* seems to have been in accordance with the definition, because the court began by saying that, looking at the premises in that case (*Lake v Bennett*), for many statutory purposes the building would be called a "house". One knows that there is, in other statutes, a definition of a "house" which is very wide indeed. The court construed the words "and reasonably so called" as, in a sense, words of limitation, starting from the assumption that, without those words, there could be no doubt that the property in *Lake v Bennett* was a house. That being the approach that this court adopted in *Lake v Bennett*, I feel considerable difficulty in adopting a different approach in this particular G case.

If that is the right way of approaching it, then I find myself unable to disagree with the learned judge's view. I accept that it is open to anybody to adopt the surveyor's description of this property and I bear in mind that in the report of *Lake v Bennett*, which includes an extract from the learned county court judge's judgment in that case, the learned judge said (reading from [1970] 1 QB 663 at p 665): "I have formed the definite opinion that it would not be called a 'house' in ordinary speech in the present century. A person looking for a house to buy would be surprised to be directed to 61 Gayton Road. On seeing it he would say 'I asked for a house and they have sent me to a shop. I have been misled' ". No doubt exactly the same H point can be made on the facts of the present case.

So I feel the greatest diffidence in distinguishing this case from *Lake v Bennett*. There are, of course, certain distinctions. First, there is the historical distinction that in *Lake v Bennett* the property had started life unquestionably as a house and had been subsequently converted to use as a shop on the ground floor. It is fair comment that both Salmon LJ and Cross LJ used language which indicated that the fact of the conversion was not irrelevant to their minds. But the learned county court judge in this case dealt with that point and dealt with it, I think, correctly. The only other distinction from *Lake v Bennett*, it seems to me, is to say that in this case the area of the property which was used as a shop was very J much greater. But that seems to introduce a quantitative consideration which could be extremely difficult. The fact which influences my mind very much is that, looking at the photographs we have been supplied with, looking at these present premises from the point of view of somebody standing in the street and comparing the appearance of these premises with the appearance of the premises which were the subject-matter of *Lake v Bennett*, it seems a little bizarre to hold that the premises at 61 Gayton Road in *Lake v Bennett* and the premises at 41 Lee High Road in another case were houses and yet that the present premises are not reasonably called a house.

I think that Parliament has left the decision on the fine distinc- K tions in this type of case to the trial judge and I am not prepared to differ from his conclusion in this court, although I agree with my brethren that, starting afresh and without *Lake v Bennett* as a guide, I think I should have come to the same conclusion as they do. But I do not think that this case can properly be distinguished from *Lake v Bennett*. Therefore, I would dismiss the appeal.

The appeal was allowed with costs. Leave was given to appeal to the House of Lords.

LANDLORD AND TENANT
RENT ACTS

A **COURT OF APPEAL**

October 22 1980

(Before Lord Justice STEPHENSON, Lord Justice DUNN and Sir David CAIRNS)

SCRIMGEOUR v WALLER

Estates Gazette January 10 1981

(1980) 257 EG 61

Landlord and tenant—Rent Act—Question whether a gardener occupied a cottage by virtue of a protected tenancy or a licence—Appeal from decision of county court judge—Gardener held a protected tenancy from the previous owner but entered into a
B **fresh agreement with new owner—Judge felt that he could disregard the documents constituting this agreement, but Court of Appeal held that it effected a surrender of the protected tenancy and the creation of a bare licence in its place—Foster v Robinson followed—Distinction as explained in Torbett v Faulkner recalled between a service occupier's licence where the servant is required to live in the house for the better performance of his duties and a licence where the servant is merely permitted to live there because it is convenient for his work—The latter was the position in the present case—The licence having been duly terminated by proper notice, the owner was entitled to possession—Appeal allowed.**

C This was an appeal by Alexander Scrimgeour, the owner of a cottage called Coach House Cottage at Hascombe Court, near Godalming, Surrey, from a decision of Judge Mark Smith at Guildford County Court, dismissing the appellant's claim for possession of the cottage and upholding a submission by the respondent, Leslie John Waller, that he was a protected tenant.

R W Seymour (instructed by Goddens & Thicknesse) appeared on behalf of the appellant; C R Semken (instructed by Macfarlanes) represented the respondent.

Giving the first judgment at the invitation of Stephenson LJ, Sir David Cairns said: This is an appeal from the judgment of His Honour Judge Mark Smith given at Guildford County Court dis-
D missing the plaintiff's claim for possession of a cottage called Coach House Cottage, Hascombe Court, near Godalming in Surrey. The plaintiff, Mr Scrimgeour, was the freehold owner of the cottage and the defendant, Mr Waller, was, from January to December 1979, employed as a gardener by Mr and Mrs Scrimgeour at their house a couple of miles away from this cottage. During that period Mr Waller was in occupation of the cottage.

On November 9 1979 a letter was written by Mr Scrimgeour's solicitors to Mr Waller giving him four weeks' notice to terminate his employment and to quit the cottage. He did not quit, he has remained in occupation and he claims to be a protected tenant there. The learned judge upheld that contention and dismissed Mr
E Scrimgeour's claim for possession. Mr Scrimgeour appeals to this court.

Mr Scrimgeour had bought the cottage towards the end of 1978 at an auction. Mr Waller had been employed as a gardener by the previous owner of the cottage, a Mrs Jacobs, and he was in occupation when Mr Scrimgeour bought the cottage. Mr Scrimgeour wrote to Mr Waller some time before December 16 1978 (we have not got the letter) offering to employ him at £40 a week. He replied in a letter dated December 16:

> I wish to confirm receipt of your letter offering employment at £40 per week, and also of accommodation in Coach Cottage owned by you. I accept your terms and look forward to commencement of employment early in the new year. F

The learned judge rightly said that not much significance attached to that letter at the trial and the reason for that was that it was superseded by something else.

On December 19 Mr Scrimgeour wrote to Mr Waller saying that he was getting out a statement of the terms of employment, which was required by law. That statement was duly prepared and some time shortly before January 3 1979 was passed to Mr Waller with a request to him to sign it. That document contains these relevant paragraphs:

> 1. Your employment is as Gardener/Handyman at Thorncombe House, Thorncombe Street, near Bramley and at Coach House Cottage, Hascombe Court. G
> 2. Your employers are Mr and Mrs A Scrimgeour.
> 3. Your employment commenced on January 1 1979.
> 4. Your wage is £36 less statutory deductions, per week, payable fortnightly in arrears, plus £4 travelling allowance.
> 9. The minimum periods of Notice by the Employer, under the terms of the Employment Protection Act are as follows:
>
> (i) One week if the period of continuous employment is four weeks or more but less than two years.

The other material clause, and most material, is clause 12. 12(a) reads:

> Service Cottage. The cottage which you occupy is free of rent and rates and is so occupied for the better performance of your duties. You will leave it in good order with vacant possession together with any equipment which has been in your charge on the date on which your employment terminates. H
> (b) The employers do not undertake to provide accommodation at Coach House Cottage for any specific period of your employment and they reserve the right to provide alternative accommodation within approximately the same distance from Thorncombe House.

After several further clauses there are the words:

> I acknowledge receipt of a copy of this statement.

It is signed "L J Waller" and dated January 3 1979.

During the year 1979 Mr Scrimgeour agreed to sell this cottage to his step-daughter and he wanted to give her vacant possession of it. He had no wish to terminate Mr Waller's employment but he J wanted to get him out of the cottage. He made offers of alternative accommodation which Mr Waller did not accept. On September 5 1979 Mr Scrimgeour wrote a letter saying:

> Dear John,
> I write in accordance with the terms of your employment to confirm that you have been offered alternative accommodation at 27 Latimer Road, Godalming, which you rejected and, although you did not go inside this house, you stated that nothing similar would be suitable.

Mr Scrimgeour went on to explain other things he had been doing to provide alternative accommodation to Mr Waller's liking, and ended with the paragraph:

> I should point out that I have no wish to terminate your employment and if you yourself can find alternative accommodation to your liking, I should be willing to consider an increase in your wages, in lieu of the rent-free accommodation which you have at present. K

That letter did not take the matter any further. There was another house proposed as alternative accommodation, a house called Palmar, which at first Mr Waller was willing to take, so much so that on October 22 Mr Scrimgeour was writing to Mr Waller saying:

> I enclose an amendment to the statement of your terms of employment. Please sign and return one copy to me.

> I will let you know as soon as I can the date of your move to Palmar—some time within the next four to five weeks I imagine.

He thought everything was in line for the move, and the amendment to the terms of employment which he was forwarding was in order to meet that new situation and was in these terms:

> The Statement signed by you on January 3 1979 is amended as follows: Delete all references to Coach House Cottage, Hascombe Court, and substitute in paragraphs 12 and 13—Palmar, Marshall Road, Farncombe, Godalming.

On receipt of that letter Mr Waller was not pleased and on October 27 he wrote a letter which, unlike the earlier one he had written, was not in his handwriting. It may be that he had taken somebody's advice by that time. In the letter of October 27 Mr Waller said this:

> The threat of impending notice in your letter dated September 5 1979 signifies total lack of security in my favour should a similar problem arise.

Mr Waller stayed in the house and, Mr Scrimgeour's solicitors having been brought into the matter by November 1, they wrote a letter treating Mr Waller's failure to go as a refusal to vacate the cottage and a breach of the terms of contract of employment, saying that alternative accommodation had been offered.

Mr Waller wrote in reply to that saying that he had not refused to leave the cottage, but it was necessary, because of his condition, to live in the country and he went on to explain that as a result of having been a prisoner of war he had a certain nervous disorder of the mind. The relevance of it in this letter was that he said it was a disorder which required him to live in the country.

Mr Scrimgeour felt that the time had now come when he would have to determine the employment of Mr Waller and with it his occupation in the cottage. Accordingly, on Mr Scrimgeour's instructions his solicitors wrote a letter dated November 9 giving four weeks' notice to terminate the employment and ending:

> Although, under your terms of employment, you are entitled to only one week's notice, our clients have asked us to give you four weeks' notice and we should be glad to hear from you therefore, that you will make the necessary arrangements to vacate the cottage at the expiration of four weeks from the date of receipt by you of this letter.

That was clearly a notice of determination of the employment and also a notice to quit the cottage.

Subject to the question of whether or not there was a protected tenancy of this cottage, the period of notice is not open to challenge and was not challenged.

The learned judge treated the documents of January 3 as being something which did not really govern the letter because, as he put it, he "doubted whether Mr Waller knew or cared what that document meant". In his defence Mr Waller had expressly admitted executing this document and neither in his defence nor in his evidence did he say that he did not understand it, or that it was in any way not binding upon him. There is no plea, no evidence of *non est factum* and there was no plea and no evidence of the parties not being *ad idem* or anything of that kind.

Because the judge felt it possible to disregard the terms of this document he treated Mr Waller as continuing to be in occupation on the same terms as applied between him and his previous employer, Mrs Jacobs, and he found that with her he had a protected tenancy.

The case on which Mr Waller was relying under his amended defence and amended further and better particulars of that defence which were drafted by his counsel, Mr Semken, who had come into the matter at a quite late stage—an amendment which was allowed by the learned judge—was that he had a protected tenancy from Mrs Jacobs and that that protected tenancy continued despite the terms of the document of January 3 or, alternatively, that a new protected tenancy was created by that document. The judge accepted, though as I have indicated he put it on the basis that he thought Mr Waller neither knew nor cared what the document contained and for that reason it was ineffective, the proposition that there was a protected tenancy and that it continued. It was on that basis that he gave judgment for the defendant.

By his notice of appeal Mr Scrimgeour contends that the judge was wrong in his finding that there was a protected tenancy with Mrs Jacobs and alternatively that the learned judge should have found that if there was such a tenancy it was surrendered and replaced by a licence, that is to say the type of contract commonly made between an employer and employee when an employee comes to reside in a house or part of a house convenient for his work. That was of course based upon the terms of clause 12(a) of the document of January 3 1979.

So far as the relationship between Mr Waller and Mrs Jacobs was concerned, Mr Waller's evidence was that it was orally agreed between them that he should be a service occupier employed at a wage of £40 a week to be paid as to £30 in cash, the difference representing the rent of £10 a week which he was in effect paying by rendering services. The learned judge accepted that evidence as to the facts of what was agreed between Mrs Jacobs and Mr Waller and against that finding it is obviously impossible for Mr Scrimgeour successfully to appeal. This court would have no grounds for disturbing it. We must treat the case on the basis that that was the contract between those two people and on that basis, if indeed it created a tenancy at all, it would be a protected tenancy. It could perhaps be arguable that, because it was apparently expressly described as a service occupation, that might have been intended as a licence rather than a tenancy, but let us assume that there was a tenancy and a protected tenancy at that stage.

It is right to point out that, although you cannot have a protected tenancy if there is no quantification of the value of any services said to represent rent (that was decided in the case of *Barnes v Barratt* [1970] 2 QB 657) it would appear from that case that if there is a quantification of the rent, which on the evidence there was in this case, you can have a protected tenancy. So I go on to assume here that there was a protected tenancy with Mrs Jacobs.

In my view the effect of the agreement of January 3 1979 was quite clearly to effect a surrender of any such tenancy, because clause 12(a) is completely inconsistent with the continuance of a tenancy at £10 a week. There is no doubt that a protected tenancy can be brought to an end by surrender (*Foster v Robinson* [1951] 1 KB 149) and it is quite clear from the report in that case that it is not necessary in order that there should be a surrender that the tenant should go out of possession and then come in again, or should hand over the keys and have them handed back to him, or any formality of that kind. It can all be derived from appropriate conduct and appropriate words. It seems to me that you could not have anything clearer than you have in this case in the document itself showing that there must be surrender of any tenancy that did exist, because either a fresh tenancy or a licence is being created.

Now, which is it? It seems to me that it is probably a licence, if full effect is given to clause 12(a), a licence such as was found to exist in *Torbett v Faulkner* [1952] 2 TLR 659. Mr Semken contends that this document contains a contract of employment only, but unless there is some way you can get rid of clause 12(a), whether you label it a contract of employment or however else you label it, it is a contract which does determine as between the parties *inter alia* the terms upon which Mr Waller was to occupy the house. Mr Semken goes on to say that the provision "... so occupied for the better performance of your duties" is not really a genuine statement of the purpose for which he was occupying at all because it was a couple of miles away from the main place where he used to do his gardening and therefore it was not necessary for him to live there for that purpose. That is in effect saying that that part of the document is a sham; and it is expressly found by the learned judge that the document is not a sham. But if it can be said that for some reason this was not a service occupancy in the sense of being a licence rather than a tenancy, but was indeed a new tenancy created between Mr Scrimgeour and Mr Waller, then the next question is "Was it a protected tenancy?", and the answer is "No," because if one thing is clear about clause 12(a) it is that there is no rent. I am not quite sure that in law you can have a tenancy with no rent. If you cannot have such a tenancy then *cadit quaestio*; if you can have such a tenancy, since it is at no rent it is at a rent which is less than two-thirds of the rateable value, so that you cannot have any protection under the Rent Act in respect of it.

I reach without doubt the conclusion that in this case the learned judge has arrived at the wrong conclusion; that this was either a licence or a tenancy which Mr Scrimgeour was entitled to bring to

A an end by the appropriate notice and four weeks' notice was more than sufficient in this case to bring it to an end. Accordingly Mr Scrimgeour was entitled to an order for possession.

He also made a claim for mesne profits. Three figures of weekly rental value were given in the course of the evidence. Mr Scrimgeour said £15: Mr Waller in evidence-in-chief said £10 and in cross-examination said £12. I think it would be fair to compromise on £12. It is 45 weeks on that basis. The judgment that, in my view, should be substituted is an order for possession and £540 by way of mesne profits. I would allow the appeal and substitute that order.

Agreeing, DUNN LJ said: I am indebted to Mr Semken in his submissions for explaining to me what I believe to be the basis of the B judgment of the learned judge. He appears to have taken the view in effect that these parties were not *ad idem* when they entered into the agreement of January 3 1979 because he came to the conclusion that the plaintiff thought that the defendant only had a service occupancy of the cottage, whereas the defendant believed that he had a protected tenancy. Because of that the judge did not really attach any importance to clause 12 of the agreement at all. In that, in my view, he fell into error. There was no plea that the parties were not *ad idem*. The document was expressly admitted in the defence and it was irrelevant whether they were *ad idem* or not because the question then was, not what they believed the document meant but what, on its true construction, it did as a matter of C law mean. The first sentence of paragraph 12(a) seems to me to give the true answer to that question because it says: "The cottage which you occupy is free of rent and rates and is so occupied for the better performance of your duties."

As Denning LJ, as he then was, said in *Torbett v Faulkner* [1952] 2 TLR 659 at p 660:

> Previously the holding of a servant was classified either as a service occupation or as a service tenancy. There was no third category. But nowadays it is recognized that there is an intermediate position. [A servant] may be a licensee. A service occupation is, in truth, only one form of licence. It is a particular kind of licence whereby a servant is required to live in the house in order the better to do his work. But it is now settled that there are D other kinds of licence which a servant may have. A servant may in some circumstances be a licensee even though he is not required to live in the house, but is only permitted to do so because of its convenience for this work—see *Ford v Langford* per Asquith LJ . . . and even though he pays the rates . . . and even though he has exclusive possession. If a servant is given a personal privilege to stay in a house for the greater convenience of his work, and it is treated as part and parcel of his remuneration, then he is a licensee, even though the value of the house is quantified in money; but if he is given an interest in the land, separate and distinct from his contract of service, at a sum properly to be regarded as a rent, then he is a tenant, and none the less a tenant because he is also a servant.

In this case the value of the house was not quantified in money, but it seems to me that it is exactly the situation envisaged by Denning LJ in *Torbett v Faulkner*. The defendant in this case was E given a personal privilege, namely, to stay in a house for the greater convenience of his work. It may not have been necessary for his work that he should occupy the cottage, because the cottage was in fact two miles from the plaintiff's house at which he was employed as a gardener, but no doubt it was a greater convenience for him. In my judgment it is quite plain that clause 12 gave the defendant no more than a bare licence to occupy the cottage and that licence governed his occupation from the date of the signature on the document of January 3 1979.

The case of *Foster v Robinson*, referred to by my Lord, shows that, assuming (as the judge found here) the defendant held under a protected tenancy from Mrs Jacobs, nevertheless it was open to the F parties to agree that that tenancy should be surrendered and that the former tenant should thereafter hold as a bare licensee. In my view this was the effect of the agreement of January 3 1979, so that from that date the defendant had no more than a licence.

Accordingly, valid notice to determine that licence having been given, he is no longer entitled to possession thereafter and I, too, would allow the appeal and grant an order for possession with mesne profits at a rate of £12 a week.

Also agreeing, STEPHENSON LJ said: I have only three things to add.

First, in cross-examination Mr Waller, according to the note of his evidence which we have, said of clause 12(a) in the agreement to which my Lord has referred: "I know what it means, but you can't G turn out someone just like that." That clearly indicates that the only defence here was a cry for help to be relieved from the plain consequences of having executed the document which included clause 12(a).

Secondly, that cry was answered by the learned judge in Mr Waller's favour, but I think it is only fair to Mr Scrimgeour to quote from the judgment the judge's opinion that "the plaintiff" (Mr Scrimgeour) "could not have behaved better." Whatever hardship there may be to Mr Waller, this is not a case of an unfeeling landlord inconsiderately evicting a good servant without any justification except the letter of the law.

Thirdly, I would like to congratulate Mr Semken on having made H some bricks, or parts of bricks, out of the very little available straw.

I would agree with the order proposed by my Lord, Sir David Cairns.

The appeal was allowed and possession ordered in 28 days, with mesne profits at £12 a week from December 9 1979 until possession. Costs were awarded in the appellant's favour in the Court of Appeal and below (scale II below).

COURT OF APPEAL J

October 17 1980

(Before Lord Justice STEPHENSON, Lord Justice DUNN and Sir David CAIRNS)

BALDOCK v MURRAY

Estates Gazette January 24 1981
(1980) 257 EG 281

Rent Act 1977, section 19 — Whether a letting at a rent which included payment for services was a restricted contract—County court judge held that it was a restricted contract and gave judgment K in favour of landlord for possession—Tenant contended that the tenancy was a regulated tenancy and thus excepted from the definition of a restricted contract by virtue of section 19(5)(a) of the Act — Held by Court of Appeal, reversing decision of judge, that there was nothing to take the tenancy out of the class of regulated tenancies — The judge had read section 19(5)(a) the wrong way round — It does not provide that if a tenancy is a restricted contract it is not a regulated tenancy; it provides that if it is a regulated tenancy it is not a restricted contract — Suggestion that the contract here constituted a licence only and not a tenancy raised too late to be considered — Appeal by tenant allowed.

L This was an appeal by the tenant, Niall Murray, from a decision of Judge Slot at Croydon County Court granting possession of a room at 25 Queens Road, Wallington, Surrey, to the landlord, Yvonne Baldock, on the ground that the letting was a restricted contract which had been duly terminated by notice to quit.

N C Nardecchia (instructed by Stoneham, Langton & Passmore, of Croydon) appeared on behalf of the appellant; P J H Ralls (instructed by Davies, Arnold & Cooper) represented the respondent.

Giving the first judgment at the invitation of Stephenson LJ, SIR DAVID CAIRNS said: This is an appeal from a decision of His Honour Judge Slot in the Croydon County Court giving judgment for possession of a room at 25 Queens Road, Wallington, in the county of M Surrey.

The plaintiff by her particulars of claim set out that this room was let by her to the defendant on a monthly tenancy, that he was given notice to quit and that he had failed to quit. It was alleged under paragraph 5 of the particulars of claim that the tenancy was a restricted contract within Part V of the Rent Act 1977. A restricted contract is defined in section 19 of the Rent Act 1977. That section provides:

> 19(1) A contract to which this section applies is, in this Act, referred to as a "restricted contract".
> (2) Subject to section 144 of this Act

A and that section is irrelevant in these proceedings,

> this section applies to a contract, whether entered into before or after the commencement of this Act, whereby one person grants to another person, in consideration of a rent which includes payment for use of furniture or for services, the right to occupy a dwelling as a residence.

The particulars of claim having alleged that this was a restricted contract, particulars were asked for of that allegation. The particulars set out that it was a contract which included services and a list of the services was added. The defendant denied by his defence that it was a restricted contract and contended that he was a protected tenant within the meaning of section 1 of the Rent Act 1977.

The learned judge, finding that it was a contract which included B payment for services, concluded that it was a restricted contract within the meaning of section 19. It was contended on the part of the defendant that it was not a restricted contract because of an exception contained in subsection (5) of section 19, which provides:

> A contract is not a restricted contract if — (a) it creates a regulated tenancy.

It is clear that this was dwelling-house within the meaning of the Rent Act, that it was within the rateable value, that the contract had all the qualities which constitute a protected tenancy and entitled it to protection as a regulated tenancy, unless it can be said that being a restricted contract took it out from that category. The learned judge held that it did. In my judgment he read section 19(5)(a) the C wrong way round. The Act does not provide that if a tenancy is a restricted contract it is not a regulated tenancy; it provides that if it is a regulated tenancy it is not a restricted contract. This clearly was a regulated tenancy and that takes it out of the definition of a restricted contract.

The learned judge said that if that interpretation were accepted it would mean that there would be very few contracts to which the definition of a restricted contract could apply. It is sufficient to say that tenancies within sections 8, 9, 10, 11 and 12 of the 1977 Act are all capable of being restricted contracts. The learned judge's suggestion that there would hardly be any restricted contracts if section 19(5) were interpreted in the way I have suggested it must be D interpreted is, it appears to me, quite wrong. I am quite satisfied that unless, by reason of some of the provisions of Part I (apart from section 19), a tenancy is taken out of the class of regulated tenancy, then if apart from that it is a regulated tenancy, it cannot be a restricted contract.

For those reasons I am satisfied that this judgment cannot be upheld on the only grounds which were considered by the learned judge in the court below.

Mr Ralls, who appeared in the county court as counsel for the plaintiff, having been only very recently instructed, has felt bound to concede that that is the position, but at the opening of his address to this court, having made that concession, he asked for leave to E raise a point which had not been raised before and which was indeed contradictory to what was alleged in the particulars of claim, namely, that this was not a tenancy at all but was a licence. I say simply this, that the whole of this case having proceeded upon the basis that this was a tenancy, and that being a fundamental allegation in the particulars of claim, it is far too late for a contention that it was a licence to be raised at this stage. I am therefore of opinion that it is inevitable that this court should refuse the application on behalf of the plaintiff for leave to serve a respondent's notice to raise that point and the appeal should be allowed.

STEPHENSON and DUNN LJJ agreed.

The appeal was allowed with costs.

F

G **QUEEN'S BENCH DIVISIONAL COURT**

October 27 1980

(Before Lord Justice DONALDSON and Mr Justice FORBES)

R v RENT OFFICER FOR LONDON BOROUGH OF CAMDEN, EX PARTE EBIRI AND ANOTHER

Estates Gazette February 14 1981

(1980) 257 EG 605

Rent Act 1977 — Question whether tenancy was a protected tenancy or excluded from protection as a holiday letting by virtue of section 9 of the Act — Duty of rent officer when faced with a jurisdictional issue — Application for judicial review by tenants seeking order of H mandamus to require rent officer to determine and register fair rent of flat — Proceedings started in county court under section 141 of Act to determine whether tenancy was protected, but county court hearing not likely to take place quickly — At a preliminary consultation rent officer expressed himself as satisfied that the letting was not a holiday letting and that he had jurisdiction to proceed — Subsequently, however, the rent officer, although still considering that he had jurisdiction, intimated that he would adjourn the consultation sine die in view of county court proceedings — Tenants urged rent officer to proceed to registration as delay might affect them adversely in view of imminent implementation of changes by Housing Act 1980 affecting the date from which excess rent could be J recovered from a landlord — Duty of rent officer analysed by court — If a rent officer was satisfied that he had jurisdiction (although his decision on this point could not be final) it was his duty to proceed — It was only if he was in doubt about his jurisdiction (or perhaps if a court decision on the issue was imminent) that he had a discretion whether or not to proceed — Observations on "Noel" case — Order of mandamus granted.

This was an application for judicial review by Cyril and Bekky Ibiri, tenants of Flat 3, 105 York Way, London N7, who sought an order of mandamus against J McW Smith, a rent officer for the London Borough of Camden Registration Area, requiring him to K determine and register a fair rent in respect of the above flat.

D Watkinson (instructed by Nic Madge, Camden Community Law Centre) appeared on behalf of the applicants; Simon Brown and John Laws (instructed by the Treasury Solicitor) represented the respondent rent officer.

Giving judgment, DONALDSON LJ said: The history of the matter is this. The two applicants took a tenancy of the premises at Flat 3, 105 York Way, London N7, in November 1979. In May 1980 they applied to have a fair rent registered. A copy of the application was forwarded by the rent officer in the usual and required way to the landlords, Sherman Securities Ltd, and the landlords by a letter dated June 18 took the point that this was not a protected tenancy but was a holiday let and so was outwith the jurisdiction of the rent L officer. As soon as that letter was received by the applicants, or very shortly afterwards, they started proceedings in the county court in order to determine whether or not this was a protected tenancy, the applicants of course contending that it was and the landlords contending that it was not.

The rent officer meanwhile had arranged for a preliminary consultation between the parties to take place on July 28, and at that meeting he went into the question of whether or not this was a protected tenancy. It was not a case, as was the position in *R v Kensington and Chelsea Rent Officer, ex parte Noel* [1978] QB1, of merely having submissions from agents or representatives of the parties. The parties gave evidence. There was a statement from the M tenants. A Mr Sherman represented the limited company landlord and Mr Sherman cross-examined the tenants.

At the end of the proceedings the rent officer said:

> In the light of what I have heard this afternoon, I have come to the conclusion that this is not a holiday let, but an ordinary regulated tenancy. In the light of the reference and Mr Ebiri's statement, I feel I have jurisdiction to proceed. I have cases where I decide straight away that there is a holiday let and adjourn them *sine die* for the tenant to prove it in court. Sometimes I have never heard any more, or the courts have confirmed it. In this instance there is no doubt in my mind, and I do feel that I can proceed.

Having done that, he then proceeded to fix August 12 as being the date for the consultation when representations are made by both parties as to the amount of the fair rent rather than proceed straight away, in order to give the landlords opportunity to apply for judicial review if they wished to prevent him proceeding further and also to prepare any arguments they might want to advance to him about the exact level of rent.

The landlords took no steps to prohibit him from proceeding, and the parties met again before the rent officer on August 12. Meanwhile the landlords had written to the rent officer on August 4 saying this:

> You will recall that despite the pending county court proceedings to decide whether the occupancy of the above flat is a protected tenancy, you made your own decision that it is so.
>
> The fact that the letting agreement describes the tenancy as ". . . a holiday letting only" is *prima facie* evidence that it is until the court decides otherwise under section 141 of the Rent Act 1977.
>
> We feel that you are pre-empting the decision of the court and acting beyond your powers in proceeding with the application for the registration of a fair rent until the court has determined whether a protected tenancy exists.
>
> In those circumstances we feel that the consultation proposed for August 12 should not be proceeded with.

On August 12 Mr Smith referred to that letter. He said that, in his view, his letter constituted "new evidence" and that in the light of that and other factors he considered that it would be wiser for him not to register a rent. He said he would adjourn the consultation *sine die*. The representative of the tenants argued that that was not the right course and that the rent officer ought to register a rent. The rent officer said that he still thought he had jurisdiction to do so but that he would not in fact register a rent. He did not say so, but clearly he thought he was acting within the scope of his discretion and he said, somewhat incorrectly, that no one could compel him to register a rent.

The proceedings then went on by Mr Madge, representing the tenants, urging the rent officer to give detailed reasons and drawing his attention to the fact that there might be rights vested in the tenants which would be lost if there was any delay as a result of orders being made bringing parts of the Housing Act 1980 into operation. I need not pursue exactly how that could operate. No orders have yet been made and it is unlikely that they will be made until the end of next month. But it clearly was a factor which he was entitled to take into consideration if and in so far as he had a discretion in the matter.

The rent officer adhered to his view that he could not register the rent, but he did proceed to hear argument on what should be the level of rent which would be fixed if and when he was prepared to register it.

The facts end with a letter from the rent officer to the tenants' representative simply saying:

> Following the meetings held at this office on July 28 and August 12 last from the information made available to me and in view of the fact that application has been made to the court for a declaration I am adjourning this case pending the decision of the court.
>
> No doubt you will let me know the outcome.

In my judgment, it is the duty of a rent officer to consider what is his jurisdiction in cases in which he has an application which depends upon a disputed question as to whether the particular contractual arrangement between landlord and tenant or landlord and licensee is or is not a protected tenancy. He has to consider it but he cannot determine it. He has no jurisdiction to determine it on a final and binding basis, subject to appeal as between the parties. His jurisdiction is like that of an arbitrator. It stems from an assumption, namely, in the case of an arbitrator that there is an agreement to refer, in the case of a rent officer determining a fair rent upon an assumption that there is a protected tenancy. Somebody else has to decide whether that assumption is valid, if the matter is to be determined in a final and binding way. But this is very far from meaning of course that every case where there is a doubt is to be referred to the courts. The duty of the rent officer, I think it is clear from such cases as *Noel*'s case to which I have referred and *Zerek*'s case which is referred to in *Noel*'s case, is to inquire as to the facts. If he is satisfied that there is not a protected tenancy he has no problem. He just does not proceed further until it has been determined by a court of competent jurisdiction that there is a protected tenancy. If he is satisfied that there is a protected tenancy (and, note, I say "satisfied" and not decided, because he has no jurisdiction to decide), in my judgment, he must proceed. It is only if he is in doubt as to whether there is or is not a protected tenancy that he has a discretion whether to proceed or not to proceed. That is putting the position, I accept, in somewhat black and white terms, and his discretion might extend to not proceeding in a case in which he is certain or virtually certain in his own mind that there is a protected tenancy but within 24 hours he knows it is going to be decided by a court of competent jurisdiction. I would not dissent from the view that in that highly unusual and marginal case he may have a discretion to postpone deciding the matter for 48 hours to get just beyond the decision. But the broad position, as I see it, is that he is obliged to proceed if he is satisfied.

What is unusual about this case is that the rent officer was satisfied. He expressed himself as being satisfied that this was a protected tenancy. He remained satisfied at all times and it is not possible to attack his satisfaction as it was possible in *Noel's* case. There it was said that the rent officer could not be satisfied because he did not have any evidence on which to be satisfied. There was evidence in this case and he was satisfied. Let me make it clear, in order that nobody should be influenced by anything that I say, that he may have been wholly wrong in being so satisfied. That is a matter which the county court judge will have to determine eventually, but he was satisfied and, being satisfied, I think it was his duty to go ahead and register this rent. If he does not do so and there is any delay in the county court deciding the matter, it looks as if these applicants will suffer a real diminution in their rights. But quite apart from that, I think that was the rent officer's duty. Accordingly, I would let an order go requiring him to determine and to register whatever fair rent he determines.

Agreeing, FORBES J said: I would only desire to add one short point because of a matter which arose in the course of argument. Of course in these cases the rent officer is often asked to determine a preliminary issue of fact: is this a protected tenancy or not? Such a determination is not conclusive, as my Lord has pointed out. The attitude which the rent officer ought to adopt is, I think, very shortly set out in the report of the case of *Ex parte Noel* at page 8 explaining the decision in *Ganatra*:

> What I understand Park J to have said was that if the rent officer, having gone into the matter, found that he was in doubt as to whether there was jurisdiction, so that he was not himself able to determine the matter, then he should leave it so that the parties could take the matter to the county court if they so chose.

Originally, on behalf of the rent officer it was argued in this case that, although the rent officer initially decided on this preliminary issue that it was in fact a protected tenancy, he subsequently changed his mind, but this argument has now been abandoned. The argument now is that the rent officer was indeed satisfied that it was a protected tenancy but nevertheless decided it was inappropriate to proceed at once to determine to register the rent. In my judgment, he was wrong about that, as my Lord has said. However, during the course of the original discussion on the argument as to whether he had changed his mind, the effect of the case of *Ex parte Noel* was said to be this, that the rent officer was precluded from deciding the question of whether it was a protected tenancy or not whenever the allegation was made that the tenancy document were a sham; and in all those cases, so it was argued or said, he should always leave it to the county court.

In my view, *Ex parte Noel* cannot be so read. It is clear to me that when one looks at the judgment of Slynn J at p 9, what he started with was saying that the rent officer was entitled to conclude that there was a protected tenancy, but in this particular case — *Ex parte Noel* — he went about it the wrong way. He attempted to decide that issue on submissions, and at F on that page Slynn J said:

> In my judgment it is quite impossible to resolve a matter of this kind merely on the basis of submissions made and arguments put before a rent officer by the solicitors for the parties. An allegation was made that this document was

a cloak or a sham, a device to avoid the effect of the Rent Acts. That is a matter which could only properly be dealt with by the hearing of oral evidence, and if required, by subsequent cross-examination.

At the top of the next page he said:

In my judgment the procedure which was adopted on this occasion is not at all satisfactory and the rent officer was not, because of the procedure adopted, in a position to come to a determination of this kind in the way that he did.

As I say, I read that as simply an indication that in that particular case the rent officer went about it the wrong way because he attempted to deal with a question which involved fraud or something of that kind — at any rate the document was a sham — not on oral evidence at all but on submissions by solicitors and one could not determine a matter of that kind merely on such submissions. But where the rent officer goes into the matter and hears evidence, as he did in this case, and hears cross-examination of the applicants, it seems to me there is nothing in *Ex parte Noel* which prevents him, if he wishes, from determining that question, as indeed he did in this case. I agree with my Lord that the order should go in the way he has indicated.

The application for an order of mandamus was granted with costs.

QUEEN'S BENCH DIVISION

November 7 1980

(Before Mr Justice GLIDEWELL)

R v RENT OFFICER FOR LONDON BOROUGH OF CAMDEN, EX PARTE PLANT AND OTHERS

Estates Gazette February 21 1981

(1980) 257 EG 713

Rent Act 1977 — Whether a letting a holiday letting for the purpose of section 9 — Case also illustrating legal questions affecting a rent officer's duty when faced with jurisdictional issues; the use of the procedure of judicial review to determine private disputes, requiring deponents to affidavits to be examined orally in court and subjected to cross-examination; and an exceptional resort to the remedy of declaration under Order 53, rule 1 (2) to decide a matter suitable for determination by the county court under section 141 of the Rent Act 1977 — Applications against rent officer for mandamus to hold a consultation and for a declaration that various applicants were protected or statutory tenants — Object to secure a registration of fair rent before November 28 1980 when it was expected that certain changes made by the Housing Act 1980 would come into operation, restricting inter alia a tenant's right to recover rent paid in excess of the registered rent to a period beginning with the date of registration instead of the date of application to the rent officer — Held on the evidence that the only tenancy with which the court was concerned was not for the purpose of occupation for a holiday within section 9 — The tenant and her fellow occupants were student nurses and students of psychology and dentistry pursuing their various courses with a view to the securing of their professional qualifications — Held also that in this quite exceptional case, where delay might lead to injustice, the court would grant an order of mandamus to the rent officer to hold a consultation and register a rent and a declaration that the applicant was a protected tenant within the meaning of the Rent Act 1977

In these proceedings five applicants occupying a flat at 72b Fleet Road, Hampstead, London NW3, sought an order of mandamus directed to a rent officer for the London Borough of Camden to direct him to take steps to hold a consultation under the provisions of the Rent Act 1977, or alternatively to register a fair rent under the Act. They also sought a declaration that the applicants were protected or statutory tenants. It was established during the hearing that the court was in fact concerned with the position of only one tenancy, that of Miss Hilary Plant, a student nurse. The issue was whether this tenancy was protected or was excluded from protection under section 9 of the 1977 Act because the purpose was to confer on the tenant the right to occupy the dwelling-house for a holiday.

Mrs T Moorhouse (instructed by Nic Madge, Camden Community Law Centre) appeared on behalf of the applicants; J Goudie (instructed by Philip Thornton & Co) represented the respondent landlord, Peter Simon; John Laws (instructed by the Treasury Solicitor) represented the rent officer.

Giving judgment, GLIDEWELL J said: These five applicants occupy a flat at 72b Fleet Road, Hampstead NW3, which is owned by the second respondent, Peter Simon. They claim that they all are, or alternatively that one of them, Miss Plant, is, protected tenants or a protected tenant of that flat under the Rent Act 1977 and thus are entitled by law to apply to the rent officer to register a fair rent. Mr Simon maintains that they are not protected tenants because, as he accurately points out, the written agreement which Miss Plant signed in October 1979 specifically says that the tenant has the right to occupy and use the flat for the purposes of a holiday.

The applicants have applied to the first respondent, the rent officer for the London Borough of Camden, to register a fair rent. For reasons which I shall explain shortly, he has declined to take the necessary steps towards doing so. The applicants therefore come before me seeking orders first of mandamus directed to the rent officer requiring him to serve a notice on the applicants and upon Mr Simon to hold a consultation under the appropriate provisions of the 1977 Rent Act, or alternatively, to register under the appropriate provisions of that Act a fair rent; and, secondly, for a declaration that the applicants are protected or statutory tenants within the meaning of the Rent Act 1977.

The issue between the applicants and Mr Simon is whether or not all the applicants are, or Miss Plant is, a tenant to whom section 9 of the 1977 Rent Act applies in relation to the tenancy. Section 9 of the Act reads:

A tenancy is not a protected tenancy if the purpose of the tenancy is to confer on the tenant the right to occupy the dwelling-house for a holiday.

There has, not unnaturally, grown up the use of the expression, "a holiday letting", but like all short expressions it is capable of being misunderstood, and what the court has to do is to look back to the wording of the section. I have to, if I grapple with this question at all, as I have already indicated I am going to, decide whether the purpose of the tenancy was to confer on the tenant the right to occupy the dwelling-house for a holiday.

Obviously, where this question is in issue, it must be determined by somebody before the rent officer has jurisdiction to proceed with the process of registering a fair rent. When the facts are put before the rent officer he may, on those facts, determine and decide himself that he either has or has not jurisdiction. Certainly if he decides that it is a letting that falls within section 9 and thus he has not got jurisdiction, he is entitled, and there are decisions of the courts to the effect that he should so decide, to decline to register the rent.

Mr Goudie urges that if the facts tend to show that the tenancy may not fall within section 9, it would be at any rate ill-advised of the rent officer to make a decision at all, but certainly he might do so. But if he finds that he has any real difficulty in deciding that matter, then the court has said that he should not do so; that he should decline to do so; and leave whichever party has made the application to take appropriate steps. Now whether or not he does make a decision, and whatever decision he makes, there is a procedure for dealing with the matter in the court, because section 141 of the 1977 Act specifically provides for an application to be made to the county court to decide a series of different questions arising under the Act, one of which is as to whether the tenancy is a protected tenancy. And thus the question whether or not a tenancy falls within section 9 of the Act, which is the point under consideration here, which may or may not result in its being a protected tenancy, is one which could, and in the ordinary case should, be determined by way of an application to the county court under section 141.

I have been referred in Mr Laws's very clear argument — and having characterised his argument in that way, may I say that all

A learned counsel have been of the greatest assistance to me in the clarity of the arguments that they have put before me — I have been referred by Mr Laws to three authorities that deal with the question "What is the proper procedure for determining an issue such as this?" They are: *R v Fulham, Hammersmith and Kensington Rent Tribunal, ex parte Zerek* [1951] 2 KB 1, *R v Rent Officer for Kensington and Chelsea, ex parte Noel* [1977] 1 All ER 356, and a very recent decision (and by recent, I mean last week) of the Divisional Court, Donaldson L J and Forbes J in the case of *R v Rent Officer for London Borough of Camden, ex parte Ebiri**, in which, taking the three decisions together, the court laid down the principles that I have just sought to set out, that is to say, that the rent officer is B perfectly entitled, if he has any real doubt about the matter, to say, "No, I am not going to decide this matter, and if you want to get it decided you must go to the county court under section 141." The rent officer does not say that in terms of course, but that effectively is the result of his not deciding it.

In this case, as far as the rent officer is concerned, I deal first with a preliminary point raised by Mr Laws. On January 7 1980, Mr Madge of the Camden Community Law Centre, on behalf of the applicants, wrote to the rent officer and enclosed an application form completed on behalf of what he described as the tenants, and signed by all of them, all the applicants and, indeed, a Mr Gray who was there at the time, for the registration of the rent, and said:

> C As you will see they have signed two holiday let agreements. Nevertheless we are of the opinion that all six tenants are joint regulated tenants. We have applied on their behalf for legal aid to obtain a declaration under section 141 of the Rent Act to this effect.

The rent officer apparently replied by telephone and then, in the letter of March 14 1980, Mr Madge set out his argument much more fully as to why the rent officer should deal with this matter; and on March 25 the rent officer replied directly to Miss Plant and, after having referred to the fact that the law centre were proposing to apply to the county court for a declaration under section 141, the rent officer said:

> D There seems little point in inspecting the premises or holding a consultation until these matters have been resolved. The inspection and consultation dates are therefore cancelled. As soon as I hear that the courts have reached a decision I will arrange other dates.

At that time it was thought, and correctly thought, that if the rent were registered at some future time and the rent that was registered proved to be less than the rent that Miss Plant was currently paying, then that decision would be back-dated to the date of the application. That was based upon what was, and indeed at this present moment still is, the law contained in section 72 of the 1977 Act, which effectively provides that in the normal case where an application has been made, the effect of the decision backdates to the date E of the application. But that provision is to be altered by section 61 of the Housing Act of 1980. Substantial parts of the Housing Act 1980 are already in force, but parts of it have not yet been brought into force. Section 61 is one of the parts that has not been brought into force, but I am told that by that process of osmosis which seems to be the process by which government now makes known the times at which statutes are to come into force, it is thought that section 72 is going to take effect or be brought into force on, I think it is the 28th of this month, but, the end of this month. That may or may not prove to be the case. Certainly there is an anticipation that it will not be too distant. The effect of the substantial alteration effected by section 61 of the 1980 Act on section 72 of the 1977 Act will most F probably be, and certainly Mrs Moorhouse argues that it will be, that when a fair rent is registered, that will take effect only on the date of registration and not be back-dated to the date of the application.

Now those provisions of what were then the Housing Bill became known to the Camden Law Centre, who know more about these things than most people in this country (the rent officers apart of course), in June of this year, and they immediately appreciated that if they did not get to the county court for some time and then got a determination in favour of the applicant or applicants, and then got their fair rent registered, there was a very real risk that while it

*Reported at (1980) 257 EG 605.

would be registered it would only take effect at the date of the G registration. And so there ensued further correspondence in June and July of this year between Mr Madge and the rent officer; effectively Mr Madge doing his best to persuade the rent officer to proceed with the procedure for registering a fair rent and the rent officer saying that he saw no reason to change his mind because legislation was likely to alter the situation.

The last letter from the rent officer is one dated July 28 1980, in which he said:

> I do not feel disposed to reconvene the consultation because your client might be disadvantaged by future legislation. The matter will remain adjourned.

H

And so the applicants then made application for judicial review and that application was initiated by notice given on August 11 1980. The application for leave was heard by McNeill J.

Perhaps I should say this before I come to McNeill J's decision. If the matter had remained there, that is to say, if all that had been in issue as it was at the time when the application for leave was made was the question whether or not the rent officer should be ordered to get on with the procedure under the Rent Act, the question under section 9 not having been determined, and no application being made to this court to determine it — if that had been the position, I would have been inclined to the view, although I am not going to decide it finally, that mandamus would not issue against the J rent officer. In other words I take the view that he was entitled to adopt the attitude he did that unless and until the matter had been decided by the county court or, as it turns out, this court, he was not going to proceed, and I do not think it would have been right to say "you must proceed" even though the question which is in issue has not been decided.

When the application for leave was made to McNeill J on August 14 1980, he pointed out, as I understand it, that under the procedure of Order 53, which was introduced in 1977 by an amendment of the rules, on an application for judicial review the applicant can now apply not merely for an order of mandamus, prohibition or certiorari or an injunction in the particular circumstances, but can K also apply for a declaration. Rule 1(2) of Order 53 sets out the circumstances in which a declaration may be sought. McNeill J referred to the particular circumstances of this case where an application had been made by Miss Plant, or on her behalf, for legal aid to pursue her section 141 application, that that application for legal aid has still not been determined. McNeill J therefore suggested that there should be added to the application for judicial review an application for a declaration in the terms which now appear, that is to say, a declaration that the applicants are protected or statutory tenants.

Mr Laws's preliminary point is that the procedure under Order 53 should not be used to determine private disputes. It should be L used only to determine public issues. He drew my attention, in relation to this matter, to the decision of the Court of Appeal in a case called *George v Secretary of State for the Environment* (1979) 38 P & CR 609, where at p 615 Lord Denning MR said:

> It seems to me that, in general, cross-examination should not be allowed in prerogative writ proceedings or in proceedings for judicial review or in applications like this for [confirmation of] compulsory purchase orders. There are very good reasons for this rule. First, usually the affidavits speak as to what took place before the determining body. It may have been before a planning inspector, or a magistrate, or someone of that kind. He may have to make an affidavit as to what took place before him. It is undesirable that M he should be subjected to cross-examination. We said so in a recent case about a certification officer.

And then he went on:

> Secondly, experience shows that on these procedural questions there is very little conflict on the affidavits. Thirdly, one party or the other might, by way of cross-examination, try to undermine the actual findings of the inspector or other officer holding the inquiry.

It should be pointed out that that particular case was not a case of judicial review at all but was an appeal to the court under the specific statutory appeal procedure under the Acquisition of Land (Authorisation Procedure) Act 1946. In other words it was one that concerned a compulsory purchase matter and it did not fall within Order 53 at all, and, with the greatest respect to his Lordship, when he was referring, as he did in that passage which I have just read, to "prerogative writ proceedings or in proceedings for judicial review or in applications like this", what he said was strictly *obiter*, and it is noteworthy that though Roskill and Cumming-Bruce LJJ, who were the other members of the court, agreed with Lord Denning on his conclusion in relation to that particular case, they did not say anything about judicial review. I say only this, that it may be, with the widening of the scope of what was formerly the prerogative orders procedure by the inclusion within Order 53 of applications for declaration, that it is right that it should become somewhat more common for matters to be dealt with by way of not merely affidavit, but cross-examination. But having said that, I do accept that it is only in an exceptional case that what is essentially a private issue should be litigated on a procedure designed to deal with public rights. To that extent I am entirely with Mr Laws.

I do not, of course, know why McNeill J took the step he did, but I imagine his thinking went something like this — and certainly having, so to speak, retrospectively tried to think myself into his mind, my thinking might have gone something like this: the law in general is opposed to delay, contrary to an impression which may prevail. The courts should decide issues as speedily as possible. Sometimes it does not greatly matter whether an issue is decided at the beginning of this year or the end of this year or even next year. If the decision is delayed it may mean that one party has to pay some more interest, or it may have its effect on costs, but apart from that it may not greatly matter, but there are many situations in which delay does work to the advantage of one party and the disadvantage of the other, and if that is the situation, then it is a general principle of law that delay is unjust and should not be allowed if it can be avoided. In a totally different field, for instance, the courts relying upon that principle will sometimes strike out an action where they take the view that a plaintiff has unduly delayed in prosecuting it, if, but only if, they come to the conclusion that the other party is prejudiced by that delay. McNeill J, I imagine, took the view, since the section 141 proceedings in the county court had not even been started, that while he was giving leave for judicial review, in relation to the mandamus, since the procedure could be used to determine the question under section 9 of the 1977 Act, that would most probably be the quickest way of getting the matter dealt with, and so it has proved, and thus he may well have thought that justice would be served in the particular circumstances of this case by suggesting the addition of an application for a declaration.

If he did think that, I agree with him, but I emphasise that the circumstances of this case are that, for reasons which I do not wholly understand, the application for legal aid for these section 141 proceedings were not dealt with for six months. For that reason, and in the particular circumstances of this case, I think it right to go on and deal with the determination of the question under section 9. But I emphasise that this should not be general practice, and if the rent officer is concerned, as I understand he is, that a decision in this case will, as it were, open the floodgates to applications for declarations, instead of section 141 determinations, it is my firm view that that should not normally be the position. I therefore propose to decide the issue as to whether or not this tenancy came within section 9. I say "this tenancy". I emphasise that I am clearly of the view that what I am concerned with is the tenancy to which the agreement dated October 28 1979, signed by Miss Plant, relates, because whether or not the tenancy is protected, that is the tenancy that is or is not protected. The facts as to the agreement of April 1979 are merely background information, which are important but not part of the actual tenancy agreement itself.

As to the facts relating to the entry into the agreement dated October 28 1979, there is in the end no dispute as to what I consider the material facts. There is a dispute about peripheral matters. I do not consider that where there is an area of dispute it relates to anything that is essential to my determination of the question. There are now five applicants because one of the two young men who were formerly in these premises has left. The four ladies are all, and were in April 1979, student nurses, and they all started their training at University College Hospital simultaneously in February 1978. It is a three-year training and, barring accidents, they will all qualify as state registered nurses in February 1981. Thereafter, I was informed, they would normally expect to stay on at the hospital as staff nurses for, at any rate, a period of time. That is as it may be, but it is quite clear that what they were concerned with was accommodation which would, they hoped, be available to them at any rate until 1981. In the spring of 1979 since they started their training, if I understand the evidence correctly, they had all been living in nurses' hostels and they had together concluded that they would quite like a change of scene, and so they decided that they would all look for a flat. They studied advertisements and they no doubt made some enquiries, and on Friday, April 27, they saw an advertisement in the *Evening Standard* which undoubtedly was an advertisement of this flat, and the cryptic message conveyed by the advertisement did not say anything about a holiday letting.

Mr Simon has put before me in his affidavit and in his oral evidence, and I accept, that the advertisement he sought to place with the *Evening Standard*, and, indeed, did place with the *Evening Standard*, which appeared in the editions of that paper earlier that week, did refer to this as being a holiday letting. How it comes about that the advertisement that appeared on the Friday did not, I have no idea. But it does not seem to me to matter whether these young ladies when they first saw the advertisement realised that what was being offered was a holiday letting or not.

On the Friday, two of them, Miss Plant and Miss Nelson, went to the premises; they saw the flat; they there met a Miss Halpen who was acting as Mr Simon's agent. She showed them around and one of them had a telephone conversation from the flat with Mr Simon. Exactly what transpired in that telephone conversation does not matter, but it is clear that they were granted what in law was an option for 24 hours to take the tenancy on payment of £20, which they did pay. It was not a deposit; it was an option to grant them the right to enter into the tenancy agreement and not to enter into such an agreement with anybody else for 24 hours. Mr Simon honoured that option and the next day they, having found two young men who were willing to occupy these premises with them, I think all six of them (certainly most of the six) went to the premises on Saturday 28 and they there met Mr Simon.

There is no doubt that he asked each of them to complete a form, of which he had copies, setting out details of their particular circumstances, and each of them did complete such a form and the copies have been produced to me. The form records their names; their present addresses which are shown as, in the case of the girls, the nurses' homes in which they were then residing; the name and address of their parents or next of kin, and in each case I think I am right in saying it is their parents; the name and address of the employers; the occupation, and this is very important in the circumstances of this case; the bankers; and another reference. Against "occupation" all the girls wrote "student nurse", and the young men wrote that they were students, one of psychology and one a dental student.

Mr Simon produced a copy of the tenancy agreement and he allowed them all — indeed, I think he invited them all — to read it, and many of them did read it. There is no suggestion here that it was simply put in front of them and they signed it without knowing what was in it. Certainly those who matter read it, and at the top of it, it says:

> THE Tenant shall have the right to occupy and use for a holiday the suite of rooms or flat consisting of 5 rooms numbered 72b . . . 72 Fleet Rd, NW3.

The evidence is that Miss Nelson, who was the lady who was going to enter into the agreement at that stage, in effect said, "What is all this about a holiday letting?" She says that Mr Simon answered that it was a protection against having bad tenants for himself and for them.

Mr Simon's account of that is somewhat different. He says that he made it clear that it was to be a holiday letting only. He also says that it was only after Miss Nelson had signed, as she undoubtedly did sign, that tenancy agreement that the detailed particulars were entered upon the forms. I find that a little surprising because, after

A all, the purpose of the forms was to enable him to find out as much as he could about the status, financial position and apparent respectability of these future tenants, and you would have thought that he would have wanted to have assured himself about that before he entered into the agreement and not afterwards. But again, it does not matter because it is October I am concerned with, not April, and what is quite clear is that from April 28 onwards on the one hand all the applicants knew that the form of agreement said in terms "This is to occupy and use for a holiday"; on the other hand it is quite clear, and Mr Simon so told me in evidence (he did not pretend otherwise — he was quite frank about it), that from then onwards he had in his possession these documents which indicated
B their occupations. He knew they were student nurses or students of psychology and dentistry and he knew they were working in that capacity and that that is what they were going to be doing when they occupied his flat. So, when we come to October that was the state of knowledge of the parties. They all knew perfectly well that in fact the occupants were going to be working as students.

What happened in the autumn seems to be this: that at some time in September the question of renewal of this agreement arose (I should have said that the April agreement was for six months), and Mr Simon indicated at some stage that he was not willing to have Miss Nelson, not that he had anything against Miss Nelson but, because of his desire to ensure that this was so far as possible
C outside section 9 of the Rent Act, he wanted to have a different tenant, a different signatory to the tenancy agreement. Moreover, he at that stage was not committed in any sense to letting to any of these six people; and indeed, he says, and I find, that he did advertise the flat again.

Having so advertised it again, in October there were some negotiations between them as to a reletting and eventually it was agreed that Miss Plant should take a new tenancy at an increased rent. There was a bit of haggling about the increased rent which seems to have gone on to the extent that one or two figures were entered on to the tenancy agreement itself and scratched out and other figures substituted. Be that as it may, in the end it was agreed
D that Miss Plant should take a tenancy for six months from October 28 at £474 per month. And so, on October 24 Mr Simon went round to the premises and I find that he had with him a copy, or rather two copies, original and counterpart, of the new tenancy agreement all ready filled in with the details, and he had filled it in, as far as Miss Plant was concerned, quite obviously by looking back at the copy of her particulars which was contained in his file (it was obvious he was meticulous about keeping these things) and there he found her occupation. It does not of course appear on the tenancy agreement but there it is; there it was in front of him; and from it he obtained the details about her, and one of the things he inserted was that her address was 4 St Pancras Way. Now that was, and he knew it was,
E inaccurate, because 4 St Pancras Way was the nurses' home that she had left in April. Mr Simon knew that, but he was putting this on, making it appear that she was a new tenant coming from outside, whereas in fact, of course, her address was Fleet Road, Hampstead, or her parents' address, whichever she cared to use. It was one or the other, but it certainly was not 4 St Pancras Way. At any rate, he took this form of agreement round; she signed, he signed; another month's rent in advance was paid and that was that.

I should have said that the way in which the rent was paid was usually that instead of the tenant paying the whole of the £425 a month as it originally was, and £474 as it became, each of them paid their appropriate share and they usually paid by way of cheques
F backed by bankers' cards, usually two cheques because the share of each of them was of course over £50 a month. There was a suggestion made that that meant, or could be interpreted as meaning, that each of them had a tenancy. I do not believe that. This was merely a convenient way of Mr Simon's getting a rent, getting it in a fairly secure way, but I think that the legal position is that Miss Plant was the tenant; Miss Nelson was the tenant from April to October; and all the others were paying their shares as agents for her. It was her responsibility; Miss Nelson's first and then Miss Plant's; and will be Miss Plant's for the future, to pay Mr Simon the rent. He would not be entitled to sue anybody else for it in my judgment. And that is how this agreement was entered into.

The law in relation to this matter, in so far as there is any decided G
authority, is contained in the decision of the Court of Appeal in a case called *Buchmann v May* [1978] 2 All ER 993. That was a case under the section of the 1974 Act which preceded section 9. The reason why section 9 came into the law in the 1974 Act is that that was when Parliament thought it right to bring furnished lettings within the control of the Rent Acts instead of the different form of control to which they had previously been subjected. Having done so, Parliament also obviously thought it right that certain types of lettings should still be excluded from the Rent Act including lettings for the purpose of a holiday; and that is the genesis of section 9.

Buchmann v May was a case in which a county court judge, on an application under the precursor of section 141, had decided the H
question whether there was a holiday letting or no, but he had apparently decided it purely on the oral evidence without any reference to the written agreement at all, and there was no reference in his judgment to the written tenancy agreement. The gist of the decision is that where there is a written agreement then the first thing the court must do is to look at the written agreement, and if the written agreement says that this is a letting for the purpose of a holiday, then *prima facie* that is the purpose of the letting. The court is entitled and, indeed, obliged to go on and consider other evidence to see whether that represents the reality of the situation, but it is only if there is clear evidence contradicting the words of the agreement that the court should decide otherwise; and that is to be found J
in passages to which I have been referred in the judgments of Sir John Pennycuick, who gave the first judgment at p 998, where he said:

> Where parties to an instrument express their purpose in entering into the transaction effected by it, or the purposes for which, in the case of a tenancy agreement, the demised property is to be used, this expression of purpose is at least *prima facie* evidence of their true purpose and as such can only be displaced by evidence that the express purpose does not represent the true purpose.

And then two sentences later: K

> When I say the express purpose does not represent the true purpose, I mean that the express purpose does not correspond to the true purpose, whether the express purpose is a deliberate sham or merely a false label in the sense of a mistake in expression of intention.

Stephenson and Megaw LJJ, who gave much shorter judgments agreeing with Sir John Pennycuick's, said somewhat the same thing in terms which Mr Goudie has suggested to me really related only to sham, rather than to the alternative of a false label in the sense of a mistake in the expression of the intention. I do not accept Mr Goudie's submission in that respect. I believe that, when read properly, all their Lordships were saying the same thing in their judgments though, not unnaturally, they expressed it in slightly L
different ways. That is not merely binding on me, but of course I entirely accept that that is the approach that I must adopt, and it is the approach that I do adopt.

Prima facie this agreement of October 28 1979 was to occupy and use this flat for a holiday, but I have clear evidence that both parties knew perfectly well in October that none of the persons who were going to occupy this flat were going to occupy it for the purposes of a holiday. I have clear evidence that all the parties knew they were going to occupy it for the purposes of their work as students. I so find and that seems to me to be conclusive of the matter.

One does not need to debate precisely what is meant by the intention of Mr Simon. As I have already said, I am satisfied that Mr M
Simon's desire was quite properly so to arrange affairs that lettings of this flat did not fall within the Rent Acts. In the particular circumstances of this case, perhaps because he found these tenants satisfactory, agreeable and respectable tenants, I know not, but for whatever reason, his letting in October 1979 failed to achieve his desire, in my view, and I find that there is clear evidence that the purpose expressed in the tenancy agreement was not the true purpose of that agreement. In arriving at that conclusion I have not found it necessary, to use the words of the present Lord Chief Justice, to bend over backwards. I have managed to remain reasonably upright while achieving that result. I have taken note, in,

A arriving at it, of the other authority to which Mr Goudie drew my attention, *Aldrington Garages v Fielder* (1978) 37 P & CR 461.

It follows that I find that Miss Plant entered into a tenancy agreement on October 28 1979 of which the purpose was not to occupy the dwelling-house for a holiday. It did not fall within section 9. It follows that, in my judgment, it was, and is, a protected tenancy.

The question then arises and has been debated, what orders I should make. Mr Laws invited me not to make any order. He said, having expressed the view I have just expressed, the rent officer would follow it and, of course, I am sure he will. I have no doubt about that at all. Mr Laws also submitted to me that it was inappro-
B priate simply to make a declaration, and on the whole I agree with that for the reasons I have already sought to express, though I hope I have made it clear that the facts of this case I regard as being quite exceptional, and I am sure, among others, those who instruct Mrs Moorhouse will recognise that coming to this court is not to be adopted as anything like the normal procedure for dealing with this particular matter.

I conclude that it really would be best if instead of leaving the matter in the air I did grant both the remedies sought. I mean no disrespect to the rent officer in doing so. I hope that he does not feel that in any sense I am being offensive to him in making an order for mandamus. It is in my judgment the best way of clearing the matter
C and I therefore grant the relief sought by Mrs Moorhouse. I grant an order of mandamus to the rent officer to hold a consultation and thereafter to proceed to register a rent and I grant a declaration that the applicant is a protected tenant within the meaning of the Rent Act 1977.

The judge made an order for costs in favour of Miss Plant against the landlord, Mr Simon, limited to the costs of a one-day hearing on the county court scale. No order for costs was made against the rent officer. At a subsequent application the judge varied his original order by directing that the rent officer should complete his consultation and registration by the end of the day of November 27 1980.

D

COURT OF APPEAL

October 27 1980

(before Lord Justice WALLER, Lord Justice ACKNER and Lord Justice WATKINS)

FLETCHER AND OTHERS v DAVIES

Estates Gazette March 21 1981

(1980) 257 EG 1149

Rent Act 1977 — Tenant's bankruptcy — Question whether tenant of Inner Temple was at date of bankruptcy a contractual or statutory tenant — If contractual, his tenancy vested in his trustee in bank-
E ruptcy, who disclaimed the tenancy — In that case the bankrupt, having ceased to be the tenant, no longer had any of the rights given to tenants by the Rent Act — If, however, the bankrupt's tenancy had been statutory by the date of bankruptcy, the statutory tenancy did not vest in the trustee and the fact of bankruptcy would not have affected the tenant's personal right to remain in occupation — Evidence that the Inn and the tenant had entered into a new contract of tenancy by the date of the adjudication in bankruptcy — Hence the tenancy vested in the trustee and, when disclaimed, left the bankrupt with no defence to a claim for possession — Appeal from decision of county court judge, who had taken the view that there had been no completed new contract, allowed

F This was an appeal from a decision of Judge Phelan at the Mayor's and City of London Court. The appellants, plaintiffs in the county court action, were the trustees of the Inner Temple and they appealed against the rejection by Judge Phelan of their claim to possession of a flat in the Temple occupied by the respondent, Mr Davies, the defendant in the county court action. The flat was subject to a regulated tenancy and the main question was whether on July 14 1978, when Mr Davies was adjudicated bankrupt, he was a contractual or a statutory tenant.

D J Nicholls QC and P Talbot (instructed by Burton, Yeates & Hart) appeared on behalf of the appellants; L J Bromley QC and C Heath (instructed by Hillearys) represented the respondent.

Giving judgment, WALLER LJ said: This is an appeal from a G
judgment of Judge Phelan at the Mayor's and City of London Court. The plaintiffs are trustees for the Inner Temple and claim possession of a flat from the defendant. The main question before the learned judge was whether the defendant was a contractual tenant on July 14 1978, or not. The judge found he was not and accordingly found against the plaintiffs. If the learned judge had found that the defendant was a contractual tenant, then because he had been adjudicated bankrupt and because his trustee in bankruptcy had disclaimed the lease, the defendant would have been a mere licensee and the plaintiffs would have been entitled to possession (see *Smalley v Quarrier* [1975] 1 WLR 938).

The facts were that the defendant had originally entered into a H
quarterly tenancy in April 1959, and the tenancy agreement was in the standard form for all residential tenancies in the Inner Temple and could be determined by a quarter's notice on either side. The defendant had continued to be a tenant, although the rent had altered marginally in the meantime, until 1977 when the first event took place. In October 1977 the Commissioners of Inland Revenue filed a notice against the defendant, and on February 6 1978 a receiving order was made. That was notified to the Inn. At that time the defendant was a contractual tenant and in March the plaintiffs were minded to increase the rent of all their residential tenants and so they sent to the defendant, among others, a notice to quit accompanied by a letter explaining that it was not intended that that J
should be taken for the purpose of getting possession of the flat, but merely for the purpose of putting up the rent. The covering letter finished with the words: "I should be glad to hear that you are willing to accept a new tenancy on the terms proposed and I attach a form for your signature and return, together with a form notifying the names of your sureties". There had been two sureties to that original tenancy agreement, which I have already mentioned.

On May 14 the defendant returned the form which read: "I agree to the new tenancy operative from March 25 in terms set out in your letter dated March 15. I enclose the form with regard to my sureties."

So that there was, at that stage, probably an offer and acceptance K
of the new tenancy. But, on May 17 after a meeting of the Estates Committee, the plaintiffs, through the sub-treasurer, wrote to the defendant thanking him for his letter and saying: "As you will appreciate, the Estates Committee can take no steps with regard to a reissue of the tenancy while the receiving order is in force". A carbon copy of that letter shows, by notes on it, what happened during the next few weeks. The first note was from the sub-treasurer to Miss Bray saying "No agreement to be issued at present". That was simply an instruction to comply with the letter, and then there is a later note, probably written on June 26, saying "Agreement to be issued", and on June 27 the standard printed forms were prepared by Miss Bray and handed to the sub-treasurer. L
The whole of the form was filled in, except for the date.

On June 27 the sub-treasurer pushed through the defendant's letterbox the two forms.

The question which arises, or which first arose for the decision of the learned judge, was whether that was a new offer which could be accepted by the defendant, or was something more required from the Inn? What in fact happened was that the next day the defendant went to the treasury. He had the two blue forms unsigned; he met Commander Flynn, the sub-treasurer, and Admiral Homan his successor. The defendant signed the forms, and he signed both copies in the treasury. He then took them away for signature by the sureties with a view to being returned to the sub-treasurer, and in M
the next two days both sureties signed — one of them in his own office — Mr Giles, the solicitor — and the other brought it to the treasury to sign.

So at that stage both copies of the tenancy agreement had been signed by the defendant and by both his sureties. They were brought to the office; Commander Flynn, the sub-treasurer, signed them where there is a printed space for him to sign.

It was the practice of the Inner Temple to send a copy to the tenant, but on this occasion it was not done, in error. The learned judge in fact found that that was a fatal error in this particular case, but I would just continue with the narrative of the facts.

A On July 14, the defendant was adjudged bankrupt, and on September 12 the official receiver disclaimed that new agreement — namely the one which I have just been describing. On October 20 the sub-treasurer wrote asking for possession. On November 15 the defendant wrote back saying that they were not entitled to possession because he had accepted the new contractual tenancy, and on November 28 his solicitors wrote a similar letter, in the course of which they mentioned the new lease.

The learned judge took the view that there was no completed contract. He said, in relation to this matter, describing the events which I have already mentioned: "Later the plaintiffs again changed their mind and in consequence a form of agreement was B prepared. This is corroborated by Miss Bray's note", and then he explains how the sub-treasurer put it in the defendant's letterbox, and said: "I hold that that constituted an offer to treat on the part of the plaintiffs". Being an offer to treat, the learned judge came to the conclusion that if there was an acceptance it was never communicated, and therefore there was no contract.

The appellants submit that there was a completed contract when the defendant signed the agreement, and the two trustees had signed. The fact of the original notice to quit and acceptance, followed by the letter and finally by the two blue forms of agreement being put through his letterbox, constituted, submit the appellants, an offer.

C On behalf of the respondent, the defendant, it is submitted that it was the defendant who made the offer; that there should have been an acceptance by the appellants, and that acceptance should have been notified to the respondent. Emphasis was laid, in argument, on the fact that there were two copies of the agreement, both copies being signed, which indicated that it was implicit that the return of one copy was necessary to show acceptance on behalf of the appellants. It was submitted, in support of that, that these two separate copies were similar to a two-part contract where one party signs one part, and the other party signs the other part and then exchange takes place, and it was submitted in this case there should have been that exchange.

D In my opinion the mode of completing a contract depends upon the intention of the parties. What was the intention of the parties in this case? On March 15 the defendant was a contractual tenant; the Inn had served him with notice to quit but indicated it was only in order to increase the rent. The defendant accepted that, and then the appellants wrote saying that while the receiving order was in force, they could not go on with the increased rent, but then they changed their mind and put those two copies through the defendant's letterbox. In my opinion, this constituted an offer of a new tenancy in the same terms as before but at an increased rent. All that was required to complete the contract was the return of the papers with the defendant's signature and the signature of the two E guarantors. When this was done, there was a completed contract. It was, of course, unfortunate that the plaintiffs omitted to send a copy to the defendant, but that in no way invalidated the acceptance.

That this was the intention of the parties is clearly shown by the defendant's letter of November 15 which I have already mentioned, because he says this: "I refer to your letter of October 20 1978, purporting to give me notice to quit the above premises by the 20th ultimo". That was the letter which was sent after the bankruptcy. "I write to say that my original contractual tenancy was recently terminated by notice to quit for the express and no other purpose of collecting an increase of rent, following which a new contractual tenancy was offered to me which I accepted. At the time this F tenancy was offered a receiving order was already in force to the certain knowledge of Commander Flynn and I therefore contend that this new tenancy cannot now be repudiated on that ground". So it is clear that the defendant was saying that the new contractual tenancy, which was offered to him, was accepted by him — in other words, that there was a completed contract and the same position is set out in the letter from the defendant's solicitors dated November 28 in reply to a letter from the appellants. They say: "Before we deal with the question of law" — that is the question of law relating to a bankrupt tenant whose lease has been disclaimed — "may we point out that the existing lease is a new lease granted to our client by the Honourable Society of the Inner Temple after they were aware that our client had a receiving order made against him." G

In my judgment the learned judge came to a wrong conclusion when he concluded that there was no completed contract. In my judgment there *was* a completed contract; it was completed when the defendant returned the two blue printed copies with his own signature and when the signature of the two guarantors were obtained.

So thereafter he was, once the contractual tenancy was disclaimed by his trustee in bankruptcy, a mere licensee. If that view is correct, the defendant submits that the plaintiffs are estopped from acting on the trustee's disclaimer. The rule upon which he relies is conveniently set out in *Snell's Principles of Equity*, 27th ed, p 563:

> Where by his words or conduct one party to a transaction makes to the other H an unambiguous promise or assurance which is intended to affect the legal relations between them (whether contractual or otherwise) and the other party acts upon it, altering his position to his detriment, the party making the promise or assurance will not be permitted to act inconsistently with it.

In my judgment, there was nothing to give rise to an estoppel in this case. Both sides knew exactly what they were doing when the new tenancy was issued. The defendant, if he had not accepted the new lease, might have been made to pay a fair rent, that is, much more than the Inner Temple were asking. I should have said the Inner Temple were asking him to increase the rent by stages, ultimately up to two-thirds of the fair rent, but starting at one-third, and so, if the matter had gone before a rent assessment committee, J the rent which would have been assessed would have been very much greater than was going to be asked for by the appellants, even after three years. The defendant might well have paid his debts by that time, and so not be made bankrupt, in which case he would have had this tenancy with all its advantages. I cannot see that any promise can be implied in the way in which it was submitted, and I agree the learned judge found as much, and I agree with his findings. For those reasons, I would allow this appeal and order that the plaintiffs are entitled to possession of the flat.

Agreeing, WATKINS LJ said: From time to time tenants of Inner Temple are invited to enter into new tenancies, primarily if not exclusively because the Inn seeks an increase in rent. There is no K doubt that the respondent wanted a fresh tenancy of his flat on terms which he knew were available for his acceptance. He had been a tenant of that flat for a considerable time. So he must have been fully aware of the usual form of negotiation adopted by the Inn for the grant of a fresh tenancy. In *Hollington Bros Ltd v Rhodes* [1951] 2 TLR 691 Harman J stated:

> It is a commonplace that in many, indeed, in most cases, where a lease is to be granted everything remains in negotiation until the lease and its counterpart are finally exchanged between the parties, and it seems to me impossible to spell out of the course of business which I have described, and which appears from the correspondence, anything more than the ordinary course of negotiations. L

But the form of negotiation adopted by the Inn was not regarded as finalised by the return to the tenant of the counterpart of the lease by the Inn.

The only safe inference, in my judgment, to be drawn from the evidence provided to the learned county court judge was that the parties had agreed to be bound once the two blue forms had been returned to the treasury at the Inn signed by the respondent and by his sureties. The absence of the signature of the sub-treasurer did not in any way affect the intention of the parties to be bound at that time; nor did the failure of the sub-treasurer to send a counterpart of the lease to the respondent. I agree with my Lord that this appeal M should be allowed.

Also agreeing, ACKNER LJ said: The correspondence at an early stage in the relevant history of this case involved a degree of informality. The defendant and the sub-treasurer were close neighbours in the Inner Temple and they were on christian name terms. But for this informality, the sub-treasurer, when he posted through the defendant's letterbox the two forms of agreement with the name not only of the defendant filled in but the proposed sureties, might well have enclosed a covering letter. If this letter had in substance stated that the Inn had changed its mind and that if the defendant (after he and his sureties signed the document) delivered

A it to the sub-treasurer, he, the sub-treasurer, would then sign on behalf of the plaintiffs, then there should clearly, in my view, be an offer by the Inn which on acceptance in the manner stated would have created a binding contract. Certainly Mr Heath, on behalf of the defendant, when he followed Mr Bromley was prepared to concede this.

Thus, the first question must be: In the light of the past history as summarised by my Lord, did the sub-treasurer, by his conduct in delivering documents by hand, represent that the Inn was making such an offer? I am quite satisfied he did and that the learned county court judge was in error in concluding that the sub-treasurer did no more than make an offer to treat.

B I am fortified in this view not only by the fact that the sub-treasurer did sign on behalf of the Inn immediately there was received by him the forms of agreement duly signed by the defendant and his sureties, but by the defendant's own letter of November 5 and that of his solicitors on November 28 written after they had been fully instructed. Those letters were referred to by my Lord in some detail. It is clear that those letters so interpreted the position. It therefore follows that the defendant, on June 28 1979 when he was adjudicated bankrupt, was a contractual tenant of the premises. That contractual tenancy vested in his trustee in bankruptcy; but after the trustee in bankruptcy disclaimed on September 12 the defendant ceased to be entitled to occupy the prem-
C ises.

Mr Bromley has submitted that the appellants are estopped, and I am bound to say I have had some difficulty in understanding this plea. Mr Bromley first submitted that the defendant acted to his detriment by entering into the tenancy agreement. I do not follow this. The facts are that until given notice to quit in March 1975 the defendant was paying one-third of the fair rent as defined by the Rent Acts. The notice to quit was designed merely to end his tenancy so that the grant of a new tenancy by the Inn could increase the rent, not up to the fair rent, but up to two-thirds of the fair rent as at 1978 — and only achieving this figure by spreading the increases over three years. Having regard to inflation, come 1980
D the defendant would probably still be paying less than one-half of the fair rent. The Inn were thus making him an extremely generous offer, as the defendant well appreciated by the expedition with which he signed the documents. He was, of course, perfectly entitled to refuse the offer and to remain in occupation as a statutory tenant, but a statutory tenant is not protected from paying the fair rent. He, the defendant, was only so protected if he entered into the new tenancy agreement on the terms offered by the Inn. If he refused to enter into that agreement, then he might well have been asked to pay the fair rent and, if he refused, required to do so if he desired to remain in occupation.

However, Mr Bromley maintained that in withdrawing their
E previous objection and making their offer by delivering by hand the printed forms of agreement, the Inn represented that they would accept the defendant as a tenant while a receiving order against him was subsisting. That may very well be, but that does not estop the Inn from relying upon a disclaimer by the trustee in bankruptcy when the defendant was thereafter adjudicated bankrupt. The question as to what the Inn would do if he was so adjudicated was never discussed. Like the learned county court judge, I conclude that there is no valid basis for the estoppel which the defendant seeks to raise. I would, accordingly, allow the appeal.

The appeal was allowed, possession of premises to be given three months from date of judgment. Legal aid taxation of the respon-
F dent's costs was ordered; not to be enforced without leave. Leave to appeal to House of Lords was refused.

COURT OF APPEAL G

February 16 1981

(Before Lord Justice ORMROD and Lord Justice GRIFFITHS)

WILLIAMS AND ANOTHER v KHAN

Estates Gazette May 9 1981

(1981) 258 EG 554

Rent Act 1977 and Housing Act 1957 — Effect of closing order on assessment of fair rent — Appeal from a decision of Stocker J, who had rejected a submission that as a matter of law the rent assessment committee should have determined the fair rent as nil or a nominal H amount — Court of Appeal agreed with judge, but it should be noted that actual ground of decision turned partly on a provision in section 72 of Rent Act 1977 which has been amended by Housing Act 1980 — The application to rent officer was received on December 14 1977 and under section 72 (before amendment) this was the normal effective date of registration of the rent, but the rent officer and committee had power to determine a later date — Closing order became operative on May 13 1978 and committee's decision was dated July 12 1978 — Committee determined that effective date should be December 14 1977 — At that date the closing order was not operative — The property was lawfully let and occupied up to May 13 1978 — Observations at end of judgment of Ormrod LJ as J to position under the amended section 72, which lays down the effective dates of registration with no discretion for rent officers or committees to alter them — It will be a question of deciding whether any rent is a fair rent for a property which is occupied by a tenant for the time being in contravention of section 27 of the 1957 Act in circumstances which expose him to a fine — Appeal dismissed

This was an appeal by Isaac Williams and Lucille Williams, former tenants of a house at 6 Clover Road, Small Heath, Birmingham, from a decision of Stocker J dismissing their appeal from a decision of a rent assessment committee of the West Midland Rent Assessment Panel. The committee had determined that the fair rent K payable to the landlord, Manir Khan, should be £5.75 per week. The facts and the relevant dates which are material are set out in the judgment of Ormrod LJ. A report of Stocker J's decision appeared at (1980) 255 EG 1208.

Andrew Arden (instructed by the Small Heath Community Law Centre, Birmingham) appeared on behalf of the appellants; the respondent landlord did not appear and was not represented.

Giving judgment, ORMROD LJ said: This is an appeal from a judgment of Stocker J on March 28 1980 sitting as an appellate tribunal under the Tribunals and Inquiries Act 1971, hearing an appeal from a rent assessment committee — a rent assessment committee of the West Midland Rent Assessment Panel. L

The facts in outline are these. The property in this case was a house known as 6 Clover Road, Small Heath, Birmingham. It was one of a number of similar houses in that area and on December 12 1977 an application was made by the tenant, Mr Williams, to the rent officer to register a fair rent. On March 23 1978 the local authority resolved to make a closing order under the Housing Act 1957 in respect of this property. On March 29 1978 the rent officer held the usual consultation on the premises and on April 10 1978 he registered the fair rent as he assessed it in the sum of £5.75 a week with effect from the date of the application to register the fair rent — though the actual date appearing is December 14 as opposed to December 12 1977, but I do not think it makes any M difference.

To that registration of fair rent objection was made by the tenant on April 18 1978. On the following day the local authority served the closing order on the landlord, who is the respondent to this appeal. That order under the provisions of the Housing Act 1957 became operative on May 13. Meanwhile the reference from the rent officer to the rent assessment committee was going through the usual machinery and it came before the rent assessment committee on July 3. They heard argument about it. They gave their decision on July 12 and they confirmed the rent officer's assessment of the fair rent and the date from which the rent was to be effective. By

A pure coincidence, on that same day — July 12 1978 — Mr Williams was rehoused by the local authority.

The matter was then taken, as I say, on March 28 1980 to appeal before the learned judge. The first point to be made is that the only ground of appeal in these matters is upon a point of law, so it has to be shown that the rent assessment committee were wrong in law in confirming the rent officer's registered fair rent. What is submitted is that they were wrong in law either because, the closing order having become operative before the rent assessment committee reached their decision, they were bound to hold that in fact the fair rent of these premises was nil or, alternatively, as Mr Arden has put it, as a matter of law they were bound to decide that the fair rent was B either nil or what he called in his notice of appeal a "nominal rent" or, alternatively, a very low rent. As between nil and a very low rent, he says it was for the rent assessment committee to decide. The second alternative way of putting the case is, I think, this. It amounts to saying that the rent assessment committee, assuming that they were not bound in law to assess the rent at nil, misdirected themselves and therefore failed to exercise their discretion in accordance with the provisions of the Act in that they came to the conclusion that the proper fair rent of this property was £5.75 a week.

The matter is one which is not very easy to deal with, partly because the respondent landlord has taken no part in these pro- C ceedings either below or here — for the very good reason, I imagine, that he would regard the expense of doing so as far outweighing any sum of money he might have to pay. Indeed, it is not very easy to see why the appellants are here because the amount involved for them is exceedingly small as it turns out, but we are told that the law centre has decided to bring this appeal at its own expense in order to obtain a decision. Whether that is within the function of a law centre or not is not for us to decide.

The appeal requires us to look a little closely at the Housing Act 1957. Mr Arden has most helpfully directed our attention to all the relevant provisions of this and the Rent Act 1977 in a very clear, lucid and succinct manner, for which we are grateful.

D It is important to bear in mind that we are talking here of "fair rent" and "houses unfit for human habitation". In both cases these are terms with legal definitions and it is very easy to forget that we are talking about terms of art. The test of unfitness for human habitation for the purposes of the Housing Act 1957 is to be found in section 4 of that Act. It reads:

> In determining for any of the purposes of this Act whether a house is unfit for human habitation, regard shall be had to its condition in respect of the following matters, that is to say:
> (a) repairs;
> (b) stability;
> E (c) freedom from damp;
> (cc) internal arrangement;
> (d) natural lighting;
> (e) ventilation;
> (f) water supply;
> (g) drainage and sanitary conveniences;
> (h) facilities for . . . preparation and cooking of food and for the disposal of waste water;
> and the house shall be deemed to be unfit for human habitation if and only if it is so far defective in one or more of the said matters that it is not reasonably suitable for occupation in that condition.

So we are talking about a habitable house which is not reasonably suitable for occupation in its present condition *as defined in the* F *Housing Act*.

There are alternative powers in the local authority. Section 9 gives them power to require the owner of the property to repair it within a reasonable time if the local authority is of the opinion that the property is capable of being repaired at reasonable expense. Section 16 gives the local authority power to accept an undertaking as to reconstruction and power to serve a notice upon him which is called a time and place notice, which is only marginally relevant to this case.

Section 17 is the operative one. It reads:

> If no such undertaking as is mentioned in the last foregoing section is accepted by the local authority, or if, in a case where they have accepted G such an undertaking —
> (a) any work to which the undertaking relates is not carried out within the specified period, or
> (b) the premises are at any time used in contravention of the term of the undertaking
> then, subject to the provisions of this section, the local authority shall forthwith make a demolition order for the demolition of the premises to which the notice given under the last foregoing section relates.

But there is power, instead of making a demolition order, to make a closing order. That comes in the proviso to subsection (1); it reads:

> Provided that if in the case of any house the local authority consider it H inexpedient to make a demoliton order having regard to the effect of the demolition of that house upon any other house or building, they may make a closing order as respect that house instead of a demolition order.

The effect of a closing order is set out in section 27. That reads:

> A closing order shall be an order prohibiting the use of the premises as respects which the order is made for any purpose other than a purpose approved by the local authority, and any person who, knowing that a closing order has become operative and applies to any premises, uses those premises in contravention of the order or permits them to be so used, shall on summary conviction be liable to a fine not exceeding [£100] and to a further fine of [£20] for every day, or part of the day, on which he so uses them, or permits them to be so used, after conviction. J

In this case the form of the closing order permitted the premises to be used for shop purposes and other purposes other than residential. The actual closing order read:

> Now therefore the council in exercise of the power conferred on them for that purpose by the proviso to section 17 (1) of the Housing Act 1957, by this order prohibits the use of the above mentioned house for any purpose other than a purpose approved by the council and hereby approve the following purposes, namely as a shop, office, warehouse or other purpose not involving human habitation.

Under a later Act, the Land Compensation Act 1973, the local K authority are under an obligation to rehouse persons rendered homeless by the operation of a closing order. In this case there was a time lag between May 13 and July 12 1978 during which the tenants continued to occupy the house pending their rehousing.

The other Act to which we must turn is the Rent Act 1977, to Part IV, section 67, which provides for the registration of a rent and for application to be made to the rent officer and it sets out the details of the matter. Subsection (3) provides:

> Subject to subsection (4) below, where a rent for a dwelling-house has been registered under this part of this Act, no application by the tenant alone or by the landlord alone for the registration of a different rent for that dwelling-house shall be entertained before the expiry of three years from the relevant L date (as defined in subsection (5) below) except on the ground that, since that date, there has been such a change in —
> (a) the condition of the dwelling-house (including the making of any improvement therein),
> (b) the terms of the tenancy,
> (c) the quantity, quality or condition of any furniture provided for use under the tenancy (deterioration by fair wear and tear excluded) or,
> (d) Any other circumstances taken into consideration when the rent was registered or confirmed, as to make the registered rent no longer a fair rent.

Section 70 deals with the determination of a fair rent. It defined M what is meant by a fair rent. Section 70(1) reads:

> In determining, for the purposes of this Part of this Act, what rent is or would be a fair rent under a regulated tenancy of a dwelling-house, regard shall be had to all the circumstances (other than personal circumstances) and in particular to —
> (a) the age, character, locality and state of repair of the dwelling-house, and
> (b) If any furniture is provided for use under the tenancy, the quantity, quality and condition of the furniture.

The other provisions of that section are not relevant to our present problem.

From that determination by the rent officer it is possible under the procedure to refer the matter to the rent assessment committee.

A The rent assessment committee's duties are set out in Schedule 10, which provides for their constitution, and Schedule 11, which provides for their function. The function of the rent assessment committee is either to confirm — if they agree with the rent officer — the fair rent which he has fixed or — if they wish to vary it — it is for them to determine the fair rent and to determine it at the time when the matter comes before them. There is power under this legislation in the rent officer and in the rent assessment committee to decide the date from which their assessment shall operate. Mr Arden tells us that, under the 1980 Housing Act, that discretion as to fixing a date has been removed and now there is no discretion as to fixing the date from which the fair rent will operate. It operates
B from the date of the determination, but we are dealing with the earlier Act.

What the rent assessment committee in this case decided is conveniently set out in the learned judge's judgment:

> Mr Bean — that is the solicitor representing the appellants — argued that because of a closing order the committee was under an obligation to fix a nominal or even a nil rent. We were unable to accept this argument. Our statutory duty is to determine a fair rent and in doing so to have regard to all the circumstances as defined in section 70 of the Rent Act 1977. If a local authority decides that a house is unfit for human habitation on grounds listed in section 4 of the Housing Act 1957, that is certainly one of the circumstances before the committee whether or not the decision is followed
> C by a closing order or some other proceedings available to the local authority under the Housing Acts. But it is not per se conclusive of the question of the fair rent.

The learned judge on appeal came to the conclusion that the rent assessment committee were correct in that approach. He said:

> In my judgment it does not follow, though there may be cases in which it would arise, that simply because a closing order has been made that the fair rent is necessarily nil. It may well be somewhat less than it would otherwise be; no doubt that is so, but for the reasons that I have endeavoured to give, in my judgment it cannot be said that there was an error of law here or that the rent assessment committee applied any false principle of law in reaching their conclusions, or failed to take into account any matter that they should
> D have done.

So he dismissed the appeal.

In spite of Mr Arden's attractive argument, I have come to the same conclusion. It seems to me that here there are two quite distinct pieces of legislation which have not in fact been linked together by any legislative provision. That always creates some difficulty, but I do not think in this case the difficulties are insuperable.

The first point to make plain is that Mr Arden has expressly conceded that, so far as the rent officer was concerned, he made his assessment of fair rent on April 10 1978 — that is, of course, before
E the closing order became operative — and he was not obliged to follow the tentative view of the local authority that this house was unfit for human habitation within the meaning of section 4. That is to say, he had to make his own assessment of the physical condition of these premises and to reflect that physical condition in his assessment of the fair rent following the principle set out in section 70 of the 1977 Act. So it could not in any event be contended, I think, that the rent officer was in any way wrong in law on April 10 1978 when he fixed this rent at £5.75. In other words, it was for him to judge what a fair rent for that property was at that time.

So the crucial question in this appeal is, what effect in law (if any) has the coming into effect of the closing order upon the assessment
F of the fair rent and how should that be dealt with by the rent assessment committee? It seems to me that the fact that a closing order has become operative is obviously a very, very important fact to be taken into consideration by the rent assessment committee, but they have a discretion to determine on what date the rent is to come into effect. They decided — and their decision seems to me plainly within their jurisdiction — that the fair rent as registered by the rent officer should operate from December 14 1977. So that they were bound, it seems to me, to take account of the fact that the property was perfectly lawfully let and lawfully occupied right up to May 13. If it is right that they have no discretion in the matter whatsoever and are bound to fix the fair rent at nil, this means either that the tenant would have no remedy against what might have been G
a very excessive rent for the period from December 14 — that is the effective date of the application for fixing the fair rent — down to May 13 — the date of the closing order, or the tenant would be living rent free in that period. It seems to me open, therefore, to the rent assessment committee, just as it was open to the rent officer, to fix a rent back-dated to the date of the application if they thought that was right. The Act provides that that shall be the operative date unless the rent officer or rent assessment committee otherwise decide. It seems to me plain that it is open to them to make the decision — which they did. I am unable to find any provision which says that, as a matter of law, the rent assessment committee must assess a nil rent because of the coming into force of a closing order. H

The reasons for this conclusion seem to me to be exactly the same as those which operated on the minds of this court in the case of *Black v Oliver* [1978] QB 870. It involved a number of quite different matters from those we have had to consider, but all three Lords Justices who decided the case, while they held that it was open in that case to the rating authority to come to the conclusion that the rateable value of the premises in question was nil, they were not prepared to hold that the rateable value of the premises in question *must* be nil. I find it impossible to distinguish the present case from that case and, accordingly, impossible to say in law that the rent assessment committee were obliged in law to fix a nil fair rent. J

That being so, I find it almost impossible to understand how, if they have a discretion to fix a nominal rent as a matter of law, it can be said that this rent assessment committee erred in fixing a rent of £5.75 unless it can be shown that they misdirected themselves in some way or failed to take into account some factor which they ought to have taken into account. It is not good enough to establish this as a point of law to argue that they did not give sufficient weight to some particular Act. As to that, it seems clear to me that they did take into account the crucial fact that, between the time the rent officer had fixed the rent and the time they came to deal with it, there had been this closing order. Of course it was obviously in the forefront of their minds. I think the only point Mr Arden could K
make on this is to say that, although they referred to the closing order, they did not treat it as having any specific effect in itself. The problem, of course, for the rent assessment committee was to decide what the fair rent should be from the date of application. This will not happen now when they have no power to decide the date from which their fair rent is going to be operative — but it will be a matter for them to decide whether any rent is a fair rent for a property which, under the provisions of the Housing Act 1957, is being occupied by the tenant for the time being in contravention of section 27 and, accordingly, in circumstances which exposes him to a fine. The question whether a landlord can enforce any rental payment after a closing order has been made is not before us. L

I can see nothing to make their decision of £5.75 a week wrong in law for the period between December and May 13 and Mr Arden concedes that. The only period, therefore, with which we are concerned — or any question of law can apply — is between May 13 and July 12. I think it was within their discretion to decide what effect to give the closing order and I think, for exactly the same reasons as this court gave in *Black v Oliver*, they were acting within their discretion. Consequently, I think the learned judge was right in his conclusion and I would dismiss this appeal.

GRIFFITHS LJ agreed.

The appeal was dismissed. There was no order as to costs. M

QUEEN'S BENCH DIVISION

November 25 1980

(Before Mr Justice WOOLF)

RE HEATHVIEW TENANTS' COOPERATIVE LTD

Estates Gazette May 16 1981

(1980) 258 EG 644

Rent Act 1977 — Appeal by "par value" tenants' cooperative from decision of rent assessment committee — Question as to whether rents should be determined as variable — Whether terms as to variation in agreements between cooperative and tenants were reasonable within meaning of section 71(4) of the Act — Committee determined fixed, not variable, rents because in two respects they were not satisfied that the terms were reasonable — Service charges to be as decided by the cooperative in general meeting — Committee considered that this did not meet the requirement of reasonableness because in their opinion an individual tenant would not have a right to see the audited service charges or to know in advance the proportion of costs allocated to his flat — Judge's analysis of constitution and rules of cooperative — Special features of a cooperative — Absence of normal rivalry between landlord and tenant — Committee held by judge to be in error in rejecting the variation terms as unreasonable — Tenants would have right to see audited service costs which, if treasurer and auditor were to perform their functions properly, would be required to be made available before the annual general meeting — Tenants would know the proportion of service costs allocated to them for a future year by the decision of the meeting — Judge, instead of remitting the case to the committee for further consideration in the light of his judgment (which was the normal practice), exercised the power given to him by Order 55, rule 7(5) of the RSC — Order that the committee's decision be varied by substituting a determination that the rent should be variable for committee's determination of a fixed rent

This was a statutory appeal under section 13 of the Tribunals and Inquiries Act 1971 against a decision of a rent assessment committee of the London Rent Assessment Panel. The appellants were Heathview Tenants' Cooperative Ltd. There were no respondents.

Stephen Sedley (instructed by Seifert, Sedley & Co) appeared on behalf of the cooperative.

Giving judgment, WOOLF J said: This is an appeal under section 13 of the Tribunals and Inquiries Act 1971 against a decision of a rent assessment committee of the London Rent Assessment Panel dated October 10 1978.

The appellants are the Heathview Tenants' Cooperative Ltd, and their appeal is against the decision of the rent assessment committee to decide to include a fixed service element in the fair rent of the block of flats occupied by members of the appellant cooperative. The dwellings which are the subject of the hearing are 18, 25, 29, 31, 33, 41 and 53 Heathview, Gordon House Road, London NW5. That is a property which was a purpose-built block of flats built in about 1937. At one time it was owned by Metropolitan Properties Ltd. In the mid-1970s that company got into financial difficulties and, as a result, the tenant cooperative was formed to acquire the property for the benefit of the members and so protect the members against the consequences of the liquidation of that company.

Because the point under appeal is in relation to the fixed service element in the fair rent which was assessed for the premises by the rent assessment committee, it is convenient that at the outset I should refer shortly to the relevant statutory provisions of the Rent Act 1977.

Section 70 of that Act deals with the determination of a fair rent, and section 71 deals with the amount to be registered as the rent. Section 71(1) says:

> The amount to be registered as the rent of any dwelling-house shall include any sums payable by the tenant to the landlord for the use of furniture or for services, whether or not those sums are separate from the sums payable for the occupation of the dwelling-house or are payable under separate agreements.

Section 71(4)(a) provides:

> Where, under a regulated tenancy, the sums payable by the tenant to the landlord include any sums varying according to the cost from time to time of —
> (a) any services provided by the landlord or a superior landlord . . .
> the amount to be registered under this Part of this Act as rent may, if the rent officer is satisfied or, as the case may be, the rent assessment committee are satisfied, that the terms as to the variation are reasonable, be entered as an amount variable in accordance with those terms.

I draw specific attention to the fact that the variation provision contained in subsection (4) requires the rent officer or the rent assessment committee to be satisfied that the terms as to variation are reasonable.

In this case the rent assessment committee had three specific matters of complaint in respect of the determination of the rent officer. First of all it was objected that the rent officer had failed to deal with the individual characteristics of the particular flats. I need say no more about that, although the cooperative were successful with regard to that, because it is not relevant to the issues before me.

The second matter which was objected to was that the rent officer had failed in his fixing of the rent to take into account the special position where the property was controlled by a body such as the appellants. With regard to that, again, the decision of the rent officer was changed in a way which reflected the contentions which were being put before the rent assessment committee by the present appellants. What the committee said was this:

> The committee considered carefully the arguments which had been put forward concerning the nature of cooperatives. It was clear to them that they were dealing with a new form of tenure which was still in its infancy and the development of which was uncertain. The evidence before them suggested that members of a cooperative would collectively find themselves under an obligation to contribute to the running of the cooperative. Because of the blurring of the traditional landlord-tenant relationship many members of a cooperative would find themselves undertaking management duties. The committee had to decide whether to abate the fair rents to make allowance for this obligation. Arguments against doing so included the fact that it would be difficult to quantify and there was no obligation in the tenancy agreement that a member should take an active part in the cooperative. Also it could be argued that members through participation in the management of the block gained advantages denied tenants in private accommodation who would have no influence on landlords' decisions on policy. However the committee were impressed by the evidence of the closely knit structure of working parties which persuaded them that member participation was an essential part of the life of the cooperative. They felt that a member on becoming a tenant would be likely to give a considerable amount of time and energy to furthering the aims of the cooperative. The initial screening system for potential members should result in only those persons becoming members who would take their obligations seriously. The sparse response from persons on the Haringey Council waiting list suggested that the public at large was sceptical of a revolutionary form of tenure and wary of the potenital obligations of membership.

And then they go on to deal with the abatement.

The third matter on which there was objection before the committee was the very matter which is the subject-matter of the present appeal; and with regard to that the committee said this:

> The committee decided to determine a fixed service charge. They were not satisfied that the terms for variation were reasonable. Although the draft tenancy agreement provided that the service charge could only be altered by the cooperative in general meeting each year, the committee decided that this charge did not meet the requirement of reasonableness under the Rent Act 1977 because an individual tenant would not have a right to see audited service costs or know in advance what proportion of the service costs would be apportioned to his flat.

As I read that decision of the committee it is quite clear, first of all that, apart from the two matters which they refer to, the committee do not, by implication, regard the suggestion that there should be a variable service charge as unreasonable; and those two reasons are quite separate; the first one being that the individual tenant would not have a right to see the audited service costs, and the second that the individual tenant would not know in advance what proportion of the service costs would be apportioned to his flat.

In order to consider the issues on this appeal, it is important that I say something more about the appellants, because it is the nature of the appellants which created the new form of tenure referred to in the committee's decision. The appellants are, first of all, a cooperative. There are different types of cooperative, as is indicated in Circular no 8/76, which is a circular issued by the Department of the Environment and the Welsh Office on January 16 1976. Paragraph 5 of that circular refers to three specific types of cooperatives; the second being non-equity cooperatives in which the tenants collectively own or lease the property but have either no individual stake in the equity or a stake limited to a share repayable on leaving at its original par value — hence the term "par value cooperative" used in the report. Of that type of cooperative the circular indicates that they are the only cooperative in the full sense of the word. As such a cooperative they come under the provisions of the Industrial and Provident Societies Act 1965, and I should make reference to that Act.

Section (1) indicates the societies which may be registered. Subsection (2)(a) of that section requires, as a condition of registration, that a society is a *bona fide* cooperative society, and, in addition, there is a provision which in effect prevents a society, if it is to be registered, from being what I will call a profit-making body.

Section 3 allows a registered society to be a body corporate, which may sue and be sued, with limited liability; and which on registration vests in the society all the property for the time being vested in any person in trust for the society, and matters of that nature. This particular society is in fact registered under the Act.

Section 14(1) provides that the rules of the society shall bind the society and all members thereof. Section 30 enables it, unless its rules otherwise direct, to hold or purchase any land and to lease that land.

The series of sections starting with section 44 deals with the books of the society, and section 46(1) provides:

> Subject to any regulations as to the time and manner of inspection which may be made from time to time by the general meetings of a registered society, any member, and any person having an interest in the funds, of the society shall be allowed to inspect at all reasonable hours —
> (a) his own account; and
> (b) all the particulars contained in the duplicate register kept under . . . this Act. . . .

And subsection (2) indicates that the registered society may, by its rules, authorise additional inspection upon such conditions as are specified in the rules.

In addition to being registered under that Act, the appellants are also registered under the Housing Act of 1974. Section 13 of that Act provides that there shall be a register of housing associations which shall be established and maintained by the Corporation which is created by that Act. Such registered housing associations are eligible for loans and grants under section 17; and section 26 limits the powers of a housing association which is registered to make any gift or pay any sum by way of dividend or bonus to any person who is or has been a member of the association, though subsection (2) contains an exception in respect of any funds paid in accordance with the rules of the association concerned, as interest on capital lent to the association or subscribed by way of shares in the association.

Subsection (2) also exempts from the effect of section 26(1):

> Any sum which —
> (i) is paid by an association whose rules restrict membership to persons who are tenants or prospective tenants of the association and preclude the granting or assignment of tenancies to persons other than members; and
> (ii) is paid to a person who has ceased to be a member of the association; and
> (iii) is due to that person under the terms of the agreement under which he became a member of the association.

Section 29 deals with grants which the Secretary of State may make to the association and the succeeding sections deal with that matter in more detail, there being the provisions contained in sections 30 and 31; and in section 32 the provision with regard to revenue deficit grants for registered housing associations. These provisions mean that there is power to advance public moneys to housing associations; that power being vested in the Secretary of State. It should be noted, however, in particular with regard to the revenue deficit grant, that there are various requirements which have to be satisfied before the Secretary of State is obliged to make a payment; and in regard to those matters it is regularly provided that the payment should be one which the Secretary of State has a discretion whether to make or not; for example, section 32(4) says:

> Notwithstanding anything in subsection (3) above, no account shall be taken for the purposes of this section of so much of any deficit as, in the opinion of the Secretary of State, arises by virtue of any such difference between income and expenditure as is specified in section 31(1) above.

And then subsection (5) says that it:

> shall be of such amount as the Secretary of State may determine in relation to that association.

I draw attention to the terms of those provisions because, although the Secretary of State will no doubt exercise his discretions reasonably and properly, whenever there is any form of discretionary requirement of the sort that I have indicated, or conditions to be performed before there is an obligation to make a payment, it is difficult to say with any certainty in advance that any particular payment would be forthcoming.

Because of the various features of a body such as the appellant which is registered under the two Acts to which I have made reference, the tenancy created by it is quite different from that which is normally dealt with under the Rent Act. No specific provision, however, is made with regard to fixing the fair rents of such bodies in a manner different from that of other tenancies. However, section 87 of the Rent Act 1977 does contain within it some recognition of the difference between this type of tenancy and others, because in dealing with the registration of rents section 87(1) provides:

> There shall be a part of the register under Part IV of this Act in which rents may be registered for dwelling-houses which are let, or are to be available for letting, under a housing association tenancy.

Quite clearly Parliament, in inserting that provision, had in mind the special features that can arise with regard to what I will call, loosely, a housing association tenancy, which may make it desirable that in the register it should be identifiable as such, as otherwise, when the register is inspected for the purpose of finding, for example, comparables, the wrong implications could be drawn.

This particular body uses the type of tenancy agreement which is included in the bundle of documents before me. That agreement starts off by saying:

> IN THIS AGREEMENT DATED then blank, "The Cooperative" is the Heathview Tenant's Cooperative, Limited . . .
>
> "The Tenant" is (and then it describes the premises).

With regard to rent the agreement says this:

> "The Rent" is £ p per calendar month exclusive of general and water rates or such amount as may be fixed from time to time by the Cooperative, upon one calendar month's Notice. Normally, this will only take place every three years and does not remove any duty, obligation or right of the Cooperative or the Tenant to refer to the Rent Officer for the District for fixing and phasing as allowed by law.

And then in regard to service charge it says:

> "The Service Charge" is £ p per calendar month or such amount as may be decided by the Cooperative in General Meeting each year and is additional to the Rent.

So far as that agreement is concerned, it makes it clear that, at the outset of the tenancy agreement, the service charge is to be a sum which is to be inserted in the agreement and therefore the tenant will know of that sum at that stage; and with regard to any increase in the sum, it has to be decided by the cooperative in general meeting each year as being additional to rent.

On the provisions of the tenancy agreement alone it does seem to me that it is difficult to understand why the committee felt that the tenant would not know in advance what proportion of the service

A costs would be apportioned to his flat. It seems to me that the very terms of the tenancy agreement presupposed that he would know, either at the outset, or because it was a matter of decision at the cooperative general meeting for a future year.

In order to consider the reasoning of the committee it is perhaps also important to look at the rules of the appellants. Those rules were before the committee and the initial rules refer to the fact that the object of the association shall be the management of the cooperative principle of housing for occupation by members of the association under the terms of an agreement granted to them by the association, which agreement shall exclude, etc; and clearly one of the objects of the association is the letting of the premises on the B terms of the agreement to which I have just made reference.

The rules provide that the business of the appellant shall be conducted by a committee and, in addition to the general powers of the committee, under rule 43(c) the committee have power to determine from time to time the terms and conditions upon which the property of the association is to be let, and at all times to enforce such terms and conditions.

There is to be a treasurer of the appellants under the rules, and rule 50(d) sets out, as one of his duties, that:

> He shall produce or give up all books documents moneys and property of the Association in his possession when required so to do by a resolution of the Committee or of the Association, and render a full and clear account at C each audit, and whenever required by a resolution of the Committee or of the Association.

There are restrictions as to the application of profits contained in rule 54; and rule 55 provides for an audit. It says:

> There shall be appointed in each year of account a qualified auditor to audit the Association's accounts and balance sheet for that year.

Rule 56 provides:

> The auditor shall in accordance with section 9 of the Friendly and Industrial and Provident Societies Act 1968 make a report to the Association on the accounts examined by him and on the revenue account or accounts and the D balance sheet of the Association for the year of account in respect of which he is appointed.

Rule 57 provides:

> (a) At least once in each quarter the Committee shall cause to be prepared a revenue account and balance sheet duly audited and signed by the auditor and incorporating the report of the auditor thereon accompanied by a report by the Committee on the position of the affairs of the Association signed by the chairman of the Committee at which the report is adopted.
> (b) In addition, the Committee shall lay a revenue account and balance sheet duly audited and signed by the auditor and incorporating the report of the auditor thereon before each annual general meeting, accompanied by a E report by the Committee on the position of the affairs of the Association signed by the chairman of the Committee meeting at which the report is adopted. The account shall be made up to such date within the period September 1 to January 31 inclusive as the Committee shall determine to be the end of the financial year.

I pause to interpose at that stage that, as a tenant will always be a member of the appellants and as such has the right to attend the annual general meeting, the member will have an opportunity to see the revenue account and balance sheet which had been audited and the report which was required to be placed before the annual general meeting.

Rule 68 gives a group of members a remedy of applying to the F registrar if in fact there is a failure to comply with the matters to which I have made reference; and of course each tenant is entitled to receive, and, indeed, has to have delivered to him, a copy of the rules, so that he should be well aware of their requirements.

The provisions I have just considered are relevant to the first ground of the decision of the rent assessment committee on the matter under appeal. Those rules, in my view, make it clear that members of this association, who are indeed the tenants, would have the right to see audited service costs. It is true that the rules do not refer specifically to audited service costs, but in accordance with the rules, if the treasurer and the auditor are to perform their functions properly, service costs, properly audited, would be required to be made available before the general meeting in each G year.

I take the view that not only is the second reason, which I have already dealt with, put forward by the committee wrong, but also the first reason put forward by the committee is wrong as a ground for saying that the terms for variation in respect of the service charges were not reasonable.

Having said that, it is right that I should indicate that I am not seeking in any way to criticise the particular committee. One of the features of this sort of body is the fact that the normal rivalry between landlord and tenant does not exist. The tenants are in fact the same collectively as the landlords and so, when the matter was before the rent assessment committee there was not the same H debate as there would normally be in a contested hearing before a committee, and one can well understand that in those circumstances the committee may not have appreciated the full implication of the rules to which I have just made reference. They, in particular, may not have realised the extent of the obligation to prepare audited accounts, which in my view, as I have already indicated, includes audited service costs, and may have been misled in that way, notwithstanding the fact, as to which I have no doubt, that those who were responsible for putting the appellants' case before the committee did their best to ensure that all material which they thought the rent assessment committee would need was made available to them. J

I am therefore quite satisfied that in this particular case the committee have come to a wrong decision. They had taken into account two matters in coming to their conclusion with regard to reasonableness which were in fact based on an entirely false factual premise. In such circumstances in the normal case, the appropriate course for this court to take on an appeal under section 13 of the Tribunals and Inquiries Act 1971 is to remit the matter to the committee so that they can reconsider the matter, questions of reasonableness being quite clearly intended by Parliament to be determined by the committee and not by this court on an appeal. However, under Order 55 rule 7(5) of the Rules of the Supreme Court, this court has power to give any judgment or decision or to K make any order which ought to have been given or made by the tribunal; and it is contended, on behalf of the appellants by Mr Sedley, that this is a case where on the particular facts here before me only one conclusion could be come to by the committee as to the question of variable service charge; and as all the facts are before me, if I was satisfied that his contention is right, I should make use of Order 55 rule 7(5) to make a proper decision, since the delays which could be caused by having the matter remitted to the committee could cause embarrassment in the special circumstances which here exist.

In considering the submission of Mr Sedley, it is right that I should point out, first of all that the provision dealing with variable L service charges is clearly one introduced into the legislation to protect tenants, though it is designed to ensure fair play between the landlords and tenants. I say that it is inserted to protect tenants because in the normal way the service charge will only be reviewed every three years, or two years if the period is changed, and in times of inflation frequently it will happen that by the end of that period the service charge will prove to be substantially inadequate from the landlord's point of view. That is important because any committee considering whether they are satisfied as to the question of the reasonableness of the provision with regard to a variable service charge would have in mind when the application is made by a body of this sort the fact that here the tenants collectively are the very M same body as the landlord, and that any service charge under this agreement is going to be fixed by the tenants themselves collectively at general meeting, and so the risk of service charges being varied inappropriately and so as to prejudice tenants is unlikely to occur.

There is, however, another factor which is very significant in the case of bodies of this sort, with regard to the reasonableness of having a variable service charge, and that is this: whereas a landlord can arrange his affairs within his own resources to deal with deficits or, indeed, even profits, which may occur in having a fixed service charge, this is singularly difficult for a body such as the appellants to do because of the statutory provisions to which I have made reference. If they run into deficit then they are dependent upon the Secretary of State being satisfied of the matters of which he is required to be satisfied before he makes good that deficit. What is more important is that whereas in the case of a normal landlord and tenant any deficit is made up out of private funds, in the case of this body the deficit has to be made up out of public funds. Where the choice is between the tenants collectively agreeing to bear the proper service charges themselves or those service charges being transformed into a deficit which has to be made good out of public funds, it seems to me that a very material matter which a committee would want to take into account in deciding whether to allow a variable service charge is the fact that by approving a provision allowing such a charge they would be avoiding obligations, which are proper obligations of the tenant, being transferred to public funds.

Looking at all the facts before me of this particular case, and I emphasise that in this respect I am looking at the facts of this particular case and not suggesting that it should in any way be a precedent in other cases in the future, it does seem to me that it would not be possible for any rent assessment committee, properly directing themselves as to the facts and as to the law, to come to a decision that the application which was obviously otherwise sympathetically considered by this committee, for the inclusion of a variable service charge, to be refused. I think, on the information which is before me, any committee would be satisfied on such a matter; and therefore, I propose to accede to Mr Sedley's submission and to make an order in this case that the rent assessment committee's decision should be varied so that there is, instead of a fixed service charge, a variable service charge as proposed by the appellants.

QUEEN'S BENCH DIVISION

March 20 1981

(Before Sir Douglas FRANK QC, sitting as a deputy judge of the division)

GUPPYS PROPERTIES LTD v KNOTT AND ANOTHER (NO 3)

Estates Gazette June 13 1981

(1981) 258 EG 1083

Rent assessment committee's duty to give reasons — Nature and extent of duty — Landlords' complaint that additional reasons supplied by committee after case had been remitted to them by the judge were still inadequate — Review of English authorities and of the conflict between them and the views of the Scottish Court of Session, who had defined the duty of committees in this respect in much more stringent terms — Scottish doctrine as expressed in Albyn Properties Ltd v Knox not applicable to England and Wales — Judge rejects proposition that it was incumbent on committees to say how they used the "comparables"; to say what would be a fair return on capital; to analyse the effect of scarcity or to carry out an arithmetical apportionment of factors such as amenity and scarcity affecting their decision — Three rules, however, formulated by judge — (1) the reasons must be intelligible; (2) although they do not have to deal with every point raised, they must deal with the substantial points, showing what matters were taken into consideration and what view was reached on them; (3) where the committee (having conformed with rule (2)) decide to rely on their own knowledge and experience they are not required to explain further how their figure is determined — Although the committee's reasons as amplified were not as explanatory as they could have been, the committee had shown that they had taken all relevant matters into account — Application to set aside and remit decision dismissed, but leave given to appeal to Court of Appeal

This is the third case reported between the same parties. The first, *Guppys Properties Ltd v Knott And Another (No 1)* (1978) 245 EG 1023, was heard, together with another case with a different respondent, by the Queen's Bench Divisional Court (Lord Widgery CJ, Cantley and Peter Pain JJ) and was an appeal from a decision of a rent assessment committee of the Devon and Cornwall Panel on an objection to a fair rent registration. The second case, *Guppys Properties Ltd v Knott And Another (No 2)* (1980) 253 EG 907, and the present case both related to the reference to a committee of the panel of an application for a certificate of fair rent. In the second case the judge (Sir Douglas Frank QC) remitted the matter to the committee for further consideration of the reasons in the light of his judgment. The present case (No 3) arose because the appellants, Guppys Properties Ltd, were dissatisfied with the additional reasons given by the committee in response to the judge's order. The application for a certificate of fair rent related to works proposed to be carried out by the appellants to a house owned by them near Truro in Cornwall. The committee determined that only certain of the works in question constituted improvements which could be taken into account in assessing the fair rent for the purpose of the certificate. The present proceedings were concerned only with the adequacy of the reasons given by the committee. The respondents did not appear and were not represented at any of the appeals.

R Pryor (instructed by Turner Peacock) appeared on behalf of the appellants; John Laws (instructed by the Treasury Solicitor) appeared as *amicus curiae*; the respondents, Mr and Mrs A Knott, did not appear and were not represented.

Giving judgment, SIR DOUGLAS FRANK QC said: This is an application by the landlords to set aside and remit a decision by a rent assessment committee sitting in Cornwall, when they determined the fair rent to be £9.75p per week after the completion of proposed works of improvement. The matter first came before me in December 1979 when I decided to remit the decision on the ground that the committee had failed to fulfil their statutory obligation to give reasons. In my judgment I said:

> I turn now to the main grounds upon which the decision is attacked and that is, that it is said that it failed to give reasons. The decision deals thoroughly with the state of the property as they must assume it to be, but when they come to the valuation of the property in that state they said:
>
> "Having considered all the evidence put before them, and having regard to the provisions of the Rent Act 1977 and to their inspection, the Committee are of the opinion that the only alterations to qualify as admissible improvements are the provision of an inside wc, of an insulating quilt to the first floor ceilings, and of a hot water system and linen cupboard, and accordingly determine that after the completion of those improvements a fair rent will be £9.75p per week exclusive of rates."
>
> Thus, all they said about the evidence is that they have considered it and, without giving any reason beyond that, they determine the fair rent to be £9.75p.
>
> . . .
>
> In the instant case it is quite impossible to discern from the decision what reasons lay behind the assessment of £9.75p per week, other than that the committee disregarded works which were not improvements. In particular, they have failed to state whether and to what extent the submissions concerning capital value and comparables were accepted as convincing or not. Consequently, the applicant knows not whether any consideration was given to his submissions and thus is left in a state of mind of dissatisfaction, which the requirement to give reasons was enacted to avoid.

On June 12 1980 the committee issued what they described as "an addendum" to their decision in these terms:

> The purpose of this additional decision is to comply with the directions in the judgment in *Guppys v Knott* (QBD 1979) by explaining the reasons leading us to set the rent at £507 a year.
>
> 2. In his application Mr Parlett proposed a rent of £936 a year and at the hearing supported this by the assertion that in his opinion the current capital value of the improved property with vacant possession would be £17,500, a figure which we accept as reasonable. In calculating the fair rent from this figure Mr Parlett did not suggest any deduction for security. On the contrary he argued that there was not local scarcity of the sort of accommodation provided by the subject property. We do not accept this argument: there is undoubtedly a sizeable demand throughout Devon and Cornwall for houses and flats. Some of the pressure is due to the number of people coming south to retire but where two and three bedroomed properties are concerned the queue is joined by local home-seekers. In our opinion the scarcity element is relevant to the present case.
>
> 3. Mr Parlett discounted the usefulness of the comparable properties listed in the case papers but we consider that of the other registered rents

A brought to our notice, Nansavallan Cottage, Kea, and 2 Chelean Cottages, Frogpool, are both relevant. Although neither is as attractive as the subject property they do give an indication of the general level of registered rents for rural properties in the area and are useful for the purposes of comparison and contrast.

4. Mr Parlett also submitted that his rental figure of £936 per annum was supported by the inherent amenity value of the property. Our inspection showed us that even when the problem of sanitation has been solved this undoubtedly attractive property still has its limitations in that its proximity to the quarry face means that on three sides there are no windows and consequently there is no through ventilation.

5. Finally we have had regard to the relevant provisions of section 70 of the Rent Act and our own knowledge and experience in making our deter-
B mination.

The matter comes before me again following a fresh notice of motion in which the landlords contend that the addendum still fails adequately to explain the committee's reasons for determining the rent at £507 a year.

Mr Pryor made the following submissions:

(1) The committee have not dealt adequately with capital value, as they have made no comment on what would be a fair return on capital, which was the whole basis of the landlords' case and they did not analyse the effect of scarcity. They failed to state the extent to which they found the submissions on capital value convincing.

C (2) The committee have not said how they used the comparables, for example, on the basis of a rent per sq ft and have not explained how they derived their rent from them. The comparables had been seriously criticised by the landlords in their written evidence and the committee did not warn themselves of their potential unreliability.

(3) One cannot derive from paragraph 4 of the decision any clear understanding of how the committee sought to balance scarcity and amenity value and do not say what they attribute to each.

(4) The decision does not convey what the committee have done with the evidence and gives no indication of how it was used or how it affected their view.

D Mr Pryor referred to *Mason v Skilling* [1974] 1 WLR 1437 and *Metropolitan Property Holdings Ltd v Finegold* [1975] 1 WLR 349, saying that those cases show that capital value is highly relevant and that the committee should have given reasons for rejecting it. Although the committee were entitled to rely on their own knowledge and experience it is not clear whether in this case they did so.

Mr Laws appeared as *amicus curiae* and most helpfully analysed what he described as the marked dichotomy between the English courts' view of the duty to give reasons and that of the Scottish courts and concluded that if the former applies the committee's decision is supportable, but not if the latter applies. He traced the English jurisprudence, starting with *R v Brighton and Area Rent*
E *Tribunal, ex parte Marine Parades Estates (1936)* [1950] 2 KB 410 where it was held, *inter alia*, that a rent committee may act on their own impression and knowledge. In the course of his judgment Lord Goddard CJ (p 420) said:

> The procedure which I have indicated can only work if the tribunal can act in that way and on their own inspection, if they like to make an inspection, of the premises, or on any information which they themselves may have. If witnesses are tendered, I have no doubt that it is the duty of the tribunal to hear them. If they are tendered and cross-examination is desired, it is the duty of the tribunal to allow cross-examination. Then, if the other side wish to call evidence to answer it, it is, again, their duty to allow it.
>
> . . .
>
> F The tribunal heard all the evidence which was submitted to them by the landlords. Having done that, they were not bound to accept the figures which were put before them. That, I think, is made clear by *R v Westminster Assessment Committee, ex parte Grosvenor House (Park Lane) Ltd* [1941] 1 KB 53.

The *Brighton* case was applied in *Crofton Investment Trust Ltd v Greater London Rent Assessment Committee* [1967] 2 QB 955 where it was held that the committee were entitled to act on their own impression and knowledge of conditions in the locality and were not obliged to act only on the evidence before it.

The *Brighton* case and the *Crofton* case were both referred to in *Metropolitan Properties Co (FGC) Ltd v Lannon* [1969] 1QB 577 and Lord Denning MR, having done so and said that the committee G
are of course entitled to use their own knowledge and experience, went on to say:

> But that does not mean that they should throw over the evidence altogether. At any rate, they should not throw over the evidence without saying why. That is what troubles me most about their written decision. They threw over the figure of the rent officer (himself an expert), the figures of the experts on each side, and the figure of the tenant himself. They fixed a much lower figure — without a word to say why they did it.
>
> If the only failure of the committee had been to give their reasons, the courts could no doubt have ordered them to do so: see *Iveagh (Earl of) v Minister of Housing and Local Government* [1964] 1 QB 395; [1963] 3 WLR 974; [1963] 3 All ER 817 CA. But no application was made for the H
> purpose. Nor was it made a ground of appeal, as it might have been.

Mr Laws said that this part of Lord Denning's judgment was obiter and I shall return to that question later. He then referred to *Mountview Court Properties Ltd v Devlin* (1970) 21 P&CR 689 and said the question there was the effect of a failure to give reasons, the failure being *ex concessis* and stated that the case was wholly consistent with the earlier and later decisions of the Divisional Court. This I regard as an important authority and shall consider it further later. Mr Laws' next case was *Tormes Property Co Ltd v Landau* [1971] 1 QB 261 and he said that three propositions can be derived from it, namely: (1) the committee may choose their own standard of comparison; (2) they are not obliged to explain their J
rejection of a different standard of comparison; and (3) they are not obliged to state how they use the comparables and what weighed with them. He then mentioned *Mason v Skilling* [1974] 1 WLR 1437 but, as he said, that case did not concern the duty of a committee to give reasons but gives authority for the proposition that any method of ascertaining a fair rent might be adopted provided that it was not unlawful or unreasonable. It was said that the nature and extent of the reasons given must in some measure be governed by the nature of the problem before the committee. In the case where the committee was not satisfied with either of the alternative figures put forward in evidence by each side and on its own expert knowledge preferred another figure, there was no need K
for it to give, nor was it bound to give, a reason, save that it thought that its figure was right. That case was very much like the later case of *Guppys (Bridport) Ltd v Sandoe* (1975) 30 P&CR 69 and *Guppys Properties Ltd v Knott* (unreported, but of which I have seen the transcript). [This case was reported at (1978) 245 EG 1023.].

Mr Laws then turned to the Scottish cases starting with *Learmonth Property Investment Co Ltd v Aitken* [1971] SLT 349 which he said demonstrates the different approach by the Scottish courts, but I do not find that the court there was concerned with the duty to give reasons. However, the next case cited, namely, *Albyn Properties Ltd v Knox* [1977] SLT 41, was concerned only with that L
question. The opinion of the court can be summarised in the following extract:

> The statutory obligation to give reasons is designed not merely to inform the parties of the result of the committee's deliberations but to make clear to them and to this court the basis on which their decision was reached, and that they have reached their result in conformity with the requirements of the statutory provisions and the principles of natural justice. In order to make clear the basis of their decision a committee must state: (i) what facts they found to be admitted or proved; (ii) whether and to what extent the submissions of parties were accepted as convincing or not; and (iii) by what method or methods of valuation applied to the facts found their determination was arrived at. In short they must explain how their figures of fair M
> rent were fixed.

Mr Laws submitted that that case places a degree of judicial consideration not justified by the English cases on a committee and to apply it to English cases would be in conflict with the authorities. He then went on to demonstrate why, for what he described as good practical reasons, the English approach should be followed. There must be a degree of informality because the committee deal with a lot of cases, often with unrepresented parties, thereby requiring loose procedures and sometimes inspections. The committees were composed of persons with expertise, including a qualified lawyer. Their task was different from that of a judge who goes as a stranger

A to a dispute. Hence for a proper working of the committees the duty to give reasons must be in conformity with the nature of the tribunal. The decisions have to take into account a great amalgam of factors, such as the age, the layout and the condition of the property. The committees must act speedily, informally and untrammelled with the duty of giving reasons on every point. In contrast, the Scottish approach demands a process of reasoning, full and objectively demonstrable, that is testable, at least as far as it is possible to be, including the quantification of the component of the rent and the detail of the method of valuation. He drew attention to the Tenants Rights, Etc (Scotland) Act 1980, section 47 of which imposes a duty on the committee to have regard to all the circum-
B stances and in particular to apply their knowledge and expertise of current rents and comparable property. However, I remark that that only disposes of one difference between England and Scotland and does not go to the question of the duty to give reasons. Further, what Mr Laws describes as compelling reasons are applicable with equal force in Scotland but apparently have left the Court of Session unmoved.

Mr Laws then turned to his submission on *stare decisis* and, having referred to a number of authorities, submitted that I am bound by the decisions of the Divisional Court. Mr Pryor argued to the contrary and cited *Kruse v Johnson* [1898] 2 QB 91. I cannot find that the question has ever been decided, but the balance of
C judicial opinion seems to favour Mr Laws, see for example Lord Goddard CJ in *Police Authority for Huddersfield v Watson* [1947] 1 KB 842 at 848. However, as will transpire, I do not have to decide this question as I find it sufficient to rely on decisions of the Court of Appeal and the Divisional Court and, where there is apparent conflict with the decisions of the Divisional Court, then to prefer the former.

Mr Laws then dealt specifically with Mr Pryor's submissions. He said that the first submission directly adopts the approach of the Scottish courts and is in conflict with the English cases and that the last sentence of the second paragraph of the committee's addendum was sufficient. The second submission failed for the same reason
D and paragraph 3 of the addendum gave a sufficient indication of the way the committee had used the comparables. Again, the third submission implied a necessity to carry out such a balancing act as was required by the Scottish approach. Finally, he said that the committee have told us what they have done with the evidence so far as they have to; and they are not required to carry out a scientific exercise involving the calculations.

I have found, in my consideration of the authorities, the case of *Mountview Court Properties Ltd v Devlin* (*supra*) to be the most helpful. I do not accept that the case proceeded on the footing that the failure to give reasons was accepted *ex concessis*, because Lord Parker CJ said:

E

> . . . the first point, as I see it, is whether the reasons given in the decision as supplemented, as they were, in January of this year are, even now, sufficient, or, in other words, whether the committee have fulfilled their duty under section 12 of the Tribunals and Inquiries Act 1958, which provides that "It shall be the duty" — now of the committee — "to furnish a statement, either written or oral, of the reasons for the decision if requested, on or before the giving or notification of the decision, to state the reasons . . ."

Then follows an important passage:

> What reasons are sufficient in any particular case must, of course, depend upon the facts of the case. I approach the matter in this way: that reasons are not deficient merely because every process of reasoning is not set out. I
> F further think that reasons are not insufficient merely because they fail to deal with every point raised before the committee at the hearing. Indeed, I would adopt the words used by Megaw J in *Re Poyser and Mills' Arbitration* [1964] 2 QB 467 at 477–478. That was dealing with an arbitrator's award, but Megaw J said:
>
> "The whole purpose of section 12 of the Tribunals and Inquiries Act 1958 was to enable persons whose property, or whose interests, were being affected by some administrative decision or some statutory arbitration to know, if the decision was against them, what the reasons for it were. Up to then, people's property and other interests might be gravely affected by a decision of some official. The decision might be perfectly right, but the person against whom it was made was left with the real grievance that he was not told why the decision had been made. The purpose of section 12 was to remedy that, and to remedy it in relation to arbitrations under this Act. G
> Parliament provided that reasons shall be given, and in my view that must be read as meaning that proper, adequate reasons must be given. The reasons that are set out must be reasons which will not only be intelligible but which deal with the substantial points that have been raised."
>
> A little further down he said:
>
> "I do not say that any minor or trivial error, or failure to give reasons in relation to every particular point that has been raised at the hearing," — and he was dealing with an error of law on the face of an award — "would be sufficient ground for invoking the jurisdiction of this court."
>
> . . .
>
> Indeed, as Lord Denning MR said in *Iveagh (Earl of) v Minister of Housing and Local Government* [1964] 1 QB 395 at 410: "The whole purpose of the enactment is to enable the parties and the courts to see what H
> matters he" — and that was the Minister of Housing in this case — "has taken into consideration and what view he has reached on the points of fact and law which arise. If he does not deal with the points that arise, he fails in his duty: and the court can order him to make good the omission."

The landlords had relied strongly on a fair rent which had been fixed for another flat (No 64) and, as to that, Lord Parker said:

> . . . the striking bit of evidence which they [the tribunal] had before them was No 64, and, beyond their saying, in the first reasons which they gave, that they had, to use their own words, "considered the evidence adduced", . . . there is no explanation at all to show why they rejected what one would think would have to be a greater market value, a greater fair rent. . . . In my judgment, that failure does result in the fact that they have J
> here given insufficient reasons; they have not complied fully with their duty under section 12 of the Tribunals and Inquiries Act 1958.

The other members of the court agreed with Lord Parker and in particular Bridge J (as he then was) said:

> It seems to me that there was here a lack of adequate reason for the decision.

Pausing there, it seems to me that from that judgment the following rules may be derived:

(1) The reasons must be intelligible;
(2) although they do not have to deal with every point raised, they must deal with the substantial points, showing what matters K
were taken into consideration and what view was reached on them.

On the other hand there is no support given for the principle apparently applied in Scotland that the committee must state what method or methods of valuation they used and apply them to the facts found.

Applying *Guppys Properties Ltd v Knott* (*supra*) I would add another rule:

(3) Where the committee (having conformed with rule (2)) decide to rely on their own knowledge and experience they are not L
required to further explain how their figure is determined.

It follows from *Mountview* that whether Lord Denning's judgment in the *Lannon case* was obiter or not, it seems fully supported by the judgment in *Mountview*, as does that part of the judgment of Edmund Davies LJ (as he then was) in *Lannon* where he said that "the committee were not obliged to accept any of the proffered evidence . . . but if they had in mind to do so (as they did) they should have made it clear why they did it . . .".

I am bound to say that I find it difficult to reconcile the *Tormes* case, particularly where Lord Parker held in terms that the committee were perfectly entitled to reject the yardstick (one based on replacement costs) put forward by the landlords without criticising M
it. However, the main argument seems to have been directed to the question whether the committee were entitled to look straight to a fair rent fixed for another property without investigating it and no consideration seems to have been given directly to the question whether there was a failure to comply with the rule that reasons must be given. Be that as it may, if *Tormes* is indistinguishable from *Mountview* and *Lannon* I am entitled to and do prefer the latter cases. Hence although I accept Mr Laws' proposition that the committee may choose their own standard of comparison I reject his other two propositions said to be derived from *Tormes*.

Therefore I conclude that the principles enunciated in *Re Poyser and Mills' Arbitration* apply to rent assessment committees as they do in other jurisdictions and particularly as they were further described in *Mountview* and *Lannon*. Those principles I have already summarised in the form of rules which I now apply to the instant case by reference to Mr Pryor's submissions.

1. It was not incumbent on the committee to say what would be the fair return on capital, nor to analyse the effect of scarcity, although the quality of the reasons would have been better had they done so. It is somewhat surprising, having regard to my previous judgment, that the committee did not state in terms whether they took capital value into account or rejected it and in not doing so they came close to failing to comply with the second rule I have already specified. However, it seems to me that either they could have said that they took it into account although mentally modified it on account of scarcity value or, alternatively, because of scarcity value they rejected it. They were entitled to take either course and so whichever they had adopted would not have vitiated their decision. Accordingly Mr Pryor's first submission fails.

2. I think it follows from the English cases, in contrast to the Scottish cases, that it was not incumbent on the committee to say how they used the comparables and that they were entitled merely to set out their conclusions in the way they did.

3. It also follows that it was not incumbent on the committee to carry out the kind of arithmetical apportionment postulated in this submission.

4. I agree that the reasons are not as explanatory as they could be, but nevertheless I think the committee have shown that they took all relevant matters into account and, having regard to the third rule I have adumbrated, this submission must also fail.

In view of the forgoing this application must be dismissed.

The appeal was dismissed. There was no order for costs. Leave was given to appeal to the Court of Appeal.

Leave has subsequently been given also to appeal on the point raised with regard to improvements in *Guppys Properties Ltd v Knott (no 1)*.

A **COURT OF APPEAL**

October 30 1980

(Before Lord Justice LAWTON, Lord Justice TEMPLEMAN and Lord Justice O'CONNOR)

NATIONAL WESTMINSTER BANK LTD v BSC FOOTWEAR LTD

Estates Gazette January 24 1981

(1980) 257 EG 277

Landlord and tenant — Lease for term of 21 years containing option for lessee to renew for a further term of 21 years — New Lease to be "at the then prevailing market rent", to be determined by arbitration in default of agreement, the new lease to be subject in all other B respects to the same covenants, provisos and conditions as the original lease — Question whether arbitrator had power to determine a rent which was subject to periodic review — Appeal from decision of judge who held that arbitrator was entitled to do so — Held by Court of Appeal that the judge's construction of the option was wrong — The arbitrator had no power to introduce a provision for periodic review — The option clause referred to the prevailing rent, not to the prevailing use and practice of the market at the date of renewal — The arbitrator had to determine a single rent payable throughout the term of the new lease — Alternative suggestion that the arbitrator might fix a differential or progressive rent also rejected — Fawke v Viscount Chelsea distinguished as being a C decision on the court's statutory power under Part II of the Landlord and Tenant Act 1954 — Appeal allowed.

This was an appeal by lessees, BSC Footwear Ltd, defendants in proceedings by originating summons brought by the plaintiff lessors, National Westminster Bank Ltd, in the Chancery Division to determine a question as to the true construction of an option to renew clause in a lease of premises at 254 High Street, Orpington, Kent. The judge, Judge Finlay, sitting as a judge of the Chancery Division, decided in favour of the plaintiff lessor that an arbitrator appointed under the provisions of the clause was entitled to determine a rent which would be subject to periodic review during the D term of the new lease.

Mark Myers QC and J R Cherryman (instructed by Titmuss, Sainer & Webb) appeared on behalf of the appellant lessees; M Essayan and V R Chapman (instructed by Linklaters & Paines) represented the respondent lessors.

Giving the first judgment at the invitation of Lawton LJ, TEMPLEMAN LJ said: This is an appeal against a decision of His Honour Judge Finlay, sitting as a judge of the Chancery Division, and given on December 3 1979. The question is whether power to determine a rent includes power to direct that the rent shall be periodically reviewed. In my judgment, that question only permits of a negative answer.

E By a lease dated June 5 1957, a shop and maisonnette were demised for a term of 21 years from March 25 1957. In due course, the plaintiffs, the National Westminster Bank Ltd, became the landlords; the defendants, BSC Footwear Ltd, became the tenants. Pursuant to the provisions of the lease, the tenants served notice requiring a renewal of the lease. The rent reserved by the lease was a peppercorn for two months "and thereafter the clear net rent of £1,250 per annum . . . payable in arrear". The lease contained covenants to pay "the said rents", meaning the annual rent I have indicated and the further or additional insurance rent for which the lease provided. There were also the usual full repairing covenants F and user and other covenants extending to some 20 in all.

Clause 4(2) contained the relevant power of renewal, which is in a not unfamiliar form and reads as follows:

> If the lessee shall be desirous of taking a further lease of the demised permises for a further term of 21 years from the expiration of the term hereby created and shall on or before 25 March 1977 give to the lessor a notice in writing of such its desire and shall pay the rents hereby reserved and subject to there being no substantial breach of the covenants on the part of the lessee and conditions herein contained which is incapable of remedy then the lessor will subject to the lessee having remedied any breach of covenant notice whereof shall have been given to the lessee grant a new lease to the lessee for a further term of 21 years from 25 March 1978 at the G then prevailing market rent (to be determined in default of agreement by a single arbitrator pursuant to the Arbitration Act 1950 or any subsisting modification or re-enactment thereof) and subject in all other respects to the same covenants provisoes and conditions as are herein contained except this subclause for renewal.

As a matter of first and last impression, that provision seems to me to empower the arbitrator to perform one task, and one task only, namely, to determine the annual rent which is to be paid throughout the 21 years of the renewed term, in lieu of the annual rent of £1,250 which was fixed by the original lease.

The tenants duly gave notice, the landlords and tenants failed to H reach agreement and they are now in process of applying to the arbitrator to fix the rent. The salient words are the requirements that the landlords, if the tenants wish, must grant a 21-year lease from March 25 1978 "at the then prevailing market rent" (namely, in 1978), to be determined by the arbitrator, and the lease must be subject, in all other respects, to the same covenants, provisos and conditions as are contained in the original lease, excluding the proviso for renewal. In my judgment, the arbitrator has no power to introduce any variations between the original lease and the renewed lease. The arbitrator must determine the rent and only the rent, and for this purpose he must determine the prevailing market rent.

J It was argued, and the learned judge declared in his order, that upon the true construction of the clause which I have read, the arbitrator is entitled, if he thinks fit, having regard to the evidence adduced before him, to determine a rent which is to be subject to periodic rent reviews during the term of the new lease. The result of that declaration is to confer on the arbitrator a discretion to decline to determine the rent payable under the renewed lease for 21 years, but only to determine the rent for a period of three or five years, or some other period which he is free to choose, having heard indeterminate evidence; and then he has power to direct that subsequent rent for subsequent parts of the term of 21 years shall be determined by such persons, at such intervals and by such machin- K ery, as the arbitrator may think fit to draft and award. Alternatively, Mr Essayan, for the landlords, submitted, if the arbitrator does not see fit to lay down any machinery, then, instead of the arbitrator determining the rent for the whole of the term, he is allowed to determine the rent for the initial period which seems to him to be proper, say three or five years, and then (if you please) to leave it to the judges of the Chancery Division to do the arbitrator's work for him and to say what the rent ought to be for subsequent periods.

In deciding whether to redraft the 1957 lease, the arbitrator is apparently to be assisted by evidence, but we do not know on what basis he could possibly form his view as to how the lease ought to be redrafted, except, I understand, that he must look and see how most leases are drafted in present times and what kind of rent review clauses was popular in the year 1978.

It is common ground that a rent review clause may take many forms. All rent review clauses enable the rent to be increased; some enable the rent to be reduced; some enable the tenant to surrender the lease if the rent review provides an amount of rent which his business cannot afford; come oblige the tenant, whether he likes the new rent or not, to keep the lease and pay up, so that he is at the mercy of galloping inflation instead of being, as the lessee of this 1957 lease was, and as we must assume he bargained to be, in receipt of the comfort of having a certain rent for a long period.

In my judgment, this lease, including clause 4(2), is, in effect, a 42-year lease with one rent review after 21 years; and what the landlords are seeking to do now, for perfectly understandable reasons in the light of history, is to redraft so that, instead of having one rent review after 21 years, they will secure for themselves rent reviews after 26, 31, or whatever other years the arbitrator thinks fit.

I fail to find in clause 4(2) anything which confers on the arbitrator the wide and discretionary powers which are claimed by the landlords. Nor, in my judgment, does the clause enable the arbitrator to decline to perform his duty, namely, to fix a rent for 21 years. He cannot say: "I can see my way clear for five years, but I am going to throw up my hands in horror and say it is quite impossible for me to fix the rent for the remaining 16 years". No one has suggested that in practice any arbitrator worth his salt will be inclined to throw up his hands. What is said is that, if the arbitrator fixes the rent at the beginning of 21 years for the whole period, then the landlord is in danger of not getting the fruits of inflation which he might otherwise get if there were rent review clauses. But what the landlords' predecessors in title gave away these landlords cannot now take back.

The learned judge conjured the power to insert a rent review clause which he ascribed to the arbitrator out of the word "prevailing". He said in his judgment that the word "prevailing" meant "that which is in general use or practice", but he omitted from that definition the fact that what has to be prevailing is the rent; it is the rent which is to be in general use or practice. It may be general use or practice to provide a rent review clause, but that is a different matter from the prevailing rent.

The learned judge appears himself to have recognised that if the arbitrator is authorised to order periodic reviews of rent, then the original lease, on this hypothesis, gave the arbitrator the power to redraft the new lease so that it does not correspond with the old lease. In his judgment the learned judge said:

> I think the right view of clause 4(2) is that there is implied in the words "at the then prevailing market rent" that the arbitrator is to determine the terms as to rent having regard to the prevailing market conditions, and if those market conditions involve a provision for review of the rent at stated intervals, then his task is to determine what, according to the practice and usage of the market at the relevant time, is the appropriate kind of review formula.

Thus the learned judge opined that the task of the arbitrator is to determine a review formula, but in my judgment it is quite impossible to read that into clause 4(2). All that the arbitrator is empowered to determine, and is bound to determine, is rent. He is not empowered to impose any formula and redraft the lease by virtue of his award. The arbitrator is concerned with an amount and not with a formula.

The alternative argument was put forward by Mr Essayan, for the landlords. He relied, in his persuasive speech in support of the rent review clause which the learned judge had been persuaded to introduce, on what he said would be an unrealistic result. In 1978 no landlord, he said — and there was evidence for this — would dream of letting premises for 21 years at a fixed rent. Every landlord would insist on having a rent review clause, and, therefore, when the arbitrator sets out to find the prevailing market rent for a 21-year lease — a fixed rent — he will find there is nothing prevailing and that there is no market. It is common ground that that does not make the task of the arbitrator impossible. Arbitrators are often faced with property which is of a peculiar nature and has, in practice, no market comparable, and yet they do not have the slightest difficulty in fulfilling the requirement that they shall fix the market rent then prevailing. In fact, although landlords may not let properties these days, voluntarily, for 21 years without rent review clauses, there is in the evidence example upon example of landlords who, like the present landlords, as a matter of history, are caught with obligations which they, or their precedessors in title, have entered into, and there are leases which require rent reviews after 10 years from now, 14 years from now, 21 years from now. The fact that a modern landlord, free from any obligations, would not enter into a lease of this nature is *nihil ad rem*. The fact is the tenants' predecessors in title were prudent and wise enough to extract in 1957 a lease which did permit and require them to stay there for 21 years and then permitted them to stay there for another 21 years if they wished to do so, at a rent which was to be fixed for that second period of 21 years by an arbitrator during the year before the first lease expired.

This landlord agreed to do that which Mr Essayan vigorously urged no landlord in 1978 would do. The answer is, this is not a 1978 lease, it is a 1957 lease. But the hypothesis which the arbitrator has to make is that this lease, with a 21-year term at a fixed rent, is on offer and he must decide the market price for a rent as between landlord and tenant if that lease came on to the market. I really think, with all respect to Mr Essayan, that was the strength of his submission, namely, that in 1978 landlords who are free from obligations would not enter into a lease of the sort which is envisaged. That does not affect the true construction of the lease made in 1957. It does not produce a terribly unrealistic result at all. It does not even produce a difficult result. What it does produce is a result which the landlords never realised might be disadvantageous, namely, the result that, whether inflation goes up or down within the next 21 years, both the landlords and the tenants will have to put up with the rent which is fixed by the arbitrator in 1978 at the beginning of the term.

As an alternative, Mr Essayan argued that, if the arbitrator cannot incorporate a rent review clause, he can at least introduce a differential rent. He could, for example, provide that the rent shall be £2,000 for the first three years, £3,000 for the next three years, and so on, as his fancy dictates. Alternatively, said Mr Essayan, if he formed the view that annual inflation was inevitable at 10 per cent compound, then he could provide a rent of (say) £2,000 in the first year increased every subsequent year by 10 per cent, and so on.

For that submission, he relied principally on the authority of the case of *Fawke* v *Viscount Chelsea* [1980] 1 QB 441 and, in particular on the judgment of Goff LJ, who came to the conclusion that section 34(1) of the Landlord and Tenant Act 1954 authorised the court, in fixing the rent payable under a business lease, to fix a rent which, so far as the decision went, was a rent which could be kept at a lower figure until the premises had been put in repair and could then be increased, and, as far as *obiter dicta* is concerned, supported the view that, in any event, a judge, determining a rent for the purposes of the Landlord and Tenant Act 1954, could provide for a differential rent possibly similar to that which Mr Essayan has urged.

To turn Mr Essayan's own argument against him, it will be a remarkable lease, even for the year 1980, which contained any differential rents of the type which he prayed in aid. Whether the evidence as regards the market would support him in that regard, I doubt. In *Fawke* v *Viscount Chelsea*, the courts were concerned with statute and with a tenant who had no right to a lease of any particular term. The court had power to grant to a tenant a lease of such period, not exceeding 14 years, as the court thought fit. In those circumstances, it seems to me that the court was faced with a different task. It could balance, on the one hand, the virtues of giving a long lease with differential rents against the alternative of giving a short lease so that the landlord could then come along and obtain the benefit of inflation, and I am not surprised that differential rents are possible under that statute. But we are dealing here with a perfectly common form lease which deliberately set out, for the first 21 years, to have all the advantages and disadvantages of a fixed rent of £1,250 per annum — a certainty which has its benefit for the landlord, in that he knows he is going to get that sum for 21 years, and which has benefits for the tenant in that he knows exactly what his obligations will be.

Against that background, when we find that the lease itself provides for a renewal and the renewal is to be on exactly the same terms as the old lease, except that there is to be substituted the prevailing market rent at the date of the beginning of the new lease, it seems to me that, on the true construction of that clause in that lease, what the arbitrator has to do is to reconsider the figure "£1,250" in the original lease and to insert the figure which he thinks proper, having regard to 1978 prevailing market rents. Then what the landlord has to do is to engross the new lease in exactly the same terms as the old lease, but instead of the figure "£1,250" there will appear the figure which the arbitrator thinks proper, payable from March 25 1978. The two months peppercorn rent does not apply.

In my judgment, the appeal should be allowed and there should be substituted the declaration for which the tenants ask, namely, that the arbitrator is bound to determine a single rent which is to be payable throughout the term of the new lease. I would leave it at that and not go on and say that he is not to be entitled to determine a rent which is subject to review, because that necessarily follows. His task is to determine the rent which is to be payable annually throughout the term of the new lease.

O'Connor LJ agreed. Also agreeing, Lawton LJ said: In 1957 the plaintiffs' predecessors in title let some shop premises in an outer suburb of London to the defendants' predecessors in title, who were a very-well-known and responsible retail trading company—Freeman Hardy & Willis. What we have to do in this case is to find out what the intention of the parties was in 1957. The learned judge took the view that the surrounding circumstances of that time, other perhaps than the fact that there was a letting of shop premises, had little bearing upon the problem of the construction of the lease.

It is manifest from the terms of the lease that the lessors were willing to give security of tenure for a longish term. That no doubt is what the lessees wanted. In 1957 there was a small cloud on the horizon. It was the small cloud of inflation. Few people in 1957 contemplated that inflation would be anything of the problem it is today. By clause 4(2) the lessors clearly were willing to grant a new lease for 21 years on the same terms as before, save that instead of having £1,250 as the anual rent there was to be another figure inserted. That figure was to be that which was the prevailing market rent.

In my judgment, part of the difficulty which has arisen in this case is reflected in the final words of Mr Essayan's submission to us, namely, as I took them down, "The arbitrator was directed to look at the prevailing practices in the market". He was not directed to do anything of the sort. He was directed to find out what in the market was the prevailing market rent. The learned judge was beguiled by the same fallacy. As Templeman LJ has pointed out, when he came to construe the word "prevailing" he said that it meant (and I quote) "that which is in general use or practice". What he was required to do was to construe the word "prevailing" as an adjective qualifying "market rent" and nothing more.

In my judgment, it is because there was introduced into this case evidence about what were the practices of the market in 1978 that the learned judge misdirected himself in construing a lease which had to be construed according to the intention of the parties at the time it was made, which was in 1957.

I agree with the order suggested by Lord Justice Templeman.

The appeal was allowed with costs in the Court of Appeal and below. Leave to appeal to the House of Lords was refused.

COURT OF APPEAL

November 21 1980

(Before Lord Justice BUCKLEY, Lord Justice EVELEIGH and Lord Justice BRIGHTMAN)

THOMAS BATES & SON LTD v WYNDHAM'S (LINGERIE) LTD

Estates Gazette January 31 1981

(1980) 257 EG 381

Landlord and tenant — Rent review clause in lease of factory premises — Lease granted in pursuance of option contained in previous lease — Lease for 14 years subject to review of rent at the end of the fifth and the tenth year — Review clause provided for the rents for the review periods to be such as shall have been agreed between the parties, with no provision for arbitration in default of agreement — Landlords claimed rectification of the lease — Held by Court of Appeal, affirming the deputy judge on this point, that the conditions required for rectification were present — The landlords erroneously believed that the review clause contained a provision for arbitration in default of agreement; the tenants were aware of the omission of such a provision and that this was due to a mistake; the tenants omitted to draw the mistake to the notice of the landlords; and the mistake was to the disadvantage of the landlords — The clause should be rectified by the insertion of a provision that in default of agreement the rent should be fixed by an arbitrator — Held further, differing in this respect from the deputy judge, that the standard to be applied by the arbitrator should be, not the market rent, but the rent that it would have been reasonable for the landlords and tenants to have agreed under the lease, having regard to all the circumstances relevant to any negotiations between them of a new rent from the review date.

This was an appeal by the defendant lessees, Wyndham's (Lingerie) Ltd, from a decision of Michael Wheeler QC, sitting as a deputy High Court judge, ordering in favour of the plaintiff landlords, Thomas Bates & Son Ltd, the rectification of a lease of factory premises at Church Road, Harold Wood, Essex.

E G Nugee QC and J C Harper (instructed by Chethams) appeared on behalf of the appellants; Robert Wakefield (instructed by Tolhurst & Fisher, of Southend-on-Sea) represented the respondents.

Giving judgment, Buckley LJ said: This is an appeal from a decision of Michael Wheeler QC, sitting as a deputy judge in the Chancery Division on June 6 1979.

The issues in the case relate to a rent review clause contained in a lease dated December 17 1970 and made between the plaintiffs, Thomas Bates & Son Ltd as lessors and the defendants, Wyndham's (Lingerie) Ltd, as lessees. The subject-matter was some factory premises at Hornchurch in Essex. In order to understand the issues it is necessary to go back in history a little while.

In the year 1956, by a lease dated August 20 1956 the plaintiffs (whom I will call "the landlords") let to predecessors of the defendants (and I will call the defendants "the tenants") the factory premises in question for a term of seven years from September 1 1956 at a yearly rent of £650. That lease contained, in clause 5, an option provision in the following terms:

> That the lessor will on the written request of the lessee made six months before the end of the term hereby created and if at the time of such request there shall not be any existing breach or non-observance of any of the covenants on the part of the lessee hereinbefore contained at the expense of the lessee grant to the lessee a lease of the demised premises for a further term of seven or fourteen years from the expiration of the said term at a rent to be agreed between the lessor and the lessee but in default of such agreement at a rent to be fixed by a single arbitrator appointed by the President for the time being of the Royal Institution of Chartered Surveyors and containing the like covenants and provisos as are herein contained.

The term under that lease was in due course assigned to the tenants, and when the time for the exercise of the option drew near, Mr Bates, the managing director of the landlords, wrote a letter to the tenants, for the attention of a Mr Avon, who was the director of the tenants who at all times has handled matters relating to this leasehold property on behalf of the tenants, a letter drawing attention to the fact that the time had come to consider a renewal of the

A lease, and Mr Bates said in that letter that the landlords would require an addition of £125 per annum, bringing the rent up to £775 per annum for the seven years from the expiration of the then current lease.

Stimulated by that communication, Mr Avon, on behalf of the tenants, gave a formal notice exercising the option on February 19 1963, requesting the landlords to grant to the tenants a renewed lease of the premises for the further term of 14 years from the expiration of the current term, "at a rent to be agreed between us" — quoting from the notice — "but in default of such an agreement, at a rent to be fixed by a single arbitrator appointed by the President of the Royal Institution."

B In response to that, Mr Bates wrote in reply saying that in fact the landlords would require rather more rent than he had stated in his earlier letter, £850 a year; in consequence of which there were some oral communications, and on March 11 1963 Mr Bates wrote to Mr Avon confirming offers which he had made orally for a further seven-year term at £800 a year and proposing that the landlords should construct certain additional buildings on the property, in consideration of which there would be a further rent of another £800 a year during the ensuing seven-year period.

Those terms were accepted by the tenants, and on November 29 1963 the parties entered into a new lease for a term of seven years from November 15 1963 at a yearly rental of £1,600; and that lease
C contained an option clause in precisely the same terms as the option clause in the 1956 lease, save that it granted an option for only a further seven years and not for seven or 14 years as had been the case in the 1956 lease. Time went by and the year 1970 arrived, when the time was approaching for the exercise of the option in the 1963 lease, and we find that on April 14 1970 the landlords wrote a letter to the tenants drawing their attention to this fact, in consequence of which Mr Avon, on May 4 signed and sent to the landlords a formal notice exercising the option contained in the 1963 lease, and by that notice he requested the landlords to grant a renewed lease of the premises for a further term of seven years from the expiration of the then current term "at a rent to be agreed
D between us, but in default of such agreement at a rent to be fixed by a single arbitrator appointed by the President for the time being of the Royal Institution," and so on. The language of that notice followed the language of the option clause contained in the 1963 lease. As has been pointed out in argument, the effect of that notice was to change the legal relationship between the parties and to bring into existence a contract for the grant of a further term of seven years at a rent to be agreed or, in default of agreement, to be fixed by an arbitrator.

The landlords wrote back to the tenants on May 7 indicating that the rent they would require would be £2,600 per annum for the first three years of the new term of seven years, the rental thereafter to
E be reviewed and agreed for the remainder of the term. Those words are taken from the letter of May 7 1970. So being then under a contractual obligation to grant a new lease for seven years at a rent to be agreed or in default of agreement to be fixed by an arbitrator, they proposed agreeing the rent for part of that term only, leaving the remainder of the term the subject of a further review and agreement at a later date.

On August 3 — that is to say, rather later than the letter I have just referred to of May 7 — there having been some oral communications in the meantime and Mr Avon having paid a visit to the landlords' offices to discuss the matter, the landlords wrote saying that they were prepared to grant a lease for a further period of 14 years from
F the expiration of the then current term, with a clause for rent reviews at the end of the fifth and 10th years of the term, the rental for the first period of five years to be £2,350 per annum exclusive of rates.

To that letter Mr Avon replied on August 17 1970 that the tenants were reluctantly prepared to accept the figure of £2,350, but he said that they were not in agreement with rent reviews after five and 10 years but were willing to accept a clause for a rent review at the end of the seventh year. That was, in effect, a counter-offer to the offer which had been put forward in the letter of August 3 1970.

The landlords replied on August 18 insisting upon rent reviews after five- and 10-year intervals, and on August 20, following some telephonic communication, they again wrote insisting upon the rent G
reviews, and that letter has a postscript: "Whilst your present lease provides for a further seven years' renewal, the question of the rent review period is something quite separate and distinct."

It seems to me that the landlords there are saying: we recognise your right under the exercise of the option; you are entitled to a term of seven years at a rent to be agreed or in default of agreement to be fixed by an arbitrator, but we are not at the moment prepared to agree any rent beyond the first five years.

On September 22 the tenants' solicitors, Nabarro Nathanson, wrote indicating that the tenants, subject to formal exchange of the lease, accepted the offer of a new lease for a term of 14 years from November 15 1970 at the exclusive rent of £2,350 per annum, H
subject to review at the expiration of the fifth and 10th years of the new term. It seems to me that it is implicit in that that the rent, at any rate in respect of the two years next following the initial five years of the term under the new lease, would be fixed by agreement or, in default of agreement, by an arbitrator appointed under the provisions to that effect in the option clause.

The lease and counterpart were then executed and exchanged. The lease had not been prepared by the landlords' legal advisers; it was prepared under the instructions of Mr Bates and was typed by Mr Bates' secretary, Miss Cannon. The lease so prepared and executed, demised the property to the tanants for a term of 14 years from November 15 1970, and now I quote from the document J
itself:

> Yielding and paying therefor during the first five years of the said term unto the lessor the yearly rent of two thousand three hundred and fifty pounds and for the next period of five years of the said term and the final period of four years of the said term such rents as shall have been agreed between the lessor and the lessee such rents to be paid clear of all deductions by equal monthly payments on the First day of each month in advance,

and then there is a provision that the tenants should also pay the costs of insurance, but we are not concerned with that.

The lease contains an option clause in, I think, the same terms as the option clause which was contained in the 1963 lease, except that K
any further lease to be granted under the option in the 1970 lease was not to be required to contain an option clause. It will be observed that in the reddendum there is no reference to arbitration in default of agreement.

At a later stage, when the time for fixing the rent from the end of the first five years of the term onwards became imminent, the landlords became aware of this and wrote a letter, signed by Mr Foley who was the property manager for the landlords, to the tenants as follows:

> As you will be aware under the terms of your lease the rent of this property is due to be reviewed, effective from November 15 1975, and this letter is intended to be a formal notice advising you of our intention to review the L
> rent. The lease has no specific note how the rent may be settled should your company and ourselves fail to agree upon a figure and I suggest that the matter be settled by an independent arbitrator. I therefore enclose an agreement in duplicate and shall be obliged if you will sign the top copy and return it to me.

As a result of that, and a hastener written on May 2 1975, a telephone conversation took place on May 6 1975 in the course of which Mr Avon said he was not prepared to sign any such agreement. Mr Foley then got into contact with the landlords' own solicitors, who advised him that because the rent review clause did not contain some means of definitely settling the new rent, eg by arbitration following appointment by the President of the RICS, M
should the parties to the lease not agree, the review clause would be unenforceable at law and the rent could remain the same until the lease expired.

Nothing further seems to have taken place immediately with regard to that, but there was later a telephone conversation between Mr Avon and Mr Foley, to which I shall have to refer again later, in which Mr Avon said — or is said by Mr Foley to have said — that he had been aware of the implications of the clause from the day the lease was signed; that he might be prepared to pay a slightly higher rent, but not nearly so high a rent as the landlords were in fact then proposing.

A It seems to me clear that the omission of any reference to arbitration in default of agreement in the reddendum of the lease of 1970 must have been due to a mistake on the part of Mr Bates, under whose instructions, as I have said, that lease was prepared. The omission was one which was clearly contrary to the landlords' interests. The only possible legal consequences which have been suggested in the course of arguments are: first, that no rents having been agreed in respect of any period after the first five years of the 14-year term, no rent would be payable after the end of that five-year period. Mr Nugee, appearing for the tenants, although he did not in the course of his argument altogether abandon the idea that that might be the legal consequence of the omission, frankly B and very properly admitted that it could not have been in the contemplation of the parties that there would be any period during the term of this lease when no rent would be payable at all, and it is inconceivable that that could have been the parties' real intention.

Secondly, it has been suggested that as no rent other than the rent of £2,350 per annum would have been agreed, that rent should continue in force until some other rent should be agreed by the parties. It seems to me that such a proposition is absolutely contrary to the clear intention of the rent review provision. The rent review provision was clearly a provision insisted upon by the landlords because they wanted the matter reviewed at the end of five years and wanted a new rent to be then arrived at. If merely by withholding C consent or agreement to a new rent the tenants could stultify the revision clause and ensure that the original rent of £2,350 should continue to be payable, the whole purpose of the rent revision clause would be destroyed or frustrated. I cannot believe that it could have been Mr Bates' intention that the lease in the form in which he framed it should have had that effect.

Thirdly, it has been suggested that in respect of the period after the first five years, and in default of agreement between the parties, the rent would have to be fixed by a process of litigation involving implication of the parties' intentions, and it has been suggested that by implication the rent ought either to be the market rent, that is to say the rent for which this property would be let in the open market, D or such rent as this particular landlord and this particular tenant would agree having regard to all the matters which would affect them in arriving mutually at a rent which the one was content to accept and the other was content to pay, which would very possibly be markedly different from a market rent, and it seems to me inconceivable that a landlord would have been content to rely on fixing a rent by a process of implication in that way without it being at all clear in what way the court would view the matter and how the rent to be so ascertained should be assessed.

The only other possible legal consequence of the omission that has been suggested at all, either in the course of argument or in the course of the evidence given before the learned trial judge, was that E the clause was entirely inoperative because of the omission and because either it amounted to no more than an agreement to agree, or it amounted to a provision which contained such defective machinery that it could not be carried into practical effect. It will be seen that advice on those lines was what was received by Mr Foley. It seems also to have been the advice received by Mr Avon.

Mr Foley gave evidence before the learned judge, but Mr Avon did not. The learned judge was somewhat critical of Mr Avon for not giving evidence. That was, no doubt, a matter which rested not so much in Mr Avon's discretion as in the discretion of those who were conducting the case on the part of the tenants. The learned judge, in the course of his judgment:

F

In evidence before me Mr Foley amplified this note — he is referring to a note written on the letter of September 23 to which I have already referred — I think it probable that in his oral evidence Mr Foley may to some extent have telescoped two telephone conversations with Mr Avon into one. His evidence—which was given with care and which I accept without hesitation — was to the following effect. He said that he quoted a new rent to Mr Avon of (he thought) about £5,000 per annum; that Mr Avon seemed amused and, when asked what he thought, said he might pay £100 or £200 more. Mr Foley said he realised that Mr Avon's figure bore no relation to the market value but he felt that he (Foley) was in a slightly ticklish position and said he would refer the position to Bates' solicitors; that Mr Avon then said that he (Foley) must be aware as he (Avon) was that the rent revision clause as drawn was inoperative and that he (Avon) had been G aware of this at the time the lease was entered into because it had been brought to his notice by Wyndham's solicitors.

The learned judge, in the course of his judgment, spoke critically of Mr Avon in that respect and said that, in his judgment, Mr Avon's conduct amounted to sharp practice. As the learned judge had not heard any evidence from Mr Avon we cannot tell in what circumstances Mr Avon acted as he did, or under what advice he acted as he did. It is clear that he was at that time in contact with and receiving advice from the tenants' solicitors, and for my part I do not feel it necessary to associate myself with that stricture on the part of the learned judge upon Mr Avon's conduct. Nevertheless, H the fact emerges that when the tenants executed the 1970 lease they did so realising the omission of any reference to arbitration in default of agreement in the review clause and without drawing the attention of the landlords to that omission in any way.

The only reasonable conclusion, it appears to me, that can be drawn from the documents is that the lease was executed in the form in which it was with regard to the terms of the review clause as a consequence of a mistake on Mr Bates' part, for at the time when the lease was prepared and put forward the tenants had a contractual right to have the rent, at any rate in respect of the sixth and seventh years of the term, agreed or, in default of agreement, determined by an arbitrator appointed by the president of the J Royal Institution of Chartered Surveyors. No doubt the grant of the lease displaced the contract which had arisen as a result of the exercise of the option, but the terms of that contract relating to fixing the rent in respect of the first seven years of the term which was granted by the 1970 lease remained in force up to the execution of the lease and that, in my judgment, affords a strong indication that until, at any rate, Mr Avon realised the omission of any reference to arbitration, it was the mutual intention of both parties that the rent to be paid under the lease after the first five years should be a rent which was agreed between the parties or, in default of agreement, ascertained by arbitration.

Mr Bates did not give evidence because, unhappily, he had died K in January 1973. That was a date a considerable number of months after the issue of the writ but before the trial. No written statement of Mr Bates made during his lifetime was adduced in evidence under the Civil Evidence Act. We are told by counsel, but we have got no other evidence of the fact, that Mr Bates did not make any written statement. So there was no evidence of any kind emanating from him.

But there was the evidence of Miss Cannon. Miss Cannon's evidence, in my view, in no way negatives the possibility or probability that the omission of a reference to arbitration was due to a mistake on Mr Bates' part. Her evidence, taking it quite shortly and generally, is to the effect that she typed the lease; that she did it in L accordance with instructions which she received from Mr Bates; but there is nothing in her evidence which establishes one way or the other whether Mr Bates, in giving his instructions, was himself labouring under a mistake. If, as Mr Avon thought was the position, the clause in the way in which it was drawn was an inoperative clause, it seems to me to be absolutely manifest that it must have been so framed as the result of a mistake, for one cannot believe that any landlord would put into a lease a clause which he intended to be inoperative.

Mr Nugee has said that there is no evidence as to what Mr Bates' intention was, and he stressed that in cases of rectification a high standard of proof is required by the court. Indeed, in some cases the M standard has been equated with the criminal standard of proof, "beyond all reasonable doubt". I think that the use of a variety of formulations used to express the degree of certainty with which a particular fact must be established in civil proceedings is not very helpful and may, indeed, be confusing. The requisite degree of cogency of proof will vary with the nature of the facts to be established and the circumstances of the case. I would say that in civil proceedings a fact must be proved with that degree of certainty which justice requires in the circumstances of the particular case. In every case the balance of probability must be discharged, but in some cases that balance may be more easily tipped than in others.

In *Hornal v Neuberger Products Ltd* [1975] 1 QB 247 Denning LJ

A (as he was at that time) said at p 258: "The more serious the allegations the higher the degree of probability that is required: but it need not, in a civil case, reach the very high standards required by the criminal law." That, in my judgment, encapsulates the law about the standard of proof required in civil proceedings applicable to all civil proceedings, and as applicable to cases of rectification as to any other kind of civil action.

The landlords claim rectification in the present case on the basis of a principle enunciated by Pennycuick J in *A Roberts & Co Ltd v Leicestershire County Council* [1961] Ch 555 where the learned judge said, at p 570:

B The second ground rests upon the principle that a party is entitled to rectification of a contract upon proof that he believed a particular term to be included in the contract, and that the other party concluded the contract with the omission or a variation of that term, in the knowledge that the first party believed the term to be included.

He says a little lower down the same page:

The principle is stated in *Snell on Equity*, 25th ed (1960) p 569, as follows: "By what appears to be a species of equitable estoppel, if one party to a transaction knows that the instrument contains a mistake in his favour but does nothing to correct it, he (and those claiming under him) will be precluded from resisting rectification on the ground that the mistake is unilateral and not common."

C Of course, if a document is executed in circumstances in which one party realises that in some respect it does not accurately reflect what down to that moment had been the common intention of the parties, it cannot be said that the document is executed under a common mistake, because the party who has realised the mistake is no longer labouring under the mistake. There may be cases in which the principle enunciated by Pennycuick J applies although there is no prior common intention, but we are not, I think, concerned with such a case here, for it seems to me, upon the facts that I have travelled through, that it is established that the parties had a common intention down to the time when Mr Avon realised the mistake in the terms of the lease, a common intention that the rent in respect
D of any period after the first five years should be agreed or, in default of agreement, fixed by an arbitrator.

The principle so enunciated by Pennycuick J was referred to, with approval, in this court in *Riverlate Properties Limited v Paul* [1975] Ch 133, where Russell LJ (as he then was) reading the judgment of the court said, at p 140:

It may be that the original conception of reformation of an instrument by rectification was based solely upon common mistake: but certainly in these days rectification may be based upon such knowledge on the part of the lessee: see, for example, *A Roberts & Co Ltd v Leicestershire County Council* [1961] Ch 555. Whether there was in any particular case knowledge
E of the intention and mistake of the other party must be a question of fact to be decided upon the evidence. Basically it appears to me that it must be such as to involve the lessee in a degree of sharp practice.

In that case the lessee against whom the lessor sought to rectify a lease was held to have had no such knowledge as would have brought the doctrine into play. The reference to "sharp practice" may thus be said to have been an *obiter dictum*. Undoubtedly, I think in any such case the conduct of the defendant must be such as to make it inequitable that he should be allowed to object to the rectification of the document. If this necessarily implies "some measure" of sharp practice, so be it; but for my part I think that the doctrine is one which depends more upon the equity of the position.
F The graver the character of the conduct involved, no doubt the heavier the burden of proof may be; but, in my view, the conduct must be such as to affect the conscience of the party who has suppressed the fact that he has recognised the presence of a mistake.

For this doctrine — that is to say the doctrine of *A Roberts & Co Ltd v Leicestershire County Council* — to apply, I think it must be shown: first, that one party (A) erroneously believed that the document sought to be rectified contained a particular term or provision, or possibly did not contain a particular term or provision which, mistakenly, it did contain; secondly, that the other party (B) was aware of the omission or the inclusion and that it was due to a mistake on the part of (A); thirdly, that (B) has omitted to draw the G
mistake to the notice of (A). And I think there must be a fourth element involved, namely that the mistake must be one calculated to benefit (B). If these requirements are satisfied, the court may regard it as inequitable to allow (B) to resist rectification to give effect to (A)'s intention on the ground that the mistake was not, at the time of execution of the document, a mutual mistake.

Mr Nugee has drawn attention to a number of other departures in the language of the 1970 lease from the language of the corresponding clauses of the 1963 lease, and he says that this lease was not, or should not be regarded as having been, granted in pursuance of the exercise of the option, but as a newly negotiated lease, the negotiations no doubt being prompted by the exercise of the option, but the H
new lease not flowing from the exercise of the option. I, with respect to Mr Nugee's argument, do not find very much force in that contention. The parties were, of course, at liberty to modify the terms of their lease in any way they mutually agreed and none of these variations to which I am now referring has any bearing upon the review clause or the language employed in it. It seems to me, as I have already said, that the omission from the review clause of any reference to arbitration was one which was clearly contrary to the landlords' interests, one which must have occurred as a result of a mistake, and one which Mr Avon, on his own evidence, recognised, and must I think be taken to have recognised, as having been the result of a mistake on the part of Mr Bates. J

The learned judge disposed of the matter on this aspect in a passage in three numbered paragraphs:

(i) that I cannot regard Miss Cannon's evidence as proving that Mr Bates was not making a mistake in omitting a longstop provision for arbitration of some sort; (ii) that although Miss Cannon said that she thought there had been other Bates' leases which contained the same type of rent review provision as the 1970 lease, none was produced, and I cannot accept that any reasonable businessmen would deliberately have adopted such a potentially defective provision; and (iii) that the provision for an option to renew the 1970 lease in clause 5 (which was in similar terms to the options in Wyndham's earlier leases) showed perfectly clearly that the parties recognised the necessity for a longstop for rent fixing purposes and that it is K
reasonable to suppose that it was a provision in these terms which Bates mistakenly omitted and which Wyndham's (through Mr Avon) deliberately allowed to go uncorrected.

I have already dealt with Miss Cannon's evidence.

With regard to the learned judge's second numbered paragraph, I agree that it is highly improbable that Mr Bates would have purposely adopted a form of clause which was so disadvantageous as the review clause is with the omission of any reference to arbitration.

With regard to the third paragraph, I would myself prefer to relate this point not to clause 5 of the 1970 lease but to the exercise L
of the option under the 1963 lease and the contract which arose from its exercise. The point is, I think, precisely the same point. It is that the parties must have had present to their minds the desirability, and indeed the obligation of the landlords in relation at any rate to the first seven years of the new term, to arrive at a rent which was not necessarily a rent which had to be at the same rate throughout the term, but they had to arrive at a rent which was agreed between them or, in default of agreement, was one determined by arbitration.

There was no precedent for a review clause contained in the 1963 lease, for the 1963 lease did not provide for any rent review; and it is not difficult to believe that a layman like Mr Bates, in preparing M
the 1970 lease, failed to detect the shortcoming of the review clause as he had framed it, and failed to apply his mind to the difficulties which would arise if no provision was made for arbitration.

On these findings to which I have just referred, the learned judge rectified the lease. The form of the order provides as follows:

This court doth declare that upon the true construction of the lease dated December 17 1970 and made between the plaintiffs and the defendants comprising factory premises at Church Road Harold Wood in the London Borough of Havering the rents during the period of five years from November 15 1975 and the period of four years from November 15 1980 should be the market rent for the said premises. And this court doth order

> A that the said lease be rectified so that in the reddendum thereof after the words "such rents as shall have been agreed between the lessor and the lessee" there shall be inserted the words "or shall in default of such agreement be determined by a single arbitrator to be appointed by the President for the time being of the Royal Institution of Chartered Surveyors".

So far as rectification is concerned, the language which the learned judge has adopted follows the language used in the option clauses in this case, except that he used the word "determined" instead of "fixed", and perhaps it would have been better if the word had been "fixed".

If the lease is so rectified the question arises: by what measure is an arbitrator to fix the rent if the parties do not agree? Mr
B Wakefield, for the landlords, initially contended that the arbitrator so-called would act not as an arbitrator but as a valuer. He based that argument upon the use of the words "shall have agreed" and the word "fixed" in the review clause. On that basis he submitted that the rent should be the market rent for the property, on the authority of a decision of the House of Lords in *Ponsford v HMS Aerosols Ltd* [1979] AC 63). Subsequently he conceded that the clause must be read as an agreement to arbitrate and not as an agreement to abide by a valuation. Upon that footing he agreed that, upon the true construction of the clause, the rent should be such as it would have been reasonable for this landlord and this tenant to have agreed under the lease. It would consequently be
C proper for the arbitrator to take into account all considerations which would affect the mind of either party in connection with the negotiation of such a rent, as, for example, past expenditure by the tenant on improvements.

In my judgment, Mr Wakefield was right to make that concession and to have accepted that the present case falls within the reasoning of the minority of the House of Lords in *Ponsford v HMS Aerosols Ltd* and not within the reasoning of the majority in that case. The review clause which was there under consideration was a review clause in a lease which provided for a yearly rent of £9,000, "during the first seven years of the said term and during the second and third seven years of the term the sum of £9,000 aforesaid or such sum,
D whichever be the higher, as shall be assessed as a reasonable rent for the demised premises for the appropriate period. Such assessment to be made in the following manner, that is to say: (a) such assessment as shall be agreed between the parties hereto in writing" — and there were certain provisions as to the date by which that agreement should be reached — "(b) in the event of the parties hereto failing to reach such agreement as aforesaid on or before the date appointed, then the reasonable rent for the second and third period shall be fixed or assessed by an independent surveyor". That form of clause, as it seems to me, focuses attention upon what is there described as "a reasonable rent for the demised premises" for the appropriate period, and that expression is first used without any reference to
E agreement between the parties to the lease at all. It then goes on to provide that such assessment — that is to say, the fixing of the amount of the rent so to be charged — shall be either agreed or, in default of agreement, arrived at by valuation by an independent surveyor. That form of wording, in my judgment, certainly affected the views of the majority in the House of Lords in that case. Lord Dilhorne at p 77 said: "The rent payable by the lessees will of course be rent for the demised premises but as I see it, the task of the surveyor is not to assess what would be a reasonable rent for the lessees to pay but what is a reasonable rent for the premises". Lord Fraser, at p 83, said: "In my opinion the words point unambiguously to the result contended for by the landlords . . . and they
F mean the reasonable rent assessed on an objective basis, without reference to the particular landlord or the particular tenant or to the history of how the premises came to be built or paid for". Lord Keith, at p 86: "In my opinion the words 'a reasonable rent for the demised premises' simply mean the rent at which the demised premises might reasonably be expected to let". The other two learned Lords, Lord Wilberforce and Lord Salmon, took a contrary view. They thought that what had to be ascertained was what would be reasonable between the particular parties to the transaction. However, they were in the minority upon the construction of that particular rent review clause. But it appears to me that the terms of the clause there under consideration were noticeably different in important respects from the clause which we have, which refers to G
nothing other than such rent as the parties shall have agreed. Consequently, I think that Mr Wakefield was well advised in making the concession which he made.

Mr Nugee, on the other hand, who had argued in the earlier stages of the appeal that in default of agreement the rent should continue after the review date at the original rate of £2,350 per annum, conceded that in the light of a decision of this court in *Beer v Bowden* which has not yet been reported but is contained in Court of Appeal transcript no 156 for 1976* — he could no longer support that argument. That again was a concession which I think he was constrained to make. The decision in *Beer v Bowden* was only brought to the attention of counsel and, through counsel, to the H
attention of the court late in the course of the argument.

So the parties are now at one, that on the true construction of the clause as rectified the rent is to be fixed by the arbitrator at such amount as it would be reasonable for the parties to agree having regard to all such considerations as I have mentioned. This was not the construction adopted by the judge, who, as appears from the terms of his order, implied a term that the rent to be agreed should be the market rent. His attention had not, of course, been drawn to the decision of this court in *Beer v Bowden*. As I understand the position, neither party now contends that the learned judge's view in that respect is right, and I myself am satisfied that the market rent would not provide a proper standard to adopt in the present case. In J
my judgment, in default of agreement between the parties, the arbitrator would have to assess what rent it would have been reasonable for these landlords and these tenants to have agreed under this lease having regard to all the circumstances relevant to any negotiations between them of a new rent from the review date.

If I were wrong on the point of rectification, then, on construction and by a process of implication, the rent to be ascertained in default of agreement must, I think, be a fair rent as between the landlords and the tenants. It would be most unjust that the landlords should receive no rent because of failure of the parties to agree. The landlords have granted a 14-year term and the court must endeavour to fill any gap in the terms of the lease by means of a fair K
and reasonable implication as to what the parties must have intended their bargain to be. See in this connection the decision of this court in the case of *F & G Sykes (Wessex) Ltd v Fine Fare Ltd* [1967] 1Lloyd's Rep 53, which was a case very different on its facts from the present, but in which the court explained the function of any court of construction where parties have embarked upon any commercial relationship but under terms that are not altogether adequate to cover the eventualities. The court would ascertain by inquiry what rent the landlord and the tenant, as willing negotiators anxious to reach agreement, would arrive at for each of the two rent review periods. In short, the standard would be the same, as I see it, as would have to be adopted by an arbitrator under the clause if it is L
rectified in the way in which I consider that it should be rectified.

For these reasons I think that the learned judge, while he came to the wrong conclusion on the matter of market rent, reached the right conclusion on the matter relating to rectification. I would accordingly uphold that part of his order which directed rectification, though I would substitute the word "fixed" for "determined", purely as a matter of pedantry, I think. It is for consideration whether, in those circumstances, any declaration is really required to be included in the order at all. That is a matter upon which, perhaps, we can hear submissions at a later stage. I would dispose of the matter in that way.

Agreeing EVELEIGH LJ said: The correspondence beginning with M
May 4 1970 contained references to rent reviews. The first reference specifically was "at a rent to be agreed between us, but in default of such agreement, at a rent to be fixed by a single arbitrator appointed by the president for the time being of the Royal Institution of Chartered Surveyors". The letter of May 7 from the defendants' solicitors accepting the proposed lease for 14 years said merely "at the exclusive rent of £2,350 per annum, subject to review at the expiration of five and ten years of the new term". Quite clearly that letter was written upon the basis that the nature

*Briefly reported at (1976) 237 EG 141.

A of the review was understood. It is inconceivable that a solicitor would confirm an agreement and ask for a draft lease which would, of course, reflect that agreement, as they in fact did, if such an important matter had not been resolved. The word "review" was clearly shorthand. I take the letter of May 7 from the plaintiffs to the defendants in the same way. The phrase there used is "thereafter to be reviewed and agreed for the remainder of the term". If anyone had asked the parties at that time how the review would take place, I am quite convinced that the answer would have been that the machinery contemplated had already been put forward by Mr Avon in his letter of May 4 to which the letter of May 7 was a reply.

B I see nothing in the words of the other letters written in the course of negotiations between the parties to indicate that the review machinery first referred to was being abandoned in favour of something else. I find it particularly difficult to conclude, as the appellants contend, that it was being replaced by a vague gentlemen's agreement. The fact that the parties ultimately agreed upon a lease of different duration from that originally agreed, and containing other terms not in the lease of 1963, in no way alters my conclusion. Certain important changes were specifically discussed. The machinery for rent review as opposed to the length of the period was treated without further discussion. The only reasonable conclusion, in my opinion must be that the parties were negotiating upon C the basis that rent review in default of agreement was to be as indicated in the letter of May 4. I therefore think that there was a common intention that the rent should be fixed by a single arbitrator in default of agreement.

I also think that the evidence established that Mr Avon knew that the lease did not contain the appropriate clause, and knew that Mr Bates intended that it should. Where a party is aware that the instrument does not give effect to the common intention of the parties as communicated each to the other, there may be an inference of sharp practice or unfair dealing. In my opinion, this will not always be so. I do not think that it is always necessary to show sharp practice. In a case like the present if one party alone knows that the D instrument does not give effect to the common intention and changes his mind without telling the other party, then he will be estopped from alleging that the common intention did not continue right up to the moment of the execution of the clause. There is no need to decide whether his conduct amounted to sharp practice. I think he might at that time have had no intention of taking advantage of the mistake of the other party. I do not think that it is necessary to show that the mistake would benefit the party who is aware of it. It is enough that the inaccuracy of the instrument as drafted would be detrimental to the other party, and this may not always mean that it is beneficial to the one who knew of the mistake.

I agree that the lease should be rectified in the way indicated by E my Lord and I agree with the order which he proposes. I should just add that I, too, regard this case as different from the case of *Ponsford v HM Aerosols Ltd* [1979] AC 63. There the reference was specifically to the demised premises and that is an important difference. At p 85 Lord Keith said:

> At first impression the words "reasonable rent for the demised premises" suggest that what has to be ascertained is simply the rent which is reasonable for the premises as such in their actual state, the situation being viewed entirely objectively. The "demised premises" must mean the demised premises as improved by virtue both of the ordinary law and the passage I have quoted from the licence agreement. So upon this view any contribution the improvements might have made to rental value would have to enter into the F assessment,

and he also clearly attached importance to the words "demised premises" in the passage which my Lord has just read.

For those reasons I agree that the lease should be rectified in the terms stated, and I further agree with the interpretation of "reasonable rent" that my Lord has given.

Also agreeing, BRIGHTMAN LJ said: I wish to say a few words only on two points. First as regards the standard of proof. The standard of proof required in an action of rectification to establish the common intention of the parties is, in my view, the civil standard of balance of probability. But as the alleged common intention *ex hypothesi* contradicts the written instrument, convincing proof is G required in order to counteract the cogent evidence of the parties' intention displayed by the instrument itself. It is not, I think, the standard of proof which is high, so differing from the normal civil standard, but the evidential requirement needed to counteract the inherent probability that the written instrument truly represents the parties' intention because it is a document signed by the parties.

The standard of proof is no different in a case of so-called unilateral mistake such as the present. The mistake in the instant case was unilateral and not mutual only because the defendants became aware of the implications of the review clause on the eve of the execution of the new lease. That consideration, as it seems to me, leads to no different conclusion in relation to the standard of proof required in a rectification action. H

The other point I want to touch on briefly is this. In his judgment the learned judge said this: ". . . the parties recognised the necessity for a longstop for rent fixing puproses and . . . it is reasonable to suppose that it was a provision in these terms which Bates mistakenly omitted and which Wyndham's (through Mr Avon) deliberately allowed to go uncorrected. In my judgment this was sharp practice . . ." I would not be prepared to assume, on the evidence, that Mr Avon was consciously guilty of sharp practice. Nor is such an assumption necessary for the plaintiffs' case. As I indicated, I take the view that there was a common intention on both sides to extend to the new lease the rent assessment arrangements contained in the covenant for renewal in the expiring lease. The discrepancy between the formula in the expiring lease and the formula in J the engrossment of the new lease was not observed by Mr Avon until it was pointed out by his solicitor. I am not willing to assume that the reputable firm of solicitors acting for him would have allowed him to execute the lease in any circumstances which they saw to be dishonest. If the judgment is intended to contain a finding of sharp practice on the part of Mr Avon, I would respectfully wish to disagree with the learned judge on such a finding. I do not think this would be justified.

As I have said, I agree that the learned judge's order should, with the slight variation mentioned, stand.

The court upheld the judge's order for rectification, substitution K of word "fixed" for "determined", but directed that the rent to be fixed by an arbitrator in default of agreement should be, not the market rent, but the rent that it would have been reasonable for these landlords and these tenants to have agreed under this lease, having regard to all the circumstances relevant to any negotiations between them of a new rent from the review date.

COURT OF APPEAL

November 4 1980

(Before Lord Justice LAWTON, Lord Justice TEMPLEMAN and Lord Justice O'CONNOR) L

AMALGAMATED ESTATES LTD v JOYSTRETCH MANUFACTURING LTD

Estates Gazette February 7 1981
(1980) 257 EG 489

Landlord and tenant — Construction of rent review clause in lease — Lease provided that the open market rental value should be arrived at, first, as specified in a notice by the landlords; secondly, if the tenants did not accept the figure specified, by agreement between the parties; thirdly, as determined by an independent surveyor M **appointed as arbitrator by agreement or, failing agreement, by the president of the RICS — Two questions arose for decision — The first was whether the figure specified in the landlords' notice had to be a bona fide and genuine pre-estimate of the open market rental value — The second was whether a letter from the tenants refusing the figure specified by the landlords, and asking for an explanation of how it had been arrived at, could be construed as a notice requiring arbitration — Held that the answer was "no" to both questions — Appeal from an order of Smith J dismissed.**

This was an appeal by tenants, Joystretch Manufacturing Ltd, from a decision of Smith J dismissing the tenants' appeals from orders of a master refusing them unconditional leave to defend and finding them liable to the present respondents, Amalgamated Estates Ltd, in the sum of £8,392.12 for arrears of rent. The lease under which the dispute arose related to property in Rugby at 14 Regent Street and 3 Bank Street. The relevant clauses in the lease are set out in the judgment of Lawton LJ.

Paul de la Piquerie (instructed by Ward Bowie, agents for H E Wynschenk, John Basset & Co, of Birmingham) appeared on behalf of the appellants; N T Hague (instructed by Frimonds) represented the respondents.

Giving judgment, LAWTON LJ said: This is yet another appeal raising the question of the construction of a rent review clause in a lease. In this case it arises by way of appeal from the order of Smith J made on July 25 1980 whereby he dismissed the defendants' appeal against the refusal of Master Ritchie to give them unconditional leave to defend, and the judgment given by the Master on July 7 1980 against the defendants in the sum of £8,392.12 for arrears of rent. There was also an order for costs.

The defendants, who were tenants, had to apply for leave to appeal out of time. That we granted; there was only a technical infringement of the rules. They asked for the judgment to be set aside on the ground that, on the true construction of the lease, the landlords were not entitled to be paid the rent which they had demanded.

The lease under consideration was made on December 7 1973. It related to some property in Rugby known as 14 Regent Street and land at 3 Bank Street, Rugby. The premises were, in fact, shop premises with a yard. The term was for 25 years at a starting rent of £3,500 per annum. For the first five years of the lease the rent was to be £3,500 per annum. Clause 1(b) of the lease provided:

> (b) For the next FIVE years of the said term either the yearly rent reserved in subclause (a) hereof or the open market rental value of the demised premises to be determined in the event of disagreement between the parties by a Chartered Surveyor to be nominated by the President for the time being of the Royal Institution of Chartered Surveyors at the review date whichever is the higher and in either case the same shall remain constant during the whole period referred to in this subclause.

Clause 1(c) was in these terms so far as it is relevant:

> (c) For each subsequent successive term of FIVE years of the said term either the yearly rent in subclause (a) hereof or that in subclause (b) hereof or the open market rental value of the demised premises at the review date to be determined as hereinbefore provided whichever is the higher and in either case the same shall remain constant during the whole period referred to in this subclause . . .

I leave out some words which are irrelevant, and the clause ended with a proviso which was as follows:

> PROVIDED THAT for the purpose of subclause (b) and (c) hereof it is hereby agreed that the following definitions and provisions shall apply . . .

Then was set out a definition of open market rental value as follows:

> The annual rental value of the demised premises in the open market which might reasonably be demanded by a willing landlord on a Lease for the term of years certain equivalent in length to the residue unexpired on the review date of the term of years hereby granted with vacant possession at the commencement of the term . . .

The second paragraph of the proviso defined what was meant by "review date". Nothing in this appeal turns on that paragraph. The third paragraph must be read in full and is as follows:

> (3) The open market rental value shall be determined in manner following that is to say (a) specified in notice in writing signed by or on behalf of the Landlord and posted by recorded delivery post in a pre-paid envelope addressed to the Tenant at the demised premises at any time provided that if a notice is posted after the review date the open market rental shall not be payable until the second quarter day following the posting the said notice and such quarter day shall be known for all purposes as the review date and such notice shall be conclusively deemed to have been received by the Tenant in due course of post or (b) agreed between the parties before the expiration of three months immediately after the date of posting of such notice as aforesaid in substitution for the said sum of

That was left blank. Then comes:

> (c) determined at the election of the Tenant (to be made by counter-notice in writing served by the Tenant upon the Landlord not later than the expiration of the said three months) by an independent surveyor appointed for the purpose by the parties jointly in writing or upon their failure to agree upon such appointment within one month immediately after the date of service of the said counter-notice then by an independent surveyor appointed for that purpose on the application of either party alone by the President for the time being of the Royal Institution of Chartered Surveyors and in either case in accordance with the provisions of the Arbitration Act 1950.

(4) is irrelevant for present purposes, and (5) is in these terms:

> (5) All stipulations as to time in the foregoing subclauses numbers (1)(2) (3) and (4) shall be of the essence of the contract and shall not be capable of enlargement save as agreed in writing by the parties.

There is reason to believe that most of the proviso was taken from a precedent in the *Encyclopaedia of Forms and Precedents* 4th ed vol 12 at p 841. What was not in that precedent was the part providing that the notice shall be conclusively deemed to have been received by the tenant in due course of post. Another provision of the lease which was not, as far as I can see, in the precedent in the *Encyclopaedia of Forms and Precedents* was clause 1(b) of this lease, which provides that, if there is a dispute as to the open market rental value of the demised premises, it is to be determined by a chartered surveyor to be nominated by the president for the time being of the Royal Institution of Chartered Surveyors.

The history of the case is as follows. Two or three days before the review date as defined in the lease the landlords wrote to the tenants by letter dated December 8 1978 in these terms:

> Dear Sirs, Re: 14 Regent Street and 3 Bank Street, Rugby. Pursuant to the terms of your Lease of the above premises dated December 7 1973, we hereby give you notice that we require a review of the yearly rent payable thereunder as from December 10 1978. We specify that the new rent payable as from December 10 1978 will be £10,000 per annum exclusive. This rent is proportionately in line with our recent lettings and reviews of other premises in the area with similar accommodation.
>
> Please will you kindly acknowledge safe receipt of this notice by signing the enclosed copy notice and returning it in the stamped addressed envelope provided.

The tenants replied to that letter on December 12 1978 in these terms:

> We are in receipt of your letter of December 8 1978 and note contents of same.
>
> We cannot agree with your rent increase, for, as you know, they are building a new Centre which is going to push Regent Street right out of position. There are shops very close to our own that are being offered for around £4,000 per annum. Perhaps, therefore, you could explain how you arrive at £10,000.

The landlords did not reply to that letter.

On April 19 1979, that is after the expiration of the three months in which a counternotice could have been served by the tenants under clause 3(c) of the proviso, the landlords demanded the payment of rent at the rate of £10,000 per annum. The rent at that rate was not paid. On February 26 1980 the landlords issued a specially indorsed writ claiming a rent at the increased rate. The defendants put in a defence to the effect that it was an implied term of the lease that any new rents specified by the landlords in a notice given under clause 3(a) of the proviso to clause 1 of the lease should be a bona fide and genuine pre-estimate of the open market rental value. Further, they said that there had in fact been a counternotice pursuant to clause 3(c) of the proviso by reason of the letter of December 12 1979. The landlords took out a summons for judgment under Order 14 and the points which had been foreshadowed in the defence were argued before Smith J. He adjudged that there was nothing in them and that, accordingly, the plaintiffs were entitled to judgment. He had before him an affidavit put in by the tenants. It was that of a surveyor deposing that, in his opinion, the maximum rent was £6,000 per annum.

There are two questions, therefore, for decision in this case: (1) should there be implied into clause 3(a) of the proviso words to the effect set out in the defence, namely, that the sum specified by

A
the landlords should be a bona fide and genuine pre-estimate of the open market rental value?; (2) was the tenants' letter of December 12 1978 a counternotice under clause 3(c) of the proviso?

As to the first point, I have no doubt at all that the words which, it was submitted, should be read into clause 3(a) cannot be so read. The proviso deals with definitions and procedure. Clause (1) of the proviso provides that the open market rental must be a rent which might reasonably be demanded by a willing landlord, etc. Clause (3) sets out how that reasonable rent is to be fixed and it is to be done in three stages. First, the landlord is to specify his figure; then there are to be negotiations for an agreement if his figure is not at once accepted. That is clear from clause 3(b). If there is no agree-
B
ment then there is to be arbitration. It is after all those steps have been taken that a reasonable rent is to be ascertained; and I can see no room at all in this lease for implying that the figure specified by the landlord must start with a bona fide and genuine pre-estimate of the rent. The whole purpose of the lease is to decide what that rent should be and, as Templeman LJ pointed out in the course of argument, if there was such an implied term to be written into clause 3(a), disputes would inevitably follow as to what was a bona fide and genuine pre-estimate of the rent. There would be no certainty at all and the lease must have intended the procedure in the proviso to deal with such uncertainties as there were.

My judgment in that respect is supported by the decision of
C
Goulding J in the case of *Davstone (Holdings) Ltd v Al-Rifai* [1976] 32 P&CR 18. The lease in that case was differently worded from the one in this case, but the issue in that case was the same as in this case; and Goulding J held that there was no room for the kind of implied term which Mr de la Piquerie on behalf of the tenants submitted should be read into this lease.

The second of the two points which were taken on behalf of the tenants I have found more difficult. The first step is to construe the letter of December 12 1978. It was a letter expressing disagreement with the figure put forward by the landlords. But it was a little more than that because, in the final paragraph, there was a request for an explanation as to how the landlords had reached the figure of
D
£10,000. On receipt of this letter the landlords would not have known with certainty what attitude the tenants were taking up. They had made it clear that they disagreed. On the other hand, by asking for particulars as to how the figure was made up, they may have been inviting the landlord to convince them that the figure of £10,000 or some figure near £10,000 was perhaps a reasonable figure in all the circumstances.

There is nothing, in my judgment, in the letter to indicate that, on December 12 1978, the tenants wanted to go to arbitration. The landlords were left without any clear indication as to what the tenants wanted to do. They might, perhaps, have got some further information about the tenants' intentions had they answered the
E
letter and given their reasons for fixing the figure at £10,000 per annum; but they did not. They were under no obligation to give any explanation. They were, in my judgment, entitled to sit back and wait to see what action the tenants took under clause 3(c) of the proviso and, when the tenants did not take any steps to call in aid the provisions of that clause, they were entitled to assume that they did not want to take advantage of it. It is worthy of note that, after the demand for rent at the new rate which was made on April 19 1979, the tenants did not seek the relief which, in my judgment, would have been available to them under section 27 of the Arbitration Act 1950. That section reads as follows:

F
> Where the terms of an agreement to refer future disputes to arbitration provide that any claims to which the agreement applies shall be barred unless notice to appoint an arbitrator is given or an arbitrator is appointed or some other step to commence arbitration proceedings is taken within a time fixed by the agreement, and a dispute arises to which the agreement applies, the High Court, if it is of opinion that in the circumstances of the case undue hardship would otherwise be caused, and notwithstanding that the time so fixed has expired, may, on such terms, if any, as the justice of the case may require, but without prejudice to the provisions of any enactment limiting the time for the commencement of arbitration proceedings, extend the time for such period as it thinks proper.

That section is of considerable importance, not only in relation to the construction and effect of the letter of December 12 but to one of Mr de la Piquerie's arguments about the need to imply a term. He G
laid great stress on the fact that the lease in subclause 3(a) fixes the tenant with notice, even though he has not in fact received the notice of the rent which the landlord wishes to charge him. But, as was pointed out by Mr Hague, when the tenant was asked to pay the increased rent after the expiration of the three months, he could have come to the court under section 27 and have applied for an extension of time. The tenants in this case accept that they did receive notice. They could have come to the court, even if they had not appreciated at first what their rights were under clause 3(c); but they did not and, in those circumstances, they cannot now complain about the operation of clause 3(c).

A more fundamental problem is the seeming difference between H
clause 1(b) and clause 3(c) of the proviso. Clause 1(b), to go back to it, provides that the open market rental value of the demised premises is to be determined in the event of disagreement between the parties by a chartered surveyor to be nominated by the president for the time being of the Royal Institution of Chartered Surveyors. Mr Hague submitted that that provision in the lease is ambiguous, because it is not clear whether, by that subclause, the rent is to be determined by an expert, in which case the Arbitration Act would not apply, or whether it is to be determined in accordance with the Arbitration Act and the provisions of clause 3(c) of the proviso.

One argument which has been canvassed is that the lease envis- J
ages two ways of finding out what the new rent should be. One is under clause 1(b), that is to say, by an expert, or under clause 3(c) of the proviso by arbitration. As there has been no request for arbitration by the tenants in this case it follows, it was submitted, that it is still open to the tenants to ask for a determination by an expert under clause 1(b). The difficulty about that argument seems to me to be this. The proviso starts by saying: ". . . for the purpose of subclause (b) and (c) hereof it is hereby agreed that the following definitions and provisions shall apply namely . . ." In other words, clause 1(b) has got to be construed in the light of what comes in the proviso, because the proviso is in the lease for the purposes of clause 1(b). K

Mr Hague put his argument in what I found an attractive manner. He submitted that the structure of this lease sets out in clause 1(b) a general proposition about the rent reviews: that, when there is a dispute, it is to be determined by an independent surveyor, and the proviso says how it is to be done. The first two clauses of the proviso deal with definitions, and clause (3) deals with the specific way the rent is to be calculated in the event of the parties not being able to agree. In other words, the determination by an independent surveyor which is envisaged in clause 1(b) is to be dealt with by the machinery provided by clause 3(c). I found that a compelling argument. An attempt to construe the lease in any other way produces this result. The parties would have provided in the lease for two L
entirely different ways of determining the new rent, one by arbitration and one by an expert. Determination by arbitration would entitle the parties to certain rights under the Arbitration Act. Determination by an expert would leave the parties bound by whatever figure the expert decides upon. It is difficult to envisage that, in this lease where the parties have specifically mentioned arbitration in clause 3(c), they, at the same time, could have envisaged that there should be a determination which had nothing whatsoever to do with arbitration.

Again, as was pointed out by Templeman LJ in the course of argument, the final words of clause 1(b)

M
> . . . to be determined in the event of disagreement between the parties by a Chartered Surveyor to be nominated by the President for the time being of the Royal Institution of Chartered Surveyors at the review date whichever is the higher and in either case the same shall remain constant during the whole period referred to in this subclause

are really surplusage. The only sense that can be made of them is to transpose them from 1(b) to a new subclause in clause (3) of the proviso. If that is done, back one comes to the situation whereby in one and the same lease there are to be two different ways with different legal consequences of determining the rent. I find that an impossible situation and, for those reasons, I would come to the

A same conclusion as Smith J, namely, that there was no defence to the landlords' claim.

Agreeing, TEMPLEMAN LJ said: I only add a few words of my own out of sympathy for the tenant, who has fallen into one of the many traps which leases lay for tenants.

The first point made by Mr de la Piquerie for the tenant is that there must be an implied term that the landlord will only specify a bona fide and genuine pre-estimate of the open market rental value. That appears to me to be an unnecessary and unworkable proposition. If a landlord puts forward a preposterous figure for rent, the tenant can always serve a counternotice. It would be
B ludicrous if the court were obliged to decide whether the landlord's figure was a bona fide and genuine pre-estimate in order to decide whether the landlord's request for an increased rent was valid or wholly void. The court does not exist to punish a landlord for being greedy, especially as the definition of "greed" varies from Shylock to Portia and from landlord to tenant.

Mr de la Piquerie submitted that the tenant could not serve a counternotice if he did not in fact receive the landlord's notice, although the tenant is conclusively deemed to have received a notice despatched by recorded delivery post. The tenant might, therefore, be bound by a proposed rent of which he had no knowledge. Therefore, so the argument runs, there must be some implied
C term which will enable the tenant to escape from a preposterous rent in those circumstances.

Apart altogether from the provisions of section 27 of the Arbitration Act 1950, which appear to give the tenant a remedy in appropriate circumstances, it seems to me that this tenant has accepted a lease under which he runs the risk of non-delivery of a landlord's notice just as under the Law of Property Act 1925 a tenant runs the risk that a notice may be delivered to the demised premises and yet never come to the attention of the tenant. At the same time, on general principles, I fail to see why a landlord cannot be content with the provisions of the Law of Property Act 1925, and I deprecate the inclusion in any lease of an express provision, such as we
D have in the present lease, whereby the tenant is conclusively deemed to have received a notice which he never had an opportunity to receive: for example, a notice which has been accidentally lost or destroyed by the Post Office authorities. That seems to me to be an unfair burden to put upon any tenant.

The tenant's present dilemma is also due to the fact that he has accepted a lease under which time is expressly made of the essence. Although that provision is applied to several subparagraphs, on analysis it can only do very real harm to the tenant who, because of that provision — if, as has happened in the present case, he accidentally disobeys the time limits — may find that he is saddled with a very onerous rent. The House of Lords has broken the implied
E tyranny of time as regards rent renewal clauses, but tenants are still harassed by express clauses making time of the essence. I think it is a great pity that any landlord should require, or any tenant should accept, a provision making time of the essence when the consequences are so onerous. In the present case, for example, if the landlord is correct as to the legal position but if the tenant's advisers are correct as regards the prevailing rent, the landlord will receive a grossly excessive rent for a period of five years simply because the tenant had either forgotten or was not sufficiently knowledgeable to understand what he had to do under the lease and made a mistake in the time factor. The result does not seem to me to be just.

Mr de la Piquerie's second point, namely that the tenant's letter
F dated December 12 1978 constitutes a counternotice, also seems to me to be untenable. By that letter the tenant refused to agree the landlord's figure of £10,000 and asked for an explanation. By the lease the tenant had three months from December 8 1978 in which to serve a counternotice electing for the rent to be determined by arbitration. I am unable to find in the tenant's letter of December 12 any words which enable me to construe it as a counternotice of the kind required by the lease. It is true that no magic formula is required but, in my judgment, the tenant must make it clear to the landlord that he proposes to have the rent decided by arbitration in accordance with the provisions of the lease. The landlord and tenant may reach a compromise before arbitration has been reached, but, nevertheless, on the plain words of the lease it seems G
to me that the tenant is directed to serve a counternotice making it clear to the landlord that he proposes to go to arbitration if that is his intention. I cannot spell that out of the tenant's letter, which simply said he did not agree with the landlord's £10,000 and asked for an explanation.

Similar views were expressed by Goulding J in *Bellinger v South London Stationers Ltd* (1979) 252 EG 699 and by Michael Wheeler QC, sitting as a deputy judge of the Chancery Division, in *Oldschool v Johns* (1980) 256 EG 381.

I was at first attracted by Mr de la Piquerie's third point that the landlord had fouled his own nest by providing in clause 1(b) for a determination by a chartered surveyor to be nominated by the H
president for the time being of the Royal Institution of Chartered Surveyors, and that this was a separate and extant provision which still bound the landlord and the tenant. Mr Hague, in a lucid exposition, has, however, persuaded me that the procedure mentioned in clause 1(b) is only a short method of referring to the machinery specified in the five provisos which are set out, those provisos expressely declared by the lease to be for the purpose *inter alia* of subclause 1(b). It would not make sense, in my judgment, to have the elaborate provisions of the provisos and to leave as some kind of long stop the similar provisions of clause 1(b).

Reading the provisions as a whole I accept Mr Hague's argument that those provisions show how open market rental value is to be J
arrived at. First of all, the landlord specifies it; secondly, if the tenant does not agree with that specification, there is the second method, namely of agreeing the rental value and, for that purpose, three months are allowed. Thirdly, if the landlord's specification is not accepted and if the landlord and the tenant cannot reach agreement, then there is the third method: determination by an independent surveyor appointed as an arbitrator in accordance with the provisions of proviso (c). Those three methods being marked out as being the methods of determining the open market value for the purposes of subclause (b), it does not, in my judgment, leave room for any fourth method of adjudication by an expert. I, too, would dismiss the appeal. K

Also agreeing, O'CONNOR LJ said: I agree with my Lords that there is no room for implying any such term as that for which the tenant contends in arriving at the figure which the landlord puts in his notice under proviso 3(a) of the lease.

I also agree that the letter of December 12 sent by the tenant to the landlord does not qualify as a counternotice under clause 3(c) of the proviso.

I have had more difficulty in coming to a conclusion on the third submission, namely that the lease provides for two different methods of determining the open market rental value at the review stage. Clause 1(b) is, in my judgment, not a submission to arbitra- L
tion. I think it clear that it is an agreement to have the open market rental value determined by a chartered surveyor in the event of disagreement between the parties; and, if it stood alone, there would be no room, in my judgment, for submitting that it was a submission or an agreement to arbitration.

What, then, is the effect of the proviso? My Lords have come to the conclusion that, even if there be any ambiguity in 1(b) (which I do not think there is), the effect of the proviso is that proviso 3(c), so to speak, provides an overriding construction of 1(b). I have difficulty with that concept, but there is another reason which has persuaded me in the end that it would not be right to dissent on that ground. The difficulty is this. There is clear machinery in clause 3 of M
the proviso for determining the open market rental value. As has already been stated, there is, first of all, a figure put upon it by the landlord; alternatively, an agreed figure (that is (b)) come to between landlord and tenant; or, alternatively, arbitration as set out in subclause (c). It will be seen that there are time-limits on that; and, however disadvantageous they may be to the tenant and however hard some of these clauses may appear, there must be some determination of the open market rental value. If all the time-limits which are dealt with in the proviso have gone by and nothing has been done other than the naming of a figure by the landlord, is it to be open to a tenant to come back, long after the event if need be,

A under clause 1(b) and ask for a determination of the rent? The answer, to my mind, comes back: "No, that will not do". Therefore, I also agree that the appeal should be dismissed.

The appeal was dismissed with costs.

QUEEN'S BENCH DIVISION

October 30 1980

(Before Mr Justice WOOLF)

B THE NEW ZEALAND GOVERNMENT PROPERTY CORPORATION v H M & S LTD

Estates Gazette February 14 1981

(1980) 257 EG 606

Landlord and tenant — Removal of fixtures — Effect of surrender of tenancy by operation of law, followed by grant of new tenancy to same tenant, on tenant's right to remove fixtures under old tenancy — Unusual case concerning a letting of Her Majesty's Theatre in the Haymarket — Question of right of removal, not directly, but in connection with the determination of rent under a rent review clause in lease — Interim award in the form of a special case for opinion of court by an official referee acting as arbit-
C **rator — In determining the "open market rental value" it was necessary to decide whether the value of tenants' fixtures, in the shape of carpets, curtains, seating and other equipment, should be taken into account — This in turn depended on whether the tenants had lost the right to remove these fixtures when they surrendered the old tenancy on entering into a new one, so that as a matter of law the fixtures reverted to the landlords — Authorities reviewed by judge, including Leschallas v Woolf and Slough Picture Hall Co (Ltd) v Wade — Held that in the present case there was a clear inference that it was not the intention of the parties that the tenant should be taken to give up his right to remove the fixtures on the determination of his old tenancy — It was not the usual practice on a surrender and**
D **regrant to include an express provision to safeguard the tenant's interest — But the judge's advice to the draftsmen of leases for the future should be noted.**

This matter came before the court for decision on an interim award made by Judge Hawser, one of the official referees, who was sitting as an arbitrator. The award was in the form of a special case setting out a number of specific questions and the arbitrator's answers thereon. The landlords were The New Zealand Government Property Corporation and the tenants were H M & S Ltd. The subject of the proceedings was Her Majesty's Theatre in the Haymarket and the relevant particulars of the leases and the questions in the interim award are set out in the judgment of Woolf J.
E R Sears QC and A Porter (instructed by Allen & Overy) appeared on behalf of the landlords; Ronald Bernstein QC and K Reynolds (instructed by Nicholson, Graham & Jones) represented the tenants.

Giving judgment, WOOLF J said: The issue in this case is as to the effect of a surrender of a tenancy by operation of law, on the grant of a new tenancy, on the tenant's right to remove fixtures under the old tenancy. The issue comes before the court in somewhat unusual circumstances and by an abnormal route.

The respondents are tenants of Her Majesty's Theatre in the Haymarket and the claimants are their landlords. The respondents
F are tenants under an underlease made on February 8 1973, which I will call the "new lease". That new lease has a rent review clause which provided that the rent for the second seven years of the term was to be the open market rental value of the demised premises, and defined "open market rental value" as:

> The annual rental value of the demised premises in the open market which might reasonably be demanded by a willing landlord on a lease for a term of years certain equivalent in length to the residue unexpired at the review date of the term of years hereby granted with vacant possession at the commencement of the term but upon the supposition (if not a fact) that the Lessee has complied with all the obligations as to repair and decoration . . . and there being disregarded (if applicable) those matters set out in paragraphs (a) (b) and (c) of section 34 of the Landlord and Tenant Act 1954.

After the procedure for rent review prescribed in the new lease G had been instituted, the parties failed to agree upon a new rent and they therefore entered into an agreement dated August 28 1978 whereby they referred their disputes and differences and all matters of fact and law arising therein to the award of a circuit judge discharging the functions of an official referee.

On August 24 1979 His Honour Judge Hawser made an interim award in the form of a special case for the decision of this court, Judge Hawser sitting as an arbitrator appointed by consent under section 11 of the Arbitration Act 1950 and Order 36 of the Rules of the Supreme Court. By the interim award the learned judge gave answers to three questions as follows: H

> Question 1: Is the open market rental value to be determined on the basis that the lease has ended, that the tenant has vacated and that the tenant has removed any tenant's fixtures that he could lawfully remove at the end of the term?
> Answer: Yes.
> Question 2: If the answer to Question 1 is yes is the open market rental value to be determined on the basis that upon vacating the tenant removed:
> (a) all the said fixtures whenever annexed or
> (b) such of the said fixtures as were annexed after July 1 1971 or
> (c) none of the said fixtures.
> Answer: (b).

Then Question 3 in an amended form: J

> In respect of each of the said fixtures which added to the letting value of the premises, is it to be treated as an improvement made immediately before July 1 1971 or as an improvement made when it was annexed to the premises?

and the answer was:

> When it was annexed to the premises.

Before me, the parties accepted that the learned judge's answer to Question 1 is correct, and most of the argument has turned on Question 2. The reference to July 1 1971 in Question 2(b) is because that was the date at which the parties agree they had agreed K the material terms for the grant of a new lease. Were it not for that agreement the critical date, subject to the provisions of the 1954 Act, would be the date on which the new lease was made, namely February 8 1973.

The reason why the answer to the second question is important in order to resolve the issues between the parties is that prior to the new lease there had been an earlier lease of the premises and the landlord contends that on the new lease being entered into the tenant lost the right which previously existed during the currency of the old tenancy to remove the tenant's fixtures. If, but only if, this were right would their value have to be taken into account in determining the increase of rent payable as a result of the review. L

The old lease was made on July 11 1899 between the Carlton Hotel Ltd, as lessors, and Herbert Beerbohm Tree and Playhouse Ltd, whereby in consideration of the expense incurred by the said Herbert Beerbohm Tree in erecting the theatre, the lessors, at the request of the said Tree, demised the theatre to Playhouse Ltd for 72 years from October 10 1898 (less 10 days) so that the term was due to expire at the end of September 1970.

By an agreement dated December 29 1952 G C Dobell & Co Ltd being the then lessees agreed first to assign to the respondent the premises demised by the lease for the residue of the term thereby created and second to sell to the respondent

> the fixtures fittings carpets curtains seating and other contents and equipment as now in and about the said Theatre and belonging to the Vendors M

for the price of £30,000.

By the date on which the term granted by the underlease was due to expire, the freehold reversion had become vested in the claimants. The claimants in due course served notice on the respondents to determine the old tenancy under section 25 of the Landlord and Tenant Act 1954. The respondent served the appropriate counter-notice and commenced proceedings in the Chancery Division for a new tenancy. Those proceedings were adjourned pending negotiations between the parties which on February 8 1973

A resulted in the new lease. The new lease was for a term of 21 years from October 1 1970, that being the date the term of the old lease expired; the old tenancy having been continued in the interim by section 24 of the 1954 Act.

The new lease described the premises demised as:

> ALL THAT piece of land situate in the Parish of St James in the said City of Westminster and being on the west side of The Haymarket and the south side of King Charles II Street Together with the buildings erected thereon and known as Her Majesty's Theatre which land and buildings are delineated on the plan annexed hereto and are thereon edged red and are hereinafter called "the demised premises".

B The new lease contained a covenant to repair which was in these terms:

> THE LESSEE . . . covenants . . . to put and keep the demised premises and the appurtenances thereof including the Landlord's fixtures doors windows and window frames and all fittings pipes and the sanitary and water apparatus and the painting papering and decoration thereof in good and substantial repair and condition and properly fitted up and decorated in a state in every respect fit for theatrical representations of a high class character (damage by fire and other insured risks save where the insurance moneys shall be irrecoverable in consequence of any act or default of the Lessee only excepted).

The tenant also covenanted:

> C Not at any time during the said term to use the demised premises for any other purposes than those of a Theatre used for the production of plays concerts, etc.

and covenanted

> To yield up the demised premises with the fixtures and fittings and additions thereto (tenants fixtures only excepted) at the expiration or sooner determination of the said term in good and substantial repair and condition (Landlords fixtures and fittings being duly renewed and replaced) in accordance with the several covenants hereinbefore contained.

There was also a collateral agreement dated February 8 1973 D whereby:

> In consideration of the Respondent having that day exchanged the New Underlease and having incurred or been about to incur substantial expenditure estimated as shown in the Schedule thereto

the respondent agreed

> to fulfil its repairing obligations under the lease the Claimants agreed that they would during the first 4 years of the term grant to the Respondent in each of those years an allowance of £5,000 as a deduction from rent subject to the production of certified receipts.

The schedule included items which would appear to be tenant's fixtures in normal circumstances and subject to the questions in E issue in this case.

The tenants have never sought to remove the tenant's fixtures which were attached to the premises during the old tenancy. However, this does not affect the importance of the answer given by the learned judge to the second question, since in carrying out the hypothetical exercises involved in fixing the rent, although those fixtures have not in fact been removed, if they are still removable by the tenants they would have to be ignored in fixing the increased rent. If they cannot be removed they would be treated as being included in the demise and would have to be taken into account subject to the correct answer to the third question, which I will deal with as a separate matter.

F In considering the effect of surrender by operation of law of the old tenancy which was continued under the 1954 Act, it is convenient to begin by stating certain general principles which are not in dispute.

Tenant's fixtures are normally only removable by the tenant during the term and for such longer period of possession which is (and I quote from *Hill & Redman's Law of Landlord and Tenant*, 16th ed, p 529) "in such circumstances that he is entitled to consider himself as tenant". So the right to remove continues while a business tenancy is continued under the 1954 Act. It is always open to the parties to make special provisions as to the right of removal, subject to such a special agreement not being inconsistent with rights of third parties. There is a qualification to these general G principles where the tenancy terminates in such circumstances that the tenant would not have time to remove fixtures. Then the right of removal continues for a reasonable time after the expiration of the term. I would refer in that connection to *V G Smith* v *City Petroleum Co Ltd* [1940] 1 All ER 260.

In accord with the general principles if there is an express surrender of a tenancy by a tenant, the right to remove subject to the terms of the surrender ends with the surrender. Furthermore, I regard myself as being bound by the authorities to which I will have to refer hereafter, to take the same view where there is an express surrender followed by an express grant of a new lease between the same parties of the same premises. There is, however, no clear authority H on what is to be the position where a tenancy comes to an end by normal effluxion of time and there is a new tenancy entered into immediately following the termination of the old tenancy. Where the tenant holds over as a yearly tenant, however, the position appears to be that the right to remove continues in accordance with the general principles which I have stated above.

Before turning to the authorities, it is right that I should make it clear that it is my view that the ordinary tenant, untutored in the intricacies of the law of landlord and tenant, would never expect that by taking a new tenancy he could, if nothing were said, lose the right, which could be of considerable value, to remove tenant's fixtures. In this respect I can see very real differences in practical J terms between an express surrender and a surrender by operation of law. In the case of an express surrender which is silent about what is to happen to fixtures, I can well understand why the courts have taken the view that the tenant has given up all his rights to what is then part of the premises. Where, however, there is no express surrender, it seems to me the position is different because what the parties have in mind is granting to the tenant new rights and not taking away old rights except in so far as they are inconsistent with the new rights. I mention this because it is clear from the reasoning set out in the case by the learned judge that he was influenced in giving the answer which he did to the second question by the fact that he regarded the authorities, as I do, as binding on him in K respect of express surrender and took the view that there was no logical basis for differentiating between the effect of an express surrender and of a surrender by the operation of law.

Turning to the authorities, I start with *Foa's General Law of Landlord and Tenant*, 8th ed (1957). I do so because Mr Bernstein, appearing on behalf of the respondent, found this his most embarrassing hurdle because in fact he was one of the editors responsible for the relevant edition of this authoritative textbook. I refer first of all to paragraph 1075:

> The mere fact that the tenant retains possession of the demised premises after the expiration of his interest does not extend the period during which he may sever fixtures, even though the reason for his holding over is that his L successor in the tenancy has failed to pay him an agreed price for them. And where the continuance in possession is under a new lease or agreement, his right to carry away the fixtures is determined, and he is in the same situation as if the landlord, being seised of the land together with the fixtures, had demised both to him; so that if a tenant taking a fresh lease of the demised premises wishes to preserve his right to remove fixtures they must be made the subject of a special agreement. So, where the old lease has come to an end by surrender, the transaction being what is usually termed one of surrender and renewal; though where such a surrender is one by operation of law upon the taking of a new lease, it may, in certain circumstances, be inferred that what are surrendered and re-demised are, not the premises in their actual condition, but the premises *minus* the tenant's fixtures. But if nothing be said about the fixtures at all, there would seem to be no ground for drawing such M an inference, and the general rule will not be displaced. It is thought, however, that the ordinary case where a tenant holds over by arrangement after the expiration of a lease, and pays rent on a tenancy from year to year, is not a continuance in possession under a "new agreement" within the meaning of the above rule, and that the right to remove fixtures when he gives up possession being one of the terms of the lease applicable to his new holding, the special agreement here spoken of would be implied.

And then paragraph 1076:

> The general principle appears equally to apply (apart from cases of fresh holding) where the tenant by his own act puts an end to the term, as where it expires by effluxion of time; so that after a surrender, for instance, the

A tenant's right to sever fixtures is gone. In the case indeed of surrender, it has been said that the right does not survive the time when the agreement upon which it is founded is made: the tenant as from that time being in the same position as if he had contracted to sell his lease. Upon a forfeiture, in spite of a somewhat guarded *dictum* of the Court of Appeal, that possibly the tenant may have a reasonable time afterwards, when he remains in possession, to sever his fixtures, it has more than once been decided that he has no such right, and that a person claiming under him is in no better position than himself.

The only part of the passages to which I have referred from which Mr Bernstein can get any comfort at all is the qualification which is made in respect of surrenders by operation of law that: "It may, in B certain circumstances, be inferred that what are surrendered and re-demised are, not the premises in their actual condition, but the premises *minus* the tenant's fixtures."

However, that passage has to be read in conjunction with the following sentence, which indicates that if nothing is said about fixtures at all there would be no ground for drawing such an inference.

The next textbook to which I shall refer is *Hill and Redman's Law of Landlord and Tenant*, 16th ed, paragraph 425 (p 529):

> Time for removal. — Where fixtures are removable by a tenant he is only entitled to exercise this right during the term, and if he omits to do so they become the absolute property of the reversioner, save that if the tenant C remains in possession after the term in such circumstances that he is entitled still to consider himself as tenant, his right to remove fixtures continues as long as this state of things lasts; and if he is a tenant holding on an uncertain tenancy, then his right to remove fixtures continues for a reasonable time after the determination of the tenancy. This rule applies in whatever manner the term comes to an end, whether by effluxion of time or by surrender or forfeiture; save that, in case of surrender or forfeiture, a third party, such as a mortgagee of the fixtures from the tenant, is entitled to a reasonable time within which to remove them. A tenant, who is entitled to remove fixtures under the stipulations of the lease, can remove them within a reasonable time after the determination of the term.
>
> From the principle that the tenant is not at liberty to remove fixtures after the determination of the term, including a determination by surrender, it D follows that where there is a surrender, followed by the grant of a new lease to the same lessee, the new lease includes the former tenant's fixtures as part of the demised premises, and, in the absence of express stipulation, any right which he had to remove them is gone. Similarly, tenant's fixtures left by a former tenant do not become tenant's fixtures of the subsequent tenant.

As I read that passage, there is not the same difficulty in Mr Bernstein's way because that passage is at least open to the interpretation that it is referring to a situation where there is an express surrender followed by a regrant; and it is not dealing with the question of surrender by operation of law.

The third textbook is *Woodfall on Landlord and Tenant*, 28th ed, and in relation to *Woodfall* I refer to paragraphs 1572 and 1573. E Without setting out the paragraphs in full in this judgment, it is sufficient if I draw particular attention to this statement in paragraph 1573:

> On negotiating a renewal, a tenant must be careful to preserve his right to fixtures, for without some express stipulation he may lose his right of removal.

It is right to say that with regard to *Woodfall*, like *Hill and Redman*, it does not expressly close the door to the right of removal continuing notwithstanding a surrender by operation of law, as opposed to an express surrender.

Mr Bernstein took me through all the authorities which were F relevant, which are cited in *Foa*, so as to support his contention that in the case of a surrender by operation of law, it is still open to the court to take the view which he submitted was in accord with commercial realities of the situation, namely, that such a surrender does not automatically bring to an end the right of removal of tenant's fixtures. In the end, the result turns on three cases which I must now examine.

The first of those cases is *Leschallas* v *Woolf* [1908] 1 Ch 641. I do not think it is necessary to extend this judgment by reading the headnote. It will suffice if I indicate that that was a case where there were three parties involved, landlord, tenant and subtenant. The tenant surrendered and there was a new tenancy granted direct to the subtenant. The decision was one of Parker J and at p 650 he said as follows: G

> Assuming, however, that I am wrong in this, I will pass to the second point in the case . . . Now it seems to me that a tenant who contracts for the surrender of his lease to his landlord is in the same position as if he had contracted to sell the lease, and cannot as against the surrenderee, any more than he could as against a purchaser, remove fixtures which were upon the freehold at the date of the contract, even though they might be of the nature of tenant's or trade fixtures. No contract for the surrender of the lease would, however, affect his subtenants or alter their rights without their consent, though, if a sublessee stepped in and rightfully removed fixtures after such a contract to surrender, the lessee might be unable to complete the surrender which he had contracted to make, or might be bound to make compensation for breach of his contract. I also think that a contract to H surrender a lease is a contract to surrender in possession free from subtenancies, and is not a contract to surrender subject to such subtenancies as may have been created by the surrenderor.
>
> This position appears to have been accepted by the solicitors for both parties.

And there he is referring to that case. I can miss out that and continue later on at p 651:

> On the construction of the surrender I am of opinion that all fixtures are included in the premises surrendered. If, as I think is at any rate arguable, the defendant was a consenting party to this surrender, his tenancy would be determined thereby, and any right he might have to remove fixtures would J be gone . . .
>
> The defendant

that is the subtenant

> claims now, not only that he had the right to remove during those three days the fixtures which had been affixed by him to the freehold during his tenancy under

the former tenant

> but that the right continued after the determination of such tenancy by the acceptance of a new tenancy under the agreement of August 7. It seems to me that there is no precise authority deciding that a tenant loses his right to K remove tenant's fixtures by the surrender of his tenancy to, and the acceptance of a new tenancy from, his landlord.
>
> It is quite clear that he loses the right by a surrender alone, but it is said that this applies only when he ceases to be the tenant, and not to cases where the tenancy is merely surrendered in order that a new tenancy on the same or different terms may be created so that he does not go out of possession of the property at all. In my opinion, however, if the tenant upon the surrender of his lease in order that a new lease may be granted makes no stipulation to the contrary, he does lose his right to remove tenant's fixtures, for the surrender of the demised premises *prima facie* includes fixtures, and the subject of the new lease is *prima facie* what is surrendered in order to be re-demised. Furthermore, it may well be that the value of the fixtures the right to remove which is thus abandoned is a material consideration in settling the terms of the new lease. The right to remove fixtures erected L during the term is, I think, a right coupled with and dependent upon the termor's interest. *Prima facie* when this interest ceases the right is gone, though there are, no doubt, exceptional cases in which, where the termor has remained in possession after the expiration of his term under such circumstances that the period of such possession can be looked upon as a mere prolongation of the term, he has been allowed to exercise the right after the term is ended.

And then he says that that seems to him to be the effect of a number of decisions, and adds:

> The law on the subject is summed up more or less by Thesiger L J in *Ex parte Brook* (1878) 10 ChD 100, 109. I will read the following passage in his M judgment: "The general presumption of law with reference to tenant's fixtures remaining affixed to the freehold when a term comes to an end is, that 'they become a gift in law to him in reversion', and are, therefore, not removable."

Pausing there: so far as an express surrender is concerned, it seems to me that that statement by Parker J is clearly binding upon me and should be given effect to by me in coming to my conclusion about this case. In saying that it is binding upon me I do have regard to the fact that that is a decision which has been referred to thereafter by the Court of Appeal, and by other courts, and has never been doubted as an expression of the law, subject to one matter which I will refer to hereafter in a later case. The learned judge later on goes on to say this:

> "But, however that may be, we are clearly of opinion that the case of a surrender of a lease by a tenant, while tenant's fixtures remain affixed to the freehold, does not, either upon principle or the authority of decided cases, give any right to the tenant subsequently to remove such fixtures. At the date of the surrender they form part of the freehold, and the law has no right to limit the effect of the surrender by excluding from it that which legally passes by it, and which has not been excluded from it by the bargain of the parties." Possibly, where the surrender is a surrender by operation of law upon the taking of a new lease, it may, under certain circumstances, be inferred that what is surrendered and re-demised is not the premises in their actual condition, but the premises minus the tenant's fixtures. Possibly also, parcel or no parcel being a question of fact, evidence might be admitted under this head. There may, too, be cases where the terms of the existing tenancy are varied only without the creation of a new tenancy: but in the present case I am of opinion that what

and he refers to the tenant

> contracted to surrender, and did surrender, included all fixtures; that what was agreed to be let to the defendant under the agreement of August 7 1907 was that [the tenant] agreed to surrender, and subsequently did surrender; and that, by accepting the tenancy commencing on August 12 1907, the defendant himself surrendered such tenancy as he might have had in the premises, and with it any right which he might have had to remove fixtures as an incident of the tenancy. This, indeed, seems to me to be the only conclusion consistent with the surrounding facts.

So far as that second passage of the judgment of Parker J is concerned, and particularly his reference to surrenders by operation of law, it could be said, and said with justification, that that passage is in fact *obiter*. None the less, I think it is right to approach this case on the basis that that passage should be regarded as setting out the appropriate approach to the law and one which I propose to adopt.

The next case to which I should refer is the case of *Slough Picture Hall Co (Ltd)* v *Wade* (1916) 32 TLR 542. The facts of that case are complicated, and again I do not propose to read those facts or indeed read the headnote. Certain of the effect of the judgment can, however, be understood from a passage at p 543 that refers to the fact that on April 14 1915 a series of events took place, the legal effect of which the learned judge, Scrutton J, said was of great importance. He then refers to that series of events and goes on:

> Both Wilson

Wilson was the tenant

> and his sublessee thus surrendered their interest in the premises to the landlords, who let to Mrs Wade, who relet to the Picture Hall Company.

What is being described in that passage is a situation whereby what was originally a letting by a landlord to a tenant, and a letting by a tenant to a subtenant, is replaced by a situation where the subtenant enters into a new agreement with a new tenant after the original tenant has given up its tenancy. The learned judge refers to the case which I have just cited, *Leschallas* v *Woolf*, and then recites what Parker J held, and goes on:

> I have come to the conclusion that this is a precise authority on the point before me and I must follow it. Wilson must have surrendered his tenancy to the brewers when he sold his interest to Mrs Wade and consented to her granting a tenancy to the Slough Picture Hall Company. The Picture Hall Company surrendered their subtenancy when they accepted a new subtenancy from Mrs Wade, who had acquired her tenancy from the brewers. As a general rule fixtures cannot be removed after the end of the term except by express agreement. In cases where the tenancy determines on an uncertain event, as in a lease for lives, the tenant may have a reasonable time to remove fixtures after the end of the last life. I have considered the suggestion of Mr Justice Parker, at p 654, that "possibly where the surrender is by operation of law upon the taking of a new lease it may under certain circumstances be inferred that what is surrendered and redemised is not the premises in their actual condition, but the premises minus the tenant's fixtures. Possibly also, parcel or no parcel being a question of fact, evidence might be admitted under this head." But in this case nothing was said about the fixtures at all, and I can find no ground for implying an exception out of the ordinary consequences of a surrender or for treating the premises demised to and sublet by Mrs Wade as anything less than the whole premises.

That case does create a difficulty from the tenant's point of view in this case, over and above that created by the words of Parker J in the first case that I cited, because it indicates that, on the facts before Scrutton J, he was not prepared to make an inference which Parker J envisaged as a possibility. However, it is to be noted that there is a distinction between that case and this in that unlike this case there was not in that case a straightforward grant of a new tenancy which operated as the surrender of the earlier tenancy, the new tenancy being between the same parties as the previous tenancy and, although on different terms, in practical effect a continuation of the old tenancy.

I propose now to approach this case relying on the same words as were considered by Scrutton J from the judgment of Parker J. Basing myself on those words I have come to the view that in this case there is a clear inference that it was not the tenant's intention to give up his right to remove the fixtures. It is correct that the demise did not refer to fixtures and the lease expressly reserved the right to remove tenant's fixtures at the end of the term granted by the new lease without indicating that tenant's fixtures meant other than tenant's fixtures attached to the premises after the commencement of that lease. However, so far as old fixtures are concerned, while I would accept that the lease can be regarded as being neutral, there is a firm inference, even in the case of a neutral lease, that the tenant was not intending to give up the right to his old fixtures when he enters into a new lease with the same landlord. If the parties turned their minds to the question I feel confident that that is what they intended. If, not having turned their minds to it, they had been asked, I feel confident the parties would have agreed that the tenant's rights should not be given up. Certainly that, in my view, would be the reaction of an objective bystander. I do not regard the collateral agreement as being inconsistent with this approach although I accept that there is a technical argument to be based on the use of the word "repair".

I regard my view as being confirmed by the fact that in Mr Bernstein's long experience in this field, he has never seen a new lease, which resulted in the surrender of an old lease by operation of law, which contained any provision to protect a tenant's right to remove fixtures. The very experienced solicitors acting for the claimants were unable to find any lease for which they had been responsible which protected a tenant's rights in such circumstances. I cannot believe that the very many tenants, of whom those practitioners can speak, whose old tenancies must have been terminated by operation of law on the grant of a new tenancy wanted to give up their right to remove the fixtures.

It may well be that there are leases where there is an express reference by the lawyers who were responsible for the drafting, to what was to happen on the taking of a new lease. Such an express reference would indeed be desirable because, as I have already indicated so far as the effects of a surrender by operation of the law is concerned on the right to remove fixtures, the law is certainly far from clear. The authorities do not deal with the matter precisely and, apart from *Foa*, the textbooks, so far as my investigations have revealed, and those of counsel who placed the relevant textbooks before me have revealed, do not deal with the matter in clear language.

It follows from what I have already said that, unlike the learned judge, I do not consider myself bound to follow alternative (b) in a situation as here where there is only a surrender by the operation of law, and not a surrender by express agreement. Accordingly, it follows that my answer to Question 2 would be (a) instead of (b) which was the answer of the learned judge.

So far as Question 3 is concerned, I can deal with the matter shortly, and I will do so by repeating as part of this judgment the reasons of the learned judge, upon which I do not feel I can improve, for rejecting the tenants' contention that the limitation contained in section 34(2) on disregarding improvements made by the tenant ran for 21 years in the case of tenant's fixtures from the date on which the tenants' fixtures became irremovable and not 21 years from the date on which they were affixed. It follows that in respect of the third question, I would reject the respondents' contentions which were argued before me.

A It was ordered that the matter be remitted to the arbitrator to continue the arbitration, the interim award to be varied in the light of the court's ruling that the open market value was to be determined on the footing that the tenants were entitled to remove all the tenants' fixtures whenever annexed. The tenants were awarded the costs of this hearing. A stay was granted pending consideration of an appeal.

CHANCERY DIVISION

B

November 20 1980

(Before Mr Justice FOSTER)

TELEGRAPH PROPERTIES (SECURITIES) LTD v COURTAULDS LTD

Estates Gazette March 21 1981

(1980) 257 EG 1153

Landlord and tenant — Rent review clause — Whether certain notices given by landlords were valid and effective for the purpose of setting the rent review machinery in motion — Also question
C **whether lease of adjacent premises between the same parties framed by reference to the original lease incorporated the rent review clause — Held as a matter of construction of the second lease that the rent review clause was not incorporated — The main question, which arose on the first lease, was whether, although time was not of the essence of the clause, a notice given more than six years after the date provided in the clause for giving notice was effective — Held that a delay of more than six years was unreasonable and that the review machinery could not be invoked after such a lapse of time — United Scientific Holdings Ltd v Burnley Borough Council cited — James v Heim Gallery (London) Ltd distinguished**

D This was an originating summons by which the plaintiffs, Telegraph Properties (Securities) Ltd, sought a declaration that certain notices were effective to operate a rent review clause. The plaintiffs were the successors in title of the original lessors and the defendants, Courtaulds Ltd, were the successors of the original lessees. The properties which were the subject of the present proceedings were portions of Lansen House, Gladstone Street, Darlington, and adjacent premises.

D M W Barnes (instructed by Radcliffes & Co) appeared on behalf of the plaintiffs; N B Primost (instructed by S B Gilinsky & Co) represented the defendants.

Giving judgment, FOSTER J said: By this originating summons the
E plaintiffs seek a declaration that certain notices given to the defendants constitute a valid and effective notice for the purpose of putting into effect a rent review clause and a declaration that the plaintiffs are entitled to have the rents payable in respect of the last seven years of two leases reviewed.

The conveyancing history. By a lease dated February 2 1966 made between A G Goldsworth and Co Ltd, the lessor, and Goldsworthy Textiles Ltd, the lessee, the lessor demised to the lessee the ground floor, mezzanine and first floor of a property, Lansen House, Gladstone Street, Darlington, for a period of 14 years from January 1 1966 for the yearly rent of £2,000. By clause 5 of that lease it was provided as follows:

F If either party so desire at the end of the first 7 years to apply for a revision of the rent and shall give 3 months notice in writing of such desire then such rent shall be revised in accordance with the rents prevailing for similar premises in the same area excluding any goodwill which shall have become attached to the demised premises by reason of the carrying on thereat of the business of the lessee and excluding any effect on rent of any improvements carried out by the lessee and in the event of the parties failing to agree then the matter should be referred to a single arbitrator to be approved by the lessor and lessee under the provisions of the Arbitration Act 1950 or any statutory modification or re-enactment for the time being in force.

The first seven years ended on January 1 1973 and three months before that was October 1 1972, so that any notice should be given previously to that date.

On March 28 1967 a deed of surrender and of variation was G
entered into between the same parties. By that deed the lessee surrendered to the lessor part of the premises demised by the lease of February 2 1966 and in consideration of that surrender the rent was reduced as from March 12 1967 to £1,250. This deed contained a clause 4 by which the lessee could by a notice given 12 months before the expiration of the lease require a grant of a further term of seven years at a rent to be determined in accordance with clause 5 of the lease.

On the same date, March 28 1967, and made between the same parties, the lessor demised unto the lessee certain other premises not being the premises surrendered but being premises adjacent to the original premises leased. The term was from March 12 1967 H
until December 31 1979 at the rent of £4,300 per annum. I must consider in much greater detail this lease later in this judgment. It also contained a provision, clause 4 (iv), for a grant of a further seven years from the expiration of the lease.

It is common ground that the plaintiffs are the present lessors of the property and the defendants are the present lessees. In fact the lessors gave no less than four notices to try to activate the rent renewal clauses. The first was dated October 11 1978, when they wrote to the tenant a letter which reads:

This notice is given in accordance with clause 5 of the lease dated February 2 1966 as referred to in the lease dated [it has got "March 28 1966" but it ought to have been "1967"] under which you occupy the premises of the J
above address. We therefore put forward the figure of £15,000 per annum as the new rental to become payable from the review date being 1st January 1973.

Then reference is made to the business rent control. I should have read the first paragraph of that letter, which is:

As authorised agents of your immediate landlords in relation to the above property we hereby give you notice of the landlords' intention to exercise their right to call for the review of the rent.

The second notice was written to Courtaulds Ltd, was dated January 24 1979, and reads as follows: K

We write with reference to our letter dated 12

it should of course be 11

October 1978 in which we stated our client's intention of operating the rent review relating to the above premises in accordance with clause 5 of your lease dated February 2 1966. For the sake of clarity and completeness we hereby give you notice that our clients also intend to call for a review of the rent payable relating to the above premises held by your company under a lease dated March 28 1967, which refers extensively to the above mentioned lease. The figure of £15,000 per annum mentioned in our letter dated October 12 relates to the whole of the premises which are let to your company under the above mentioned two leases. L

A letter was sent to Goldsworthy Textiles Ltd in exactly the same terms.

Finally, on July 27 1979 the solicitors wrote to the receiver, who had then been appointed for the property, and said this:

Hereby give you notice that your landlords desire to apply for a revision of the rent payable in respect of the premises let to you (a) by a lease dated February 2 1966 and (b) a lease dated March 28 1967 made between the same parties. The said notice is given pursuant to clause 5 of the lease of 1966 and clause 2 of the lease dated March 28 1967.

Two questions, in my judgment, arise. The first is: Does the lease of March 28 1967 of the additional premises contain a clause for a rent review; and, secondly, in any event can the plaintiff invoke the rent M
review clause by giving notice some six years and 10 days after the required date?

The lease of March 28 1967. I have come to the conclusion that this lease does not include a clause for a rent review. In the lease of February 2 1966 the date for a review was, as I have said, January 1 1973 and the notice had to be given before October 1 1972. Paragraph 2 of the second lease reads as follows:

Subject as to the premises demised the terms of years granted, the rent reserved, and the matters specifically referred to hereafter, this demise is made upon the same covenants on the part of the lessor and lessee and guarantor respectively and to the same stipulations and conditions as were

> A expressed and contained in the lease dated February 2 1966 and made between Goldsworth and Company and Goldsworthy Textiles Limited and Bell Nixon of the third part, as if the same were herein set forth at length with such modifications only as are necessary to make the same applicable to the present demise save that it is hereby agreed and declared that sub-clauses (vi), (viii), and (ix) of clause 2 and sub-clauses (iii) and (iv) of clause 3, and clause 8 of the said lease of February 2 1966 shall not apply to this demise.

In the opening the words "the rent reserved" refer in my judgment to the sum of £4,300 and do not include any additional rent which might be payable if there was a rent review clause. The opening words "subject as to" exclude in my judgment the term of years
B granted, the rent reserved, and matters specifically referred to, from the terms of the lease of February 2 1966 to be implied. The lease was to run from March 12 1967 to December 31 1979 and it would have been extraordinary if there was implied a rent review on January 1 1973, that is some five years nine months and ten days from its commencement. It was pointed out that clause 5 of the original lease was not specifically excluded; but the reason for this is that it is used for the possible extension of the term under paragraph 4(4). It is also to be noticed that in the original lease of February 2 1966 the lessee covenanted by a clause 2(i) "To pay the rent hereinbefore reserved and to make the payment by way of additional rent hereinafter provided". In the lease of March 28 1967 the
C lessee covenanted by clause 3(i) "To pay the reserved rent at the times and in the manner aforesaid". There is no mention of any additional rent. In my judgment the clause for the rent review is not to be included in the lease of March 28 1967. That does not, however, mean that I should not deal with the second question, because whether I am right or wrong on the construction which I have given to the 1967 deed the question still arises as to whether the rent review clause is or is not effective.

The defendant levelled several criticisms of the four notices given: that there should have been two notices; that the proposed rental should have been put forward for each of the demised premises; and that in the final notice no figure was mentioned. In my
D judgment there is nothing in any of those criticisms.

Delay. Assuming that the first notice was given on October 11 1978, this amounted to a delay of some six years and 10 days since its due date. Can the plaintiff succeed after such a delay? The plaintiff relies on the decision of the House of Lords in *United Scientific Holdings Ltd v Burnley Borough Council* [1978] AC 904. The decision was actually given in March 1977. It had always been thought before that case that in rent review clauses time was of the essence, so that if the notice was one day late the landlord could not enforce the provision. There were two appeals before the House of Lords and in neither was there any undue delay. Lord Simon, at p 942, says this:

> E The Court of Chancery would therefore decree specific performance of the conveyance notwithstanding that various steps leading to completion had not been observed timeously, provided that B had been guilty of no such delay as to make it unreasonable for him to call on A to complete out of time or that it would otherwise be unfair to A.

At p 951 Lord Salmon says this:

> Nevertheless any unreasonable delay caused by the landlords and which is to the tenants' prejudice would prevent the rent being revised after the review date.

In that case the House of Lords was not dealing with a case where
F there had been long delay in serving the notice. In the recent case of *James v Heim Gallery (London) Ltd* (1980) 256 EG 819 the notice which had to be given was not less than six months before June 24 1972, ie before December 24 1971. Notice in fact was given on June 5 1972, a delay of some five months and ten days, and before the expiry of the first seven years. In that case the defendants relied on certain facts which they said amounted to promissory estoppel and the learned judge found for the defendants on that ground. That decision was reversed on October 7 1980 and I have a transcript of that which is numbered 80/660 from the Law Library. That decision was based on the fact that the delay in giving the notice was not great and that the acts relied on by the defendants were insufficient to raise a plea of promissory estoppel. Again, that court was G
not faced with long delay in giving the notice.

Counsel for the plaintiff went so far as to say that any such delay would be valid not only after the expiration of the lease but at any time in the future, a submission which I find fanciful. In a case such as the present, where the delay is over six years before the notice itself is given, I do not think that the doctrine of promissory estoppel has to be considered at all. In my judgment the plaintiff has been guilty of such a delay as to make it unreasonable for it to call on the defendant for a rent review and to do so would be of necessity unfair for the defendant. I also refer to the fact that in the *Heim* case the matter was raised some six months after the House of Lords' decision. But in this case the matter was not raised for a year and a H
half. That again seems to me to be delay to which the court will give heed. I therefore need not go into the facts in regard to the matters which have been put in evidence as to the effect both on the plaintiff and the defendant if the notice is not held to be good.

One further matter I should deal with is that in the *James v Heim Gallery* case no mention whatever is made of the legislation which controlled rents and that legislation was not repealed until April 1 1976. The plaintiff does not seek to say that the rent could be increased before that date. Exactly the same position must have occurred in the *James v Heim Gallery* case and I for my part cannot see that the plaintiff gains any comfort from that fact. It follows that in my judgment I do not propose to make either of the two declar- J
ations for which the plaintiff asks.

The summons was dismissed with costs.

CHANCERY DIVISION

November 21 1980

(Before Mr Justice FOX)

WELLER v AKEHURST AND ANOTHER K

Estates Gazette March 28 1981
(1980) 257 EG 1259

**Landlord and tenant — Rent review clause in lease — Time of the essence — Lease providing for rent of £850 per annum for first seven years of 14 years' term — Rent for next seven years at "open market value" as defined in the lease — Such value to be (a) the sum specified in a notice by the landlord sent to the tenant before the beginning of a clear period of two quarters of a year beginning on one of the usual quarter days preceding the review date, or (b) agreed between the parties within three months after the posting of such notice, or (c) determined by the election of the tenant by an independent surveyor according to a procedure laid down — Landlord's L
notice under (a) not sent within the required time — Submitted by landlord that nevertheless the rent from review date should be at the open market value and that the court should order an inquiry as to that amount — Submission rejected — As a result of the failure to adhere to the time-table, in which time was expressly made of the essence of the contract, the rent would continue at the old figure of £850 per annum**

The plaintiff in these proceedings by originating summons was Florence Lily Weller, the lessor of a shop, store and dwelling-house with which the proceedings were concerned, and the defendants, Richard Charles Akehurst and Jill Susan Akehurst, his wife, were M
the lessees. The case raised questions as to the correct interpretation of the rent review clause in the lease in the events which had happened.

N L V Mears (instructed by Mayo & Perkins, of Eastbourne) appeared on behalf of the plaintiff; J S Colyer QC and Jonathan Brock (instructed by Burton & Ramsden) represented the defendants.

Giving judgment, Fox J said: This case is concerned with the interpretation of a rent review clause in a lease. The lease is dated August 14 1972 and is a lease of a shop, store and dwelling-house. The term demised was 14 years from July 1 1972.

A Clause 1 of the lease makes the following provisions as to rent:

(a) For the first Seven Years of the said term a yearly rent of £850
(b) For the next Seven Years of the said term the open market value of the demised premises at the review date which revised rent shall remain constant during the whole period referred to in this subclause. . . .
Provided that for the purpose of subclause (b) hereof it is hereby agreed that the following definitions and provisions shall apply namely:
(1) The expression "open market value" means the annual rental value of the demised premises on the open market which might reasonably be demanded by a willing Landlord on a lease for a term of years certain equivalent to the length of the residue unexpired at the review date of the term of years hereby granted with vacant possession at the commencement of the term but upon the supposition (if not a fact) that the tenant has B complied with all the obligations as to repair and decoration herein imposed (but without prejudice to any rights or remedies of the lessor in regard thereto) and there being disregarded (if the section is applicable) those matters set out in paragraphs (a) (b) and (c) of Section 34 of the Landlord and Tenant Act 1954 and there being disregarded (so far as may be permitted by law) all restrictions whatever relating to rent or to security of tenure contained in any statute or order or regulation thereunder . . . such lease being on the same terms and conditions as this present demise without the payment of any fine or premium.
(2) The expression "review date" means the expiration of the Seventh year of the said term for the purpose of ascertainment of the open market rental value under sub-clause (b) hereof.
(3) The open market rental value shall be determined in manner following C that is to say it shall be such annual sum as shall be
(a) specified in a notice in writing signed by or on behalf of the Lessor and posted in a prepaid envelope addressed to the Tenant at the demised premises at any time before the beginning of a clear period of two quarters of a year (commencing on one of the usual quarter days hereinbefore mentioned) immediately preceding the review date or
(b) agreed between the parties before the expiration of three months immediately after the posting of such notice as aforesaid in substitution for the said sum or
(c) determined by the election of the Tenant (to be made by notice in writing to be served by the Tenant upon the Lessor not later than the expiration of the said three months) by an independent surveyor appointed for that purpose by the parties jointly in writing or upon their failure to agree upon D such appointment within one month immediately after the service of the said counter-notice then by an independent surveyor appointed for that purpose on the application of either party alone by the President for the time being of the Royal Institution of Chartered Surveyors and in either case in accordance with the provisions of the Arbitration Act 1950.
(4) (1) In the event of the determination of such independent surveyor not having been published prior to the review date for any reason whatever then in respect of the period of time (hereinafter called the said interval) beginning with the review date and ending on the quarter day immediately following the date on which such determination shall have been published the Tenant shall pay to the Lessor in manner hereinbefore provided rent at the yearly rate payable immediately before the review date PROVIDED that at the expiration of the said interval there shall be due as a debt payable by the E Tenant to the Lessor on demand a sum of money equal to the amount whereby the yearly rent determined by such independent surveyor shall exceed the yearly rent at the yearly rate aforesaid but duly apportioned on a daily basis in respect of the said interval.

Subparagraph (2) I can omit.
Then:

(5) All stipulations as to time in the foregoing subclauses numbered (1) (2) (3) and (4) shall be of the essence of the contract and shall not be capable of enlargement save as agreed in writing by the parties.

By a letter to the tenant dated January 25 1979 on behalf of the F lessor, it was specified that the annual rent payable under the lease for the period of seven years commencing on July 1 1979 should be £2,500 per annum. The notice contained in that letter purported to be given under the provisions of clause 1 (3) (a) of the lease. In fact, however, the clear period of two quarters of a year commencing on one of the usual quarter days and immediately preceding the review date commenced on December 25 1978 (not on January 25 1979). Clause 1 (3) (a) requires the notice to be sent at any time before the commencement of the said clear period to which I have referred.

In these circumstances, the originating summons asks:

(1) For a declaration that the letter of January 25 1979 was a valid notice for the purpose of clause 1 (3) of the lease.

(2) Alternatively, that it may be determined whether upon the true construction of the lease the rent reserved for the seven-year period from the review date is (a) the open market value (as defined in the lease) of the premises at the review date or (b) £850 per annum or (c) some other and if so what rent. G

In the event, it was conceded by the lessor (the plaintiff) that the letter of January 25 1979 was *not* a good notice for the purposes of clause 1 (3) of the lease as it was not given in time. The question then is what rent is payable for the seven years of the lease commencing on the review date.

It is contended on behalf of the lessor that such rent is the open market value (as defined in the lease) of the premises at the review date and that the court should now order an inquiry to determine H that. It is submitted by the tenant that such rent is £850 per annum (ie the original rent).

The argument on behalf of the lessor, put shortly, is this. The lease expressly provides that for the second seven years of the term (ie the seven years from the review date) the rent payable shall be the open market value. That expression is defined in the lease in terms sufficiently certain to be capable of ascertainment. Accordingly, while it is true that the particular machinery contained in clause 1 (3) of the lease for ascertaining the open market value cannot be operated (because the lessor failed to serve in due time the notice which would bring clause 1 (3) into operation), nevertheless the open market value is capable of ascertainment and the court J should direct an inquiry for that purpose. The rent so ascertained would then be the rent payable for the second seven years of the term.

I do not feel able to accept that argument. The open market value as defined in the lease is, it is true (as indeed is conceded by the lessor) expressed with sufficient certainty to be capable of ascertainment. But the lease expressly provides in clause 1 (3) that the open market value shall be determined "in manner following that is to say it shall be such annual sum as shall be . . ." There then follows the series of provisions, all of which hinge upon the giving by the lessor of such a notice as is specified in clause 1 (3) (a). The lessor failed to give such notice within the time specified in the lease. K Further, clause 1 (5) provides that all stipulations as to time in the relevant provisions of the lease shall be of the essence of the contract and shall not be capable of enlargement save by agreement between the parties.

The general presumption is that strict adherence to the timetables specified in the rent review clause is not of the essence of the contract (see the decision of the House of Lords in *United Scientific Holdings Ltd v Burnley Borough Council* [1978] AC 904). But as Lord Diplock observes at p 923:

It is not disputed that the parties to a lease may provide expressly that time is or time is not of the essence of the contract in respect of all or any of the steps L required to be taken by the landlord to obtain the determination of an increased rent, and that if they do so the court will give effect to their expressed intention.

And at p 936 he said:

. . . the best way of eliminating all uncertainty in future rent review clauses is to state expressly whether or not stipulations as to the time by which any step provided for by the clause is to be taken shall be treated as being of the essence.

The position therefore in the present case is that although the M lease provides that the rent shall be the open market value and defines that term with sufficient precision, the lease also requires that the ascertainment of the open market value shall be effected in a particular way and stipulates that time shall be of the essence of the provisions in that procedure.

In my judgment, the parties cannot have intended that if the lessor failed to observe the time provisions she could still be free to obtain the determination of the open market value by the court. The parties chose the procedure for rent review with precision and made it plain that the procedure was only to be available on a strict timetable. The lessor having failed to observe that timetable, it seems to me that there is thereafter no room for the operation of

A any procedure to determine the open market value. If the open market value cannot "be determined in manner following" (to use the language in clause 1 (3)) it cannot, in my view, be determined at all. Clause 1 (3) states in terms that the open market value shall be "such annual sum as shall be . . .". There then follows (a) (b) and (c). That language, coupled with the direction for determination "in manner following" excludes, I think, the possibility of it being any other sum or determined in any other way. For the court now to take upon itself the determination of the open market value would, I think, be to fly in the face of the clearly expressed requirement of strict compliance with the time provisions. I think that the rent revision procedure was intended to be carried out timeously or not
B at all. The provision for revising the rent was in the lessor's favour; only the lessor could invoke it. But the lessor could only invoke it upon strict terms. I see no reason to suppose that the parties intended that the lessor should get a rent increase in any other way.

I was referred to the decision of the Court of Appeal in *Kenilworth Industrial Sites Ltd v EC Little & Co Ltd* [1975] 1 WLR 148, but I do not think that that decision assists me on any aspect of the present case, since in *Kenilworth* it was held that time was not of the essence of the provisions. And, in so far as the case proceeded upon the basis of a dichotomy between "option" clauses on the one hand and "obligation" or "machinery" provisions on the other, it is now inconsistent with the decision of the House of Lords in the *United*
C *Scientific* case (see per Lord Diplock in the latter case at p 936). I was also referred to *Brown v Gould* [1972] Ch 53. In that case the lease contained a formula for ascertaining the rent but no machinery for working out the formula. The question was whether the provision was void for uncertainty. It was held that it was not; the court could itself work out the formula. In the present case the lease provides both a formula and a specific mode of working out the formula. The question is whether, in view of the provisions that time is of the essence, any other mode of working out the formula is permissible. For the reasons which I have given, I do not think that it is.

I should add this. If the lessor gave a *bona fide* notice under clause
D 1 (3) (a) stating an amount which was in fact above the open market value and the tenant failed to give a counternotice in due time, the tenant could not, I think, invoke the definition to redress the position. The tenant would have to accept the amount stated in the landlord's notice. It would be a curious result if the tenant, having failed to observe the timetable, is bound by the clause 1 (3) procedure, but if the lessor fails to serve notice in time, the lessor can still get the open market value determined by the court.

I should mention that the expression used in the opening sentence of clause 1 (3) is "open market *rental* value" not "open market value", but that, I think was merely a slip and the contrary was not contended.

E The result in my view is that the provisions of clause 1 (3) are wholly inoperative and that the provisions of the lease do not permit the open market value to be determined in any other manner than that specified in clause 1 (3).

In those circumstances, the tenant asserts that the rent for the second seven years is £850 per annum. It is contended that the position is expressly dealt with by clause 4 (1). It is said that the language of that provision is very wide. It begins: "In the event of the determination of such independent surveyor not having been published prior to the review date for any reason whatever . . ." While I agree that that language is indeed wide, it seems to me that clause 4 (1) is assuming that an independent surveyor has been duly
F appointed under clause 1 (3). The provision, I think, is simply dealing with the period which elapsed after the review date until the quarter day after the determination of the rent by the surveyor. The whole provision, I think, assumes that the surveyor will issue a determination, but that there will be an interval before he does so. Thus the proviso begins "PROVIDED that at the expiration of the said interval . . ." I do not think that clause (4) (1) is dealing at all with the present circumstances where no surveyor has been or can now be appointed. In my view, therefore, the lease contains no express provisions to deal with the present case.

What then is the position? That the parties should, in the circumstances, have contemplated that no rent at all should be payable for the residue of the term, I reject altogether. The lease grants a term G
for 14 years and it is inconceivable that the parties contemplated that no rent might be payable during the second period of seven years. A liability on the part of the tenant to pay rent must have been intended. The question is how much? In my opinion, if the parties had been asked at the time of the grant of the lease what was to happen if the lessor chose not to serve a notice under clause 1 (3), or failed to serve such notice in time, it is wholly reasonable to suppose that they would have said that the old rent of £850 per annum was to continue. Thus, I do not think that the parties could have been contemplating any reduction of the rent at all. The lease was granted during a period of inflation, and it is clear that only the lessor could set the clause 1 (3) procedure in motion. It was a H
provision for the lessor's benefit.

On the other hand, I see no reason to suppose that they would have contemplated that the rent should be increased in the event of no valid notice being given under clause 1 (3) by the lessor. The lessor was adequately protected by clause 1 (3) if he wished to increase the rent in line with the market. But, it having been agreed that time was of the essence of those provisions, it cannot, in my view, have been contemplated that, having failed to operate clause 1 (3), the lessor should still be free to get the rent increased by a quite different machinery not expressed in the lease at all. Clause 1 (3) was in general for the benefit of the lessor, but the time provisions in clause 1 (3) (a) were for the protection of the tenant. J

What I am dealing with is a rent review provision containing strict requirements for its operation. It did not operate because the person at whose option the rent review was to be made did not comply with the requirements. I think that in those circumstances the reasonable and necessary inference is that the old rent continues. I will so declare.

CHANCERY DIVISION

November 3 1980 K

(Before Vivian PRICE QC, sitting as a deputy judge of the division)

VINCE AND ANOTHER v ALPS HOTEL LTD AND ANOTHER

Estates Gazette April 25 1981

(1980) 258 EG 330

Landlord and tenant — Rent review clause in lease of hotel — Lease for 21 years with provisions for review at 7 and 14 years — Agreement on variation of rent, or application for appointment of arbitrator, to be made before a date mentioned in review L
clause — No agreement reached or application made before date in question — Decision of House of Lords in United Scientific Holdings Ltd v Burnley Borough Council subsequently drawn to the attention of the lessors, nearly two years after the first review date — Lessors claimed declaration that review provisions could be operated and any increase of rent recovered retrospectively — No evidence in present case that time was of the essence of the rent review machinery — Declaration in favour of lessors granted — Defence based on estoppel rejected

In these proceedings Edmund David Vince and Maria Vince, freehold owners of the Malvern Hotel, 3 and 5 Petersham Road, M
Richmond, West London, sought a declaration as plaintiffs against Alps Hotel Ltd, the lessees of the Malvern Hotel, and Krishan Chand Chhabra, a director of Alps Hotel Ltd and a guarantor of the lessees' liabilities under the lease. The declaration sought was that the plaintiffs were entitled to operate the provisions of the rent review clause and that a variation of rent thereunder would be valid retrospectively to December 10 1975, the first of the review dates under the clause.

P J Martin (instructed by Stone Rowe Brewer & Devane) appeared on behalf of the plaintiffs; N Patterson (instructed by Duke-Cohan & Co) represented the defendants.

Giving judgment, MR VIVIAN PRICE QC said: Edmund David Vince and Maria Vince are and have been at all relevant material times the freehold owners of a property known as The Malvern Hotel, 3 and 5 Petersham Road, Richmond, in West London. Their predecessor in title to the property was a company called Muncip Developments Ltd, who by a lease dated December 10 1968 let the property to a person named Brian Middler for a term of 21 years from December 10 1968 at an initial rent of £4,000 per annum. In 1973 by a deed of licence made on June 20 1973 between Edmund David Vince and Maria Vince of the first part, Brian Middler of the second part, Alps Hotel Ltd of the third part and Krishan Chand Chhabra of the fourth part, Edmund David Vince and Maria Vince granted a licence to Brian Middler to assign the lease to Alps Hotel Ltd in consideration of which Alps Hotel Ltd and Mr Chhabra covenanted with the Vinces to perform and observe the several covenants and conditions contained in the lease. Thereafter, by an assignment dated June 22 1973, Brian Middler assigned the lease to Alps Hotel Ltd so that at all relevant times the Vinces have been the freehold owners of the property; Alps Hotel Ltd has been the leaseholder of the property; and Mr Chhabra (a director of Alps Hotel Ltd) has been in effect the guarantor that Alps Hotel Ltd will perform and observe the covenants and conditions contained in the lease.

The lease dated December 10 1968 includes provision for the review of rent at seven-year intervals. Clause 1 of the lease is in the following terms:

> In consideration of the rents and covenants by and on the part of the lessee hereinafter reserved and contained the landlord hereby demises and releases unto the lessee all that premises known as The Malvern Hotel, 3 and 5 Petersham Road, Richmond in the London Borough of Richmond-upon-Thames to hold the demised premises unto the lessee from the 10th day of December 1968 for a term of 21 years paying therefor yearly during the first seven years of the said term the rent of £4,000 and during the residue of the said term a yearly rent of £4,000 or such amount whichever be the greater as may be agreed between the landlord and the lessee before the 10th day of December 1975 and 1982 or in the absence of such agreement as may be determined by an arbitrator to be nominated by the President for the time being of the Royal Institution of Chartered Surveyors on the application of the landlord made before but not more than two quarters before the said 10th day of December 1975 and 1982 and so that in the case of such arbitration the amount to be determined by the arbitrator shall be the amount which shall in his opinion represent a fair yearly rent for the demised premises having regard to rental values then current for property let without a premium with vacant possession and to the provisions of this lease other than the rent hereby reserved all such respective rents to be paid without any deduction except for the landlord's property tax by equal quarterly payments in advance on the four usual quarter days in each year the first of such quarterly payments to be made on the execution hereof.

It will be seen that in order for there to be a change of the rent amount chargeable during the second seven-year period the lease provides that there must either be agreement between the parties before the expiry of the first seven-year period, ie December 10 1975 or the landlord has to make an application to the president for the time being of the Royal Institution of Chartered Surveyors before December 10 1975 but not more than two quarters before. In fact in this case no agreement was reached between the Vinces and Alps Hotel Ltd before December 10 1975 about the rent to be paid during the second seven-year period of the lease, even though there had been some desultory negotiations which came to no conclusion; nor had the Vinces by that date (that is December 10 1975) made any application to the president for the time being of the Royal Institution of Chartered Surveyors for the appointment of an arbitrator to decide the rent payable.

The correspondence has been exhibited to various affidavits that have been read before me and has all been put together in a convenient bundle. I do not think it is necessary for me to go through that correspondence in detail, since there can be no doubt, and it is agreed between the parties, that the conditions laid down in terms in clause 1 of the lease for the agreement of the new rent or application for the appointment of an arbitrator were not complied with within the times specified.

It will be seen from the correspondence, and I do not again need refer to the details, that Alps Hotel Ltd and Mr Chhabra, its director and guarantor, maintained the position from an early stage that since no agreement between the parties as to an increased rent had been reached prior to December 10 1975 and the landlords had failed to apply by that date to the president of the Royal Institution of Chartered Surveyors for the appointment of an arbitrator then the rent payable for the property must remain at £4,000 per annum until the next review date, ie December 10 1982. In other words, time was of the essence of the rent review provisions of the lease. The landlords never explicitly acknowledged that they were out of time, they never explicitly acknowledged that their rights had been lost for a review of the rents, but I think it is true to say that quite apart from the advice that they had received later on as to the apparent failure of their then solicitors to advise them properly, they seem to have accepted, so far as can be seen, that there was little that they could do at that stage to secure an increased rent. Correspondence between the parties on the subject of an increased rent seems to have ceased.

In 1977, however, the position changed. In *United Scientific Holdings Ltd v Burnley Borough Council* and *Cheapside Land Developments Co Ltd v Messels Service Co Ltd* [1978] AC 904, both cases being heard together by the House of Lords, it was decided that in relation to the rent review clauses in those cases there was nothing in either of the leases in question to displace the presumption that strict adherence to the timetables specified in their respective rent review clauses was not of the essence of the contract and that therefore the new rents should be determined in accordance with the procedures specified in the respective leases. It was also decided that the rents fixed by the valuations would be payable retrospectively from the respective review dates.

Although the House of Lords gave their decisions in March 1977, it was not until October 1977 that their decisions in these cases were drawn to the attention of the Vinces. Mr Vince was advised that the effect of the change in the law (and it was quite clearly a change in the law as then understood both so far as the textbooks are concerned and as far as the courts up to the Court of Appeal were concerned) was that it was possible that time was not of the essence of the rent review provisions in the lease held by Alps Hotel Ltd and that, despite the failure to invoke the procedure under clause 1 of that lease by applying before December 10 1975 for the appointment of an arbitrator under that clause, it might still be possible to apply for such an appointment and to obtain a rent increase retrospective to December 10 1975. Accordingly, on January 19 1978 application was made to the president of the Royal Institution of Chartered Surveyors to appoint an arbitrator and at the same time Alps Hotel Ltd and Mr Chhabra were informed through their solicitors. Further correspondence ensued, but the president of the Royal Institution seemed reluctant to act in the absence of agreement between the contending parties, even though the lease itself did not require agreement between the parties either to go to arbitration or upon the appointment of an arbitrator. Ultimately the present originating summons was issued between Edmund David Vince and Maria Vince as plaintiffs, Alps Hotel Ltd as the first defendant and Mr Chhabra as the second defendant. At one stage Mr Middler was added as the third defendant, but he has played no part in the proceedings before me.

The relief claimed by the plaintiffs by the summons against the defendants is for a declaration that on the true construction of the above-mentioned lease, that is the lease dated December 10 1968 and made between Muncip Developments Ltd and Brian Middler, the valuation of the rent by an arbitrator appointed by the president of the Royal Institution of Chartered Surveyors pursuant to clause 1 of the said lease and the plaintiffs' application made to the said president in a letter dated January 19 1978 would be valid and binding on the defendants if higher than the figure of £4,000 per year and recoverable retrospectively to December 10 1975.

The defendants do not accept that time is not of the essence of the rent review provisions, although they accept that the burden of displacing the presumption is upon them. The defendants further assert that by reason of the plaintiffs' conduct in allowing them to incur substantial expenditure in the belief that the rent would remain at £4,000 per annum the plaintiffs should be estopped from being allowed to assert that time is not of the essence of the rent

A review provisions of the lease. I have very carefully considered clause 1 of the lease, and the submissions of counsel in relation thereto which necessarily rely by example on other cases. In construing the words of this clause I can see no indication anywhere that the timetable there set out is other than procedural. I can see no indication, still less firm evidence, that time was ever intended to be the essence of the bargain between the parties. To take an extreme case which was discussed by counsel before me, suppose that negotiations between the parties took up the whole two quarters before December 10 1975 and yet no agreement was reached upon a new rent, and the landlord then made his application to the president of the Royal Institution of Chartered Surveyors on
B December 11 1975 (or even on December 10 1975 itself, since it is to be observed that the timetable set out in clause 1 itself is that the application should be made before December 10), can it really be said that an application on such a date was too late and that the landlord must wait another seven years for a rent review? I cannot accept that there is any indication in the clause whatsoever that such was the intention of the parties, and in my judgment there is nothing in clause 1 to displace the presumption that time in relation to the rent review provisions was not of the essence of the bargain.

I then turn to the other side of the defendants' case, namely that in some way the plaintiffs are estopped by their conduct from asserting that time is not of the essence. Essentially it falls into two
C heads, which I shall deal with together. First of all the plaintiffs' delay in applying by their letter dated January 19 1978 for the appointment of an arbitrator, upon which they rely in these proceedings, and secondly that by that delay they caused the defendants to act to their prejudice by incurring expense at a time when they thought the rent was going to remain at £4,000 for the next seven years, that is until the next rent review moment in December 1982. I think it is important that I should read at this stage the evidence upon which the defendants rely and which is also answered by Mr Vince on behalf of the plaintiffs.

I first of all turn to Mr Chhabra's affidavit dated May 25 1979, and I pick it up at paragraph 8:

D

> The Plaintiffs did not make any application to the President of the Royal Institution of Chartered Surveyors before December 10 1975 asking him to appoint an independent arbitrator to assess the rent payable under clause 1 of the lease of The Malvern Hotel.

That of course is accepted. Then we pick up the delay that is relied upon in support of the plea of estoppel.

> The Plaintiffs have delayed in applying to the President of the Royal Institution of Chartered Surveyors until January 19

and it should be 1978, not 1976 as it is expressed in the affidavit,

> as appears from the letter marked EDV3 and exhibited to the affidavit of
> E Edmund David Vince sworn on June 19 1978. Since December 1975 I have paid rent to the Plaintiffs at the rate of £4,000 per annum. Since December 1975 I have been able to execute works of improvement to The Malvern Hotel which I would not have been able to execute if the rent payable under the terms of clause 1 had been more than £4,000 per annum. In particular, central heating has been installed to the value of £3,800, an extension at the rear of the hotel has been built at a cost of £1,360, additional rooms have been created by partitioning at a cost of £385, and new toilets, baths and showers have been installed at the hotel at a cost of £660.
>
> I have also been able to execute the works of repairs required by clause 2.4 and 2.5 of the lease to a higher standard than I would have been able to if the rent payable under clause 1 of the lease had been more than £4,000 per
> F annum. In particular £650 has been spent on the decoration of the hallway and £350 on the decoration of the lounge of The Malvern Hotel. All the rooms of The Malvern Hotel have been newly painted, which has cost £845.
>
> I am informed by my solicitors that the Plaintiffs have been guilty of unreasonable delay which has caused prejudice and hardship to the Defendants in that the several sums set out at paragraphs 9 and 10 above would not have been spent if the rent review provisions had been implemented by the Plaintiffs in accordance with the timetable provided in the lease.

In reply to that Mr Vince, by his third affidavit of March 10 1980, in paragraph 7 says:

> In relation to paragraphs 9 and 10 of Mr Chhabra's affidavit and the executed works of improvement: under clause 2.16 of the said lease the First Defendants were prohibited from executing such works without my consent. This consent was never granted and copies of the correspondence G

and he sets them out in EDV28.

> Clauses 2.3, 2.4 and 2.5 of the said lease require the First Defendants to observe the therein set out repairing covenants. Since 1975 their schedules of dilapidations and notices have been served on the First and Second Defendants requiring them to remedy the dilapidations set out. Since 1975 these dilapidations have not been repaired within the said clause 2 and on June 11 1979 this Honourable Court granted the Plaintiffs leave as required by the Leasehold Property Repairs Act 1938 to bring forfeiture proceedings in relation to such dilapidations and failures to repair, and trial in relation to this matter is still pending, the action having been commenced on July 31 1979 by writ of summons for dilapidations and failure to repair in 1975, H
> 1976, 1977, 1978 and 1979,

and then he refers to further letters. Again I need not refer to them, and nor do I need refer to the copies of the schedules of dilapidations and notices requiring repairs.

> The schedule of dilapidations and notice dated June 27 1978 being relied upon by the Plaintiffs in the said pending action relates to the state of the said premises.

Then he refers to arrears of rent.

> I therefore contest the First and Second Defendants' claim in paragraph 11 of Chhabra's affidavit that the sums referred to in paragraphs 9 and 10 J
> therein would not have been spent if the rent review provisions had been implemented by the Plaintiffs. The installation of the central heating, rear extension and additional rooms, toilets, baths and showers increased the value to the First and Second Defendants of the said premises at 3 and 5 Petersham Road as an hotel. Secondly, the decoration referred to increased again the value to the First and Second Defendants of the premises of the said hotel and/or was required by the various schedules of dilapidations and notices herein variously referred to. Thirdly, the First and Second Defendants prefer this type of expenditure to the expenditure required of them under the lease.

Then finally Mr Chhabra in his affidavit of May 9 1980 says at paragraph 3 when dealing with delay: K

> In paragraph 5 of Mr Vince's affidavit he purports to show that in the circumstances set out therein he was not guilty of unreasonable delay. He does not say when he first instructed Messrs Perry & Company but in any event whatever his discussions and rights may have been as against his former solicitors I know nothing of any claim as against them. So far as I was concerned nothing further was written or discussed as between the Plaintiffs and the Defendants between September 1976 and January 1978. In any event, in April 1976 I had been advised by my solicitors and verily believed that no rent increase was payable to the Plaintiffs under the terms of the lease. Further, the judgment of the House of Lords in the cases of *United Scientific Holdings* and *Cheapside Land Development Company* was pronounced on March 23 1977, almost nine months before the Plaintiffs' solicitors wrote to the Royal Institution of Chartered Surveyors and then to L
> my solicitors. On the question of delay I further mention that the Plaintiffs' affidavit under reply is dated March 10 1980 and purports to reply to my affidavit of some nine months earlier.

Then he produces the letter from Mr Vince's present solicitors of January 19 1978, together with a copy of his solicitors' reply of January 24 1978, which I do not need to read. Then paragraph 5:

> As to paragraph 7 of Mr Vince's affidavit he has purported to deal with a number of matters and I will try and deal with these individually as follows. As to exhibit EDV28

that is the correspondence between June 8 1977 and August 17 M
1977 between solicitors acting on behalf of the defendants, the then solicitors acting on behalf of the plaintiffs and the present solicitors acting on behalf of the plaintiffs.

> which relates to the failure by the Plaintiffs to give consent for the construction of the extension to the rear of the hotel I am informed by my solicitors and verily believe that no reply was ever received to the letter of August 17 1977 in which they informed the Plaintiffs' solicitors that they considered the Plaintiffs were unreasonable in withholding consent for those works and alterations which were in my view an improvement to the property and which enured to the benefit of both the Plaintiffs and the Defendants. (b), the Plaintiffs have exhibited certain letters under EDV29 and the schedule of dilapidations under EDV30 which in my humble opinion give an incom-

plete picture of the situation. Now produced and shown to me marked KCC8 true copies of the originating application which was issued by the Plaintiffs herein for leave to take proceedings for damages for breach of covenant together with true copies of the affidavits sworn in those proceedings but I have not included the exhibits. It is correct that leave to issue proceedings as against the Defendants was granted by Master Chamberlain on June 11 1979 but in so doing he stated, according to the note made by my solicitors, as follows: "I think that this is a close one and may very well be that the Plaintiffs will go no further. I find that at the date of the summons the Plaintiff has brought himself within the Act and is therefore entitled to leave as asked under direction in paragraph 4." The point which the learned Master was making I am informed and verily believe is that the works to the roof and the rear elevation had not been completed at the time of issuing of the summons so that in law the Plaintiffs were therefore entitled to the order which was therefore requested. There is now produced and shown to me

and he then produces the pleadings.

(c) As to the payment of rent so far as I am aware at all times the rent was paid within 21 days of the due date.

(d) As to the suggestion that the Defendants would in any event have executed the works set out in paragraphs 9 and 10 of my affidavit of May 25 1979 the position is that if the Defendants had known that there was likely to be an increase in rent of any substantial nature as from December 25 1975 they would not have executed the improvements such as the provision of central heating and the extension at the rear of the hotel. I cannot at this stage state whether they would have executed the works in relation to the toilets, baths and showers, but they would not have spent more than was necessary in order to comply with the repairing obligations under the provisions of the lease.

The essential factors giving rise to an estoppel in pais or estoppel by conduct were defined by Lord Tomlin in *Greenwood v Martins Bank Ltd* [1933] AC 51. I do not need to deal with the facts of that case, which are of course very remote from the facts here. The factors defined by Lord Tomlin are set out at p 57, where he says:

> The essential factors giving rise to an estoppel are, I think, (1) a representation or conduct amounting to a representation intended to induce a course of conduct on the part of the person to whom the representation is made. (2) An act or omission resulting from the representation, whether actual or by conduct, by the person to whom their representation is made. (3) Detriment to such person as a consequence of the act or omission. Mere silence cannot amount to a representation, but when there is a duty to disclose deliberate silence may become significant and amount to a representation.

Have the defendants in the proceedings before me adduced evidence before me to satisfy all those factors there set out? I think that the defendants' essential difficulty in establishing an estoppel by conduct in the present case is to show by their evidence that at any time they relied upon the plaintiffs' representations, either express or implied, in any way to their detriment or prejudice or otherwise. It was submitted to me that the defendants acted to their prejudice by reason of the plaintiffs' failure to do anything from 1976 onwards. The plaintiffs, so it is asserted, seem to have pursued the matter as far as they could and then apparently cried off. But the answer to that assertion is firstly that the plaintiffs could not pursue their right until they knew of it, and that the date when they knew of their rights cannot on the evidence before me be earlier than October 1977 when they were advised of the decision in *United Scientific Holdings* case and the possibility that they could go to court if necessary and get a declaration of the sort that is sought before me.

Secondly, there is no evidence that the defendants relied in any way upon the plaintiffs' silence rather than upon their solicitors' advice, and indeed the evidence that I have read establishes that.

Thirdly, there is no real evidence before me as to when the works were done or what exactly they amounted to.

In those circumstances on the evidence before me it does not seem to me that the defendants have established any prejudice or hardship to themselves, still less that it was caused by any representation, express or implied, by conduct or absence of action, silence or otherwise by the plaintiffs. In my judgment, therefore, the plaintiffs are entitled to the declaration that they seek.

CHANCERY DIVISION

March 4 1981

(Before His Honour Judge Mervyn DAVIES QC)

AVON COUNTY COUNCIL v ALLIANCE PROPERTY CO LTD

Estates Gazette June 20 1981

(1981) 258 EG 1181

Landlord and tenant — Construction of rent review clause in lease — Whether the "rack rental market value" was to be determined on the basis of the deemed vacant possession of the whole of the demised premises or having regard to the fact that the premises were subject to a number of leases and licences — Held by the judge that the arguments in favour of a vacant possession basis of valuation of the rent were stronger than the contrary arguments — Summons accordingly answered in favour of the landlords' (county council's) construction

This was an originating summons for the construction of the rent review clause in a lease of office premises known as Northernhay House, Exeter. The plaintiffs were the landlords, Avon County Council, and the defendants were the tenants, Alliance Property Co Ltd.

Gerald Godfrey QC and E Prince (instructed by the solicitor, Avon County Council) appeared on behalf of the plaintiffs; A L Price QC and D M Burton (instructed by Alfillé & Co) represented the defendants.

Giving judgment, JUDGE MERVYN DAVIES QC said: This is an originating summons which raises a question about the true construction of a rent review clause in a lease. The plaintiff, the Avon County Council, is the freeholder of office premises known as Northernhay House, Exeter. The defendant is Alliance Property Co Ltd who own a lease of the premises. The lease (hereinafter called "lease") is dated March 5 1975. It was made between the Avon County Council ("the council") of the first part, Copthall Holdings (Midlands) Ltd of the second part and Cornwall Property (Holdings) Ltd of the third part. The council is the landlord under this lease and the Copthall company were the tenants. The lease was on June 5 1978 assigned to the Alliance Property Co Ltd. The Cornwall company was a party to the lease simply for the purpose of guaranteeing the payment of rent and takes no part in the proceedings.

By the lease the council demised to the tenants the plot of land, buildings and premises described in the First Schedule to the lease and it was declared that the said plot of land, buildings and premises "with any alterations or additions" were in the lease referred to as "the demised premises". The term of the lease was 125 years from December 21 1974. The demise was made subject to and with the benefit of the leases and licences mentioned in the Third Schedule to the lease. These leases and licences are hereinafter referred to as "the scheduled leases and licences". The lease reserved "the yearly rent following that is to say during the first five years of the said term the yearly rent of £55,200-00 and during the twenty-four successive periods of five years of the said term each a yearly rent to be assessed in accordance with Clause 6(iv) hereof" together with a further rent in respect of insurance expenditure.

The clause 6(iv) already mentioned is in these terms:

> 6(iv)(a) The Council shall have the right to review the yearly rent for the time-being payable hereunder at or after the end of the fifth year and also at the end of each subsequent fifth year of the term hereby granted (the date when such review takes place being hereinafter referred to as "the date of review") on giving to the Tenants not more than twelve nor less than three months notice in writing prior to the date of review and after the date of review this Lease shall be read and construed and shall take effect in all respects as if the yearly rent for the time-being payable hereunder instead of being the sum of £55,200-00 had been 70 per cent of the rack-rental market value (as hereinafter defined) of the demised premises but without prejudice to any of the other terms and conditions contained in the Lease and so that in no event shall the yearly rent payable by the Tenants to the Council after the date of review be less than the yearly rent payable by the Tenants to the Council immediately before the date of review.

(b) The rack-rental market value of the demised premises shall (subject as hereinafter provided) by the amount which shall be agreed between the Council and the Tenants to be the best rent at which the whole of the demised premises might reasonably be expected to be let in the open market by a willing landlord to a willing single tenant for a term of years equal to the unexpired residue of the term granted by this Lease and subject to similar covenants and conditions (other than the amount of the rent) to those contained in this Lease there being disregarded:
(i) any effect on rent of the fact that the Tenants or their predecessors in Title have been in occupation of the demised premises or any part thereof.
(ii) Any goodwill then attaching to the demised premises or any part thereof by reason of the carrying on thereon whether by the Tenants or by their predecessors in Title or by an under-Tenant of the Tenants or their predecessor in Title of any business or businesses.
(iii) Any improvements made to the demised premises by the Tenants with the consent of the Council or the Court.
(c) In the event of the Council and the Tenants failing to agree as to the rack-rental market value of the demised premises at the date of review within three months of such date of review, then and in any such case the dispute shall as soon as practicable be referred to the award of some competent person to be agreed upon by the Council and the Tenants (or in default of agreement) to be nominated by the President for the time-being of the Royal Institute of Chartered Surveyors . . .

The First Schedule already mentioned is in these terms:

All that plot of land situate in and having a frontage to Northernhay Place, Exeter, in the County of Devon containing in the whole 0.606 of an acre or thereabouts which said plot of land is delineated and coloured pink on the plan annexed hereto together with the office premises erected thereon together also with all appurtenances thereto.

The plan referred to shows a pink area marked "car park" with on the edges two small shops and a few other buildings. I was told that at the date of the lease two large office blocks had been built on the pink land, the buildings together being known as Northernhay House.

The Third Schedule gives particulars of the schedules leases and licences as follows:

(1) A lease dated March 29 1974 by the Bristol Corporation to ECC Quarries Ltd for 35 years from March 29 1974.
(2) A lease dated May 29 1974 by the Avon County Council to the Provincial Insurance Co Ltd for 25 years from March 25 1974.
(3) A lease dated June 6 1974 by the Avon County Council to Harvey McGill & Hayes for 25 years from March 25 1974.
(4) A licence dated July 18 1974 granted by the Avon County Council to Harvey McGill & Hayes for 25 years from March 25 1974.
(5) A licence dated August 5 1974 granted by the Avon County Council to the Provincial Insurance Co Ltd for one year from June 24 1974.

As I understand, the three scheduled leases took up all the lettable space in the buildings upon the demised land, while the licences took up some but not all of the parking spaces available on the demised land.

When the first date of review arrived on December 21 1979, the freeholders (that is the council) and the tenants (that is Alliance) failed to agree a new rent. The council took the view that clause 6(iv) means that the "rack-rental market value" is to be determined on the footing that the whole of the demised premises are to be deemed in vacant possession at the date of the determination. The tenants took the view that the determination ought not to be on that footing; and that the determination ought to be made with a recognition of the fact that the demised premises are subject to the provisions of the scheduled leases and licences. The affidavit evidence gives an indication of the issues that then arose, but the issues were more precisely defined when the originating summons was amended in the course of the hearing. The amended originating summons in effect puts the question in this way:

Whether "the rack-rental market value of the demised premises" within the meaning of clause 6(iv)(b) of the said lease shall be determined for the purpose of that clause (A) on the assumption that the "willing single Tenant" referred to therein is given vacant possession of the demised premises on the date of review *or* (B) without making that assumption so that as regards the current date of review December 21 1979 such "willing single Tenant" is not assumed to be entitled to vacant possession of the demised premises but (as was the fact) is entitled to possession thereof subject to the then existing subtenancies and licences.

Put shortly, the question is whether the rent is to be fixed on the basis of vacant possession or on the basis that the scheduled leases and licences subsist.

Now that I have set out the material terms of the lease and stated the question arising I must mention some of the circumstances in which the lease came to be granted. If an agreement is to be understood it must be placed in its context. In *Prenn v Simmonds* [1971] 1 WLR 1381, Lord Wilberforce said at p 1383 H:

In order for the agreement of July 6 1960 to be understood, it must be placed in its context. The time has long passed when agreements, even those under seal, were isolated from the matrix of facts in which they were set and interpreted purely on internal linguistic considerations. There is no need to appeal here to any modern, anti-literal, tendencies, for Lord Blackburn's well known judgment in *River Wear Commissioners v Adamson* (1877) 2 App Cas 743, 763 provides ample warrant for a liberal approach. We must, as he said, inquire beyond the language and see what the circumstances were with reference to which the words were used, and the object, appearing from those circumstances, which the person using them had in view.

So despite the fact that the original parties to the lease are not before me it is, I think, permissible to take into account the circumstances that existed at the date of the lease. What I will say in this regard does not come wholly from the evidence filed. It comes in part from what I was told by counsel as being, as I understand, accepted facts.

The story begins with an agreement dated February 4 1972 made between Northernhay Car Park Ltd ("Northernhay") of the first part, Copthall Holdings (Midlands) Ltd ("the developer") of the second part, Cornwall Property (Holdings) Limited ("the surety") of the third part and the Bristol Corporation of the fourth part. The agreement is exhibit "WE2" to an affidavit of Mr William Evans sworn on September 15 1980 and filed herein. By the agreement the Bristol Corporation agreed to buy the land comprised in the lease (then a car park with two shops) from Northernhay. Then the developer at the cost of the Bristol Corporation agreed to develop the land. It was to be an office development. It was then agreed that at the completion of the development the land would be let to the developer for 125 years "at an exclusive annual rent (hereinafter referred to as 'the Initial Rent') to be calculated prior to the grant of the Lease as hereinafter provided the Initial Rent to be subject to review as set out in the form of draft Lease hereinafter referred to ... and the Lease shall be in all respects according to the form of draft Lease as set out in the Second Schedule hereto". Clause 22 provided for the Initial Rent being fixed by reference to the cost incurred in developing. Clause 17(b) of the agreement was in these terms:

At the request of the Developer at any time on or after the date of the Developer's Architect's Certificate of Practical Completion prior to the grant of the lease to the Developer herein as hereinafter mentioned the [Bristol Corporation] shall grant leases or tenancy agreements of parts of the building comprising the said works for such periods not exceeding 21 years and at such rents (payable by the direction of the [Bristol Corporation] to the Developer) as the Developer (consistently with the provisions as to permitted underleases contained in the Lease) shall approve such leases or tenancy agreements being prepared by and at the expense of the Developer and absolving the [Bristol Corporation] from all responsibility whatsoever thereunder and in such case the Lease granted to the Developer shall be subject to and with the benefit of the leases or tenancy agreements so granted.

I note that the rents were to be approved by the developer and, as well, the rents were to be paid to the developer by the direction of the corporation.

It was pursuant to this agreement that the lease (that is the lease dated March 5 1975) was granted. The fact that the lease was granted by Avon County Council and not by the Bristol

A Corporation was due, as I understand, to the fact that the county council succeeded under the Local Government Act 1972 to certain of the corporation's rights and obligations. I was not told whether or not the initial rent in the lease, ie £55,200-00, was fixed in the way provided for in the clause 22 I have mentioned, but I have no reason to suppose that it was not fixed in that way.

Between the date of the agreement (February 4 1972) and the date of the lease (March 5 1975) the scheduled leases and licences were granted on the dates already mentioned. No doubt all five were executed pursuant to clause 17(b) of the agreement. The initial rents payable under the scheduled leases were (1) £34,500-00, (2) £23,450-00 and (3) £15,830-00. I note that the
B three scheduled leases begin with the words "This underlease" but they are in fact leases. I suppose they were looked upon as underleases because the rents were to be handed over to the developer and had the lease been in existence the terms of years granted by the schedule of leases would have been granted by the developer and so would have been true underleases. The terms of the scheduled leases are, as I have said, 35, 25 and 25 years. The lease refers to the scheduled lease dated March 29 1974 as having a term of 25 years. This is a mistake. In fact its term is 35 years. The scheduled leases contain rent review provisions. All three provide for new rents at successive intervals being fixed by reference to an assumed letting "with vacant possession".

C So I come to the question whether the phrase "rack-rental market value" in clause 6(iv) of the lease means a value determined on a vacant possession basis or on some other basis such as that indicated in the second limb of paragraph 1 of the amended originating summons.

Mr Price appeared for the tenants. He contended for a non-vacant possession value. The main points of his argument were:

(1) clause (iv) does not contain the words "with vacant possession"; such words could be read into the clause only if it was necessary in order to make sense of the valuing provision
D and that was not the case. He contrasted the rent review provisions of the scheduled leases where the words "with vacant possession" do appear. There is perhaps some justification in the circumstances of this case for supposing that the words of the scheduled leases may throw some light on the lease and vice versa.

(2) clause (iv) came into operation at a time when the demised premises were not in vacant possession. The scheduled leases had already been granted. They will continue for many years and may well be extended by virtue of the Landlord and Tenant Act 1954. Accordingly, the lease was drafted on the footing that the rent arising under the lease would be a rent based on a
E recognition of the fact that the scheduled leases existed; and that meant that the rent would not be fixed on any vacant possession basis. The lease was, he said, a reversionary lease.

(3) It was suggested that the main reason for the constitution of the (reversionary) lease was to bring into being an intermediate single managing tenant. This single tenant would relieve the council from the risk and trouble of management in that the single head lessee would attend to the superintendence of the common parts, etc.

In light of these considerations, Mr Price said that the meaning of clause 6(iv) was that a new rent would be fixed not on any
F vacant possession basis but with a recognition of the fact that the council's interest in the land was burdened by the scheduled leases and licences.

In my view the absence from subclause (iv) of the words "with vacant possession" is not fatal to the claim that subclause (iv) contemplates a vacant possession valuation. The words quoted may be implicit in the clause. I see the strength of Mr Price's G
second point but it must be tested by reference to the actual words of subclause (iv) in the whole context of the lease.

I therefore turn to subclause (iv). First, paragraph (a) says that the new rent is to be "70 per cent of the rack-rental market value (as hereinafter defined) of the demised premises". Then paragraph (b) defines "rack-rental market value of the demised premises"; I have already read paragraph (b) in full.

Mr Godfrey, for the council, pointed out that "(b)" is concerned with a rent of the "demised premises", a phrase defined in the lease as being the buildings and land described in the First Schedule to the lease. Thus, it was submitted, subclause (iv) does not call for a rent that is in any way related to the H
scheduled leasehold interests that subsist; the rent is to be related to the land and buildings and that implies a rent based on vacant possession. I see great force in that submission. In conjunction with this submission, Mr Godfrey said that if Mr Price's construction is to be upheld one has to alter or disregard the words "of the demised premises". Mr Godfrey said that the Price construction required the deletion from paragraph (b) of the words "the whole of the demised premises" and the insertion of some such words as "the reversion immediately expectant on the determination of the leases to which the demised premises are subject at the date of the review". I note too the use of the phrase "*the whole* of the demised premises" in connection with J
the phrase "a willing *single* tenant". Those phrases together certainly seem to suggest that no account is to be taken of the scheduled leases. Continuing with the wording of (iv)(b) I come to the words "and subject to similar covenants and conditions (other than the amount of the rent) to those contained in this lease". So paragraph (b) does not say specifically that the notional letting is to be regarded as subject to the scheduled leases and licences. Then come the "disregard" provisions of subparagraphs (i), (ii) and (iii). These seem to me to be appropriate to a rent fixed by reference to vacant possession rather than by taking account of the scheduled leases.

These considerations taken together lead me to the conclusion K
that the words "with vacant possession" are implicit in subclause (iv)(b). It is to be read as though those words followed the words "expected to be let". I have reached this conclusion simply by considering the words of subclause (iv). The circumstances that existed at the date of the lease do not seem to me to make this construction in any way doubtful. I add that the conclusion that it is right to be able to construe "(iv)" as I have is I think satisfactory for three reasons. In the first place, the phrase "rack-rental market value" of itself suggests to me a vacant possession valuation. I say that because a rack-rent is the best rent that can be reasonably obtained and the best rent would ordinarily be obtained in respect of premises in vacant posses- L
sion. Secondly, the initial rent under the lease was not intended to be fixed by any reference to the scheduled leases: see agreement dated February 4 1972, clause 22. In making this comment I appreciate that such rents were not intended to be fixed by reference to vacant possession either. Thirdly, the rents of the schedule leases had to be approved by the developer, ie the person about to become the tenant under the lease: see 1972 agreement, clause 17. That being so, it is difficult to see why the rent of the lease should be tied in any necessary fashion to the rents of the scheduled leases.

Accordingly the originating summons is answered as to paragraph 1 in the vacant possession sense. M

As to paragraph 2 of the summons, I was told that on a vacant possession basis there was agreement that the new rent of the demised premises is to be £92,000 a year.

The defendants were ordered to pay the costs of the application.

RATING

COURT OF APPEAL

July 23 1980

(Before Lord Justice MEGAW, Lord Justice EVELEIGH and Lord Justice OLIVER)

LONDON BOROUGH OF CAMDEN v CIVIL AVIATION AUTHORITY AND ANOTHER

Estates Gazette January 24 1981

(1980) 257 EG 273

Rating — General Rate Act 1967 — Appeal by rating authority against decision of Lands Tribunal reducing the rating assessment of Space House, Kingsway, London WC2, two office blocks occupied by the Civil Aviation Authority, to the nominal figure of £1 — At the date of a proposal by the CAA works were necessary to complete the building, to carry out repairs because of deterioration during previous years, and to remedy structural defects due to the use of high alumina cement concrete (HAC) in beams and to the existence of stresses and strains in "Y" columns — The Lands Tribunal assumed for the purpose of this decision that the building was in a completed state and that, with the exception of the remedial work required for the HAC beams and the "Y" columns, the rest of the work was repairs — The tribunal concluded that, by reason of this remedial work, which differed from ordinary repairs within the meaning of section 19, and the general doubt and uncertainty about the state of the building, the premises were for rating purposes unlettable — Held that the appellants had failed to establish any error of law in the tribunal's conclusions — Appeal dismissed — Observations by Megaw LJ on the need for appellants to formulate with clarity and precision the questions of law which the court is asked to consider.

This was an appeal by the London Borough of Camden against the decision of the Lands Tribunal (1978) 248 EG 957, 1019, whereby the assessment for rating of the two large office blocks known as Space House in Kingsway, which had been shown as gross value £1,200,000, rateable value £999,972, was reduced to the nominal figure of £1. The first respondents to the appeal were the Civil Aviation Authority, the lessees of the hereditament, and the second respondent was the valuation officer, A J Langford.

G E Moriarty QC and G R G Roots (instructed by the Solicitor, London Borough of Camden) appeared on behalf of the appellants; J C Taylor and P H Morgan (instructed by Richards, Butler & Co) represented the first respondents; A P Fletcher (instructed by the Solicitor of Inland Revenue) represented the second respondent.

Giving the first judgment at the invitation of Megaw LJ, EVELEIGH LJ said: This is an appeal from the decision of the Lands Tribunal, where the assessment of a hereditament described as "Offices, Showrooms and Premises, Sub-basement and 1st-5th Floors, Space House ,43-59 Kingsway, London WC2" was reduced to the nominal value of £1. The premises which are the subject of the appeal comprise two large office blocks known as Space House situated in Kingsway in the London Borough of Camden. Those offices were completed as a shell in or about 1966. There then remained a substantial amount of work to be done before the building was in its final completed state and, by 1968, the work was far from complete. In consequence of that a notice was served upon the owners by virtue of Schedule 1 to the General Rate Act of 1967. By that notice and with the agreement of the parties concerned, December 1968 was fixed as the date for completion. The property was assessed and rates were in fact paid thereafter; and, in 1973, in the new valuation list which came into force on April 1 of that year, the premises were shown with an assessment of gross value £1,200,000, rateable value £999,972.

At about the end of 1974 the Civil Aviation Authority took an interest in the building and finally entered into a lease to begin in February 1975. The premises were still incomplete. There remained a great deal of work to be done and, in addition to that, the premises had deteriorated, having stood empty, and substantial repairs were needed. Furthermore, it had been discovered that the concrete referred to as HAC was suspect; and, also, the reinforced concrete columns, which were in the shape of a "Y" and called in this case "Y" beams, had been subjected to strain and some of them showed minor cracks which led to some anxiety as to their future stability and duration.

The authority, none the less, were prepared to take the property and they entered into occupation on June 5 1975, that is to say, of part of the premises, and occupied other parts of the premises from time to time as work was done upon the building, making them fit for occupation.

On March 21 1975, some two months before the occupation began, a proposal for the alteration of the valuation list was made on behalf of the Civil Aviation Authority, and it was in relation to that proposal that the decision of the Lands Tribunal was given and with which this court is concerned. The principles applicable to a case of this nature were referred to in the decision of the Lands Tribunal. The first principle, as to which there is no dispute in this court, is that the premises have to be assessed as at the date of the proposal, namely March 21 1975. The valuation is governed by section 19(6) of the General Rate Act 1967, which provides as follows:

> "Gross value", in relation to a hereditament, means the rent at which the hereditament might reasonably be expected to let from year to year if the tenant undertook to pay all usual tenant's rates and taxes and the landlord undertook to bear the cost of the repairs and insurance and the other expenses, if any, necessary to maintain the hereditament in a state to command that rent.

Schedule 1 to the Act was also relevant to the determination of the case and, in the decision of the tribunal, that schedule is referred to in detail. There is no need to repeat it here in this judgment.

Now, the condition of the premises as at the date of the proposal is of considerable importance. It is agreed as a general principle that the valuation should be arrived at *rebus sic stantibus*, that is to say, taking into consideration the existing state of affairs as at that date. The existing state of affairs in so far as is relevant for consideration of this appeal was the condition of the property. There were three categories of work that required to be attended to: firstly, the work necessary to complete the building, the work that should have been done by December 1968; secondly, the work necessary to make good the decay that had occurred in the intervening period, which can unequivocally be called repair work; then a third category of work which, at the moment, I forbear from giving any particular name, the work that related to the concrete and the "Y" beams.

The actual extent to which work might be done, or might be thought advisable, in respect of those two matters — the cement and the "Y" beams — was not precisely known. What was certainly known was that attention ought to be given to it; but it was work

A which, in the submission of counsel for the appellants, was certainly not required to be done there and then. It has been submitted in this court, indeed, that, in so far as that work would produce a condition that made the building unlettable, it need not be done for a very considerable time; but I shall return in greater detail to that matter in a moment.

Faced with that position — the condition of the building on March 21 1975 — the member decided that the premises were in fact unlettable and, accordingly, made the nominal assessment. Referring to the work of the "Y" beams or columns and the work of the HAC beams, he said this:

B I do not consider that these two items can in any sense be deemed to be work of repair, still less that they can be considered as repairs which the hypothetical landlord is to be deemed to have carried out. They both represent substantial structural defects going far beyond what would ordinarily be considered to be repair. They had their origin in defects of design or construction, and the remedying of these defects has the result of producing a building which was in a material and significant way better than and different from that which existed before.

That being so, he came to the conclusion that this work, which was not to be regarded as work of repair, was such, however, as made the building unlettable.

It was submitted on behalf of the appellants that the building should be treated as though it had been completed and all necessary
C works done. It was said that their failure to comply with the notice specifying December 1968 as the completion date meant that the building should be deemed to have been completed on that date and that it should be deemed to be in a lettable condition thereafter. The member assumed, for the purpose of his decision, that the building was in a completed state; in other words, he ignored the work that was done for the purpose of completing the building and, furthermore, he ignored the effect of work done for carrying out the repairs in respect of those defects that were the result of deterioration. He however regarded the HAC beams and the "Y" columns as in a category of their own and held that they were responsible for the building being unlettable.

D In the notice of appeal the grounds stated were as follows:

That the Lands Tribunal was wrong in law in deciding that the Y column work and the work on the HAC beams were not works of repair which were to be deemed for the purposes of section 19 of the General Rate Act 1967 to have been done at the date of the proposal; and

(2) That there was no evidence that at any time after the date of deemed completion the hereditament was for any substantial period incapable of occupation by reason only of the Y column work or the work on the HAC beams.

There have been various reports received in relation to the "Y" columns and evidence was given to the effect that the cracks in those columns indicated to the expert, a Mr D A S Lloyd, that there was movement in the building which, to some extent, might jeop-
E ardise its stability and work eventually was done, that is to say, subsequent to March 21 1975, the total cost of which work to the "Y" columns amounted to £330,000. The cracks were sealed, struts were inserted between the columns and, as an extra precaution, an extra ring or collar was added to the building to prevent movement. In so far as the cement is concerned, there had been considerable interest as a result of the fall in the roof of the swimming pool at the Sir John Cass and Red Coat School in Stepney, and the Department of the Environment had circulated local properties, drawing attention to the suspicion that surrounded this particular concrete, high alumina cement (HAC), and one passage in that circular reads:

F It follows from this that there is now greater cause for concern, and that all buildings with roofs of a type similar to the Stepney school, that is to say having isolated pre-stressed beams of high alumina cement concrete . . . must be regarded as suspect.

A report obtained from consulting engineers dated September 20 1974 contained this sentence:

To date we have not yet found any reasons which would cause us to believe that serious manufacturing faults have occurred in the production of the HAC floors in this building.

A subsequent report on March 18 1975 contained this statement. It related to the fact that test cores or holes had been drilled and samples taken from examination:

Since of the 60 cores tested as many as 32 per cent have an equivalent cube strength below 20 N/mm² (2,900 lbs/sq in) and 13 per cent have a strength lower than 15 N/mm² (2,175 lbs/sq in) we are seriously concerned with the problems which may arise due to high porosity of the concrete which would accompany such low strengths. The possibility of corrosion of reinforcement should be given more serious consideration than merely the slight likelihood of further deterioration of the strength of the already highly converted concrete. G

I draw particular attention to the passages quoted from these reports because they do not state a positive condition in the building but, in effect, are cautioning against the possibility of the concrete being defective to a serious degree. This is a matter that will be of importance in considering the nature of the repairs that were contemplated and when and whether they might be done. H

I say "repairs"; that is, in fact, begging the question. It is important to have this in mind when considering whether the work in relation to the HAC beams and "Y" columns can truly be described as work of repair. At March 21 a considerable amount of work was being done and it is agreed on all sides that, as a result of that work, the premises could not be let and they could not be occupied for the purpose for which they were intended. Against those facts and the precise details of the work being carried on which has been stated with great particularity in the very thorough decision of the tribunal that is before this court — I do not propose to reiterate them here—one can now consider the two grounds of appeal.

As to ground 1, namely that the work to the "Y" columns and the HAC work should be treated as works of repair, this, to my mind, is J
a question of fact. Are they works of repair as contemplated by section 19 of the General Rate Act? The work that would be done — or was, in fact, ultimately done — was work which resulted in the building being a stronger and better building than it had been before that work was done. It cannot, in my opinion, be said that the fact that the building must be deemed to be completed in December 1968 also means that the building must be deemed to be in the condition in which it would be after the remedial work to the "Y" columns and the HAC beams was carried out. It can only mean that the building would be deemed to be completed to the extent that those responsible for its design had contemplated that it would be K
completed in December 1968, and that certainly did not include this particular work. Therefore, one asks: Is this work to be regarded as repairs? That is to say, is it to be regarded as repairs for which the landlord would make himself responsible? Whether or not something is to be treated as repair is a question usually of fact and degree. In the case of *Brew Brothers v Snax* [1970] 1 QB 612, which was dealing with the tenant's liability under a repairing covenant but, *mutatis mutandis* the decision would apply clearly to a landlord's liability, the headnote at p 614 reads:

. . . that whether the end-product of work requiring to be done properly constituted "repair" was a question of degree in every case; that (Harman LJ dissenting) the correct approach was for the court to conclude, on a fair interpretation of the precise terms of the lease in relation to the state of the property at the date of the lease whether the work could fairly be called repair or not; and that in coming to its conclusion the court must look at the work as a whole and not at individual component parts thereof . . . L

Phillimore LJ at p 645 said this:

Megaw J had of course specifically held that with the exception of the foundation to the flank wall the underpinning of the other two main walls of the premises would be necessary to render the building safe in accordance with the Pynford report and the notice of dilapidations served by the landlords.

I agree with the judge, who concluded after reviewing the authorities and particularly the observations of Lord Esher MR in *Lister v Lane & Nesham* [1893] 2 QB 212, 216 and those of Buckley LJ in *Lurcott v Wakeley & Wheeler* [1911] 1 KB 905, 924, that the vital question in each case is whether the total work to be done can properly be described as repair since it involves no more than renewal or replacement of defective parts, or whether it is in effect renewal or replacement of substantially the whole. It is, as Megaw J held, a question of degree in each case. It is well established that a tenant is not liable to produce a different thing from that which he took when he entered into the lease or to remedy the results of bad design. M

In the present case I would have come to the same conclusion as that of the Lands Tribunal. This was work that, in effect, produced a building of a different quality from that which was contemplated in 1968 and was work which, to my mind, would not be said to fall within the landlord's repairing covenant. For one thing it was not

A work that positively had to be done. The condition of the building was not such that, as a result of the "Y" columns and the HAC beams, it could not be used as offices. The condition of those members was such that a prudent landlord might well, at some time, decide, for the continuing and ultimate stability and duration of the building, to carry out such work as those investigations indicated was desirable — at some time and a time of his choosing. It happened in this case, by agreement between the tenant and the landlords, the work was begun — the investigatory work anyway — before March 21 and was in the process of being done during March 21. But the effect of the uncertainty that then overshadowed the building was to make that building unlettable in the view of the Lands Tribunal.
B So, in so far as the first ground of appeal is concerned, I would reject that contention for those reasons. It is a question of fact, but, for myself, I would have come, for the reasons I have stated, to the same conclusion.

I now turn to the second ground. The way the member approached the matter was to assume, for the purpose of his conclusion, that the works of completion should be deemed to have been done, and that any disturbance that existed from the fact that those works were in the course of being done was to be ignored. He further assumed that, with the exception of the "Y" column work and the HAC beams, the other work was to be regarded as work of repair. He reached his conclusion, therefore, upon the basis that the
C disturbance and, as a result, the fact that the premises were made unlettable, was due to the "Y" columns and the HAC beams. He referred to the fact that work was going on at the relevant date in relation to those two items. He said:

> The fact is that it — that is to say, the work — was going on at the relevant date and is a material circumstance in valuing the hereditament at that date *rebus sic stantibus*. This extends not only to the work in the beams and the work in preparation for the Y columns. It also includes, in my judgment, all the necessary consequential disturbance involving the making good of damaged decorations, the delivery of scaffolding and the erection of the hoarding. Finally, and probably as important a factor as any, there was the general doubt and uncertainty.

D The hoarding was the hoarding that was put around the building to enable this work to be carried out without interference. So what he did was to regard all the circumstances and the "Y" columns and the HAC beams as if they would have affected a prospective tenant and the landlord at that time and to come to the conclusion that all those circumstances, and in particular the general doubt and uncertainty, would have made the premises unlettable.

Before this court it has been contended that, although it is right from the point of view of assessing the gross value of the premises to ignore the fact that repairs were going on, the fact that repairs were going on are relevant for the court's consideration in deciding what is the cause of the unlettability — if one can use that expression — of
E this hereditament. Mr Moriarty has submitted that the actual repair work being carried on was so extensive as, of itself, to make the premises unlettable; and, if not that work, then the work of completion that was actually being carried on. He says that, although for the purpose of deciding whether or not the premises are to be treated as occupied, the deeming provision must be applied and the premises must be deemed to be complete. He also says that in so far as the valuation side of the problem is concerned, the actual state of affairs must be considered and, if the works being done are having an effect upon the value of the property, then it cannot be said that the work on the "Y" columns and the HAC beams was the only cause of the property being unlettable. In that event, he has asked
F the court to remit the matter to the member for further consideration, submitting to this court that there should have been a detailed examination of the amount of disturbance attributable to each of these factors and, as I understand it, for how long that disturbance was likely to persist.

That argument, to my mind, does not take proper account of section 19. That section requires assessment to be made on the basis that the landlord is responsible for repairs, and it is assumed that he will do them and that the tenant will be content to pay the rent that the premises will, in consequence, command. Therefore, the fact that repair will be required will not affect the assessment so as to produce a lower rent. See *Wexler v Playle* [1960] 1 QB 217. The same principle must apply if the repairs are actually done, so that G
the normal use of the building is being interrupted.

Now, this does not mean that repair activity is outside the principle that the matter must be considered *rebus sic stantibus* at the date of the proposal. The condition of the premises is not regarded as an adverse factor affecting the tenant's willingness to pay rent and the landlord's acceptance of it, so long as the condition is remediable by the landlord fulfilling his obligation to repair. So the court is not precluded from having regard to what is going on, to looking at what is going on and then deciding this is repair for which the landlord will be responsible. The conclusion from a valuation point of view is that the value is in no way diminished on that account. The valuation *for rating purposes* (and I would emphasise those words) H
is not affected by repairs and, therefore, the premises are not, for rating purposes, to be treated as unlettable on that account.

On all sides in the present case it is agreed that these premises were unlettable; and, if they are not to be treated as unlettable because of the repairs, they can, in my opinion, only be so regarded as a result of the other work, namely, the "Y" columns and the HAC beams. In so far as the work of completion is concerned, again, they have no effect upon the valuation. They have to be treated as being done, and it follows from this, in my opinion, that the only relevant cause of the premises being unlettable from a rating valuation point of view in accordance with section 19 would be structural conditions relating to HAC beams and the "Y" J
columns.

For those reasons I would reject the second ground of appeal also and would dismiss this appeal.

OLIVER LJ agreed.

Also agreeing, MEGAW LJ said I agree so completely with the conclusions expressed by Eveleigh LJ, with the reasons which he has given and with the clarity with which he has expressed those reasons that it would be otiose verbiage if I were to try to express any of the reasons in words of my own. I shall, therefore, avoid the conventional judicial meiosis expressed in the phrase "adding a few words of my own". But I qualify that with two observations.

The first is that I would like to pay tribute to the attractive K
presentation and the ingenuity with which Mr Moriarty for the appellants sought to demonstrate that questions which were essentially questions of fact were questions of law. That leads up to the second observation: while I appreciate that there may be difficulties in the way of dealing with the matter in the way which I think would be desirable, it does seem to me that there is grave danger of waste of time and money if, in matters of this sort coming on appeal from the Lands Tribunal, the questions of law which the court are asked to decide are not stated in the case to the court with reasonable clarity and precision. The actual case stated to this court in the present case was:

> . . . whether upon the findings of fact I came to a correct decision in law. L

That very carefully reasoned decision setting out the views of the member of the Lands Tribunal on facts and on law, without any unnecessary verbiage, occupies some 31 pages. It is apparent that, in those 31 pages, there may well be a large number of questions which could be described as questions of law. It is not, therefore, very helpful in this court to be asked to decide whether the member of the tribunal came to a correct decision in law. It would certainly have very great advantages in saving time and money for all concerned if it were possible that the true questions of law which it is sought to bring before this court could be defined with reasonable precision in the case stated for the consideration of this court. It M
should, I think, be for the intended appellant to formulate with precision the question or questions of law which he regards as arising in the decision of the tribunal or which he wishes to contend have been wrongly decided. It would then be for the tribunal to consider whether or not that question or those questions is or are properly formulated as a question or questions of law and whether or not it or they can properly be said to arise in the tribunal's decision.

The appeal was dismissed, the first respondents, the Civil Aviation Authority, being given the costs of the appeal. No order was made as to the costs of the second respondent, the valuation officer.

A **COURT OF APPEAL**

November 19 1980

(Before Lord Justice WALLER, Lord Justice ACKNER and Lord Justice WATKINS)

CORSER (VALUATION OFFICER) v GLOUCESTERSHIRE MARKETING SOCIETY LTD

Estates Gazette February 28 1981

(1980) 257 EG 825

B **Rating — Whether auction hall belonging to a farmers' co-operative marketing society qualified for agricultural derating — Whether auction hall an "agricultural building" within the meaning of section 26 of the General Rate Act 1967 as amended — Description of arrangements for the auction of fruit and vegetables — Bidding by recognised buyers only — A substantial proportion of goods put in for sale were from non-members called "members on probation" who might succeed in due course to membership — A minor use of the hall was for private treaty sales — Held by Court of Appeal, reversing decision of Lands Tribunal, that the auction use of the hall, which involved the marketing of non-members' goods as well as those of the society's members,** C **amounted to merchandising in an independent commercial and business sense — The hall was not, therefore, used "in connection with agricultural operations" — Held further that in any case the hall was not "solely used" for the purposes set out in section 4 (1) of the Rating Act 1971 because a substantial proportion of the use of the hall was attributable to the sale by auction of the produce of non-members — Appeal by valuation officer against decision of Lands Tribunal, which had accepted the derating of the auction hall, allowed**

This was an appeal by case stated from the dismissal by the Lands Tribunal of three appeals by Haden Howard Corser, the valuation D officer, from decisions of the Gloucestershire Local Valuation Court. The matter in dispute concerned the rateability of an auction hall owned by the respondents, the Gloucestershire Marketing Society Ltd, and used for sales of vegetables and fruit. The Lands Tribunal (R C Walmsley FRICS), upholding the decisions of the local valuation court, had accepted that the auction hall was entitled to agricultural derating as an "agricultural building".

Alan Fletcher (instructed by the Solicitor of Inland Revenue) appeared on behalf of the appellant valuation officer; Malcolm Spence QC and Hugh Merry (instructed by Ivens, Thompson & Green, of Cheltenham) represented the respondents.

Giving the first judgment at the invitation of Waller LJ, Ackner E LJ said: This is an appeal by way of case stated from the Lands Tribunal, who on December 4 1978 dismissed three appeals by the appellant, the valuation officer, against decisions of the Gloucestershire Local Valuation Court dated November 4 1976. These decisions determined the assessment of a hereditament described as "fruit and vegetable market and premises at Alstone Lane, Cheltenham". The appeals arose out of originating proposals lodged in 1973, 1974 and 1976. The first proposal lodged by the ratepayers, Gloucestershire Marketing Society Ltd, the respondents to this appeal, sought a reduction from gross value £28,250 to gross value £19,600 on the grounds that the existing assessment was excessive and incorrect and not according to law. The second proposal was F lodged by the valuation officer seeking an increase from gross value £28,250 to gross value £29,700 on the grounds that structural alterations had been carried out. The third proposal was likewise lodged by the valuation officer on the same grounds seeking an increase to gross value £36,225.

Although the hereditament comprises three substantial buildings, this appeal is concerned with the rateability of the auction hall (including the flower room). The issue which thus appeal raises is whether or not the hall should receive agricultural derating. This depends upon whether or not the hall was an "agricultural building" within the meaning of section 26 of the General Rate Act 1967 as amended.

The respondent society was founded in 1919 by a group of local G farmers with the object of marketing their produce collectively. By the late 1950s their then premises had become cramped and they purchased the site in Alstone Lane on which the present market was built later. The membership of the society fluctuated between 430 and 450; about 65 per cent of the members are in Gloucestershire, 15 per cent in Worcestershire, 5 per cent in Herefordshire and the remaining 15 per cent are spread over a further 20 counties in England. The total area of land occupied by the members extends to some 60,000 acres.

It is common ground that the main aim of the society is marketing. The use which they make of the hall comes under three heads, viz sales by auction, sales by private treaty and sales by contract. H

Although Mr Fletcher had argued before the Lands Tribunal that, in the case of sales by contract, the hall was used only for loading into or for unloading from a cold store, before us he accepted that this movement was ancillary to the agricultural operation and he therefore did not pursue his objection. We are therefore concerned only with sales by auction and sales by private treaty. These two uses should be considered separately.

Sales by auction. Growers bring their produce into the hall and place it on the floor in a special section. This is done either on the afternoon before an auction or early on the day of the auction. Auctioning begins each day at 9 am. The busier days are Tuesdays, Thursdays and Saturdays, and on those days the auction time may J extend to about 4 pm in the peak growing period but otherwise until 1 pm. Mondays, Wednesdays and Fridays are the slacker days; on those days the auction may be completed by 10.30 am or by 3 pm in the peak growing period.

No member of the public is allowed to bid. There are about 150 recognised buyers, of whom up to 100 may be present on any one day. A few of the buyers are wholesalers but the vast majority are retailers. Each grower's produce is sold separately. The successful bidder puts his ticket on the bought produce and takes it away as soon as he has completed all his purchases.

Sales by private treaty. These sales are arranged, usually by telephone, in one of the general offices, a part of the hereditament K which it is agreed is not "agricultural". After the sale has been arranged the produce is brought into the hall as to 50 per cent by the growers, and as to the other 50 per cent, this is collected by the respondents. The produce is put on the floor in a different special section of the hall. The buyers are secondary wholesalers and the produce which they have thus bought is cleared from the hall by the hauliers working for the respondents on the same day as it is brought in.

Two other matters need separate consideration. These are:

Non-members. In the agreed statement of facts put before the Lands Tribunal there is a category of persons referred to as "non- L members being members on probation". They were referred to on the appeal for brevity's sake as "probationers". When an existing member of the society drops out, be it by removal, retirement or death, the succeeding occupier of the member's land often wishes to continue sending produce for sale. The respondents' practice was to put such an applicant on trial for a period of 6-12 months in order that the respondents might satisfy themselves, *inter alia*, as to the quality of the produce being sent. A further reason for such non-members arises from the respondents' need for increased turnover. With this in view, they approach a new grower to become a member after a similar trial period. About 75 per cent of probationers become accepted as members and the other 25 per cent drop out. M No distinction is made as far as the use of the hall is concerned between produce coming from members and produce coming from probationers.

Produce bought for resale. A small proportion of the total produce handled and sold in the hall comes neither from members nor from probationers but has been bought by the respondents for resale. The case stated shows that although the percentage turnover relating to produce bought for resale is increasing, in the last year for which figures are given (1976) it was under 7 per cent. The reason for buying and reselling such produce is that the buyers at the auction, who are in the main retailers, usually themselves stock

A a full range of fruit and vegetables. If such buyers have to go to a competing organisation to obtain fruit and vegetables not grown by the members or probationers, the risk exists of their buying all their requirements from such a wholesaler. Moreover, in addition to buying fruit, the respondents may in the event of local shortage also buy home-grown fruit or vegetables.

The bought-in produce is only taken into the hall on the three busy days of the week. It is put on the floor in a special section first thing on a Tuesday, Thursday or Saturday morning, taking up about 1 per cent of the floor space of the hall and it is offered for sale at fixed prices before the auctioning of the other produce begins at 9 am. The fixed prices are 10 per cent higher than the cost price to B the respondents. The selling of these fixed-priced items usually took up to no more than 15 minutes of the auctioneers' time.

Having dealt with the facts it is now convenient to turn to the relevant statutory provisions.

General Rate Act 1967. Section 26(1): "No agricultural land or agricultural buildings shall be liable to be rated or be included in any valuation list or in any rate". (2) I do not read. "(3) In this section the expression 'agricultural land' — (a) means any land used as arable meadow or pasture ground only, land used for a plantation or a wood or for the growth of saleable underwood," . . . (b) I do not read. "(4) In this section the expression C 'agricultural buildings' — (a) means buildings (other than dwellings) occupied together with agricultural land . . . and . . . used solely in connection with agricultural operations thereon . . ." (b) I do not read.

Rating Act 1971: Section 1(1): "In section 26 of the General Rate Act 1967 (in this Part of this Act referred to as 'the principal section') — (a) the expression 'agricultural buildings' shall include any building which is an agricultural building by virtue of section . . . 4 of this Act . . . (b) the expression 'agricultural land' shall include land occupied with, and used solely in connection with the use of, one or more such buildings.

"(2) In this Part of this Act and in the principal section 'building' D includes a separate part of a building; and in determining for the purposes of this Part and that section whether a building used in any way is solely so used no account shall be taken of any time during which it is used in any other way, if that time does not amount to a substantial part of the time during which the building is used."

(3) (4) and (5) I do not read.

Section 4(1): "Subject to subsection (3) of this section, a building, other than a dwelling, is an agricultural building by virtue of this section if — (a) it is used in connection with agricultural operations carried on on agricultural land (as defined in the principal section); and (b) it is occupied by a body corporate any of whose members are, or are together with the body, the occupiers of the land." (2) I E do not read. "(3) A building used as mentioned in either of the preceding sub-sections is not an agricultural building by virtue of this section unless that use . . . is its sole use."

The rival contentions were succinctly summarised in the admirably drawn case stated. Save in one relatively minor respect these contentions have been repeated and, before turning to the authorities, it would be convenient now to refer to them quite briefly.

Mr Fletcher for the valuation officer, contended that the hall was not used "in connection with agricultural operations". The process of making the growers' produce marketable had been completed either on the farm or in the grading and packing section, which F section was agreed to be derateable. He thus contended that the whole of the marketing in the hall was an independent business. The most important use was for the holding of sales by auction. Selling by auction was a form of "merchandising" or of commerce and was not a process that could be expected to be used by a farmer. Moreover, a substantial quantity of the produce sold by auction had been grown by non-members, the probationers. As for the sales by private treaty, the hall was used as little more than a transit depot.

Mr Fletcher further contended that if his main submission was not acceptable, then the hall was not "solely used" as defined, because for a substantial part of the time it was used for the sale of produce that had been grown by the non-members.

Mr Fletcher successfully submitted to the Lands Tribunal that the G sale of produce bought by the society for resale was a non-qualifying use. The tribunal, however, held that it could be disregarded because the time that the hall was used for the handling of that produce did not amount to a substantial part of the time during which the hall was used (see section 1(2) of the 1971 Act). Although in the notice of appeal he had alleged that the Lands Tribunal had erred in law in reaching this conclusion, Mr Fletcher was content to accept the decision, so long as we were satisfied that the tribunal had taken into account not only the time involved in the resale of this produce but also the whole of the time which the produce was in the hall. I am quite satisfied the tribunal did take into account the whole of the time and not just the limited time H spent in the reselling of the produce and accordingly this point features no further in the appeal.

Mr Spence for the society contended that the use of the hall was simply to enable growers to dispose of their produce in the most effective way by making use of a cooperative society. The produce was unloaded, stored, auctioned and then loaded again, all these uses being in connection with agricultural operations. There was no question of the society running a shop and selling to the public. They sold only to wholesalers or retailers and it made no difference whether the sales took place at fixed prices or by inviting bids.

As to Mr Fletcher's alternative submission, Mr Spence contended that the "use" of the hall as referred to in section 4(3) was J the same use as that referred to in section 4(1)(a). This subsection was unqualified by section 4(1)(b), which related to the occupation of the agricultural land on which the produce had been grown. Accordingly it was immaterial to the question of sole "use" that some of the produce handled in the hall came from non-members, ie the probationers.

The only relevant English authority on this subject is *W & J B Eastwood Ltd v Herrod (Valuation Officer)* [1971] AC 160. The case was concerned with a different Act, but the relevant section contained the words "used solely in connection with agricultural operations thereon". In commenting on the words "used in connection with" agricultural operations on the land, Lord Reid stated K that: "ordinary usage of the English language suggests that the buildings must be subsidiary or ancillary to the agricultural operations" (p 168d). He went on to state (p 169d): "the whole object of producing a crop on the agricultural land is to market it in one form or another, and I think that anything done in the farm buildings, including storage and treatment, must be held to be done in connection with the agricultural operations on the land". But he added this caution: "Everything is saleable at a price, so even storage for a time or very simple treatment is not strictly necessary. One must have regard to ordinary and reasonable practice. There comes a stage when further operations cannot reasonably be said to L be consequential on the agricultural operations of producing the crop".

Lord Reid agreed with Lord Hunter when he said in *Midlothian Assessor v Buccleuch Estates Ltd* 1962 SC 453 at p 459: "I would agree that agricultural and pasturage do not cease when crops are grown or beasts raised, but may properly include operations reasonably necessary to make the product marketable or disposable for profit."

Lord Reid also agreed with Lord Sorn when he said in *Perth and Kinross Assessor v Scottish Milk Marketing Board* [1963] SC 95, that if a farmer set up a butcher's shop on his land to sell his fatstock M as meat no one would suggest that it be derated, for the shop would be used for an independent purpose distinct from the farming operations.

It seems to me clear that in reviewing the Scottish cases Lord Reid accepted that one test to apply in order to decide whether the use was or was not consequential on or ancillary to the agricultural operations was to ask whether or not the use, whether agricultural or not, was in connection with an operation which was an independent business or enterprise. Thus, in the first instance you look at the agricultural land to the exclusion of the buildings to ascertain what are the agricultural operations undertaken on that land. Next you look at the activities carried on in the buildings and determine

A whether those activities are connected with, viz used consequential on or ancillary to, the agricultural operations undertaken on the agricultural land and do not comprise any other activities (whether agricultural or not), which are not so connected.

Not surprisingly, fine distinctions have been drawn. In the words of Lord Avonside in *Roxburgh Assessor v West Cumberland Farmers' Trading Society Ltd* [1977] RA 298 at 307: "It is notoriously difficult to draw the line between disposal of farm products in a manner consequential on or ancillary to agricultural operations and disposal of these products in a manner which involves 'merchandising' in an independent commercial and business sense."

I can deal quite shortly with the sales by private treaty. As stated
B above, Mr Fletcher's contention was that the hall was used as little more than a transit depot. So be it, but to my mind the existence of a building to which goods which have been sold can be brought, for the purpose of their being collected by the purchasers, cannot amount to the disposal of those goods in a manner which involves "merchandising in an independent commercial and business sense". Mr Fletcher did not criticise the judgment of Lord Keith in the Court of Session in the case of *Ayrshire Assessor v Macster Poultry Ltd* [1977] RA 189, where he stated at p 198: "I do not consider the mere fact that a farmer regularly delivered his produce or part of it to a purchaser or purchasers would afford proper ground for holding that he had gone beyond what was reasonably
C necessary for marketing the produce and had entered into some sort of separate enterprise".

To my mind there can be no question of the use of the hall for the reception of the goods after they have been sold and pending their subsequent collection, producing some sort of separate enterprise.

The position with regard to the sales by auction raises difficult questions. I do not accept Mr Fletcher's submission that just because it may be unusual for individual farmers to auction their produce, there is anything unusual in a cooperative society disposing of their members' goods by auction. If the cooperative society had graded and packed the goods and then, as the final act in its disposal of the products, had auctioned them, I would not have
D accepted that such a use of the building amounted to a separate or independent enterprise. However, the society's functions in relation to the sales by auction is purely a marketing function and this may well involve crossing the line. Moreover, what to my mind clearly tips the balance is that the society is marketing not only its own members' goods, but non-members' goods, and such sales form a substantial part of the total auction sales. In 1973 the total of the auction sales amounted to just over £400,000, of which £118,000 was attributable to auction sales of probationers' products. In 1974 the total was just over £500,000, the probationers' sales accounting for £153,000. Approximately the same pattern applies to 1975. Thus a substantial part of the use made of the hall
E was devoted to the marketing of non-members' products. It cannot be suggested that such use was consequential on or ancillary to the agricultural operations. I therefore conclude that the auction use of the hall did involve the society disposing of the products in a manner which involved "merchandising in an independent commercial and business sense". To my mind it matters not that such sales were made not to the public in general but to wholesalers and retailers. I would allow the appeal on this ground.

It is accordingly unnecessary to deal with Mr Fletcher's alternative ground, but for the sake of completeness I will deal with this shortly. In my judgment the words "that use" in section 4(3) of the 1971 Act refer back only to section 4(1)(a). They do not refer back
F as well to 4(1)(b), which is concerned not with use but with extent of the occupation of the agricultural land referred to in 4(1)(a). That is all of that land and not merely part of it. However, the use has to be the "sole use" (section 4(3)). It is common ground that a substantial proportion of the use of the hall was attributable to the sale by auction of non-members' produce. This produce was grown by them on land occupied by them. Accordingly, the answer to the second question is that in the true construction of section 4(1)(a) and (3), the hall was not solely used for the purpose set out in section 4(1).

Agreeing, WATKINS LJ said: In recent times groups of farmers have found it convenient and profitable to sell the produce of their lands through the agency of societies which are bodies corporate G
specially created for that purpose and of which they are members. These societies have headquarters consisting of offices and buildings in which the produce of members is stored, prepared for sale and in various ways sold to wholesalers and retailers.

Before 1971 none of these buildings qualified for exemption from rating. This was because none of them was an agricultural building as defined in section 26 of the General Rate Act 1967. Parliament decided to assist these societies to reduce their liability to pay rates. It do so by sections 1 and 4 of the Rating Act 1971, which were expressly enacted to extend the provisions relating to exemption from rating of land and buildings used in connection with agriculture. H

These extended provisions permit exemption from rating of a building provided it is established that: (1) The building is occupied by a body corporate, ie the society, and (2) the building is solely used in connection with agricultural operations, and (3) those agricultural operations are carried on upon land used as arable, meadow or pasture ground only, for a plantation or a wood or for the growth of saleable underwood, and (4) that land is occupied either by any members of the society or by any members together with the society.

If all these conditions are satisfied the building is properly called "an agricultural building" as defined in the Act of 1971.

I agree that the Gloucestershire Marketing Society Ltd clearly J
failed to satisfy the second of these conditions, since on a substantial scale they regularly sold by auction in their hall the produce of non-members, called "probationers", which had been grown on land which was not occupied either by members or by members together with the society. Therefore, they are not entitled to exemption from rates.

I would add a word or two about one method of sale in particular adopted by the society, since much was said of it in argument. It was submitted that the holding of a sale although only of members' produce by auction and although confined to wholesalers and retailers would of itself disable the society from satisfying the second condition. I do not understand the numerous authorities to K
which reference was made to lead to that conclusion, and I cannot think of any other reason which would justify it.

Would an auction by a sole farmer in a building on his farm of the produce from it to wholesalers and retailers only disentitle him from exemption? I think not. If that be right I see no reason why a society when it does a like thing on behalf of its members and is in conformity with all other conditions should not be exempted from paying rates.

If a society sells directly to the public or sells the produce of non-members or sells produce which it has bought for resale, the building where this activity goes on will inevitably be rated unless possibly the insignificance of that activity invokes the *de minimis* L
principle.

I also would allow this appeal.

Also agreeing, WALLER LJ said: Ackner LJ has already referred to the speech of Lord Reid in *Eastwood v Herrod* and I do not repeat it. In coming to a conclusion in this case I have found it helpful to consider a number of Scottish decisions showing where the line has been drawn between an agricultural building on the one hand and activities which are not derated. The argument before us has been on the part of the valuation officer that the line is drawn between that which is part of the normal process for making the product merchantable and activities which are activities of merchants. On the other hand the ratepayers submit that the line should be drawn M
between activities of a farmer which are activities of a wholesale nature and activities of selling which are of a retail nature.

In *Perth and Kinross Assessor v Scottish Milk Marketing Board* [1963] RA 24, the court decided that a service court for the production of semen and a laboratory were used for the business of handling and purveying semen and were not agricultural buildings. Lord Patrick at p 30 said: "In the use the ratepayers make of these buildings they have I consider passed from the sphere of their farming activity into the sphere of their business which consists of handling and purveying semen" and then he went on to instance a butcher's shop on a farm saying it was clearly not an agricultural building, and he went on: "The reason would be that the building was truly being used for an independent purpose distinct from the farming operations." In *Rimmer v Perth and Kinross Assessor* [1965] RA 506 the court held that a potato store was not of the agricultural process because the ratepayers also were in business as potato growers and seed and ware merchants. And Lord Kilbrandon, at p 509, said: "It is not decisive that each potato in the store is a product of the farmer's business because it is also part of the stock of the merchant and held in the premises as such. If that be so then the building is being used both as a farm store and as a merchant's warehouse and it therefore cannot claim . . . that it is occupied solely in connection with the former type of occupation to the exclusion of the latter." And Lord Avonside, at p 513, said: "While a producer may sell his crop and still be within the confines of agricultural operation it is plain that if he moves into the field of merchandising or sales promotion, whether wholesale or retail, he is removing himself from the field of agriculture as such." In *Moray and Nairn Assessor v Meldrum* [1968] RA 562, where the farmer ran a business as a retail milk salesman it was held that the dairy was not within the exemption, and Lord Avonside, at p 569, said: "The ratepayers have moved into the field of merchandising or dealing, they have entered into retail trade in the market and that part of their activities cannot be described as an agricultural operation." In *Roxburgh Assessor v West Cumberland Farmers Trading Society Ltd* [1977] RA 298, where the Scottish Court was considering section 7 of the Rating Act 1971, the court held that egg-grading and packing buildings were not agricultural, and Lord Fraser at p 305 said that the enterprise was partly for the benefit of the society which was a separate entity from its members and gave three reasons for saying that it was separate: firstly, the ratepayers were taking active steps to market the eggs advantageously, secondly, they used two trade names for their eggs, and thirdly, and I quote, "In this case it seems to me that the operations carried on in the subject of appeal are in effect operations carried on by a merchant because the ratepayers are acting as merchants of the eggs produced by their members."

In this case we have to consider whether or not the operation of the auction hall was part of the agricultural process or whether it had passed into the realm of merchandising. In my judgment the daily auction taking place every morning of the week, sometimes running into the afternoon, was a separate enterprise for merchanting. It would be wrong in my opinion to say that it was still part of the agricultural process. The normal practice of many farmers or agriculturists would be to send their produce to market for it to be auctioned at market and that is done by merchants who do the auctioneering. Furthermore, in this case, a substantial proportion of the goods sold by auction belonged to probationers. The proportion varied each year but, for example, it was over 25 per cent in 1973. It is, in my view, clear that in such circumstances the ratepayers had passed beyond the stage of an agricultural process because they were in the business of marketing produce of those who were not members of the association.

This would be sufficient to dispose of this appeal, but a second question was raised and argued before us, and I think it right to express my opinion. The question was: "Whether on a true construction of subsections 4(1)(a) and (b) and subsection (3) the said hall was not solely used for the purpose set out in subsection 4(1) by virtue of the fact that it was used for the sale by auction of crops grown by non-members on agricultural land occupied by them." In considering the answer it is necessary to look first at the General Rate Act 1967.

Section 26(4) of the 1967 Act defines "agricultural buildings" as "buildings . . . occupied together with agricultural land . . . used solely in connection with agricultural operations thereon." It is the common occupation of agricultural land and buildings which provided the basis but the use had to be solely in connection with agricultural operations on *that* land.

The 1971 Act, by section 4, extended that original definition to buildings operated by bodies corporate and certain associations. In my opinion the land referred to in section 4(1)(b) is the same as the "agricultural land" mentioned in section 4(1)(a). I agree with the Lands Tribunal decision in *Home Grown Fruits* [1974] RA 329 on this point. Subparagraph (b) therefore should be understood as saying ". . . any of whose members are or are together with the body occupiers of the agricultural land in connection with which the agricultural operation mentioned in (a) above are used". To qualify as an "agricultural building" it is therefore necessary that: (1) the building be occupied by a body corporate, (2) the land on which the agricultural operations are carried on be occupied by any of the members of the body corporate either alone or with the body corporate, and (3) the building be solely used in connection with agricultural operations on that land. I think that this results from the words of section 4(1)(a) and (b) and section 4(3).

Applying these principles to this case, under subparagraph (b) the building is occupied by the respondents and the occupiers of the land have to be either the respondents or their members or both. In the *Home Grown Fruits* case the Lands Tribunal dealt with the construction of 4(1)(b). The arguments supporting a liberal interpretation are at p 341 and the answer by the rating authority at p 344. Although he does not specifically say so, it is, I think, implicit that the Lands Tribunal member in his decision adopts the interpretation at p 344. I believe this to be right. Were it not to be so, the amending statute would be departing far from the original provisions of the 1967 Act.

Now, coming to subparagraph (a) the use has to be in connection with agricultural operations carried on on agricultural land occupied by the respondents or their members. So far, so good. But then subsection (3) says that that use has to be the sole use. It is not the sole use because a substantial proportion of use is from non-members whose agricultural connection is with land not owned by members of the respondent body. It will be seen from what I have said that the question of whether subsection (3) qualifies subsection (1)(b) does not arise. Subsection (1)(b) is already limited to the society or its members.

My answer to the question posed is that on a true construction of sections 4(1)(a) and (3) the said hall was not solely used for the purposes set out in subsection 4(1) by virtue of the fact that it was used for the sale by auction of crops grown by non-members on agricultural land occupied by them.

The appeal was allowed, the assessments to be entered in accordance with agreed figures. The appellant was awarded the costs in the Court of Appeal and Lands Tribunal.

COURT OF APPEAL

March 2 1981

(Before Lord DENNING MR, Lord Justice SHAW and Lord Justice OLIVER)

WESTMINSTER CITY COUNCIL v HAYMARKET PUBLISHING LTD

Estates Gazette May 16 1981

(1981) 258 EG 641

Rating — General Rate Act 1967, sections 17A and 17B — Rating surcharge on unused commercial building — Commercial building mortgaged "up to the hilt" — Question of priority between surcharge and mortgage debt — Bank as mortgagee had sold the property to the present occupiers — Surcharge registered as land charge in local land charges register — Bank claimed that mortgage debt due to them (of over £6m) took priority over rating authority's surcharge of about £17,000 — Rating authority claimed the reverse — Held, affirming decision of Dillon J, that the surcharge was a charge on the entirety of the interests in the property, the whole proprietary interests in the premises — Accordingly the surcharge in favour of the authority took priority over the mortgage debt to the bank — Appeal dismissed

In this case Haymarket Publishing Ltd appealed against a decision of Dillon J that a building at 22 Lancaster Gate, London, which the company had purchased from the National Westminster Bank, selling as mortgagee, was charged in favour of Westminster City Council with a surcharge of £16,940.93 for a period as an

A unused commercial building. Dillon J's judgment was reported at (1980) 253 EG 805.

W J Mowbray QC and Colin Braham (instructed by the Solicitor, Westminster City Council) appeared on behalf of the respondents (plaintiffs before Dillon J); Peter Millett QC and Gregory Hill (instructed by Wilde Sapte) represented the appellants (defendants before Dillon J).

Giving judgment, LORD DENNING MR said: On the north side of Hyde Park there is an elegant period property, 22 Lancaster Gate. It was very suitable for use as offices by high-class organisations. But it remained empty for many months — like the notorious Centre Point. The owners did not try to let it to anyone. It suited
B them better — in days of inflation — to let it remain empty, not paying rates.

Parliament did not approve of such goings on. So in 1974 they enacted provisions under which the local authority could impose a surcharge on commercial buildings left empty in this way. If a commercial building was left empty for 12 months the surcharge would be double the normal rates. Treble in the second 12 months. Quadrupled in the third 12 months. And so forth.

The question in this case is whether the surcharge applies to commercial buildings which are mortgaged up to the hilt. Is the surcharge payable by the mortgagor, or by the mortgagee? If it is not paid, how can it be enforced?

C This building at 22 Lancaster Gate was bought on February 3 1974 by a company called Shop Investments Ltd. That company was a subsidiary of a holding company, Amalgamated Investment & Property Co Ltd. Shop Investments guaranteed the indebtedness of the holding company to the National Westminster Bank. It came to over £6m. As security Shop Investments mortgaged 22 Lancaster Gate to the bank. It was by way of a charge of legal mortgage on February 4 1974. A few days later, on February 8 1974, the Local Government Act 1974 was passed bringing the surcharge into effect for empty properties left unused.

Despite the new Act, Shop Investments Ltd left 22 Lancaster Gate empty. They left it empty for nearly two years. They made no
D effort to let it. This came to the notice of the Westminster City Council. They decided to levy a surcharge for the period from February 8 1974 — when the Act came into force — to October 24 1975: that being the period when the mortgagors had made no effort to let it at all. The surcharge came to £16,940.93. Westminster City Council registered it in the local land charges register.

In 1977 the holding company went into liquidation — owing the bank £6m. The bank, as mortgagees, sold the property to Haymarket Publishing Ltd for £220,000. The Westminster City Council claim that this surcharge of £16,940.93 takes priority over the mortgage debt. But the bank say that the mortgage debt of £6m takes priority over the surcharge of £16,940.93. So the practical
E question is: Who gets the £16,940.93? The Westminster City Council or the bank?

It all comes down to the correct interpretation of two sections which were inserted in the General Rate Act 1967 by section 16 of the Local Government Act 1974. It is written in by two supplemental sections, 17A and 17B. Section 17A says who is to pay the surcharge. It says:

> If for a continuous period exceeding six months a commercial building is not used for the purpose for which it was constructed or has been adapted, its owner shall pay in respect of that period . . . a surcharge additional to the rates (if any) payable apart from this section.

F So there is a liability to pay put upon the owner. The meaning of "owner" is defined in section 17B(7):

> . . . "owner" means the person entitled to possession.

If the owner does not pay, then section 17B provides for how it can be enforced. Subsection (3) of that section says:

> A surcharge imposed under section 17A of this Act in respect of a hereditament shall until recovered be a charge on the land comprised in the hereditament.

The first question is: Who is liable to pay the rates? Who is the "owner"? Who is "entitled to possession" when it is a question between mortgagor and mortgagee? It is quite plain to my mind G
that unless and until the mortgagee enters into possession, in accordance with the mortgage law, and takes the rents and profits, the mortgagor is the person entitled to possession. He is the person who is entitled to put up advertisements, to let and do all that is necessary to let — unless and until the mortgagee interferes. It seems to me that the "owner" under section 17A is not two, three or more people. It is the one who really has control of the letting. In this case the "owner" means the mortgagor, Shop Investments Ltd. They were liable to pay the surcharge. No doubt Westminster City Council could sue them for it if they were worth suing — which unfortunately they are not.

Then under section 17B we get the provision that the surcharge is H
"a charge on the land". What does that mean? I agree that it is ambiguous. The meaning is difficult to discover from the words of the statute. But fortunately there has been a series of cases in the courts which lead inevitably to the interpretation that it means a charge on "all the estates and interests in the land". I will go through the series of cases which lead to this result. The first is the *Corporation of Birmingham v Baker* (1881) 17 Ch D 782. In that case there was a charge on the "premises". Sir George Jessel MR said at p 786:

> Now, the houses per se being inanimate, they cannot bear the burden. If there be a charge on the houses it is a charge on the total ownership — if I J
> may call it so, on the proprietorship; not on any particular section or portion of the proprietorship, but on the whole.

The next case is the *Guardians of Tendring Union v Dowton* [1891] 3 Ch 265. That again was a question of a charge on the "premises". I would pick out the words of Fry LJ at p 269:

> All the Act does is to create a charge on the premises — that is, on the land — that is, on all the interests of the owners of the land.

To finish the citations, I would refer to *Paddington Borough Council v Finucane* [1928] Ch 567. That was a decision of Russell J (as he then was), who said of the Public Health Act 1875:

> . . . when it confers a charge on the premises, [it] means not a charge only on K
> the interests of the rack-rent owner in the premises, but a charge upon the entirety of the interests of the premises, the whole of the proprietary interests of the premises.

Having regard to that line of authorities, it seems to me that we should interpret the words in this statute, "a charge on the land comprised in the hereditament", in the same sense.

It was suggested that it should be confined to cases where the charge was in respect of improvements or expenses incurred upon the land — as happened in some of those cases — and that it should not apply to a charge in respect of a penal provision such as this. That distinction is not well-founded. The issue depends simply on L
the interpretation of the words in the statute. It seems to me that "a charge on the land" means a charge on "all the estates and interests in the land".

This conforms with the intention of Parliament. Any other view would mean that it would be open to companies, by manipulation of their affairs between holding and subsidiary companies, to avoid the charge altogether. The land could be bought in the name of a subsidiary company with money obtained on mortgage from a holding company: and, lo and behold, the charge would be gone. That cannot be right. The judge was quite right in holding that there was a charge on the land in favour of Westminster City Council: and that it took priority over the mortgage debt which was owing to the M
bank.

I would dismiss the appeal accordingly.

Agreeing, SHAW LJ said: I would only add that support is to be found in the subsection itself to limiting the application of the phrase "person entitled to possession" where it appears in section 17B(7) of the Act to the person who is at material times actually exercising that right. It provides for apportionment between persons who become successively entitled to possession during a period of non-use. This could hardly be apt in relation to persons who are in law collaterally and contemporaneously entitled to possession.

I would dismiss the appeal.

Also agreeing OLIVER LJ said: It is quite clear, in my judgment, that section 17A contemplates, contrary to Mr Mowbray's submission, only one owner — a single individual or persons holding jointly — on whom the surcharge would be levied. But that, of course, is not conclusive of the matter because the question at issue is the ambit of the charge which is imposed by subsection (3) of section 17B. In subsection (2) of section 17B it is said that: "References in section 17A of this Act and this section to a commercial building are references to a hereditament . . . whose net annual value falls to be ascertained under section 19(2) of this Act". So quite clearly the hereditament there is the commercial building.

Then one comes to subsection (3), where it is provided that: "A surcharge imposed under section 17A of this Act in respect of a hereditament shall until recovered be a charge on the land comprised in the hereditament". This is a formula which is familiar in other statutes, for instance the Public Health Act and Housing Acts, and my Lord the Master of the Rolls has already referred to cases in which similar words have been construed.

What is there in this statute to restrict the words in section 17B(3) in such a way as to lead to the conclusion that they do not mean exactly what they say? There is the fact, first of all, that this is a charge imposed by way of a penalty. It may be argued that the statute should be construed restrictively, but that by itself does not, I think, justify the court in concluding that the section does not mean exactly what it says. Then it is said that it is very unfair that the effect of the charge is to make the rates recoverable from persons such as mortgagees, tenants and owners who are not in possession and were not under an original liability. But the words of the statute are, as I think, plain and one must start from the position that Parliament must have been aware of how similar words in previous statutes had been construed. If it had been intended to confine the charge to the interest of the owner in the premises, nothing would have been simpler than to have said so. The legislature did not do that.

I agree that the learned judge came to the right conclusion, and I too would dismiss the appeal.

The appeal was dismissed with costs. Leave to appeal to the House of Lords was refused.

COURT OF APPEAL

November 18 1980

(Before Lord Justice WALLER, Lord Justice WATKINS and Lord Justice GRIFFITHS)

BRENT LONDON BOROUGH v LADBROKE RENTALS LTD

Estates Gazette May 30 1981
(1980) 258 EG 857

General Rate Act 1967, Schedule 1 — Rating of unoccupied property — Dispute as to date from which rating authority could properly charge the unoccupied property rate on the appellants in respect of two floors of a building — The building as a whole was deemed to be completed in 1974, but appellants contended that the two floors only became a separate hereditament when they were leased to the appellants with effect from June 24 1975 — It was therefore argued that the rate liability did not arise until three months from that date – Held, rejecting this argument, that the date of separation in the valuation list was not the test — As part of the whole building the two floors had been unoccupied for more than three months before June 24 1975, when the lease began, and it followed that the appellants were liable to pay the unoccupied property rate from that date — Appeal from decision of Brown J dismissed

This was an appeal from a decision of Brown J, holding that the appellants had been properly charged with the unoccupied property rate under the General Rate Act 1967, section 17 and Schedule 1, in respect of the fourth and fifth floors of Chancel House, Neasden, in the London Borough of Brent, as from June 24 1975, the date when the lease to the appellants of these floors commenced.

C S Lewsley (instructed by P W Moore, of legal staff of Ladbroke Group) appeared on behalf of the appellants; Viscount Colville of Culross (instructed by K B Betts, Town Clerk, Brent) represented the respondent rating authority.

Giving judgment, WALLER LJ said: This is an appeal against a decision of Brown J given on July 25 1979, when he held that the appellants were properly rated as owners of the 4th and 5th floors of a building known as Chancel House, Neasden Lane in the London Borough of Brent.

The facts were that Chancel House was a new building which was nearing completion in 1974. Notices were served by the rating authority and it was not disputed before us that the building was deemed to be completed (within the meaning of paragraph 7 of Schedule 1 to the General Rate Act 1967) on November 15 1974. The appropriate completion notices had been served separately in relation to each of the six floors of the building.

By a lease dated May 23 1975 the appellants became tenants of the 4th and 5th floors for a term of 25 years from June 24 1975. The respondent rating authority had passed a resolution in 1968 empowering themselves to rate unoccupied property. Furthermore in 1974 they had passed another resolution empowering themselves to rate unoccupied property at 100 per cent of the proper rate. Accordingly, it being accepted that the building was deemed to be completed on November 15 1974, the rating authority was empowered to charge rates if the building remained unoccupied after three months from that date. After being informed that the two floors had been let, the rating authority demanded rates from June 24. The appellants denied liability for the period before they went into occupation on September 1 1975.

The relevant provisions are contained in the General Rate Act 1967.

Paragraph 7 of Schedule 1 provides:

Completion of newly erected or altered buildings. 7. For the purposes of paragraph 1 of this Schedule, a newly erected building which is not occupied on the date determined under the subsequent provisions of this Schedule as the date on which the erection of the building is completed shall be deemed to become unoccupied on that date.

Paragraph 8 lays down the procedure and provides:

8.— "(1) Where a rating authority are of opinion — (a) that the erection of a building within their area has been completed; or . . . and that the building is, or when completed will be, comprised in a relevant hereditament, the authority may serve on the owner of the building a notice (hereafter in this paragraph referred to as 'a completion notice') stating that the erection of the building is to be treated for the purposes of this Schedule as completed on the date of service of the notice or on such later date as may be specified by the notice."

Paragraph 1(1) provides:

Liability to be rated in respect of certain unoccupied property. 1.— (1) Where, in the case of any rating area in which, by virtue of a resolution under section 17 of this Act, this Schedule is in operation, any relevant hereditament in that area is unoccupied for a continuous period exceeding three months, the owner shall, subject to the provisions of this Schedule, be rated in respect of that hereditament for any relevant period of vacancy; and the provisions of this Act shall apply accordingly as if the hereditament were occupied during that relevant period of vacancy by the owner.

The definition of "hereditament" is in section 115 and is as follows:

"hereditament" means property which is or may become liable to a rate, being a unit of such property which is, or would fall to be, shown as a separate item in the valuation list.

By paragraph 15 of Schedule 1 "relevant hereditament" means "any hereditament consisting of, or of part of, . . . office, factory, . . . or other building whatsoever. . . ."

Mr Lewsley submits that the top two floors of Chancel House became a separate hereditament from June 23 1975 and that paragraph 1(1) of Schedule 1 would start to apply on that date. In other words that the relevant hereditament, ie the top two floors, would be unoccupied for a continuous period exceeding three months from that date, and that only after the three-month date would the hereditament be rated and he submits that anything that had happened before that date is irrelevant because there was no separate hereditament.

On the other hand Lord Colville submits that in the case of a newly erected building a completion notice relates at the same time to the whole as well as to each individual part of the building with the result that the three-month nil rate could only occur once. The corollary to Mr Lewsley's argument would be that whenever a fresh part of the building was let not only that part but the part remaining unoccupied in the hands of the owner would be a new hereditament and would be entitled to three months of unoccupation before being chargeable with the unoccupied rate. It would be surprising if this were the intention of Parliament and therefore one must examine the phrases carefully to see their exact meaning.

Paragraph 7 of the Schedule does not mention hereditament, it is "*the building*" which is deemed to become unoccupied. The building may consist of one hereditament or many. The valuation list may divide the building into several hereditaments or not divide it at all as in this case up to the time of the lease to the appellants. In this case more than three months had passed since the specified date. Although the two floors of the building had been unoccupied for more than three months, were they entitled to another three months from the beginning of their lease?

In my opinion the date of separation in the valuation list is not the test. "Hereditament" is "property which is . . . *or would fall to be* shown as a separate item in the valuation list". On the specified date these two floors comprised a unit of property which "would fall to be" shown as a separate item.

Although the case of *Camden London Borough Council v Post Office* [1977] 1 WLR 892 does not directly cover the present facts the judgment of Bridge LJ is helpful in its general approach. In the last sentence of his judgment on p 899 he says:

> The procedure under paragraphs 7 and 8 having fixed the date when the whole of it was deemed to become unoccupied, the owner, as defined in this Schedule, of any hereditament carved out of that entire building was liable to pay the unoccupied rate from the appropriate date.

In that case both the completion date and the appropriate date occurred after the date of the commencement of the tenancy. Paragraph 7 of Schedule 1 defines the date when the building is deemed to become unoccupied. And paragraph 8(1) indicates that the rating authority has to be of opinion that the building "is or when completed will be comprised in a relevant hereditament". The judgment in the *Camden* case indicates that "the relevant hereditament" will also include any hereditament carved out of the entire building.

In my opinion the building referred to in paragraph 8 includes the hereditament of the whole building as well as any separate hereditaments carved out of that whole building. In this case the top two floors comprised a unit of property which would fall to be shown as a separate item. It follows that in paragraph 1 "the relevant hereditament", although only leased on a date after the date of completion, nevertheless has been unoccupied from the date of completion. The hereditament referred to in paragraph 1 is a unit of property which is or would fall to be shown as a separate item in the valuation list. The fact that it was not shown as a separate item up to the date of the commencement of the lease does not affect the matter because it must be included in the relevant hereditament in paragraph 8 which, as I have already said, must be deemed to include any hereditaments carved out of the whole. The same consequence would follow with relation to the remainder of the building. It would comprise one or more units of property which would fall to be shown as separate items.

It follows that in my opinion, therefore, the date of completion is the date when the hereditament including the whole building was deemed to be completed and the date when any separate hereditaments which were afterwards carved out of that whole building became unoccupied would be the same date.

I would dismiss this appeal.

WATKINS LJ agreed

Also agreeing that the appeal should be dismissed GRIFFITHS LJ said: Ladbrokes submit that the top two floors of the office building cannot be at one and the same time a part of one hereditament, namely the entire office building, and also a separate hereditament, namely the top two floors. Until the lease of the top two floors to Ladbrokes it was proposed to rate the entire building as one hereditament, but when the two top floors were let to Ladbrokes they became a separate hereditament for rating purposes. So it is submitted there was no hereditament comprising the top two floors until it was created by the lease to Ladbrokes on June 24 1975, and therefore there was no hereditament upon which Schedule 1, paragraph 1, could bite until June 24 1975. Therefore Ladbrokes say they are not liable for rates until *that* hereditament has been unoccupied for three months, namely September 23 1975, or until they entered into occupation of it.

I confess that I was for a time beguiled by this argument. But on reflection I am convinced that it cannot be correct, for it leads to a result that goes far to frustrate the power to rate unoccupied property. In this case we are concerned with a newly constructed building, but the position would be no different in the case of an existing building. Suppose that a large office block had been leased for occupation to one large company, let us say ICI, and accordingly was rated as a single hereditament. At the end of their lease, ICI decide to move elsewhere and the whole office block becomes empty. The owner of the block will become liable for an unoccupied rate on the whole building after three months. A month later he lets the top floor of the building as offices to a small company. Apply the appellants' argument; the top floor becomes a separate hereditament upon which no unoccupied rates are payable for another three months and it follows that the rest of the building becomes a different hereditament which is also exempt from the unoccupied rate for a further three months. So the owner can keep postponing his liability for the unoccupied rate by successive lettings of small parts of the building, for each time he lets a part capable of being a separate hereditament it alters the hereditament remaining in the hands of the owner.

Because the problem is identical whether considering a newly constructed or an existing building I do not myself find any assistance to its solution in the provisions of Schedule 1, which deal with the date upon which a newly erected building is deemed to become unoccupied. Save only this, that they do show that Parliament was alert to one stratagem that a building owner might employ to avoid paying an unoccupied rate by deliberately delaying the completion of his building and it is a pointer to the conclusion that Parliament could not have intended the alternative stratagem of successive lettings to achieve the same objective.

I think the clue to the problem is to be found in the appreciation of the fact that the term "hereditament" in rating law is a term of are used to identify a property that has two attributes: (1) that it is rateable and (2) that it is or can be shown as a separate unit in the valuation list (see section 115). The property does not gain or lose any physical attribute or change in any way by calling it a hereditament; the term hereditament is a legal shorthand used to denote the fact that that property is or will be separately rated. With this in mind I turn to Schedule 1, paragraph 1, and bearing in mind the definition of "hereditament" in section 115 and the definition of "relevant hereditament" in Schedule 1, paragraph 15, I read for "any relevant hereditament" "any property consisting of an office liable to be rated and shown in the valuation list as a separate item" and I find that where such a property is unoccupied for a continuous period exceeding three months it becomes liable to an unoccupied rate. The definition covers the top two floors of this building, they had been unoccupied for more than three months before June 24 1975 and it follows that Ladbrokes are liable to pay the unoccupied rate upon them from that date.

Accordingly I would dismiss this appeal.

The appeal was dismissed with costs. Leave to appeal to the House of Lords was refused.

QUEEN'S BENCH DIVISION

January 12 1981

(Before Mr Justice WOOLF)

LEFORE HOLDINGS LTD v CROYDON LONDON BOROUGH

Estates Gazette June 6 1981

(1981) 258 EG 970

Rating — Case stated by magistrates — Whether company liable to rates in respect of piece of land — Whether company in beneficial occupation — Land had been intended at one time to be used as a car park — Company had done some work to make the land suitable for a car park, such as laying down hogging (but no tarmacadam), but land had never been used as a car park — Submission that carrying out of work to render land fit for beneficial occupation was not sufficient in itself to amount to beneficial occupation — Arbuckle Smith & Co Ltd v Greenock Corporation — Held that the finding of the magistrates that the land was in beneficial occupation was wholly unreasonable on the material before them — Decision of magistrates that company was liable to rates in respect of the land in question set aside

This was a case stated by magistrates arising from a complaint by the London Borough of Croydon in respect of alleged non-payment of rates by the appellant company, Lefore Holdings Ltd. The land which was the subject of the claim was described as a car park situated at the rear of 77 to 81 High Street, London SE25.

Graham Clark (instructed by W G R Saunders & Son) appeared on behalf of the appellant company; Brian Jubb (instructed by the town clerk, London Borough of Croydon) represented the respondent borough.

Giving judgment, WOOLF J said: This is an appeal by Lefore Holdings Ltd. It is an appeal which has had a rather chequered history. The matter has previously been before the Divisional Court and the Court of Appeal because the justices were refusing to state a case. The justices did, however, as a result of those earlier proceedings, state a case and the case which is stated is an admirable one. It clearly outlines the issues, indicates the facts as found by the justices and goes on to give three reasons for their decision.

The matter arises because on August 11 1978 a complaint was preferred by the London Borough of Croydon as a rating authority against the appellant company, Lefore Holdings Ltd, requiring the company to appear before the court to show why it had not paid the rates specified in the complaint in accordance with section 97(1) of the General Rate Act 1967, a rate amounting to £377.05 with regard to land which was described as a car park at the rear of 77 to 81 High Street, London SE25, for the period from September 8 1977 to September 30 1978, which period was subsequently amended so that the amount was reduced to £240.26 for the period from September 8 1977 to May 7 1978.

Mr Clark, who has represented the appellant before me, was also representing the appellant before the justices and he indicates, as appears from the case, that certain concessions were made by the appellant before the justices, the consequence of which was that the only issue before the justices was whether the undoubted possession by the appellant of the land in question was of some value or benefit to the company. If it was of some value or benefit, then it is, as I understand the position, conceded now and has at all times been conceded by the appellant that he would be liable for the appropriate amount in respect of rates.

The concession had the effect that the only witness who was called before the justices to give evidence was Mr Trundell, a director of the appellant company. He gave evidence which is set out in the case in detail, and I can summarise it in this way. It was to the effect that at one time the appellant intended to use the land as a car park, but they themselves had never in fact so used it and therefore they had not been in beneficial occupation. It was agreed that in respect of the car park for a period work had been carried out to the land to make it more suitable for use as a car park, and in particular hogging had been laid on the land, but no tarmacadam. However, in respect of the carrying out of that work, reliance is placed upon previous authority which makes it clear that the mere carrying out of work to render land fit for beneficial occupation is not sufficient in itself to amount to beneficial occupation. The authority relied upon for that proposition is the decision of the House of Lords in *Arbuckle Smith & Co Ltd v Greenock Corporation* (1960) 4 RRC 95, where, according to the headnote, it was held that the "appellants were not in rateable occupation of premises during the year in question because their use for the purpose of effecting alterations was a preparing of them for future occupation but was not occupation for the purpose for which they were bought".

In the case, as part of the findings of fact made by the justices, it is recorded that it was possible during the relevant period for motor vehicles to be driven on and off the hogging and to remain parked on it. It is also recorded that on occasions motorists drove on to the land without the appellant's permission and that the appellant's staff ordered them to leave and that the appellant's staff and customers could have parked motor vehicles on the land. It is also recorded that there were parking restrictions in the area and the appellant could have obtained income from the land by charging for parking spaces.

As I understand those findings, they could be perfectly proper findings for the justices to make in a case of this sort in so far as they were setting the scene in order to decide the question whether or not the land was in fact beneficially used, since clearly if they had come to the conclusion that the land could not be used that was a matter which would be relevant so far as the appellant was concerned because that would be inconsistent with the use. It also follows, in my view, that it was important for them to come to a conclusion about this because it meant that they were able to approach the vital question at least on the basis that the land was land which could be used for the purpose which was contended for and therefore one had to go on to consider if it was in fact used.

I make those comments about that paragraph of the case because, as part of his submission on the appellant's behalf, Mr Clark submitted that it may be in relation to that paragraph the justices were misdirecting themselves in making the assumption that because land could be used for a particular purpose, it was so used; or alternatively, they were misdirecting themselves in assuming that the fact that land could be used for a particular purpose was sufficient for a finding that the land was being beneficially used for the purpose of rating.

Although I understand the reasoning behind Mr Clark's submission, I do not think that that particular criticism of the justices' approach is warranted. I think that the probable explanation of that finding in the case is merely that the justices were doing what I would regard as perfectly proper and making preliminary findings of fact before going on to consider the vital question.

In relation to that vital question, having set out the evidence of Mr Trundell and the submissions of the parties, the justices went on in the case to say:

> We decided that [Mr Trundell] was exaggerating when he suggested that the land could not be used for parking motor vehicles because it only had a surface of hogging. This was not our belief. Some motorists had driven vehicles on to the land without permission. We accepted that in August 1977 he informed the respondent, on behalf of the appellant, that the appellant intended to use the land as a car park in about two weeks' time. We did not accept Mr Trundell's testimony that the land was not used during the relevant period by the company or its staff, or by other persons with motor vehicles with the appellant's consent. At the conclusion of our deliberations, we believed that the appellant had not discharged the burden of proof resting upon him. We decided that the appellant's possession of the land had been of value and benefit.

When the evidence of Mr Trundell as recorded in the case is looked at with care, it is quite apparent that if any weight was attached to that evidence, it clearly established a situation where, on the uncontradicted evidence before the justices, the only conclusion that one could come to as a matter of law was that this particular land at the relevant time was not rateable on the basis contended for by the appellant. I do not propose in this case to go so

A far as to suggest that it is never open to the justices in a matter of this sort to reject out of hand the only evidence which is put before them on behalf of a party in the position of the present appellant. Before a decision can be reached as to whether or not that is a course which justices are entitled to adopt one has to consider all the circumstances of the particular case. However, in this case I am satisfied that the reasoning of the justices in rejecting Mr Trundell's evidence in its entirety was not justified.

First, the whole basis of their decision appears to rest upon the fact that they came to the conclusion that Mr Trundell was "exaggerating". So far as that conclusion is concerned, they were in error because they based it upon the fact that they understood Mr Trun-
B dell as giving evidence that the land could not be used for parking vehicles because it only had a surface of hogging. As I understand Mr Trundell's evidence as recorded in the case, it is clearly to the effect that the land *should* not be used for parking motor vehicles and not that it *could* not be used. Indeed, he was accepting that it had been used, but he was saying that it was objectionable use because it would churn up the hogging and cause damage. It does not, therefore, seem to me that the criticism of exaggeration was justified.

The remainder of the reasoning of the justices all depends upon that premise, on the finding of exaggeration. They accepted Mr Trundell's evidence that he informed the respondent "that the
C appellant intended to use the land as a car park in about two weeks' time", but they apparently, because of the exaggeration, rejected the evidence that he gave at the same time that he did not do so because of the approach by the bank.

A selective approach to the evidence of a witness of that sort again can be justified in certain cases, but in my view it is only understandable in this case because of the misunderstanding by the justices of the evidence of Mr Trundell which caused them to come to the conclusion that he was exaggerating.

It seems to me, bearing in mind that the only evidence before the justices was that of Mr Trundell, taking into account that that evidence was not contradicted in detail in any way at all, that it was
D not in this case open to the justices to come to the finding of fact which they purported to do that this land was in beneficial occupation. Their finding was one which was wholly unreasonable on the material before them, and therefore one which was not open to them as a matter of law.

That being my conclusion on the material placed before the justices, it follows that this appeal should be allowed.

Mr Jubb, who has helpfully drawn my attention to all relevant matters on behalf of the respondent, accepts that if I come to the decision that on the evidence the justices were not entitled to come to the conclusion that they did, there would be no purpose served by
E remitting the case to the justices for rehearing and I therefore allow the appeal and set aside the decision of the justices that the appellant is liable for the amended amount by way of rates.

The appeal was allowed with costs of the present hearing in the High Court; the appellant also to receive £50 towards costs in the magistrates' court.

QUEEN'S BENCH DIVISION

February 6 1981

F (Before Mr Justice GLIDEWELL)

DEBENHAMS LTD v EALING LONDON BOROUGH

Estates Gazette June 20 1981
(1981) 258 EG 1184

Rating — General Rate Act 1967, section 6 — Proposal by valuation officer for alteration in valuation list following changes in the occupation of a hereditament — Question whether ratepayers were liable to pay rates based on proposed alteration in the list before their objection to the proposal had been resolved or whether, pending such resolution, they should pay rates on the basis of the existing entries (subject in either case to any necessary eventual adjustments) — No specific resolution by rating authority G
under section 6 in this case, but there was a general resolution to make the rate — Held that in principle a rating authority could lawfully demand payment of rates on the basis of a proposal made by the valuation officer which has not yet taken effect or been entered in the list — Case of B Kettle Ltd v Newcastle under Lyme Borough Council considered — In the present case, however, the authority's general resolution was not so worded as to cover altered hereditaments which were the subject of proposals not yet resolved — Authorities would have to employ different wording if they wanted a general resolution to empower officials to operate section 6 — Magistrates therefore in error in granting a distress warrant in respect of rates which had been unlawfully H
demanded — Decision of magistrates quashed

This was an appeal by case stated by justices for the Middlesex area of Greater London. Debenhams Ltd, who were in rateable occupation of warehouse premises in Trident Way, Southall, in the London Borough of Ealing, challenged a distress warrant granted by the justices on the basis that Debenhams were liable to pay rates calculated on a proposed alteration to the valuation list the objection to which had not yet been resolved.

Matthew Horton (instructed by Forsyte, Kerman) appeared on behalf of the appellants; G Seward (instructed by N L Green, chief solicitor, London Borough of Ealing) represented the J
respondent rating authority.

Giving judgment, GLIDEWELL J said: This is a case stated by justices for the Middlesex commission area of Greater London, acting for the petty sessional division of Brentford, relating to a decision made by them on July 24 1980. Debenhams Ltd (whom I shall call "Debenhams") are in rateable occupation of certain premises in the London Borough of Ealing. The valuation officer has made a proposal for the alteration of the valuation list which affects those premises. Debenhams have objected to the proposal, but the objection has not yet been resolved. Put shortly, the issue is whether, until the decision is made on Debenhams' objection, Debenhams should pay an amount of rates based on K
the existing entry in the valuation list or on the entry as proposed by the valuation officer.

I then deal first with the legal provisions. These are to be found in the General Rate Act 1967. Section 2(1) of the Act requires every rating authority from time to time to make such rates as will be sufficient to provide for so much of their total estimated expenditure during a given future period (usually one year) as is not to be met by other means.

Section 2(3) reads:

> Any rate made by a rating authority under subsection (1) of this section shall be made and levied as a single consolidated rate for the whole of their rating area which shall be termed "the general rate"... L

The relevant part of section 2(4) provides that the general rate for any rating area:

> (a) shall be a rate at a uniform amount per pound on the rateable value of each hereditament in that area ... and (b) shall be made and levied in accordance with the valuation list in force for the time being ...

Ealing London Borough Council ("the council") is a rating authority for the purposes of the Act.

Section 67 of the Act requires that the valuation officer shall maintain for each rating area a valuation list, which he shall cause to be altered from time to time. Section 67(6) provides: M

> ... the valuation list in accordance with which, under section 2(4)(b) of this Act, any rate falls or fell to be made, as in force ... at the date of the making of the rate, shall be conclusive evidence for the purposes of the levying of that rate of the values of the several hereditaments included in the list.

The major argument in this case has centred on section 6 of the Act. The relevant parts of that section read as follows:

> 6(1) Subject to the provisions of this section, the rating authority may at any time make such amendment in a rate (being either the current or the last preceding rate) as appear to them necessary in order to make the rate conform with the enactments relating thereto, and in particular may

A — (a) correct any clerical or arithmetical error in the rate; or (b) correct any erroneous insertions or omissions or any misdescriptons; or (c) make such additions to or corrections in the rate as appear to the authority to be necessary by reason of — (i) the coming into occupation of any hereditament which has been newly erected or which was unoccupied at the time of the making of the rate; or (ii) any change in the occupation of any hereditament; or (iii) any property previously rated as a single hereditament becoming liable to be rated in parts.

(2) Where the effect of the amendment would be either — (a) to alter, otherwise than by way of correction of a clerical or arithmetical error, the value on which a hereditament is rated; or (b) to charge to the rate a hereditament not shown, or not separately shown, in the valuation list, the rating authority shall not make any amendment of the rate unless B either the amendment is necessary to bring the rate into conformity with the valuation list or a proposal for a corresponding alteration to the valuation list has been made by the valuation officer. . . .

(3) In the foregoing provisions of this section (other than subsection 1(c)(i)) reference to a rate shall be construed as references to that rate as it has been applied in relation to particular hereditaments. . . .

In general terms, therefore, the scheme of the Act is that the valuation officer makes a valuation list which describes and attributes a rateable value to each rateable hereditament in the rating district; the rating authority makes each year an estimate of the total rate income it is likely to require to meet its expenses for the ensuing year, and it then makes a "rate in the pound" C which, when applied to the aggregate of the rateable values of all hereditaments in its area, will produce that sum; the sum payable by any particular person in rateable occupation of property will be the product of the rateable value of his property multiplied by the rate in the pound for the relevant year. Section 6, as I read it, is intended to deal with two separate problems: (i) the correction of errors in the rate, however they have arisen, and (ii) changes in the rate as it applies to a particular hereditament which are necessitated in any of the circumstances set out in section 6(1)(c). Such a change may only be made if it is necessary either to bring the rate into conformity with the valuation list or at least to a proposal made by the valuation officer for an alteration to that D list.

It is clear that, when a proposal is made for an alteration to the valuation list, and any objection to that proposal has been resolved, the rates payable on a particular hereditament for the relevant period will fall to be calculated in accordance with the altered entry in the valuation list. If during the period between the making of the proposal and the resolution of the objection the ratepayer has paid more than in the event proves to have been justified, he will receive a refund. If on the other hand he has paid less, he will then have to pay the excess. I am not concerned with the question whether the valuation officer's proposal was justified or not. That will be determined in due E course by a local valuation court, if the question cannot be resolved by agreement. What the magistrates were concerned with, and I am now concerned with, is the question whether, pending that resolution, Debenhams can be required to pay rates based on the proposed alteration to the valuation list, with the right to a refund if in the end they pay more than proves to be justified, or whether pending that resolution they simply make payments based on the existing valuation list entries, with the prospect of having to make up an excess when the problem is resolved.

One other section of the Act is of importance in this case, namely, section 5. That provides that a demand note on which F the rate is levied shall include information with respect to a number of matters, the relevant matters here being: "(a) the situation of the hereditament in respect of which the demand note is issued and such description thereof reasonably necessary for purposes of the identification as may be prescribed; and (b) the rateable value and, where it differs from the rateable value, also the net annual value of the hereditament. . . ." Although under the Act the Rate-demands Rules 1974 were made, those rules do not prescribe any further details necessary for the purpose of identification.

Finally, section 96 of the Act provides that a magistrates' court may issue a distress warrant to enforce the payment of rates by any person who fails to pay any sum legally assessed on and due G from him in respect of a rate within seven days after it has been legally demanded of him.

The magistrates dealt carefully with a complicated matter. In particular, they have set out the facts which they found in their case stated with admirable clarity, on which they are to be congratulated. I also had before me a bundle of relevant documents, most of which are referred to in the case stated, which has been agreed between the parties. In summary, the relevant facts are:

(a) at all material times Debenhams have held under a lease warehouse units nos 12A and 12B Trident Way, Southall, in the H Borough of Ealing. Debenhams themselves have at all times occupied no 12A, but until June 25 1978 no 12B was occupied by Messrs Rosser & Russell, who also occupied the unit adjoining 12B, namely, 11 Trident Way.

(b) On the valuation list in force on June 25 1978 these properties appear as follows: (i) 11 and 12B Trident Way — as one hereditament with a rateable value of £31,513, (ii) 12A Trident Way — with a rateable value of £11,222. I was informed by counsel that no alteration has yet been made to those entries in the valuation list.

(c) On June 25 1978 Rosser & Russell terminated their lease of no 12B, whereupon Debenhams entered into occupation of that J property.

(d) On July 6 1978 the valuation officer made a proposal for the alteration of the valuation list by substituting for the two hereditaments shown in it two new hereditaments, namely (i) 11 Trident Way — with a rateable value of £20,763, and (ii) 12 Trident Way — with a rateable value of £24,722. At that time there was a dividing wall between 12A and 12B, with no means of communication through it. On July 28 1978 Debenhams objected to the proposal, contending that 12A and 12B did not constitute a single hereditament. That objection has not yet been resolved.

(e) In the summer of 1979 Debenhams pierced the dividing K wall between 12A and 12B and inserted a door, so that there is now communication between the two units. On May 21 1980 the valuation officer, lest he should fail in the July 1978 proposal, made a further proposal in the same terms. Debenhams objected to that proposal also, on June 9 1980, again alleging that 12A and 12B do not form part of one hereditament. That objection has also not yet been resolved.

(f) Meanwhile, the council had been sending to Debenhams or their agents demands for rates, and statements of account of unpaid rates, as follows:

(i) In April 1978, a demand for rates on 12A Trident Way, based on the rateable value of £11,222 for the year 1978-79. L

(ii) In April 1979, a demand for rates on 12A Trident Way, based on the rateable value of £11,222 for the year 1979-80.

(iii) On May 18 1979, a demand note headed "General Rate 1979-80", relating to a hereditament described as "Unit 12B, Trident Way", demanding rates based on a proposed rateable value of £24,722, both for the year 1979-80 and for the total period from June 25 1978 to March 31 1979, making a total of £36,918.84.

(iv) On November 16 1979, a statement of account expressed to relate to the hereditament of 12B Trident Way, showing an amount outstanding to March 31 next as £36,918.84. M

(v) In December 1979, a document in the form of a demand for rates for 1979-80, which is addressed to Debenhams Ltd, Unit 12 Trident Way, but which is not expressed to relate to any defined hereditament. The document bears in manuscript the words, "Amended provisional account as attached — £25,182.24". There was indeed attached to the demand note another document in the form of a provisional account, which did relate to Unit 12, Trident Way, and which showed how the sum of £25,182.24 was calculated as being the sum of £36,918.84 less amounts paid by Debenhams.

(vi) In April 1980, a demand note in respect of the heredita-

A ment 12A Trident Way, based on the rateable value of £11,222, for the year 1980-81.

(vii) According to the case stated, "A further demand for rates in respect of the year 1980-81 had been issued by [the council] to [Debenhams] on May 13 1980, although [Debenhams] denied having received any such demand". No copy of this demand note is among the documents before me, and neither the magistrates' court nor I can therefore tell to what hereditament it was expressed to relate, or on which rateable value it was based.

(viii) Thereafter, a series of statements of account and reminders were sent by the council to Debenhams, culminating in a document which calculates Debenhams' liability based on B the proposed rateable value of the proposed hereditament "12 Trident Way", after giving credit for payments made by Debenhams, as at March 31 1980 — £20,160.39, and for the year April 1 1980 to March 31 1981 — £27,812.25. These sums were the subject of the application for the distress warrant.

(g) There was no specific resolution of the council authorising the amendment to the rate made under section 6 of the Act. The authority of the council's officials to demand an amount of rates based on the assessments proposed, and not on those contained in the valuation list, therefore depends on the general resolution to make the rate, passed by the council in March of each year. I will refer to the wording of this resolution when I deal with the C submissions on this point.

I turn to consider decided authority on the points in issue. I was referred by counsel to the decisions of the Queen's Bench Divisional Court and of the Court of Appeal in *B Kettle Ltd v Newcastle under Lyme Borough Council*, which is reported in (1977) RA 181 (DC) and (1979) RA 223 (CA).* In that case Kettles, the ratepayers, were in occupation of a new hereditament, which had no assessment in the valuation list at April 1 1975. In September 1975 the valuation officer proposed an entry in the valuation list showing the hereditament with a rateable value of £13,000, to take effect from April 1 1975. Kettles objected, but the objection (which resulted in the rateable value D being reduced to £4,000) was not heard until October 1977, after the Divisional Court hearing. The borough council made a demand for rates for the year 1975-76 based on the rateable value of £13,000. Kettles did not pay, so the borough council sought a distress warrant from the magistrates, who dismissed the complaint. The borough council appealed successfully to the Divisional Court, and Kettles' further appeal to the Court of Appeal was dismissed.

In both courts, two issues were raised: (1) Whether under section 6 of the Act a rate can be amended, and the rating authority lawfully require the ratepayer to pay rates, based on a proposal by a valuation officer to which there is an objection and E which has thus not yet taken effect? Both the Divisional Court, in which Lord Widgery CJ gave the principal judgment, and the Court of Appeal (Lord Denning MR and Lawton and Geoffrey Lane LJJ) answered this question "Yes". (2) Whether a specific resolution of the rating authority is needed in order to apply section 6 to a particular hereditament and, if not, whether there was a general resolution which was sufficient to authorise the rating authority's officers to apply section 6?

Again the Divisional Court and the Court of Appeal answered the first part of this question "No", although Lawton LJ expressed some doubts about it. Lord Denning, with whom Geoffrey Lane LJ agreed, said:

> F It seems to me, in the circumstances, that it is not contemplated that this matter should have been brought before the rating authority every time an error is discovered or an omission has to be corrected or after a new proposal is made in the course of a year. It seems to me that it can be done by a general resolution in advance, enabling the staff in the office to make these corrections to errors or to deal with proposals during the course of the year.

Then further down on the same page he said: "As I have said — and counsel for the rating authority agrees — there should be a general resolution at least to authorise the staff in the office to make these clerical amendments and matters of that kind."

*See also (1979) 251 EG 59.

Lawton LJ put it more dubiously thus: G

> I have not found the problem of construing section 6 in the way Lord Widgery CJ construed it as easy as he did because there are parts of section 6 which are not easy to reconcile with other parts. For example, it is easy to see that it would be nonsensical for the rating authority as such — and that would not include a committee of the rating authority but the rating authority itself — to have a resolution every time a clerical or arithmetical error in the rate was discovered. On the other hand, under section 6(1)(c), the amendment to a rate can only be made as appears to the authority to be necessary. That is more akin to a matter requiring a specific resolution of the rating authority.
>
> I have had grave doubts about this matter; but, having regard to the view which Lord Denning MR has taken and I understand Geoffrey Lane LJ will take, I feel that I should accept that the overall intention of H section 6 is that a specific resolution is not required every time the rating authority decides to make an amendment.

The second part of the question, that is to say whether there was a general resolution which was sufficient to authorise the rating authority's officers to apply section 6, created difficulties in *Kettle*'s case in the Divisional Court, because no resolution of the borough council had been put in evidence before the magistrates, and the court thus had no means of knowing what it said. Lord Widgery dealt with the matter in this way at p 187 of the (1977) RA report:

> But more substantial than that is the argument put forward by counsel J for the ratepayers on the absence of any resolution of the rating authority to do what was done. I have had some difficulty with this, speaking for myself, because this is a case stated and we can only work on the facts which are disclosed, and no word has been said in the case about whether there was a resolution or not. I do not myself consider that any resolution was necessary except such resolution as in the end must accompany all activities of the rating authority. In the end, almost everything which a local authority do has to be sanctioned by the resolution of some committee, if not of the authority themselves. But leaving that kind of thing aside, I feel quite confident that it was not the intention of Parliament that a specific resolution should be made every time it was desired to bring a new hereditament into the rate under the powers relied on in this case by the rating authority. Section 6 is drafted in clear and K specific terms, and I feel confident that it would be said if any further resolution was required.

It seems to me that his Lordship must have assumed, in the absence of evidence to the contrary, that the wording of the general resolution was wide enough to authorise the relevant officials to apply section 6. In the event, that assumption proved to be correct.

When the matter came before the Court of Appeal, by agreement between the parties, a copy was produced of the resolution of the borough council making the rate for 1975-76. It provided that ". . . a general rate for the borough at the rates in the pound on the rateable value of each hereditament chargeable L thereto as specified below is hereby made and approved . . ."

The question thus was whether that wording was wide enough to authorise the application of section 6. On that question Lord Denning said at p 227 of the (1979) RA report, reading on from the point which I had reached in my earlier citation:

> But the whole question comes to this — whether the resolution made in advance on March 18 1975 was sufficient for the purpose. As I have already said, it empowers and authorises the rates at so much in the pound on each hereditament chargeable thereto. It seems to me not only to be chargeable on the existing valuation list but also chargeable in the course of the year by reason of correction of errors and by reason of proposals which have been made: and that general resolution, it seems to M me, is sufficient to satisfy the statute. Otherwise it seems to me to be adding quite an unnecessary amount of detail for the rating authority which Parliament cannot have contemplated. It cannot have contemplated that every small amendment, clerical or otherwise, is to be brought up every time the council meets.

Lawton LJ said on p 229:

> Counsel for the rating authority submitted that that resolution was in wide enough terms to authorise the officials of the rating authority to make amendments to the rate in accordance with the provisions of section 6 of the General Rate Act 1967. The words are indeed very general. With some hesitation and doubt, I have come to the conclusion

A that that resolution was wide enough to authorise what was done in this case, so I, too, would dismiss the appeal.

Geoffrey Lane LJ said on p 230:

> Whether that passage is right or not,

that is a passage from the judgment of Lord Widgery in the Divisional Court.

> by a very sensible agreement between the parties here, we now have before us a copy of the relevant resolution which was made in advance on March 18 1975. If the argument of counsel for the ratepayers is correct, it seems that every single clerical error which might occur would have to be amended by a specific resolution of the authority in respect of each error.
> B I cannot think that that was the intention of Parliament. In my view, the resolution in advance of March 18 1975 was enough to satisfy the requirements of section 6, and for those reasons I, too, would dismiss this appeal.

I turn to the questions posed in this case. The case stated by the magistrates helpfully sets out five questions on which the opinion of the court is asked. The answer to these questions in turn determines whether the magistrates were correct in granting the application for a distress warrant. I will deal with each in turn.

(1) Whether the respondent rating authority could lawfully
C demand the payment of rates on the basis of a proposal made by the valuation officer which has not yet taken effect nor been entered in the valuation list?

Mr Horton accepted that I am bound by the decision of the Court of Appeal in *B Kettle Ltd v Newcastle under Lyme Borough Council* to answer this question "Yes". He reserves the question for possible future argument elsewhere.

(2) In the alternative to (1) above, whether in respect of a building already the subject of an entry in the valuation list either as a separate hereditament or as a part of a larger hereditament the respondent rating authority could lawfully demand the payment of rates on the basis of a proposal made by the
D valuation officer which had not yet taken effect nor been entered in the valuation list?

On this question, Mr Horton made two submissions on behalf of Debenhams: (a) The general rule is to be found in section 2(4) of the 1967 Act, ie that the general rate shall be made and levied in accordance with the valuation list in force for the time being. Only if the particular circumstances of a case bring it within the terms of section 6 can a rate be demanded on the basis of a proposed amendment. Mr Seward, for the council, accepted, as I do, that this general proposition is correct.

(b) The facts found by the magistrates in this case do not establish any of the necessary conditions for the application of
E section 6 as set out in section 6(1). Mr Seward, to the contrary, pointed out that section 6(1)(c)(ii) authorises the council to "make such additions to or corrections in the rate as appear to the authority to be necessary by reason of . . . any change in the occupation of any hereditament . . ." He says that the changes proposed by the valuation officer have come about as a result of a change in the occupation of a hereditament, namely, no 12B Trident Way.

I agree with Mr Seward. Thus, in my judgment, the answer to this question is also "Yes".

(3) In the alternative to (1) and (2) above, whether the respondent rating authority had made such a resolution as to
F permit it lawfully to demand the payment of rates on the basis of a proposal made by the valuation officer which had not yet taken effect nor been entered in the valuation list?

As I have said, there was no specific resolution of the council authorising any of its officers to apply section 6. The general resolutions passed in March of each year, making the rate for the ensuing years commencing April 1 1979 and April 1 1980 were, so far as this is concerned, in identical terms. The relevant part of the 1979 resolution reads:

> Notice is hereby given that the Council of the London Borough of Ealing, being the Rating Authority, by resolution dated the 8th day of March 1979 made a general rate of 89.5 pence in the pound on the rateable value of each hereditament as shown in the Valuation List in G respect of the period commencing on the 1st day of April 1979 and terminating on the 31st day of March 1980 . . .

Mr Horton contrasted the wording of this resolution, and in particular the words "on the rateable value of each hereditament as shown in the Valuation List", with the words in the Newcastle under Lyme resolution in *Kettle*'s case, namely, "on the rateable value of each hereditament chargeable thereto". He submitted that, while the words "each hereditament chargeable thereto" had been held to be wide enough to comprehend not merely the hereditaments shown in the valuation list at the time when the resolution was passed but also such altered or added hereditaments as might be the subject of a proposal and thus fall within H section 6, that interpretation could not apply to the resolution of the Ealing Council in the present case.

Mr Seward contended that there was no difference of any substance between the two resolutions. He thus invited me to reach the same conclusion as the Court of Appeal had reached in *Kettle*'s case.

In my view this is very much a matter of impression for me. The question is that posed by Lord Denning in *Kettle*'s case, namely, whether the resolution made in advance on March 8 1979 was sufficient for the purpose? In my judgment it was not. The resolution in the present case, in my view, means what it says, ie it applies to the hereditaments shown in the valuation J list. Other altered hereditaments the subject of proposals are not covered by the resolution. If the Ealing Council wishes to have a general resolution which empowers its officers to make use of section 6, the wording of the resolution will have to be altered. Thus my answer to this question is "No".

This means that I conclude that the magistrates were wrong to grant a distress warrant. Strictly, therefore, I need not go on to consider the other questions posed. However, I have thought it right to do so.

(4) In the alternative to (1), (2) and (3) above, whether the respondent rating authority were precluded from lawfully demanding the payment of rates on the basis of a proposal by virtue of K the fact that some or all of the said demands were made in addition to demands made in the same rate years on the basis of existing entries in the valuation list in respect of the buildings the subject of the said proposal?

In argument, Mr Horton put this point rather more succinctly. He contended that it is unlawful for a rating authority to issue a demand based on a proposed value after it has issued an earlier demand based on an existing entry. Alternatively he submitted that the second demand can only lawfully be issued if it is accompanied by an express notification that the previous demand is withdrawn, but he quoted no authority for this proposition.

As a matter of practice, I have no doubt that when it does L issue demands for rates a council ought to take every step open to it to make clear to the ratepayer what is properly being demanded and why, and if a later demand is intended to be made in place of an earlier one, it should make that clear too. Fortunately, in the present case I do not for a moment think that these ratepayers, with their resources, advised as they were by most distinguished rating surveyors, were at all misled. But other individual ratepayers, without the benefit of such advice, might very well be misled, and indeed seriously worried if they received a series of apparently conflicting demands such as those in the present case.

Having said this, however, it is my view that as a matter of law M the council are entitled to do what they did. A demand is a preliminary to steps to enforce payment of the sums demanded, and, provided that the council do not seek to enforce payment of the same sum twice (which of course they have not done here), the fact that they have served a series of apparently conflicting demand notes does not prevent them from recovering a sum which is in fact due. My answer to this question is therefore "No".

(5) Whether there had been issued any or any lawful demands for the rates in respect of which a warrant of distress was sought?

The distress warrant was sought in respect of the hereditament

A which will appear in the valuation list if the valuation officer's proposals, or one of them, is approved by the local valuation court, namely, no 12 Trident Way. As I have said earlier, most of the demand notes based upon the proposed rateable value were said to relate to a hereditament described as "Unit 12B Trident Way". In my judgment, this is not a compliance with section 5(1) of the Act, in that it is not a description of the situation of the hereditament described as "Unit 12B Trident Way". In my judgment, this is not a compliance with section 5(1) of the Act, in that it is not a description of the situation of the hereditament 12 Trident Way. There is, on the other hand, no proposal in relation to, nor any entry in the valuation list relating B solely to, no 12B. Thus, in my view, the council are not entitled to recover any sums based upon demand notes relating to a hereditament 12B Trident Way.

The demand note which the magistrates found had been issued on May 13 1980 was not before the court and no copy of it is before me. As I have already said, the magistrates could not find, and I cannot tell, to what hereditament it was expressed to relate. Thus in respect of that demand note the council have C failed to prove that they have made a lawful demand in respect of the property no 12 Trident Way.

The only demand note relating to no 12 which does, in my judgment, satisfy section 5 is the document dated December 20 1979. I appreciate that this, on the face of the document itself, was not expressed to refer to any defined hereditament, but it did expressly refer to the "amended provisional account as attached", and that document indeed identified the hereditament to which it related as being Unit 12 Trident Way. Thus in my judgment those two documents together did constitute a lawful demand note relating to no 12 Trident Way and to the proposed rateable value of that hereditament which demand note satisfies D section 5. If, therefore, I were wrong in my answer to Question (3), my answer to Question (5) would be that a lawful demand for rates in the sum of £25,182.24p was made on December 20 1979, but that no other or later lawful demand has been made.

The appeal was allowed with costs and the decision of the magistrates quashed.

REAL PROPERTY AND CONVEYANCING

COURT OF APPEAL

February 26 1981

(Before Lord Justice STEPHENSON, Lord Justice DUNN and Sir David CAIRNS)

SOLLOWAY v HAMPSHIRE COUNTY COUNCIL

Estates Gazette May 30 1981

(1981) 258 EG 858

Claim against highway authority for damage caused to house by encroachment of tree roots — Only issue on appeal was whether encroachment gave rise to a reasonably foreseeable risk of subsidence — Trial judge decided in favour of plaintiff and highway authority appealed — Vital question was whether the authority should reasonably have foreseen that there might be pockets of clay under the house so that the foundations rested partly on plateau gravel and partly on clay, with consequent risk of subsidence — After a detailed consideration of the evidence before the judge the Court of Appeal did not accept that the risk was reasonably foreseeable — The possibility of an intrusion of clay under the respondent's house was no more than a vague possibility, not a real risk — Even assuming that there was a real risk, it was an outside chance and had as such to be balanced against the practical steps which could reasonably have been taken by the appellants to minimise the damage — There was no breach of duty and consequently no nuisance — Authority's appeal allowed

This was an appeal by Hampshire County Council from a judgment of Stocker J in favour of Brian Thomas Solloway, the freehold owner of a house at 72 Shirley Avenue, Shirley, Southampton, which had suffered structural damage. Mr Solloway, the plaintiff in the action, complained, and the judge found that subsidence at the house, caused by the encroachment of the roots of a horse chestnut tree on the adjacent highway, was a nuisance for which the defendants, the present appellants, were responsible.

John J Smyth QC and Richard Dening (instructed by Hepherd, Winstanley & Pugh, of Southampton) appeared on behalf of the respondent, the plaintiff in the action; John Foy (instructed by Theodore Goddard & Co, agents for R A Leyland, solicitor and county secretary, Winchester) represented the appellants, defendants in the action.

Giving the first judgment at the invitation of Stephenson LJ, DUNN LJ said: This is an appeal from a judgment of Stocker J given on February 20 1980. The learned judge held that the subsidence at the plaintiff's house, caused by the encroachment of roots of a horse chestnut tree on the highway adjacent to the house constituted a nuisance for which the defendants, as the highway authority, were responsible.

There were numerous issues at the trial. There was an issue of causation, there was an issue of contributory negligence, there was an issue of law and there was an issue as to the responsibility of a highway authority for a nuisance of this kind; but the only issue on the appeal has been whether the encroachment by the roots gave rise to a reasonably foreseeable risk of subsidence. The judge held that it did; the defendant appellants contend that it did not.

The case has been well argued and I personally am indebted to all counsel, both for their arguments in this court and in the way in which the evidence was presented in the court below.

The primary facts as found by the judge were accepted by Mr Foy, for the appellants. Those facts are fully set out in the judgment and it is not necessary for me to do more than to summarise them. The plaintiff is the freehold owner of the house, which he bought in 1967. Outside the house there is a horse chestnut tree, probably planted in about 1874 and long before Shirley Avenue became a residential street. No 72, which is the plaintiff's house, was built in about 1922. The foundations of it lie at a depth of about 2 ft below the surface of the ground. Such foundations conformed to local building requirements at the time when the house was built, but would not be considered adequate for a house of this size if built today. There was no evidence of any significant structural damage to the house over the 50 years or more of its existence, until the events giving rise to this claim.

Immediately outside the boundary of the plaintiff's house stands the horse chestnut tree. It is about 60 ft high and stands in the pavement a few feet from the front fence of the property: to be exact, at 8.3 metres from the nearest point of the house, which is the window at the bay of the house. Over the years there was some trouble with the drainage, owing to blockage of the drains by roots from the horse chestnut tree. This occurred in 1968, 1970 and 1971. On the first occasion the plaintiff wrote to the local council asking them to top the tree and see to its roots, but he received no reply. It is accepted that the blockage of the drains by the roots has no relevance to the question of foreseeability of the subsidence.

The position, therefore, as found by the judge was that the plaintiff lived in the house from 1967 to 1976 unaware of any threat to the structure from the chestnut tree on the roadway outside.

In August 1976 the plaintiff and his wife went on holiday for some three weeks. Some time before doing so, substantial redecoration and repair work had been done to the house, including rebuilding of the chimney stacks and work to the bay windows. No structural damage was observed at the time of those works.

In 1976 occurred the second successive very hot summer, with drought conditions, which had lasted since the summer of 1975. On return from his holiday the plaintiff observed substantial cracks in the fabric of the house. These cracks were in the area of the bay window and the side walls of the building. Rectification of the building was carried out by means of underpinning. The agreed cost of this work was £5,656.04, and that was the amount of the damage which the judge ordered for the nuisance.

The judge made crucial findings as to the subsoil under this house and they are so important that I read that passage of his judgment in full. He said:

> Shirley Avenue is situated in an area, the top sub-soil of which is plateau gravel. In so far as small pockets of clay outcrop may exist, they are of insufficient size to appear at all on the geological plan. A house founded on plateau gravel is unlikely to be affected by intrusion from the roots of trees. The plaintiff's house was founded upon plateau gravel at the rear and the greater part of the flank wall, but upon an outcrop of Bracklesham-based clay at the front. Differential settlement was therefore likely to occur in any event, even if uninfluenced by external factors, such as dehydration by the roots. However, such settlement would be likely to have revealed its full effect within 10 years of the erection of the house. No structural damage in fact occurred during this period, and there was no structural damage of any significance prior to the events of 1976.

On the basis of those findings the judge then dealt with the law and he expressed it in this way:

> A I therefore find that the law which I have to apply to the facts of the present case to be, that the duty in respect of the nuisance created by the roots arises if the encroachment of those roots is known, or ought to be known, to the owner, occupier or other person responsible for the tree and its maintenance, if the encroachment is such as to give rise to a reasonably foreseeable risk that such encroachment will cause damage.

That formulation of the law, although it had been contested below, was accepted in this court by Mr Smyth, for the plaintiff. I am bound to say that at first I had doubts whether it was an accurate formulation of the law because of the case of *Davey v Harrow Corporation*, a decision of this court reported in [1958] 1 QB 60, which makes no reference to foreseeability as an element in the tort B of nuisance, where the nuisance is caused by the extrusion of tree roots or branches, but on further consideration I agree that Mr Smyth was right to accept the formulation of the law as made by the judge.

In *Leakey v National Trust* [1980] 1 QB 485 Megaw LJ reviewed the whole of the law of nuisance not brought about by human agency, and at p 522 he dealt with the case of *Davey v Harrow Corporation*. He said (at D):

> I have no hesitation in preferring [that] decision as stating the law as it now is: subject to the proviso that the duty arising from a nuisance which is not brought about by human agency does not arise unless and until the defen- C dant has, or ought to have had, knowledge of the existence of the defect and the danger thereby created.

Shaw LJ, with some doubt, and Cumming-Bruce LJ both agreed with the judgment of Megaw LJ. Although it was not necessary for the decision, the proviso suggested by Megaw LJ puts nuisance by encroachment of tree roots and branches into the same category as any other nuisance not brought about by human agency. It is consistent with *Sedleigh-Denfield v O'Callaghan* [1940] AC 880 and confines the strict liability for nuisance to cases where there has been some non-natural user of the land as stated in *Rylands v Fletcher* (1868) 3 HL 330. I would accept the proviso as stated by D Megaw LJ as applicable to the facts of this case.

Mr Smyth's basic submission was that there was ample evidence to support the judge's finding that the risk of subsidence to the plaintiff's house was reasonably foreseeable by the defendants. Mr Smyth submitted that it was a finding of fact and that this court should not interfere with it. I do not think that this is the right approach. The finding as to foreseeability was not a finding of primary fact. It did not depend on an assessment by the judge of the reliability of the witnesses, or upon a resolution of a conflict of evidence. It was an inference which the judge drew from the whole of the evidence. Having read the whole transcript, this court is in as good a position to draw inferences of that kind as was the judge, and E in some ways better.

I approach the case on the basis of asking the question, "What is the proper inference to be drawn from the evidence?"

One of Mr Foy's principal criticisms of the judge's finding was that, in considering foreseeability, he had concentrated on the likelihood of pockets of clay being present in the subsoil and had virtually ignored the long period of drought from the early summer of 1975 to September 1976. The evidence was that this was quite exceptional and abnormal, as Mr Baker, the expert witness called by the plaintiff, put it, and the probability of such a long period of drought (Mr Baker said) was extremely remote as it had occurred only once in the last 200 years.

F Mr Foy said that the drought was an important factor which should be taken together with the possible presence of clay in the subsoil and with the shallowness of the foundations and the proximity of the tree to the house in considering the question of foreseeability. He submitted that it was the combination of those factors which caused the subsidence and that all four factors should be taken into account when considering the foreseeability of subsidence.

The difficulty with that approach, it seems to me, is that Mr Baker said that, even if there had not been a drought, the tree roots would eventually have caused a subsidence because of the existence of the outcrops of clay under the foundations and that the drought merely accelerated this process. And Dr Biddle, the expert called by the G defendants, said that as there was clay under the house there was a probability of subsidence that ought to be considered in any event.

In my view, in considering the question of foreseeability the judge was right to concentrate on the likelihood of pockets of clay under the house and on the proximity of the tree to the house in the light of the evidence that the roots of chestnut trees grow as far as their height. Indeed it was accepted by Dr Biddle that he would expect the roots of this tree to extend under the foundations of the house and the roots would be likely to dehydrate the clay and cause the subsidence.

So the vital question is whether the defendants should reasonably have foreseen that there might be pockets of clay under this house. H If there were, and the foundations were resting partly on plateau gravel and partly on clay, as was the case, there would be a greater risk of subsidence than if the foundations were resting wholly on clay. If the foundations were resting wholly on gravel there would be no substantial risk of subsidence, even though the tree roots were extruding under the foundations.

It was common ground that the geological map showed the subsoil in the Shirley area to be almost entirely plateau gravel with clay underneath. The section shows that the layer of plateau gravel must have been comparatively thin, although it was accepted that it extended probably to 10 ft to 12 ft, which would have been sufficient to bring it well below the foundations. In any event, it is the J map itself that is of importance for this purpose, and not the section. It was accepted that geological maps are in general very accurate and that those concerned with foundations are normally entitled to rely on them. But there was also evidence that, from the knowledge of surveyors, civil engineers and others working in Southampton, there were pockets of clay in the Shirley area. These were quite small; perhaps between five and ten sq yds in extent. But this only added to their potential danger if they were under a house because the foundations would not be resting on a homogeneous subsoil.

The clay found under no 72 consisted of a uniform, fairly stiff, yellow clay with below it a yellow sandy clay, and below that a dark mottled clay. K

The pockets of clay in the Shirley area were said by Mr Baker to be very variable and completely unpredictable in this location. It was common ground that the only sure method of ascertaining the existence of a particular pocket would be by sinking a borehole, and because of the small size of the pockets as at no 72 it would be necessary to sink one borehole at each side of the house where clay was suspected.

Although there were passages in the evidence of all the experts to the effect that pockets of clay might be found under any house in Shirley Avenue and that that was a risk that ought to be considered, they all agreed that it was unlikely that clay might be found under any particular house. In his evidence Mr Baker was asked: L

> Q. Unless you happened to have done excavations in the particular area, you would have no reason for suspecting that there was clay under any particular house. — A. Correct.
>
> Q. As I understand your report, and tell me if this is right, before you started these excavations, you had expected that it would be plateau gravel. I rather get the impression you were slightly surprised to find that it was clay. — A. Well, with this particular property, we had done the one next door previously, and we found clay there. So we were not so surprised on this particular one.
>
> Q. Next door — that put you on guard as it were, but before you did next door, it would certainly be a surprise to you to learn there was clay underneath these houses? — A. It was, yes. M

Dr Biddle said this:

> Q. If you had been consulted in 1974, before there was any question of drought, and asked to advise whether trees in this area were likely to damage adjoining houses, what would have been your advice? What would you have done, and what conclusions would you have drawn? — A. Well, when asked this sort of question routinely by local authorities in the main, I think you have got to get the best available information of the risks pertinent to the particular area, and the source of this information is a geological survey map. This will tell you basically, whether you are on a highly shrinkable soil, such as London clay, which outcrops in various areas in Southampton, whether you are on a

soil of substantially or slightly lower shrinkability, such as Bracklesham beds, or whether on particular sands or gravels, and I think one has to put the problem into one of those sort of categories, and clearly this would fall into the last category — an area where I would certainly not anticipate any problems.

Q. Would that have been your advice if you had been asked in, say, 1974? — A. Yes. I would have said there was no problem in this area. It might be thought to be necessary to take action where there are settlement risks.

He was asked a direct question in cross-examination:

Q. Did it surprise you to find this clay here at this depth? — A. It certainly surprised me, yes, because my experience of geological survey maps shows they are right in 99 per cent of cases I have investigated, and would normally give you a pretty good idea of what you would find.

Taking the evidence of Mr Baker and Dr Biddle as a whole, and they were the two independent experts, a fair reading of it seems to me to show that both were surprised at the presence of pockets of clay under no 72, and Dr Biddle, at any rate, thought that no subsidence problem in that area was likely.

The judge relied strongly on the evidence of Mr Cocain, the defendants' engineer. He dealt with it in his judgment where he said:

In fact, however, the possibility of an intrusion of clay was known to Mr Cocain, the city structural engineer of Southampton who, for this purpose, are the agents of the defendants. He said that — "In some places in the gravel there is a high, and in some places low, clay content, and in the Shirley library area you might find five to twelve sq yds of clay."

Mr Foy said that, in that passage of the judgment, the judge appears to have misunderstood the evidence of Mr Cocain. His evidence on this part of the case appears as follows:

Q Were you aware before this matter came to light, that part of Mr Solloway's house rested on clay? — A. No.

Q. Did it surprise you to find that out? — A. Yes, very much. In the past years I have worked for the Southampton Council. I have been concerned with a number of projects in this particular area — public developments, the new library and housing projects. Generally the subsoil has shown to be gravelly, of a gravelly nature. I should say that it varies in that in places it has more clay content and at others less clay content. It is difficult to classify, but it is generally very good bearing strata throughout the area. I do not recall any particular foundation problems that we have had.

Q. Have you looked at the geological drift map in this case, for this area? — A. Yes, obviously when we had this problem crop up, it is the well known first routine action.

Q. What was the conclusion you drew from looking at the map? — A. I was surprised at the foundation damage. Obviously I had known that it was a good gravelly area, and as events have shown the size of the clay pocket, this has rather been a surprise to me.

Then he was asked to look at the map and asked this question:

Q. Would you understand from looking at that map there was a likelihood, or possibility even, of clay being on the surface in the vicinity of this house? — A. Well, I think it has already been said that part of Southampton is very variable. The conditions can vary from yard to yard, but I would not expect in this area, to find any very bad conditions. I would not have thought there would be any large pockets of clay. That is based on the experience I have of the area.

In cross-examination he was asked:

Q. Well, whether there is a problem or not, one is concerned about its existence in any quantities in the sort of areas you would concede you have found in the past. When you talk about areas where there was a high clay content, what is the largest you think in Shirley for example — if you are talking about Shirley — you have found of a high clay content? — A. I have seen areas where the development I am referring to was at the edge of the gravel outcrop area. In fact, the site at the Shirley library; on that particular site I would say that the clay content would have been more than 50 per cent.

Q. I am not concerned about the content, but the area you found the clay content. What sort of area are you talking about — an area of 100 sq yds, or less? — A. A handful of sq yds. Five to ten sq yds perhaps — very localised.

Q. How far on average, would it be before you found the next area of clay content? — A. It is impossible to say.

Q. In your experience? — A. It might occur twice in the building area. It might do. It might occur once. It might not occur at all; but in no case in those places would it constitute a foundation problem, because there was always a large stone content, which made the ground a fairly stable bearing strata.

Q. So you would say that had not caused you to have any worries about the foundations you were putting in? — A. Correct, yes.

Mr Foy said that Mr Cocain, in those answers, was not talking about pockets of homogeneous clay, he was talking about pockets of clay mixed with gravel or gravel mixed with clay, and that in each case the gravel content was such that he had no worries about the foundation; whereas, said Mr Foy, the judge seemed to think that he was talking about pockets of homogeneous clay.

I find force in that submission. I can well understand that the judge might have viewed Mr Cocain's evidence with caution because, as Mr Smyth said, it was not until cross-examination that Mr Cocain admitted that the subsidence was caused by the tree roots. But the judge did not deal with his evidence in that way. He dealt with it on the basis that he accepted it. I think that Mr Foy was right in saying that the judge did not understand the full implication of it. Mr Cocain, like Mr Baker and Dr Biddle, was surprised at the presence of homogeneous clay under no 72, but his evidence as to the mixture of clay and gravel in the pocket, in my view, is crucial on the question of foreseeability.

The judge dealt with the question of foreseeability in two parts, by dealing first with what he called "the nature of the duty" and then with its scope. In reality the scope of the duty depends on the extent of the risk of damage and the two should be considered together.

In *Leakey's* case [1980] 1 QB 485 at p 524 Megaw LJ dealt with this question in this way:

This leads on to the question of the scope of the duty. This is discussed, and the nature and extent of the duty is explained, in the judgment in *Goldman v Hargrave* [1967] 1 AC 645 at pp 663, 664. The duty is a duty to do that which is reasonable in all the circumstances, and no more than what, if anything, is reasonable, to prevent or minimise the known risk of damage or injury to one's neighbour or to his property. The considerations with which the law is familiar are all to be taken into account in deciding whether there has been a breach of duty, and, if so, what that breach is, and whether it is causative of the damage in respect of which the claim is made. Thus, there will fall to be considered the extent of the risk; what, so far as reasonably can be foreseen, are the chances that anything untoward will happen or that any damage will be caused? What is to be foreseen as to the possible extent of the damage if the risk becomes a reality? Is it practicable to prevent, or to minimise, the happening of any damage? If it is practicable, how simple or how difficult are the measures which could be taken, how much and how lengthy work do they involve, and what is the probable cost of such works? Was there sufficient time for preventive action to have been taken, by persons acting reasonably in relation to the known risk, between the time when it became known to, or should have been realised by, the defendant, and the time when the damage occurred? Factors such as these, so far as they apply in a particular case, fall to be weighed in deciding whether the defendant's duty of care requires, or required him to do anything, and, if so, what.

In considering whether there is a breach of duty, the extent of the risk and the foreseeable consequences of it have to be balanced against the practicable measures to be taken to minimise the damage and its consequences.

The judge held that intrusions of clay were likely in the Shirley area. With all respect to the judge, I cannot, on the evidence, accept that finding. I think that the possibility of an intrusion of clay under no 72, which is the real question, was unlikely. I would hold that it was no more than a vague possibility, not a real risk in the words of Lord Reid in the *Wagon Mound (No 2)* [1967] 1 AC 617, but assuming that there was a real risk or chance, I would say that it was an outside chance and that outside chance has to be balanced against the practical steps which could reasonably have been taken by the defendants to minimise the damage. At first the judge found this an insuperable difficulty. He said:

In these circumstances, Mr Foy for the defendants submits that this would present the defendants with a problem which would be insuperable without a very great expense since in practice, every house in the street, and no doubt other houses in similar tree-lined roads where similar circumstances

A existed, would have to be considered, and the roots of the adjacent trees located, or at least the nature of the subsoil ascertained.

He further submitted that the alternative would be to cut down all of the trees which were potential candidates for causing root damage, and refrain from planting trees in streets in case they gave rise to liability. He submitted that the proper conclusion of these facts was that it would not be reasonable in all the circumstances to require the defendants to abate the nuisance or to make them liable for the damage consequent from the failure to abate, and that in accordance with the principle enunciated in the *Leakey* case by Megaw LJ, this gave rise to a situation in which the proper consequence was that the individual householder should each accept such burden as might exist in respect of their own property.

B The judge eventually dealt with the problem in this way. He said:

> It may well be, and I would so assume in the absence of evidence, that many of the houses could be eliminated as the subject of risk, either by reason of their position in relation to the adjacent trees, or by reason of the nature and size of the tree — or for other reasons. It may also be that in discharge of the duty which I have found to exist, the appropriate steps to eliminate or reduce that risk might involve some form of co-operation between the house owner and the defendants. Had the defendants considered the problem, and approached the plaintiff in an attempt to find an agreed solution, it may be that this would have been sufficient to discharge the duty. In the event, they did no such thing, and in failing to take any steps I find they were in breach of their duty.

C In my view there is no reason to suppose that many of the houses in Shirley Avenue could be eliminated from this risk. We were told that there is an avenue of trees all along that road and the evidence was that pockets of clay might exist anywhere in Shirley Avenue. All the householders, it seems to me, would have to be approached, not only in Shirley Avenue but in any other street in Hampshire where there are trees adjacent to houses.

The judge said that such costs as might be involved should be regarded as part of the cost of maintenance of trees to the public good. "In any event," he said, "trees require to be examined, and from time to time lopped or topped, and generally kept in good order, and such additional costs as there may be in respect of trees
D whose roots may present potential risk to adjacent property owners are in my view, in the absence of any specific evidence on the subject, properly to be regarded as part of the general maintenance costs, and are unlikely to add significantly to such costs, having regard to the comparatively rare instances in which the problem is likely to arise."

Felling and lopping of trees, as suggested by the judge, might give rise to all sorts of complaints and difficulties, quite apart from its interference with public amenities, and in the end such action might well not be appropriate, because the evidence was that the only effective method of locating the pockets of clay would be by sinking boreholes at every house adjacent to a tree.

E Mr Smyth did not go so far as the judge. He submitted that the defendants having done nothing, the onus was on them to lead evidence to the effect that remedial work was not practicable. He submitted that, at the very least, the defendants should have considered and discussed the question, and there was no evidence that they did; that they should have then circularised all householders whose houses were adjacent to trees, pointing out the dangers and asking them to pay for, or share, the costs of boreholes.

It seems to me that that evidence is not necessary in a matter of this kind. The expense would be considerable. Common sense tells that action of the kind suggested by Mr Smyth would be likely to cause more problems than it solved. Householders circulated in
F that way would be likely to suffer apprehension for the value of their property, and probably with no reason. The consequences of such circulation, it seems to me, are unpredictable and unsatisfactory. Balancing the risk of pockets of clay under no 72 with the steps necessary to deal with that risk, in my judgment there was no breach of duty by the defendants in this case and so no nuisance.

I reach this conclusion with regret.

At the end of his judgment, the judge, having considered the question of foreseeability and the reasonableness of the steps needed to abate or remove the nuisance, said that "the real question was on whom should the loss fall?" and concluded that "the loss should properly be a charge on the public, for whose benefit the tree was maintained and not on the individual who has suffered the damage." G

There is much to be said for that approach, but it is not the law. There being no breach of duty by the defendant, the plaintiff's claim must, in my judgment, fail and accordingly I would allow the appeal.

Agreeing, SIR DAVID CAIRNS said: I add a few words because we are taking a different view from that taken by Stocker J.

Since the subsoil at Shirley was almost entirely gravel, since there was no indication on the geological map even of pockets of clay, and since there was no experience of damage caused to the structure of any of the houses in Shirley Avenue by the roots of trees since the H
houses were built in 1922, it appears to me not surprising that the county council clearly never directed their minds to the question of whether any particular house in that road was liable to be damaged by such roots. If they had considered the matter, I am of opinion that they could have reasonably taken the view that any risk of such damage was so remote that it could be disregarded. To say that a risk of damage is reasonably foreseeable means that it is foreseeable not merely as a theoretical possibility but as something the chance of which occurring is such that a reasonable man would consider it necessary to take account of it. The risk of being struck by lightning when one goes for a walk is not a reasonably foreseeable risk. I should be prepared to hold that the risk in this case was J
not a reasonably foreseeable risk.

If, however, it could be said to be a reasonably foreseeable risk, I am satisfied that it was a risk such that the cost and inconvenience of taking any effective steps to remove it or reduce it would be quite out of proportion to that risk. There is nothing in the evidence to show that 72 Shirley Avenue was any more at risk than any other house in the avenue. Nor is there anything to show that any operation on the trees, short of felling them, would have made the roots safe if there were exceptionally dry weather and if the roots of any particular tree were passing through clay.

Clearly it would have been unreasonable for the council to remove all the trees in the avenue and it is unlikely that Shirley K
Avenue was the only road in the area where there were trees in proximity to houses. The alternative would have been to make an examination of the subsoil of every house which might possibly be at risk by boreholes in order to see whether the ground there was of clay, gravel, or a mixture of both. Obviously a number of boreholes would have been needed close to each of the houses. It would all have been of great inconvenience to householders and the making of any proposal for it would have been likely to cause considerable anxiety to them. The value of the houses might well have been reduced by reason of the operation.

For those reasons and those given by my Lord, I agree that the appeal should be allowed. L

Also agreeing, STEPHENSON LJ said: I have reluctantly come to the conclusion, after the great assistance received both by the judge and by this court from counsel on both sides, that the appeal must be allowed for the reasons given by my Lords.

If, as is suggested by the language of Lord Goddard CJ, giving the judgment of this court in *Davey v Harrow Corporation* [1958] 1 QB 60, all the property owner had to prove was a nuisance causing actual damage by the encroachment of the appellant council's tree roots, he would be entitled to retain his damages. But in this court Mr Smyth, for the respondent, has abandoned that contention and concedes, rightly in my judgment, that the statement of the law in M
Davey's case is subject to the proviso put upon it by Megaw LJ in *Leakey v National Trust*, and rightly explained and applied by Stocker J. So there are two questions to be considered: (1) was there a foreseeable risk that the encroachment of these tree roots would cause damage to the respondent's house, and (2) were there any reasonable precautions which the appellant could have taken to prevent or minimise that risk? The second question takes that form because the appellant took no precautions.

It is common ground that the respondent cannot succeed in getting a favourable answer to the first question unless he can prove that the appellant ought to have foreseen that there was a real risk

A of a pocket or outcrop of clay in the gravel under the respondent's house.

On the evidence analysed by my Lord, Dunn LJ, I am willing to assume that the expert evidence entitled the judge to find that that risk was more than a mere possibility which would never occur to the mind of a reasonable man, in the oft-quoted words of Lord Dunedin in *Fardon v Harcourt-Rivington* (1932) 146 LJ 391 at p 392, and which would therefore require no extraordinary precautions. But granted there was a real risk, reasonably apparent to the appellant's engineers, if they had thought about it, of clay under the respondent's house, I cannot agree with what I think was the judge's opinion that the appellant could effectively discharge its
B duty by investigating comparatively few properties and trees, and perhaps approaching a few property owners, or felling or lopping a few trees, at little expense. I do not think it needed evidence to decide that to have prevented or minimised the plaintiff's damage would have required correspondence with many property owners in the county, wholesale inspection of and boring under properties, C
and the felling of many trees, which would have put the appellant to very great expense and produced many complaints from property owners upset by warnings of dangers which might not exist and from a wide public deprived, perhaps needlessly, of the amenity provided by trees like those lining Shirley Avenue.

Such precautions would not, in my opinion, be reasonable and no less would have been required, without exceptional good luck, to discover and deal with the pocket of clay which unhappily underlay a crucial part of the respondent's house before the abnormally long, dry season of 1975-1976 made it too late to stop the subsidence and damage. D

I therefore feel, as did the judge at first, that this submission of Mr Foy's, for the appellants, is not only cogent but insuperable, and I accordingly agree that the appeal succeeds.

The appeal was allowed with costs.

TOWN AND COUNTRY PLANNING

COURT OF APPEAL

October 31 1980

(Before Lord Justice STEPHENSON, Lord Justice DUNN and Sir David CAIRNS)

KWIK SAVE DISCOUNT GROUP LTD v SECRETARY OF STATE FOR WALES AND ANOTHER

KWIK SAVE DISCOUNT GROUP LTD AND OTHERS v SECRETARY OF STATE FOR THE ENVIRONMENT AND ANOTHER

Estates Gazette January 17 1981

(1980) 257 EG 169

Town and country planning — Appeals from Queen's Bench Divisional Court — Decisions of Secretaries of State upholding enforcement notices challenged — The question in both appeals was whether the appellants, owners of discount stores, were in breach of planning control, and thus subject to enforcement notices, by use of premises in the first case as a retail supermarket and in the second case as a retail discount store — In the first appeal planning permission for an improved motor service station included permission to convert an existing workshop to a "retail showroom" — Held that there was an implied limitation that the showroom should be used for the sale of cars and automobile goods and accessories — Use as a retail supermarket for the sale of groceries, bread, confectionery, meat, cigarettes and other like goods was a breach — In the second appeal the planning permission as regards the showroom was specifically limited to "car showroom" — It was sought to overcome this difficulty by granting a licence to another company, as concessionaire for a short period, to use the showroom for the sale of cars, after which the appellants used the building as a retail discount store — Held that such minimal or token use for the sale of cars did not enable the appellants to take advantage of the Town and Country Planning (Use Classes) Order to justify use as a retail store — Enforcement notices upheld in both cases and appeals dismissed.

In the first appeal the appellants were the Kwik Save Discount Group Ltd and the premises which were the subject of the enforcement notice were situated at Erw Cross Roads, Holyhead; the respondents were the Secretary of State for Wales and Ynys Mon Borough Council. In the second appeal the appellants were the Kwik Save Discount Group Ltd and five other companies, Status Discount Ltd, Courtesy Shoes Ltd, North West Vintners (Retail) Ltd, Hirsel Ltd and Supermarket Tyre Services Ltd; the respondents were the Secretary of State for the Environment and Oldham Borough Council. The enforcement notice in this appeal related to a site in Oldham. The decisions of the Divisional Court were reported at (1978) 247 EG 561 and 562.

Andrew Rankin QC and W Rankin (instructed by Lee, Bolton & Lee, agents for Bullivant & Co, of Liverpool) appeared on behalf of all the appellants; Simon Brown (instructed by the Treasury Solicitor) represented the respondent Secretaries of State; the planning authorities were not represented and took no part in the proceedings.

Giving the judgment of the court, STEPHENSON LJ said: We have before us two appeals, with leave, from two decisions of the Divisional Court given on June 26 1978. They dismissed appeals from (1) a decision of the Secretary of State for Wales given on July 22 1976, whereby he upheld an enforcement notice issued by the Anglesey County Council against the appellants on February 8 1974; (2) a decision of the Secretary of State for the Environment given on December 15 1976, whereby he upheld an enforcement notice issued by the Oldham Borough Council against the appellants on August 5 1975. The first notice relates to a site in Holyhead, the second to a site in Oldham. We shall take the Holyhead appeal first.

On September 1 1972 Esso Petroleum Co Ltd applied to the Anglesey County Council for permission to develop land at Erw Cross Roads, Holyhead. As a result of previous planning applications and permissions there was on the site a number of buildings comprising Swift's Service Station. In December 1952 permission had been granted for use of land there, which did not include the present appeal site, as a public service garage. In April 1955 a detailed application for the erection of a petrol filling station and showroom on the same land was approved. In April 1963 permission was granted for additions to the service station, comprising a new store and showroom. On March 12 1965 permission was granted for the erection of a workshop on land adjoining the service station, and that is the present appeal building. On January 7 1969 conditional permission was granted for a new access to the workshop from the trunk road adjoining the site.

In September 1972 Esso were prospective purchasers of the site and buildings comprising Swift's Service Station. In their form of application they stated:

(4) Describe briefly the proposed development including the purpose for which the land and/or buildings are to be used. If they are to be used for more than one purpose, give details. See note (a):

Conversion of existing workshop to retail showroom including new display windows.

Erection of specialist service bays. Installation of canopy, new tanks and new self service pumps.

Alterations to existing service station building. Use of existing waste ground for car sales display and parking.

Against (5):

(5) State the purpose for which the land and/or buildings are now used, and if used for more than one purpose, give details:
Petrol service station and workshop.

Against (10):

(10) If the building is to be used wholly or partly for industrial or commercial use, state:

(i) the nature of the proposed industry or business, including, if for industrial use, a brief description of the type of processes to be carried on;

(ii) the total floor area:

. . . are the words:

(i) Existing petrol service station to self service. Existing workshop to retail showroom. New specialist service bays — car diagnostic — tune-up, etc. Existing showroom to shop for retail of motor vehicle accessories and petroleum products.

(ii) Existing service station building — 2,500 sq ft.
Workshop to be retail showroom — 6,750 sq ft.
New Service bays — 1,725 sq ft.

On January 2 1973 the council granted conditional permission in these terms:

. . . in accordance with the plan and application submitted to the Council on October 4 1972 subject to the condition(s) specified hereunder:

(i) The development to which this permission relates shall be begun not later than the expiration of five years beginning with the date of this permission.

A (ii) The type and colour of the facing bricks proposed to be used in the construction of the building shall be submitted to and approved in writing by the local planning authority before any development is commenced.

(iii) The type and colour of the exterior finishing materials is reserved for further consideration and approval before any development is commenced on the erection of the building.

(iv) The development permitted by this consent shall be carried out strictly in accordance with the plans attached to this permission.

On March 19 1973 the appellants' solicitors wrote to the council:

We are acting for a proposed purchaser of this site and the vendors solicitors have supplied us with a copy of a planning application dated September 1 1972 by E C Blades on behalf of Esso Petroleum Company Limited. The application is for the conversion of the existing workshop to a retail shop
B and showroom including the retail of motor vehicle accessories and petroleum products. The existing petrol service station is to be converted to self service and there are to be specialist service bays. The existing service station building has an area of two thousand five hundred square feet while the area to be used as a retail shop and showroom will have an area of six thousand seven hundred and fifty square feet; the new service bays having an area of one thousand seven hundred and twenty five square feet.

Omitting the next paragraph, the letter goes on:

We should be grateful for your confirmation that the planning permission given by your council on January 2 1973 gives permission for use, *inter alia*, as a retail shop and showroom in accordance with the plan and application submitted to the council on September 1 1972 as well as in accordance with
C the further plan submitted to the council on October 4 1972.

On March 21 1973 came the prompt reply:

With reference to your letter of March 19 in connection with the above, I have looked at the planning consent issued and I would confirm that this permission ref no V/607j dated January 2 1973 gives permission for a retail shop and showroom in accordance with the plan and application which was submitted to the planning authority on October 4 1972.

The "proposed purchaser" on whose behalf this request for confirmation was made was not Esso but the appellants, and on March 27 1973 they entered into a contract to purchase the site from a Mr Williams, who still owned the adjoining land occupied by Swift's Service Station. On May 18 1973 the purchase was com-
D pleted and after certain alterations the appellants opened the present appeal building not as a retail showroom for motor vehicles or accessories but as a supermarket, the Kwik Save Discount Store.

The council took the view that this was a breach of planning control in that "the said land has been developed by the making of a material change in the use thereof to a use for the purpose of a retail supermarket for the sale of groceries, bread, confectionery, meat, cigarettes and other like goods without the grant of permission required in that behalf under Part III of the Town and Country Planning Act 1962 or Part III of the Act of 1971". They therefore gave notice:

E That in exercise of the powers contained in the said section 87 of the Act of 1971 the Council HEREBY REQUIRE YOU within the period of twelve calendar months beginning with the date on which this notice takes effect to discontinue the use of the said land for the purpose of a retail supermarket for the sale of groceries, bread, confectionery, meat, cigarettes and other like goods and to restore the said land to its condition before the said development took place.

The appellants appealed under section 88 of the Town and Country Planning Act 1971. The inspector who held an inquiry into their appeal reached no conclusion on their appeal under subsection (1)(b), as it depended, in his view, expressed in paragraph 154 of his report of October 1975, upon an interpretation of law and the documents, but on their appeal under subsection (1)(a) he recom-
F mended that planning permission should not be granted if it was required. As we have said, the Secretary of State (who is the only respondent to this appeal, the council not being a party to it) decided on June 22 1976 in favour of the council, and the Divisional Court dismissed the appellants' appeal from his decision.

The appeal has been pursued on two grounds, at least as far as this court. But on the second day of its hearing before us Mr Rankin on behalf of the appellants conceded that as there was in fact no use of the appeal building as "a shop for the sale of motor vehicles" he could not rely on any change of its use to another purpose within Class III of the Schedules to the Town and Country Planning General Development Order 1973 and the Town and Country Planning (Use Classes) Order 1972, as requiring no planning per- G
mission by virtue of article 3 in those orders. He accordingly abandoned the second ground in his notice of appeal. We need therefore read only the first ground:

1 The Divisional Court was wrong in law in holding:

(a) that the planning permission granted on January 2 1973 in respect of the said land and in particular the building on the appeal site restricted the use thereof to an activity involving automobile goods;

(b) that the said planning permission created a planning unit with a defined scope and purpose and that parts of the said land the subject matter of the said planning permission could not be developed in isolation for purposes and uses not within the original scope and purpose of the planning unit without express planning permission being obtained for the relevant H
particular part of the said land.

Why, asked Mr Rankin, should this permission to convert the existing workshop to a retail showroom, including new display windows in accordance with the plans attached, and to use it as a retail showroom be a permission restricting its use to the display and retail sale of automobile goods? Nothing would have been simpler than to limit the use to use for a retail motor car showroom, or for a car sale showroom, which is how the Minister of Housing and Local Government described "use for the sale of motor vehicles" in Circular no 10/60, paragraph 7, when that was introduced as a fifth exception into what was then Class 1 in the Schedule to the Town and Country Planning (Use Classes) (Amendment) J
Order 1960. (Compare also the permission for a "car showroom" which is the subject of the Oldham appeal.) But the permitted use was not so restricted expressly by any such qualification of the words "retail showroom" or by a condition. In circular after circular, beginning in May 1948 and including Welsh Office Circular no 215/72 issued on September 25 1972, within a fortnight of this application, the responsible minister had drawn the attention of planning authorities to the fact that where it was desirable to restrict the use of a building to a single named activity this could be done by way of a condition of the kind set out in Circular no 5/68 (paragraph 23 and appendix, paragraph 7).

K
The evidence of the council's development control officer had been that the possibility of imposing a condition restricting the use of the workshop had *not* been overlooked; indeed, that he had given careful consideration to the need for such a condition but had concluded that it was unnecessary.

Furthermore (said Mr Rankin) the four conditions imposed under the power given by section 29(1)(a) of the Act of 1971 related to entirely different matters. Section 33(2) conferred a power to specify the purposes for which a building permitted to be erected could be used, and section 34 required prescribed information about planning applications and decisions to be registered and available for inspection. The specified purpose for which the existing appeal building was permitted to be used, and could be seen by L
the public to be permitted to be used, was simply for a retail showroom. That was contrasted with the "use of existing waste ground for car sales display and parking" and the use of "existing showroom to shop for retail of motor vehicle accessories and petroleum products" — both uses plainly restricted as the use "for retail showroom" was not. So also a subsequent permission of June 29 1973 restricted the use of a shop on the site to the sale of tyres, batteries and accessories only and for no other purpose whatsoever. One further submission was that a retail showroom meant a retail shop and showroom, and that meaning was confirmed by the council in their letter of March 21 1973.

We do not consider that the letter carries the matter any further M
in either direction. If the writer misinterpreted the grant of permission, his opinion would be irrelevant to its true construction and would not stop the council — or the Secretary of State — from relying on that construction. But there was no misinterpretation of the grant. It did give permission for a retail shop — from the exisiting showroom, and for a showroom — from the existing workshop, whatever the true nature and extent of the permission for a retail showroom.

We doubt whether permission to use a building for a retail showroom could, without more, mean permission to use it for a supermarket and we would have been willing to dismiss this appeal

A on the ground that "retail showroom" did not naturally mean or include "supermarket" but for a concession by Mr Brown for the respondent that in a planning context it might. We should have thought it difficult to argue that a tenant, for instance, who convenanted with his landlord to use business premises for a retail showroom was not in breach of covenant if he used them for a supermarket. But the Divisional Court heard no argument and gave no decision on this simple point and we therefore turn to their decision and the reasons given in their judgments for holding that in this grant of planning permission "retail showroom" did not mean or include "supermarket".

The judgment of Lord Widgery CJ (with which Talbot J agreed) B linked Mr Rankin's argument that there was no express restriction on the activities in the appeal building with his argument that the building was a separate planning unit, and preferred the respondent's argument that there was no separate planning unit and there was an implied restriction to automobile goods. The Lord Chief Justice regarded the appeal building as one of a complex of buildings supplementary one to another and was of the opinion that a new planning unit could not be created either by the owner selecting one of those buildings and conveying it to a purchaser or by a purchaser by his own say-so whipping it out of the planning unit formed by the whole complex or what he called "the main site".

C Ackner J agreed that the appeal be dismissed. He added:

> The essential question seems to be to be what was the nature of the permission granted on January 2 1973. In the context of the application made by Esso Petroleum, to which my Lord has referred, and in the context of the actual approval, which was in these terms: "Alterations and extensions to Swifts Service Station, Erw Cross Roads, Holyhead, in accordance with the plan and application submitted to the Council", it seems to me clear that the permission was limited to the use of the retail showroom for the sale of cars and was not without limitation.

We do not find the question of a separate planning unit, or the authorities relating to it, important or helpful in deciding this appeal. We agree with Ackner J (as he then was) that the essential D question is the nature of the permission granted and that it must be construed in the context of the application. That means, as Mr Brown has argued for the respondent, in the context of the whole of the proposed alterations and extensions to the service station buildings on the whole site which were the subject of this composite application. From the application as a whole and the other parts of it we have no hesitation in implying the limitation to the use of the showroom for the sale of cars and automobile goods and accessories. If that means that the whole site or complex is being treated as one planning unit, so be it. It does not mean that no site can be developed "piecemeal", as Mr Rankin contended was the effect of the argument for the respondent, only that this application related E to a number of separate but connected "pieces".

Mr Brown conceded that the appeal building was not to be used for a purpose ancillary to the service station but contended that it was to be used as a constituent part of it or of the overall site. For that contention, however described, he relied on five pointers; the application was made by Esso; for use of the appeal building as a retail showroom; with the physical characteristics shown in the deposited plans and drawings of display windows, double doors and a ramp showing that it was designed as a car showroom; what was permitted was "alterations and extensions to Swift's Service Station"; one of which was to one building on a site which was part F of a larger site with other buildings being altered and extended.

In our judgment these items in the application and grant do point to the limitation or restriction implied by the Divisional Court. The express restrictions to cars and motor vehicle access in other parts of the application support rather than contradict the restriction, as does the fact that one existing showroom (we do not know what it displayed) is being converted to a shop for motor vehicle accessories at the same time as an existing workshop (for repair of motor vehicles presumably) is being converted to a showroom.

We would accordingly dismiss the Holyhead appeal.

The Oldham Appeal relates to a grant by the Oldham Borough Council to a company called Transport & General Investments Ltd G on December 10 1973 of planning permission for the erection on a site in Oldham of a self-service petrol-filling station, carwash, car showroom, and tyre fitting bay. Soon after the grant the company entered into negotiations for the sale of the site to the appellants, a company with which the vendor company had no connection, ending in a sale to the appellants which was completed on April 16 1974. The showroom was erected and handed over to the appellants about the end of January 1975. Reading from the papers:

> On advice received at that time the company understood that the showroom building would have to be used for the sale of cars, and so a concession for the sale of cars was offered to Barkers of Wortley Ltd, motor retailers. They were informed that only a short-term licence would be granted, and that this would be terminated when the company was ready to undertake other forms H of retail sales. Terms were agreed, and the concessionaire took possession on February 25 1975 although the agreement was actually signed on March 13 1975. The concession continued for about one month, and was terminated on March 27 1975. On April 10 1975 the building was opened as a retail store, and the form of retail trading which then began is substantially the same as at present.

The concession agreement provided for the premises to be rent free and for the appellants to supply capital funds or guarantee up to £25,000 and to bear any loss and be entitled to any profit. Five cars were brought to the showroom for sale in February 1975 and were advertised in two papers on March 13 1975. On August 5 1975 the respondent council served an enforcement notice on the appel- J lants stating that there appeared to have been a breach of planning control in "the making of a material change in the use of the said land from use as vacant land and buildings intended for development and use as a self service petrol filling station, car wash, car showroom and tyre fitting bay to use for the purpose of a retail discount store with ancillary car parking" . . . and requiring the appellants to "(i) Cease the use of the said land as a retail discount store with ancillary car parking", and "(ii) remove all merchandise from the said land".

In April 1976 the appellants' appeals (only one of which is relevant) were heard by an inspector, who stated his conclusions beginning with a passage which Talbot J, giving the first judgment K in the Divisional Court, cited with implied approval, as follows:

> The offering of five cars for sale for a period of about one month in a building with a floor space of about 20,000 sq ft amounted to no more than a token use of the appeal premises as a shop for the sale of motor vehicles, so minimal as to be of no planning significance. Article 3 and Class III(b)(v) of Schedule I of the General Development Order 1973 refers to a change of use of premises from "use as a shop . . ." and not from a permitted use of premises which have not actually been brought into use. There was therefore no effective use of the new building on the appeal site until the appellant's use of it as a discount store, which constituted a material change of use from a non-use, involving development for which specific planning permission was required. L

The inspector went on to recommend that if it was decided that development requiring planning permission was involved planning permission should not be granted. The Secretary of State accepted his recommendation, upheld the relevant enforcement notice and refused planning permission.

I again read from Talbot J's judgment in the Divisional Court:

> It is against the Secretary of State's acceptance of that conclusion that Mr Rankin moves this court to find that the Secretary of State was wrong in law, and that the matter should be remitted to him for guidance. He relies upon the terms of the relevant order, which I take from Volume 3 of the *Encyclopaedia of Planning*. The relevant order from which I read is M the Town and Country Planning General Development Order 1977, though the terms of the previous order then applicable we are told were precisely the same. The relevant part reads as follows: "Development consisting of a change of use to (b) use as a shop for any purpose included in Class I of the Schedule to the Town and Country Planning (Use Classes) Order 1972 from use as: . . . (v) a shop for the sale of motor vehicles." So therefore it is argued by Mr Rankin that development is permitted by reason of that part of the order, because the appellants in changing to their discount store use did so from use of this building as a shop for the sale of motor vehicles. If that is right, then the appellants' present use is a permitted use under the terms of the relevant legislation.
>
> The question therefore is, was there a use of these premises for the sale of motor vehicles? Mr Rankin accepts that what the appellants did was in fact

A to involve themselves in what can only be described as a device in order to bring themselves within the terms of the order. But accepting that, his argument is that what in fact took place — and I have given it from the citation of the facts found by the inspector — was in fact a use of these premises as a motor car showroom. He submits that it does not matter how many cars there are. It does not matter that the terms of the agreement were in effect to guarantee Barkers of Wortley Ltd against any loss, and to subsidise them if necessary. It does not matter that the period of time was in fact short. What he said is, looking at what happened, can there be any other description than that these premises were in fact used for the sale of motor vehicles?

Though there is no qualifying adjective in the order of the word "use" Mr Woolf has submitted that one ought to qualify it by the word "material". To
B this end he sought support from section 22 (1) of the Act. The relevant part of that subsection is this: "In this Act, 'development' means the making of any material change in the use of any buildings or other land." Therefore, as the change of use referred to in the subsection is qualified by "material", so also, he argues, that word in the order must also be qualified in like manner. He points out that what in fact took place was only a token activity, that when one comes to consider the activity of selling motor cars this is a case where there were the trappings only. No motor cars were concerned in this short period of time, and were being advertised twice.

What is the answer to these submissions? In my judgment, the very fact that a device was resorted to by the appellants makes me suspect the use to which it is said the land was put. The inspector and the Secretary of State found that it was *de minimis* on the facts. I would not disagree with that view, and in my judgment if a use is a *de minimis* use it is not a use within the order.
C But there is yet a wider ground to which I would refer, which in my judgment leads me to the conclusion that the appellants are wrong and that the Secretary of State came to the right conclusion. Here there was the grant of a planning permission, as I have referred to, in December 1972 for the activity involving motor cars, a self service petrol filling station, a car wash, a car showroom, tyre fitting bays, and so on. That was the permission, a grant for that activity, and in my view it is not within that grant permission to take part of the site covered by the permisssion and to make use of it for some business which is quite different to that which is covered by the grant of planning permission. For those reasons, in my judgment this appeal fails.

With that judgment Lord Widgery CJ and Ackner J agreed.

In the appellants' notice of appeal that decision is challenged on
D three grounds, the last of which is not pursued in this court.

It will be seen that there is common ground between this appeal and the Holyhead appeal, but the express restriction to a car showroom avoids any need to imply a restriction to cars and compels the appellants to rely in this appeal on the General Development Order and the Use Classes Order, which were ultimately no help to the appellants in the other appeal. On the view which we take it becomes unnecessary to consider the wider ground on which the Divisional Court rested its decision at the end of Talbot J's judgment. It is conceded that if we have to support the decision on that ground only we should have to remit the case to the respondent Secretary of State. But in our judgment the Divisional Court came
E to a right decision on the first point and the appeal from it must fail.

There is no finding that the device of getting a reputable car sales company to "use" the showroom for car sales was improper. It was done openly on legal advice, and the managing director of the company gave evidence that it was a genuine effort to find another outlet for the sale of cars without a time-limit and if the business had been successful he had hoped to continue, although he could offer no explanation of the unusual terms of the concession. But it was admittedly a device to bring the benefits of the General Development Order into operation by creating a use for showing and selling motor vehicles from which a use for a retail store could legitimately spring.
F The only question was therefore one of mixed law and fact:

(1) Was there in fact a use for the sale of motor vehicles?
(2) If there was, was it minimal?
(3) If minimal, should it be disregarded?

The answer to the first question was admittedly "yes". Mr Rankin, for the appellants, submitted that (2) and (3) should each be answered "no". (2) is, in our opinion, a question of fact which a reasonable Secretary of State was entitled to answer "yes", as the inspector answered it and as we think we would answer it. (3) is, in our judgment, a question of law to which the answer is "yes". We can see no reason to exclude the commonsense *de minimis* principle from such an area of life and law as this. To introduce the question G
whether the use was "material" is not, we think, helpful and may be misleading.

We would therefore dismiss this appeal on that ground.

Both appeals were dismissed with costs. Leave to appeal to the House of Lords was refused.

QUEEN'S BENCH DIVISIONAL COURT

December 15 1980

(Before Lord Justice DONALDSON and Mr Justice KILNER BROWN) H

SYKES v SECRETARY OF STATE FOR THE ENVIRONMENT AND ANOTHER
SOUTH OXFORDSHIRE DISTRICT COUNCIL v SECRETARY OF STATE FOR THE ENVIRONMENT AND OTHERS

Estates Gazette February 28 1981

(1980) 257 EG 821

Town and Country Planning Act 1971 — Appeals against decisions of Secretary of State raising points of general interest as to the need for planning permission where land is used in connection with horses — Planning authorities said to be concerned about proliferation of J
uses of small areas of land for schooling horses, teaching young riders, practising for gymkhanas and other recreational activities connected with horses — Judgments seek to clarify position — Although the use of land as grazing land, being a use for "agriculture", is not subject to planning control, even if the use is for the grazing of "non-agricultural horses" (ie horses used for recreation), it is necessary to define what is meant by such use — The use must be for the purpose of grazing — Grazing must be the predominant or substantial use — Incidental grazing by horses fed otherwise would not be enough — Common sense rather than reference to legal precedents should be the guide for inspectors and planning officers — In the main appeal the finding of fact that the land was used for K
grazing could not be attacked — Point raised in other appeal as to need for clarity and precision in enforcement notice — Appellant was admittedly using land only for exercising ponies, not for grazing, but notice required him to desist from using the land as "a paddock" — Notice ambiguous — Strictly speaking, the concept of a paddock is merely that of an enclosure, not a use — In any case, interpreted as a use it could have meant use for grazing, which was not a breach of planning control, just as easily as use for the keeping of ponies — Planning authority's appeal on the grazing point dismissed — Landowner's appeal on enforcement notice point allowed

The first-named appeal was by David John Sykes against the L
Secretary of State and the South Oxfordshire District Council as respondents, the issue being the validity of an enforcement notice served by the second respondents. In the second-named appeal the South Oxfordshire District Council as appellants challenged a decision of the Secretary of State that, in view of a finding of fact that land was used for grazing of horses, there was no breach of planning control. It was argued on behalf of the planning authority that the grazing did not escape planning control unless the horses were themselves used for agricultural purposes. The respondents to this appeal were the Secretary of State and Timothy B Underwood, Elizabeth Underwood and Peter Brian Lance.

D E W Turriff (instructed by Bircham & Co) appeared on behalf M
of David John Sykes in the first-named appeal and on behalf of Timothy B Underwood and Elizabeth Underwood, respondents in the second-named appeal; Simon Brown (instructed by the Treasury Solicitor) represented the Secretary of State in both appeals; D N R Latham (instructed by Sherwood & Co, agents for J B Chirnside, chief executive and solicitor, South Oxfordshire District Council) represented the council in both appeals.

Giving judgment, DONALDSON LJ said: Today we have been concerned with two appeals against decisions of the Secretary of State in his planning jurisdiction. Both appeals have something in common in that they raise the question of whether and to what

A extent it is necessary to have planning permission if you are using land in connection with horses which are themselves kept for recreational purposes.

Both cases arise in South Oxfordshire. The exact details of the appeal sites are irrelevant for present purposes. In the case of Mr and Mrs Underwood they kept three racehorses, two point-to-point horses, a driving pony, two family ponies and one retired mare which had been badly injured, all on the 2½ acres of land concerned. In the case of Mr Sykes, he kept two show ponies on his rather smaller piece of land.

The Underwoods were successful in their appeal to the Secretary of State, who took the view that what they were doing, as found by
B the inspector, did not constitute a breach of planning control. In the case of Mr Sykes, he reached the opposite conclusion. In the case of the Underwoods the South Oxfordshire District Council appeals and in the other case Mr Sykes is the appellant.

Let me put Mr Sykes' appeal on one side for the moment because he is only concerned with the second of the two points which arise in the Underwood appeal. He cannot contend in his case that what he was doing was a permitted development because it has been found that he was not using the land for grazing purposes in any way at all. He was using it merely for exercising the ponies. He has, however, a point on the form of the notice to which I will return.

In Mr and Mrs Underwood's case, as I say, the Secretary of State
C took the view that no planning permission was needed because there was no breach of planning control in the use which they were making of their land, and it is that point which I think has to be examined and it is that point which is of general interest, as I understand it, to the Secretary of State and to planning officers throughout the country.

The matter starts with section 22(2)(e) of the Town and Country Planning Act 1971, which provides:

> The following operations or uses of land shall not be taken for the purposes of this Act to involve development of the land, that is to say — . . . (e) the use of any land for the purposes of agriculture . . .

D I think I can omit the rest of the words as being immaterial for present purposes.

"Agriculture" is defined, somewhat indigestibly, in section 290(1) of the Act as follows:

> "Agriculture" includes horticulture, fruit growing, seed growing, dairy farming, the breeding and keeping of livestock (including any creature kept for the production of food, wool, skins or fur, or for the purpose of its use in the farming of land), the use of land as grazing land, meadow land, osier land, market gardens and nursery grounds, and the use of land for woodlands where that use is ancillary to the farming of land for other agricultural purposes, and "agricultural" shall be construed accordingly.

E The Secretary of State in his decision letter [Ref APP/5355/C/77/5117 dated May 15 1980] dealt with Mr Underwood's appeal in the following terms. He said:

> In regard to ground (b) of the appeal against Notice A, it was argued on behalf of your clients, citing the judgment of the Court in *Rutherford v Maurer* [1962] 1 QB 16 and *McClinton v McFall* (1974) 232 EG 707, that the use of the appeal site for the grazing of horses was an agricultural use and, as such, did not amount to development within the meaning of the 1971 Act. The inspector found as facts, which are accepted, that Mr Underwood purchased the appeal site in 1975 and, since then, it has been used for grazing his horses as an alternative to their accommodation in the stable buildings at "The Well House". There was no evidence to show that the site had been used for any other purpose in connection with these horses.

F The Secretary of State's decision letter continued:

> The inspector concluded: "Site A is a well defined field. Although now open to the extreme rear part of the land purchased as the property 'The Well House', that area is not embraced by Notice A and is physically separated from the land containing the buildings concerned in Notice B. Site A should therefore be considered as an isolated planning unit.
>
> "The judgment given in *Belmont Farm Ltd v Minister of Housing and Local Government* (1962) 13 P & CR 417 was concerned with land and buildings which as a whole were used as a stud farm, a situation not pertaining at Site A.
>
> "The other two cases cited by the appellant's advocate, *McClinton v McFall* and *Rutherford v Maurer*, although not concerned with planning law, were concerned with circumstances very similar to those of Notice A. In G
> both cases the courts appeared to have had no doubt that the grazing of land by horses, whatever the purposes of those horses, came within a definition of agriculture similar to that given in section 290 of the 1971 Act.
>
> "Following those last two judgments rather than that given in *Belmont Farm Ltd v Minister of Housing and Local Government*, as Site A has only been used for the grazing of horses, which can aptly be described as use for the purpose of a horse-paddock and which comes within the definition of agriculture, no development requiring planning permission has occurred by reason of section 22(2)(e) of the 1971 Act, and the appeal succeeds on ground 88(1)(b)."

The Secretary of State continued:

> These conclusions have been considered. The inspector's view that Site A is H
> a separate planning unit is accepted, subject to the qualification that that unit is seen as including parts of "The Well House" land to the north of the fence, behind the stables. Though occupied with the remainder of "The Well House" land — comprising the house and its curtilage, including the stables — Site A, together with that northern part of "The Well House" land is seen as being a separate planning unit, as a physically distinct area which has a separate use; namely for grazing horses. Following the inspector's view of the matter, it is further considered that, as a "use of land as grazing land", this use is within the definition of "agriculture" in section 290(1) of the 1971 Act and that, by virtue of section 22(2)(e) of that Act, it is consequently a use which is not to be taken as involving the development of land. From this it follows that the introduction of that use did not constitute a breach of planning control, and the appeal succeeds on
> ground (b). J

Mr Latham for the South Oxfordshire District Council submits that this is wrong and that, properly construed, section 22(2)(e) does not permit the use of land for the grazing of horses unless those horses are themselves being used for agricultural purposes. If you have, for instance, a carthorse, you could graze that on the land assuming that the carthorse would be used for agricultural purposes, although one can of course use carthorses for other purposes. On the other hand, horses which are used purely for recreational purposes are not, he submits, within the definition. If you read the definition literally that clearly is not correct. But Mr Latham relies heavily upon the decision of this court in *Belmont Farm Ltd v* K
Minister of Housing and Local Government (1962) 13 P & CR 417.

In that case this court was concerned with a different limb of the definition, namely, the words "keeping of livestock" with its parenthetical qualification relating to creatures kept for the production of food, wool, skins or fur, or for the purpose of its use in the farming of land. This court held that, taking account of that parenthesis, the words "keeping of livestock" had to be restrictively construed and did not cover the keeping of horses for purposes other than agricultural purposes. In Mr Latham's submission, the same approach should be adopted in relation to the words "use of land as grazing land".

For my part, I see no reason why the words should be construed L
restrictively in that way. There are no qualifying words such as exist in relation to the words "keeping of livestock". There are no such qualifying words as exist at the end of the definition in relation to the use of land or woodlands. It is thus quite clear that, if Parliament had intended to qualify the apparent width of the words "use of land as grazing land", it could have done so, and I see absolutely no reason why we should imply any such limitation.

The whole of the decision in the *Belmont Farm* case, as appears from the judgment of Lord Parker CJ, turned upon the qualifying words and what was to be implied from them. There is nothing, as I see it, in that judgment which would have any application to the words with which we are concerned. Accordingly I would give them M
their natural meaning. As it seems to me, faced with the inspector's conclusion of fact that this land — the Underwoods' land — was used for the grazing of horses, the Secretary of State's decision was wholly correct.

The form of the notice of appeal does not permit Mr Latham to attack the findings of fact, and indeed he might have had some difficulties in view of the semi-sacrosanct nature of findings of fact in this field. But it is, I think, fair to say that the Underwoods may perhaps have been fortunate in their findings of fact in this case because it is not, as I see it, every grazing of land by horses which enables an owner of land to say that he does not need planning

A permission. Section 22(2)(e) says in terms that what is permitted is the use of the land, and I substitute the relevant part of the definition, for the purposes of using that land as grazing land. The concept of there being more than one cause or more than one purpose is well known to the law.

What an inspector in these circumstances has to decide is: what was *the* purpose — and I stress the word "the" — for which the land is being used? If horses are simply turned out on to the land with a view to feeding them from the land, clearly the land is being used for grazing. But if horses are being kept on the land and are being fed wholly or primarily by other means so that such grazing as they do is completely incidental and perhaps achieved merely because there B are no convenient ways of stopping them doing it, then plainly the land is not being used for grazing but merely being used for keeping the animals. On the other hand, of course, if the animals are put on to a field with a view to their grazing and are kept there for 24 hours a day, seven days a week over a period, it would not, I would have thought, be possible to say that as they were being kept there, they were not being grazed. It is quite possible for horses to be both grazed and kept in the same place.

The predominant question here is: what use was being made of the land? Was it for the purpose of grazing? I do not find any particular difficulty in deciding what is a predominant use. To take an example which I mentioned in argument, if somebody goes to a C restaurant and smokes after the meal they do not go to the restaurant in order to smoke; they go for the meal. There is no difficulty about that. I cannot see any difficulty in most cases in recognising whether the land is being used for grazing or for the keeping of non-agricultural horses. It is only if it is being used for the purpose of grazing that no planning permission is required.

Let me now turn to the second problem, which arises out of the fact that in the case of the Underwoods' enforcement notice they were required to desist from using the land as "a horse paddock", and in the case of Mr Sykes' enforcement notice he was required to desist from using the land as "a paddock". It is unnecessary to say anything about the Underwoods' case since, as I have already said, I D would support the Secretary of State's decision in allowing the appeal. Therefore that notice is dead. But as far as the Sykes' notice is concerned, it is crucial because this is the only ground upon which he can have the Secretary of State's decision set aside.

For my part, I think that he is entitled to have it set aside. I say that for this reason. Section 87(6) of the Act provides:

> An enforcement notice shall specify — (a) the matters alleged to constitute a breach of planning control; (b) the steps required by the authority to be taken in order to remedy the breach . . .

It follows from that, and indeed there is ample authority to support the proposition, that there must be a clear indication to the E addressee of the notice as to what it is that he is doing wrong and what he must do in order to stop doing it. Both those propositions call for some clarity of expression on the part of those who prepare the enforcement notice. To require somebody to stop using land as a paddock seems to me to be wholly lacking in any clarity whatever. It is open, to start with, to the comment that you do not use land as a paddock; it either is or is not a paddock. A paddock is not a concept of use at all; it is a concept perhaps of enclosure. But Mr Brown says that in this context it must mean "for the keeping of horses". It seemed to me that it could equally well mean "for the grazing of horses", which is not a breach of planning control at all. It is a thoroughly unsatisfactory term, and as it is ambiguous as well as F being unsatisfactory I think that the section is not complied with. In the case of Mr Sykes' appeal, too, I would allow the appeal and remit the matter to the Secretary of State for further consideration.

Agreeing, Kilner Brown J said: With regard to the question of principle which this court was asked to consider, it seems to me that in this situation, as we are told and it may well be so, there is growing anxiety felt by a number of local authorities as to the proliferation of small areas of land used for the purpose either of keeping horses for recreational purposes or alternatively for using them for schooling or for teaching young riders, particularly in the more difficult art of show jumping and performing in gymkhanas. But in the end it is always a question of common sense. I would have thought that planning officers in the first place should apply the sort G of test which my Lord has indicated and to see really what is the land used for, and, as Mr Brown for the Secretary of State rightly said, you look to see what is its substantial use. It is easy enough to detect the situation where a piece of land — a paddock, a small meadow, call it what you like — is simply used for the purpose of schooling horses or training young riders. The amount of grazing which occurs on that land is merely incidental while each horse and rider is waiting his or her turn to be trained. On the other hand, there may be, as the inspector found in the case of the Underwoods, normally speaking a degree of actual and substantial grazing which was carried out.

Again I would deprecate the use of reference to authorities such H as those which were cited before the inspector. In view of the judgment given by my Lord, I would hope that henceforth inspectors and planning officers would be able to approach the growing problem using their common sense, assessing each situation upon the facts as they appear to be. Having said that, I agree with the judgment of my Lord.

The appeal by Mr Sykes was allowed with costs against the district council. The appeal by the district council against the Secretary of State and others was dismissed with costs.

J

QUEEN'S BENCH DIVISION

October 24 1980

(Before Mr Justice FORBES)

BERNARD WHEATCROFT LTD v SECRETARY OF STATE FOR THE ENVIRONMENT AND ANOTHER

Estates Gazette March 7 1981

(1980) 257 EG 934

Town and Country Planning Act 1971 — Application to quash order K of Secretary of State for the Environment dismissing appeal against refusal by local planning authority of permission for development — Original application for permission related to 35 acres and 420 dwelling units — On the refusal of permission for this development the applicants appealed to the Secretary of State and at the public inquiry put forward an alternative proposal for development of 25 acres with 250 dwelling units — The Secretary of State dismissed the appeal and expressed the view that he had no power to grant permission for development on a smaller site and at a lower density than those indicated on the application — Secretary of State misdirected himself by acting on the principle that conditions could only be used to reduce development below that proposed in the application where L the application was severable — The correct test was whether the development permitted was in substance different from that for which application was made — As the Secretary of State fell into error in this respect his decision had to be quashed

This was an application by Bernard Wheatcroft Ltd to quash an order by the Secretary of State for the Environment dismissing an appeal by the applicants against refusal of planning permission by Harborough District Council. The site concerned in the proceedings was a 35-acre portion of agricultural land forming part of a larger area owned by the applicants at Bitterswell Road, Lutterworth, in the District of Harborough. The first respondent, the M Secretary of State, was not represented, as he had formed the opinion that his decision could not be defended. The district council, however, decided to contest the application.

J C Harper (instructed by R G Frisby & Small, of Leicester) appeared on behalf of the applicants; J Sullivan (instructed by the solicitor to Harborough District Council) represented the second respondents, the district council; the first respondent, the Secretary of State, was not represented and took no part in the proceedings.

Giving judgment, Forbes J said: In this case Mr Harper moves to quash an order of the Secretary of State for the Environment whereby he dismissed an appeal against refusal of planning permission by the second respondent, the Harborough District Council.

Despite the fact that it is concerned solely with the extent of his powers the Secretary of State is not represented.

The facts may be set out briefly as follows. The applicants own a large area of agricultural land at Bitterswell Road, Lutterworth, in the District of Harborough. The site with which we are concerned was a 35-acre portion of that land lying to the north of Lutterworth and immediately adjacent to a developed area of that town, a large portion of which in fact was a previously developed estate of the applicants. The applicants also own a still further area of 7 acres of land which was not included in the application. In 1972, despite objection by the local planning authority, the Secretary of State decided that 25 acres of the 35 now applied for were suitable for development if two problems could be overcome. The first was surface water disposal and the second was access to Bitterswell Road. He said, as I understand it, that the access problem alone would not have been sufficient to prevent planning permission being granted. After that appeal, and encouraged by the Secretary of State's decision, the applicants purchased some further land, which enabled, in their view, the access problem to be overcome, and that land was included in the current application. But the 35 acres included further land, beyond the original 25 and not included in the access land, which extended into open meadow with no particular natural boundaries. The application was made on the appropriate form on April 3 1978. After giving the address of the site and identifying the 35 acres on a plan the applicants went on to answer an invitation on the form to state the number of dwelling units proposed by filling in "approximately 420 dwellings". It is pointed out that this is the mathematical result of taking a density of 12 houses to the acre over the whole of the 35-acre site. That application was refused by the second respondents, the local planning authority, on July 12 for a variety of reasons including amenity, population, access, traffic and surface water disposal. The applicants appealed to the Secretary of State on December 22 1978, and a public inquiry was held on January 22 to 24 1980. On January 4, that is less than three weeks before the inquiry was due to be held, the applicants wrote to the local planning authority indicating that they were proposing to put forward another proposal and submitted what they described as a schematic layout showing about 250 dwellings on a reduced area of 25 acres. The letter emphasised that the applicants "would wish the schematic layout to be considered as a viable alternative proposal to the application as originally submitted only if the issue of scale of development is deemed to be critical to the determination of the appeal and without prejudice to the proposals contained in the original application".

On January 11 the planning authority wrote back: "My council is of the opinion that this is a new application and should be considered in the normal way, that is determined by the council after consultation with interested parties", etc. At the inquiry the applicants called a planning consultant who said that he could not support the development of 420 units on the 35-acre site, and he produced three alternative plans. Two of them provided for 250 dwellings on 25 acres and differed only in their proposals for access and internal roads. The third included another 6 acres, making 31 in all, and provided for 330 to 350 dwellings. The planning authority's case was almost wholly concerned to argue that any development on this site would have undesirable consequences, although it is clear that the impact of the development reduced to 250 houses had been examined by the traffic experts of the county council who appear to have given evidence that even this reduced number was unacceptable on traffic grounds. The planning authority maintained at the inquiry that the Secretary of State could not legally reduce the area of the appeal site by 10 acres, and that he only had power to deal with the application as submitted.

It was accepted that the surface water objection could be adequately resolved by using a balancing reservoir scheme, and this reason for refusal was abandoned.

Various other parties appeared at the inquiry. A fair reading of their evidence and arguments recorded in the inspector's report is that they objected to any development on the site. One of them clearly stated that even 250 houses would be objectionable. The inspector reported on March 6 1980. It is unnecessary to refer to his report other than to summarise his conclusions and recommendations. His conclusions were, first, that it was a legal matter for the Secretary of State to determine whether it was possible to restrict any planning permission granted on that appeal to an area smaller than 35 acres and to fewer than 420 dwellings. Secondly, if the appeal was restricted to consideration of 420 dwellings on 35 acres, he felt that it should be dismissed. And thirdly, if it was permissible to restrict the area to 25 acres and for the number of dwellings to be reduced, then such development would not be objectionable. He recommended that, on the assumption that there was no legal bar to such action, permission should be granted for the erection of dwellings on 25 acres at a density of 10 to the acre.

The Secretary of State gave his decision by a letter dated April 24 1980. After setting out the inspector's conclusions and recommendations he went on in paragraphs 4 and 5:

> 4. Having regard to the inspector's conclusions concerning a smaller site than that proposed in the application under appeal, while it is accepted that there are circumstances where a split decision would be appropriate, the opinion is held that where an appeal results from an application for permission to erect a specified number of dwellings without any indication at all of their sizes or of the individual plots, the proposed development is not severable and it would be improper to purport to grant permission in respect of part of the site or for a lesser number of houses. In this particular case it must be noted that although plans D, E and F illustrate a possible layout and a reduced approximately 25-acre area of the appeal site for about 250 dwellings which your clients agree would be an acceptable alternative development, it was clearly indicated at the inquiry that these plans, which were submitted after the appeal had been made, were not provided as replacements for the original appeal proposals. Consequently the view is held that it would not be appropriate for the appeal proposal to be severed or reduced, and the Secretary of State has therefore considered the appeal on the basis of the original application before him.
>
> 5. The Secretary of State agrees with the inspector's conclusions regarding the proposal on this appeal and concurs with his opinion that the appeal should be dismissed. Any proposal for a smaller development would have to be the subject of a further application which would lead to consideration by the local planning authority in the first instance. In the circumstances the Secretary of State does not propose to comment on any of the inspector's conclusions regarding a reduced development. For the reasons given he does not accept the inspector's recommendations and thereby dismisses the appeal.

The real question in this case is whether the Secretary of State was right in considering that he had no power to grant planning permission for development on a smaller site and with houses at a lower density than were indicated on the application form originally submitted to the local planning authority.

Mr Sullivan, however, had an argument that on a true reading of the decision the Secretary of State was in fact exercising his planning discretion. It will be convenient to deal with this argument first. The inspector in his conclusions and recommendations clearly poses a legal question. I have no doubt that in paragraph 4 the Secretary of State was attempting to answer it. When he uses the term "improper" in the first sentence of this paragraph, he refers, it seems to me, to an improper, that is an illegal, use of powers. This first sentence sets out in general terms the legal proposition to which the Secretary of State commits himself. One can expand it in the context of the appeal in this way. If the application indicates a number of sites for development, each with a single house, then it can be severed by as it were lopping off individual sites. In such a way permission can be granted for a reduced area or for a lesser number of houses. But if all you have is an area covered by the application and a number of houses proposed to be built on it, such severance is impossible and therefore reduction in the area or the number of houses is improper, because no power is given to achieve a reduction by this means. Put simply, the Secretary of State is saying, "The only way in which I can properly exercise my powers and achieve a reduction in the area or the number of houses is if the application can be regarded as severable. If it cannot be so regarded I have no power to achieve this end." The second sentence in paragraph 4 does no more than set out those circumstances in the current appeal which led the Secretary of State to say that despite the other proposals put forward what he is dealing with is a non-severable application. The last sentence is the conclusion to the other two. The three sentences of this paragraph, properly read, amount to my mind to a logically unimpeachable syllogism: only severable applications can result in planning permission for a reduced area: this is not a severable application: therefore it cannot result in planning permission for a reduced area. Like all syllogisms the conclusion is valid only if the premises are sound. It remains to be seen whether the major premise here is a valid statement of the law.

The powers of the Secretary of State are derived from section 36(3) of the Town and Country Planning Act 1971. They are well known but I should refer to them.

> Where an appeal is brought under this section from a decision of a local planning authority, the Secretary of State, subject to the following provisions of this section, may allow or dismiss the appeal, or may reverse or vary any part of the decision of the local planning authority, whether the appeal relates to that part thereof or not, and may deal with the application as if it had been made to him in the first instance.

What can be done when the application is made in the first instance is to be found in section 29(1):

> Subject to the provisions of sections 26 to 28 of this Act, and to the following provisions of this Act, where an application is made to a local planning authority for planning permission, that authority, in dealing with the application, shall have regard to the provisions of the development plan, so far as material to the application, and to any other material considerations, and

subject to certain sections of this Act

> may grant planning permission, either unconditionally or subject to such conditions as they think fit.

At this point I can I think go straight to the judgment of Lord Widgery CJ in *Kingston-Upon-Thames Royal London Borough Council v Secretary of State for the Environment* [1973] 1 WLR 1549, the passage I wish to refer to being at p 1552:

> . . . one has got to look at the learning on the question of what conditions can properly be attached to planning permissions. The attachment of conditions to planning permissions is as old as the planning legislation itself, and is now to be found in section 30(1) of the Town and Country Planning Act 1971: "Without prejudice to the generality of section 29(1) of this Act, conditions may be imposed on the grant of planning permission thereunder — (a) for regulating the development or use of any land under the control of the applicant (whether or not it is land in respect of which the application was made) or requiring the carrying out of works on any such land, so far as appears to the local planning authority to be expedient for the purposes of or in connection with the development authorised by the permission. . . ." Those are wide words; they clearly on their face entitle the local planning authority to impose conditions which affect land not the subject of the application itself, and which go to the restriction of the past user or the removal of existing works. Although they are wide it has been recognised for a very long time that they are subject to certain restrictions. The two principal restrictions which the courts have placed on those words are first that a condition is invalid as being contrary to law unless it is reasonably related to the development in the planning permission which has been granted. It must not be used for an ulterior purpose, and must, in the well-known words of Lord Denning MR in *Pyx Granite Co Ltd v Minister of Housing and Local Government* [1958] 1 QB 554, 572, "fairly and reasonably relate to the permitted development". The second restriction on those words which the courts have adopted in recent years is that a condition which is so clearly unreasonable that no reasonable planning authority could have imposed it may be regarded as *ultra vires* and contrary to law and treated as such in proceedings in this court. But as far as I know those are the only two general limitations on the wide powers in section 30 of the Town and Country Planning Act 1971.

Mr Sullivan initially argued that the Secretary of State was right and that severability was the only test. In his subsequent submissions, however, he seemed to have abandoned that stance, because they proceed on the basis that the proper test is whether the development permitted is in substance different from that applied for. The extent to which this latter formulation is incompatible with the former I shall deal with in a moment. Although therefore Mr Harper and Mr Sullivan put forward a number of propositions, in the end I do not think they differ markedly from each other on the essential principles governing the question of when conditions can be regarded as *intra vires*. Both I think accept as a starting point the passage in Lord Widgery's judgment which I have just quoted. In the context of that passage the question here is whether it is permissible to grant a planning permission subject to a condition that only what I might call a "reduced development" is carried out. Both counsel, I think, accept that it is permissible to grant planning permission subject to such a condition, and both I think would seek to limit such conditions to those which do not alter the substance of the application: and both agree that in considering whether it is right to grant planning permission subject to such a condition the planning authority should, among other things, have regard to one of the underlying purposes of Part III of the Act of 1971, which is to ensure that before planning permission is granted there should be adequate consultation with the appropriate authorities and a proper opportunity for public comment and participation. The broad proposition therefore, as I see it, to which both counsel would give assent is that a condition, the effect of which is to allow the development but which amounts to a reduction on that proposed in the application, can legitimately be imposed so long as it does not alter the substance of the development for which permission was applied. If it does alter the substance, so the argument goes on, it cannot legitimately be imposed, because there has been no opportunity for consultation and so on about what would be a substantially different proposal. Parliament cannot have intended conditional planning permission to be used to circumvent the provisions for consultation and public participation contained in this part of the Act.

Now the test of substantial difference is not at all the same thing as the test of severability. It is possible to imagine an application for two related developments on the same piece of land, say a major and a minor development, which is clearly severable into these two portions. To give planning permission subject to a condition that the minor development were not carried out might well not alter the substance of the application. On the other hand, if the condition prevented the major development being carried out, that might well amount to a permission substantially different from the application. Thus the application of the severability test alone could result in planning permission being given for development which is substantially different from that applied for. The proposition that conditions can only be used to reduce the development below that proposed in the application where the application is severable is derived from a decision of Sir Douglas Frank QC, sitting as a deputy High Court judge in a case called *Kent County Council v Secretary of State for the Environment* (1976) 33 P & CR 70. That decision itself clearly arose from the argument put forward by counsel for the Secretary of State, which was in these terms and recorded in the judgment at p 75:

> 1. Where an application contained a number of separate and divisible elements it was lawful for them to be separately dealt with. 2. Alternatively, if the elements were not divisible there was power to modify the application providing that (a) the scope of the development was not enlarged; (b) the essential nature of the development was not altered; and (c) any persons affected were given a chance to make representations.

It can be seen that the second alternative formulation looks remarkably like the proposition to which I have just referred and to which both counsel would assent. In giving judgment, Sir Douglas Frank acceded to the first part of this argument and presumably thought it in consequence unnecessary to deal with the second. The Secretary of State in the case I am dealing with has clearly directed himself that it is only if the application is severable that he can by condition reduce the ambit of the planning permission granted. He has had no regard to the question whether the planning permission if granted subject to a condition would be substantially different from that applied for.

For my part I cannot accept that the proper test is whether the development proposed in the application was severable or not. Unless coupled with a requirement that the result must not be substantially different from the development applied for, it would be possible, as I have just indicated, for local planning authorities to grant planning permission for developments which were in fact substantially different and thus defeat the consultative objects of Part III of the 1971 Act. The severability test, therefore, could only be a proper one if combined with a test of substantial difference. But I can see no justification for the severability test at all. It should be remembered that we are dealing here with applications for outline planning permission. Many of these applications are no

A doubt for multiple purposes, some of them severable, some of them perhaps not. But many applications, as here, are for single purposes, for instance residential development. Why should it be impossible for the local planning authority to say on an application for outline planning permission: we think 35 acres is too much but 25 will be all right; and similarly with a reduction in density? As long as the reduction passes the test of not altering the substance of the application what vice is there in that? It is clearly a condition fairly and reasonably related to the permitted development (see *Pyx Granite, supra*) and it is not unreasonable under the *Wednesbury* doctrine. To give permission for a substantially different development would, on the other hand, be unreasonable as that
B word is understood in these cases (see, for instance, a passage from the judgment of Diplock LJ, as he then was, in *Mixnam's Properties v Chertsey Urban District Council* [1964] 1 QB 214 at p 237) because it would not be what Parliament intended for a consultation process to comprehend. The test of substantial difference is thus firmly based on the broad principles of *Wednesbury*. The severability test, on the other hand, seems to me to have no particular validity. To grant a planning permission for part only of an application which was not severable does not appear, merely on that fact, to run counter to either of the two general limitations referred to by Lord Widgery in the *Kingston* case. Perhaps the argument on severability put forward by the Secretary of State in
C the *Kent County Council* case and accepted by Sir Douglas Frank had its origin in the fact that the application in that case clearly was severable. But that does not seem to me to justify its elevation into a matter of general principle.

I conclude for my part that there is no principle of law which prevents the Secretary of State from imposing conditions which have the effect of reducing the permitted development below the development applied for except where the application is severable. The Secretary of State clearly directed himself that there was such a principle and thus fell into error, and his decision must be quashed.

I should add a rider. The true test is, I feel sure, that accepted by both counsel: is the effect of the conditional planning permission to
D allow development which is in substance not that which was applied for? Of course in deciding whether or not there is a substantial difference the local planning authority or the Secretary of State will be exercising a judgment and a judgment with which the courts will not ordinarily interfere unless it is manifestly unreasonably exercised. The main, but not the only, criterion upon which that judgment should be exercised is whether the development is so changed that to grant it would be to deprive those who should have been consulted on the changed development of the opportunity of such consultation. And I use these words to cover all the matters of this kind with which Part III of the Act deals.

There may of course be in addition purely planning reasons for
E concluding that a change makes a substantial difference; but I find it difficult to believe that where a proposed development has been the subject of such consultation and has produced a root and branch opposition to any development at all, whether larger or smaller, it should be necessary in all cases to go again through the process of consultation about the smaller development. It is clear that in this case the processes of consultation had resulted in such root and branch opposition and further consultation could not result in more opposition but only, if there was any change in public attitudes, in less. In those circumstances Mr Harper invites me to say that only an unreasonable Secretary of State could conclude that the course recommended by the inspector would result in a development
F substantially different from the application. In consequence, he says I should make an order, the effect of which would be to substitute for the dismissal of his client's appeal planning permission as recommended by the inspector. But as I understand it all I have power to do under section 245 of the 1971 Act is to quash the order, and that is all in fact that Mr Harper's notice of motion asks me to do. The court cannot grant planning permission. I must decline his invitation and merely order that the Secretary of State's decision should be quashed.

I might add that I have come to my general conclusion with a certain feeling of satisfaction, as it seems to me to permit a welcome degree of flexibility in the conduct of planning applications and appeals while at the same time maintaining adequate safeguards for G
the interests of those in whose favour the provisions for consultation were enacted.

The Secretary of State's order was quashed, the Secretary of State to pay such costs as would ordinarily have been incurred by the applicants if the Secretary of State had submitted to judgment, the balance of costs to be borne by the district council.

QUEEN'S BENCH DIVISION

October 16 1980 H

(Before Mr Justice FORBES)

GRANADA THEATRES LTD v SECRETARY OF STATE FOR THE ENVIRONMENT AND ANOTHER

Estates Gazette March 21 1981
(1980) 257 EG 1154

Town and Country Planning Act 1971 — Application to quash decision of Secretary of State dismissing appeal against refusal of planning authority to grant permission for change of use of a cinema to a bingo social club — Fate of last cinema in Chichester — Decline J
in cinema attendances, although a significant upturn in 1978 — Owing to unprofitability of cinema use owners wished to change to a "top-class social club premises based on cash bingo" — Local public pressure for cinema use to be continued, the nearest other cinema being at Bognor Regis — Relevance of desirability of preserving existing use when there was no possibility of a refusal of permission resulting in a continuation of that use in its present form — Secretary of State accepted that continuation of use of whole building as a cinema was not commercially viable, but was not satisfied that the applicants had exhausted the possibilities of retaining a cinema facility in the building combined with some other use — Applying the principles stated in Associated Provincial Picture Houses Ltd v K
Wednesbury Corporation, the judge held that the Secretary of State had not taken into account matters which he should not have taken into account, or the converse, and that he had not come to a conclusion at which no reasonable authority could have arrived — Application to quash accordingly dismissed

This was an application by Granada Theatres Ltd to quash a decision of the Secretary of State for the Environment whereby he dismissed the applicants' appeal against the refusal of Chichester District Council to grant planning permission for the change of use of the Granada Theatre, East Street, Chichester, from a cinema to a bingo social club. L

M S Rich QC and J F Mummery (instructed by Turner, Peacock) appeared on behalf of the applicants; Simon Brown (instructed by the Treasury Solicitor) appeared on behalf of the first respondent, the Secretary of State; the second respondents, Chichester District Council, were not represented and took no part in the proceedings.

Giving judgment, FORBES J said: In this case Mr Rich moves to quash a decision of the Secretary of State dated November 30 1979 whereby he dismissed an appeal by the applicants against a refusal of the Chichester District Council to grant planning permission for change of use of the Granada Theatre, East Street, Chichester, from a cinema to a bingo social club. Although at the start of the argument this appeared to be a matter of some complexity, it has M
now been resolved through the efficiency of counsel into what is really a very short point.

It is necessary nevertheless to set the scene. The applicants, a well-known concern, own the Granada Theatre in East Street, Chichester. The building was erected in 1832 as a Corn Exchange. It is a two-storey building and it fronts on to East Street, Chichester. I take its description from the inspector's report.

> The stepped front entrance from East Street with two pairs of doors is set back 17 to 20 ft from the carriageway and the wide footpath is spanned by an entablature supported by six Doric columns close to the kerb. The premises are used as the Granada Cinema with 820 seats in an auditorium of some 4,600 sq ft and a balcony of 820 sq ft. It is a Grade II starred listing building.

A Now cinema-going has declined in the last year or so, and the Granada, Chichester, is now the second most unprofitable cinema in the Granada group. I take the further facts again from the inspector's findings of fact.

> The appellants [that is the present applicants] decided in early 1977 to convert the enterprise to a bingo social club. Gaming Board consent was obtained in January 1978 and a bingo licence in May 1978. The popularity of cinema-going has declined enormously in the last 30 years, but there was a significant increase in 1978. Twenty-six Granada cinemas were closed in the last 10 years; two other cinemas in Chichester were closed by the end of 1960. Admissions to the Granada cinema were approximately halved between 1965 and 1976 since when they have improved considerably due to B increased performances and attractive films. In 1976 admissions were equivalent to five visits by each person in Chichester.

Pausing there, that is a mathematical computation, the validity of which I am not certain about, but the facts were that there were just over 100,000 admissions and the population of Chichester was 20,000 persons. The finding goes on:

> For the last four years under 20 per cent of the seats have been filled on average. The possibilities of better attendances are limited. Operating costs of the cinema have increased greatly and profitability is critically low. Admissions might need to double for the cinema to pay its way. The appellant intends to operate a top-class social club premises based on cash C bingo.

The applicants applied for planning permission for this change of use and were refused by the local planning authority. They appealed and a public inquiry was held on February 6 and 7 1979. The inspector's report is dated February 23 1979 and the Secretary of State's decision letter November 30 1979. One of the points raised by the applicants at the inquiry was the question of the desirability of preserving the existing use. As I understand what was their argument there, as it is Mr Rich's argument now, it was that the desirability of preserving the existing use is not a material consideration in a planning application such as this where, it is said, D there is no possibility that refusal of planning permission would result in continuance of the existing use. The authority for this proposition is said to be the case of *Clyde & Co v Secretary of State for the Environment* [1977] 1 WLR 926. It was argued at the inquiry that there was no such possibility and that therefore any question of preserving the existing use was immaterial. The inspector recorded this as a question of legal interpretation. He put it in this way in his findings of fact: "Legal representations have been made as to whether the desirability of preserving the possibility of an existing use would be material." In these circumstances his conclusions proceeded on alternative hypotheses as did his recommendations. I shall read those.

E

> 74. It is for the Secretary of State to consider whether the desirability of retaining the existing use of the appeal premises as a cinema is material or not (paragraphs 25 and 50 referred). In the former case the remarkable volume of public protest against loss of the cinema mainly voiced through a petition organised and heavily supported by young people, the wide appeal of cinema to all kinds of people and the comparative lack of other easily accessible forms of public entertainment in the area combine to suggest that it is very desirable to retain the existing use for social reasons. The poor profitability of the cinema in recent years implies that it may be too large as at present organised, and although the appellant now sees commercial advantage only in converting the premises to a bingo social club at considerable expense, it may be that the discernible increase in popularity of cinema going or further examination of alternative ways of using the building could result in the retention of a viable cinema facility. In these circumstances it F would be wrong to allow the appeal and so lose the existing use. 75. If, on the other hand, the desirability of retaining the existing use were not material, the situation is quite different. Preservation of the special quality of the premises themselves as a listed building and of their historic surrounds in an outstanding conservation area must be a matter of the greatest importance, but there is no conclusive evidence that they would be materially harmed by the proposed change of use providing that advertising on the outside of the building were to be restricted as proposed by the appellant. Changed circumstances could lead to attempts to introduce more eye-catching advertising of a kind associated elsewhere with bingo clubs which could lead to most undesirable loss of visual amenity, perhaps for a considerable time until control could be re-established. But if planning permission were made personal to the appellant, as invited, the danger of unacceptable harm to the amenity and character of the area could not be so conclusive by itself to justify dismissal of the appeal.

And his recommendations in 76: G

> Depending upon what decision is taken upon the legal point at issue, I recommend either that the appeal be dismissed or that it be allowed subject to a condition that the permitted use of the premises as a bingo social club should inure only for the benefit of Granada Theatres Ltd and not for the benefit of the land.

The Secretary of State's decision letter sets out those conclusions and refers to the inspector's recommendations and continues, so far as is material, in paragraphs 5 and 6 which again I think I should read.

> 5. The inspector put forward alternative conclusions and recommendations depending on whether or not the Secretary of State took the view that the H desirability of retaining the existing use of the appeal premises as a cinema is material to the decision on the appeal. The Secretary of State has examined the legal submissions which have been made on this issue. He does not regard as material the alleged demerits such as they may be (he holds no view on the point) of bingo as an activity, and he has not been influenced by unfavourable comments made about it in connection with the appeal. He does, however, consider that public demand for the retention of a cinema facility in Chichester on the one hand and for the introduction of commercial bingo on the other as well as the availability of alternative suitable premises and ways and means of providing these facilities are all material considerations in relation to the present appeal. 6. He notes that the appeal building is the last cinema in Chichester, that the next nearest is in Bognor 6 miles away, and that petitions and correspondence received show there to be massive public pressure for the appeal building to continue as a cinema and point to its being much valued as a social amenity catering for all age J groups including the young. He accepts that notwithstanding this expression of public opinion in relation to the present appeal, the appellants face difficulties in running the cinema as it stands, but they have already given some thought to the possibility of finding an alternative use which would still incorporate a cinema element, and that the proposed change of use to bingo is likely to be much more profitable than the present use. However, it may well be that a cinema and bingo operation or a multi-unit cinema are not the only possibilities and, despite the appellant's declared intention of closing the present cinema soon, the Secretary of State cannot rule out the hope that as the inspector suggested an increased interest in the cinema or further examination of alternative ways of using the building could yet result in the retention of a viable cinema facility.

Mr Rich's first argument is, as already stated, that the desirability of preserving the existing use is not a material consideration where K there is no possibility that refusal of planning permission would result in the continuation of the existing use. He says that the inspector in paragraph 74 misunderstood this argument and thought it was an argument simply that the desirability of retaining existing use was not material, that he therefore asked himself the wrong question and never considered whether it was possible or not that refusal of planning permission would result in such retention. I do not think that the inspector so misdirected himself. Both in his recording of the appellant's argument at paragraph 25 of his report and in his finding of fact which I have already read he refers to the desirability of preserving the possibility of an existing use. It is true that in the first sentence of paragraph 74 he omits the words "the L possibility", but this I apprehend is mere shorthand for a concept he has correctly recorded elsewhere. The tenor of the rest of that paragraph and the use of the expression "it may be" at the end of it clearly show that it was the possibility that he had in mind.

The case of *Clyde & Co v Secretary of State for the Environment*, to which I have already referred, has been relied on by both counsel. I confess that I was surprised that it should have been necessary for there to be an authority for the proposition that in a change of use case the desirability of preserving the existing use was a material consideration, because it seemed to be so self-evident a proposition. Properly looked at, however, I think that at first instance M Willis J was really dealing with a different argument, as it seems the case was being put on the basis that the Secretary of State was trying, by an improper use of planning powers, to force an owner to retain an existing use against his will. Sir David Cairns in the Court of Appeal was indicating that there was no doubt that the desirability of preserving the existing use was a material consideration, and that in the case before him there was ample evidence that refusal of planning permission might well result in such retention. Read in this way there is nothing in the *Clyde & Co* case to suggest that I am going too far in considering that the desirability of preserving the existing use must always be a material consideration when deciding whether to grant planning permission for a change of use. Of course there

A may be facts which make it unnecessary to consider this question very deeply or beyond a certain point. If all the parties agree that the continuation of an existing use is undesirable, one need not consider the question of desirability further. It is a concluded question. It is nevertheless a material question, but one which has been easily answered. If both parties agree that though desirable it is impossible to preserve the existing use, again it is a concluded question. But wherever there is a dispute between the parties as to whether an existing use should be retained or not, it seems to me inevitable that the desirability of retaining it is a material question. That is why I prefer to see Mr Rich's main proposition (and it is Mr Brown's main proposition also) restated in this way: the desirability B of retaining an existing use is always a material consideration where planning permission is sought to change it; there may, however, be facts which make it unnecessary to consider it, as where there is no possibility that refusal of planning permission will result in its retention. Looked at in this way the sole relevant issue which the Secretary of State had to decide was whether there was a possibility, a reasonable possibility perhaps, that the existing use would be preserved if planning permission were refused. And the sole issue which I have to decide is whether in answering that question it was reasonable, in the context of *Associated Provincial Picture Houses Ltd v Wednesbury Corporation* [1948] 1KB 223, for the Secretary of State to come to the conclusion on the evidence before him, that C for either of the reasons he advanced there was a reasonable possibility that a cinema facility might be retained. Mr Rich says no, there was no material before him on which he could reasonably have come to that conclusion, and Mr Brown says yes, there was.

It is perhaps useful to summarise what, as I understand it, the Secretary of State's letter was saying before dealing with the possibility of retaining the cinema use. It seems clear to me that he was saying that for the purposes of this decision he accepted the appellants' contention that the use of the whole building as a single cinema was not commercially viable and therefore not likely to continue. He also accepted that the applicants had investigated two other methods of preserving the cinema use; one was combining the D cinema with bingo and the other was turning the building into a multi-unit cinema; and that they had properly concluded that those alternatives were unworkable. The vital sentence then follows:

> However, it may well be that a cinema and bingo operation or a multi-unit cinema are not the only possibilities, and despite the appellant's declared intention of closing the present cinema soon, the Secretary of State cannot rule out the hope that as the inspector suggested an increased interest in the cinema or further examination of alternative ways of using the building could yet result in the retention of a viable cinema facility.

The question then as I indicated earlier is simply: could the Secretary of State reasonably have come to that conclusion on such E material as he had? What material did he have? As to the first possibility, an increased interest in the cinema, he had the inspector's findings of fact in the body of his report, and I refer to G and H. I have read them already but I shall refer to them again.

> The appellant decided in early 1977 to convert the enterprise to a bingo social club. The popularity of cinema going has declined enormously in the last 30 years but there was a significant increase in 1978.

In other words, the applicants' decision to abandon the cinema preceded the upturn in the popularity of cinema-going. As to the second possibility the evidence appears to have been this. The applicants had investigated the two possibilities already mentioned. One of their witnesses had conceded that the building could be put F to other uses than a bingo social club (that is paragraph 49 of the report). He had apparently also conceded that they were considering offers from firms wanting to use it for other purposes, even punk rock (that is paragraph 66E). On this the inspector had found as a fact that the possibility of other uses exist that have not been fully explored. Some uses have disadvantages (that is finding of fact T). There appears to be no particular evidence that any of the possible uses which had not been fully explored included dual or multiple use which would feature a cinema facility. As the Secretary of State had accepted that use as a single cinema was not a viable possibility, it follows that to be in any way material to the question of retaining the existing use, any possible use would both have to include a cinema and be a dual or multiple use. There was thus no direct G evidence before the inspector or the Secretary of State that there existed any alternative viable and relevant use.

That being the state of the evidence, was it reasonable for the Secretary of State to come to the conclusion that there was still a possibility that the existing use might be retained. He expressed this as a hope, but it is clear that it was the possibility that he was considering. Now in considering whether the Secretary of State has acted reasonably in this way it is important not to fall into the trap of equating the reasonableness of his decision with the reasonableness of that well-known common law figure the man on the Clapham omnibus. The test is not whether the Secretary of State has acted reasonably in that sense at all. And perhaps it is well to remind H oneself of the *ipsissima verba* of Lord Greene MR in the *Wednesbury* case. It is *Associated Provincial Picture Houses Ltd v Wednesbury Corporation* [1948] 1 KB 223. The passage I have in mind is in the judgment of Lord Greene at p 233:

> The court is entitled to investigate the action of the local authority with a view to seeing whether they have taken into account matters which they ought not to take into account, or, conversely, have refused to take into account or neglected to take into account matters which they ought to take into account. Once that question is answered in favour of the local authority, it may be still possible to say that, although the local authority have kept within the four corners of the matters which they ought to consider, they have nevertheless come to a conclusion so unreasonable that no reasonable J authority could ever have come to it. In such a case, again, I think the court can interfere.

A further explanation of this kind of reasonableness is to be found in a passage from the judgment of Diplock LJ, as he then was, in the case of *Mixnam's Properties Ltd v Chertsey Urban District Council* [1964] 1 QB 214 CA at p 237. It is very short.

Thus, he says, and he is, of course, dealing with byelaws:

> the kind of unreasonableness which invalidates a bye-law is not the antonym of "reasonableness" in the sense in which that expression is used in the common law, but such manifest arbitrariness, injustice or partiality that a court would say: "Parliament never intended to give authority to make such K rules; they are unreasonable and ultra vires".

It seems to me that one can paraphrase what the Secretary of State is saying in his decision letter like this: "I consider that it is important to retain the existing use as a cinema. I know that you decided it was wholly unprofitable to continue, but that was before the upturn in cinema attendance. I am not satisfied that that was a valid decision and you should look at it again. Similarly I know that you have explored two possible multi-uses each involving the retention of a cinema facility. But there are other uses which you have not explored, and I am not satisfied that you have exhausted the possibilities. In those circumstances I find myself unconvinced that the retention of the existing use is out of the question". There are only L two reasons for a court to overturn a decision based on such a chain of reasoning: that the Secretary of State took into account matters which ought not to have been taken into account or the converse on the one hand, and unreasonableness on the other. The only matter which Mr Rich urges as something which the Secretary of State should not have taken into account is the possibility of retaining the existing use, because he says there was realistically no such possibility. But this question is itself entirely dependent on the other question: could the Secretary of State reasonably, as defined in *Wednesbury*, have come to the conclusion that there was a realistic possibility that the use might be retained? There is therefore only one question: did the Secretary of State act reasonably? I find M myself quite unable to say that the conclusion to which the Secretary of State came on the material before him was so unreasonable that no reasonable authority could ever have come to it.

On that basis it seems to me that this appeal must fail.

The application was dismissed with costs.

A
QUEEN'S BENCH DIVISIONAL COURT

November 10 1980

(Before Lord Justice DONALDSON and Mr Justice HODGSON)

EMMA HOTELS LTD v SECRETARY OF STATE FOR THE ENVIRONMENT AND ANOTHER

Estates Gazette April 4 1981

(1980) 258 EG 64

**Town and Country Planning Act 1971, section 246 — Appeal against decision of Secretary of State for the Environment upholding
B enforcement notice served by planning authority — Question as to application of Class XI of the Schedule to the Town and Country Planning (Use Classes) Order 1972 — "Use as a boarding or guest house, or an hotel providing sleeping accommodation" — The planning unit in this case consisted of an hotel, formed out of three private houses, with drinking facilities in a bar open to non-residents — The authority served an enforcement notice requiring the hotel company to cease using the enlarged bar area for the sale and consumption of intoxicating liquor and other drinks to the general public — The Secretary of State on appeal held that the enforcement notice was right — The case came before the Divisional Court on two occasions, having been sent back on the first occasion to the Secretary of State for further consideration — The Secretary
C of State adhered to his original opinion on the ground that the use of the enlarged area as a non-residents' bar was not an incident of the hotel use, but, because of its scale, advertisement, manner of operation etc, was a separate use in breach of planning control — Held that the Secretary of State was in error as none of the features on which he relied showed that the use was not an incident of the hotel use — Appeal allowed and decision sent back a second time to the Secretary of State for further consideration**

This was the second appeal against a decision of the Secretary of State to uphold an enforcement notice served by the Southend-on-Sea Borough Council in respect of the Boston Hall Hotel on The
D Leas, Westcliff-on-Sea. The first appeal was reported at (1979) 250 EG 157.

C Cochrane (instructed by Kingsford, Dorman, agents for Drysdales & Janes, of Southend-on-Sea) appeared on behalf of the appellants; Simon Brown (instructed by the Treasury Solicitor) represented the Secretary of State; the borough council was not represented and took no part in the proceedings.

Giving judgment, DONALDSON LJ said: Emma Hotels appeal under section 246 of the Town and Country Planning Act 1971 against the decision of the Secretary of State for the Environment to uphold an enforcement notice served by the Southend-on-Sea Borough Council. It is the second decision by the Secretary of State
E to uphold the enforcement notice and it is the second appeal to this court.

The planning unit comprises what was once three private houses, namely, 25, 26 and 27 The Leas, on the sea-front at Westcliff-on-Sea. In 1969 permission was granted for the conversion of 25 and 26 into a private hotel and that produced the Boston Hall Hotel. In 1974 permission was obtained to incorporate 27 into the hotel as six bedrooms, a lounge and a bar. Shortly after that, permission was obtained to change the use of the bar in 27 into a use as a non-residents' bar. A question was raised as to whether that permission was necessary, but I do not think it matters now. The position, historically, is that from three private dwelling-houses these prem-
F ises have changed, bit by bit, into a larger hotel with drinking facilities for non-residents.

Following the enlargement of the bar area the appellants applied to the local licensing justices for a licence — I take it to have been a full publican's on-licence — and were refused it by the licensing justices but obtained it on appeal to the crown court.

The enforcement notice was dated June 3 1976 and it required Emma Hotels Ltd to cease using this enlarged area for the sale and consumption of intoxicating liquor and other drinks to the general public. There was a public inquiry and the Secretary of State upheld the enforcement notice, concluding that there had been a change from use as a private hotel into a composite use partly for the
G purposes of an hotel and partly for the purposes of a fully licensed non-residents' bar and lounge: in other words, something in the nature of a public house. There being a material change of use, he had to consider whether planning permission should be granted and he refused planning permission and upheld the enforcement notice.

That decision was appealed to this court which, on that occasion, consisted of Lord Widgery CJ, Bridge LJ and Caulfield J, and the judgment of the court was given by Bridge LJ on January 23 1979. He recited the facts of the matter and the decision of the Secretary of State that the premises were in composite use for the purposes of an hotel and a non-residents' lounge and bar. He went on to say:

> H
> At an early stage in the argument, Mr Cochrane for the appellants conceded, and in my judgment rightly conceded, that, viewed independently of any provision in the Use Classes Order to which I shall have to turn in a minute, it could properly be said that use of premises as a licensed public house was a materially different use from use of premises as a private residential hotel. That concession having been made, it would certainly seem to me that, if one had to decide this appeal independently of the Use Classes Order, one would be bound to say that there was material on which the Secretary of State could properly conclude that here there was a composite use of a single planning unit in relation to which a change in the proportion of the elements making up the composite use had taken place, which was capable of affording material on which the Secretary of State could conclude as a matter of fact and degree that a material change of use of the whole had taken place.
>
> However, the matter does not rest there, for Mr Cochrane relies on the
> J well known provisions of Article 3 of the Town and Country Planning (Use Classes) Order 1972, which it will be recalled provides: "Where a building or other land is used for a purpose of any class specified in the Schedule to this order, the use of such building or other land for any other purpose of the same class shall not be deemed for the purposes of the Act to involve development of the land."
>
> The class in question here is Class XI, which refers to: "use as a boarding or guest house, or an hotel providing sleeping accommodation."
>
> Before the extension of the non-residents' licensed bar area involved in which is now alleged to have been a material change of use for the purposes of planning law took place, there cannot be the slightest doubt that the whole unit, 25, 26 and 27, constituted an hotel providing sleeping accommodation, and was within that class.
>
> The question I ask myself is whether the fact that the permitted small area included in 27 for use as a non-residents' bar has been enlarged to take in the
> K area to which the enforcement notice relates for some reason takes the entire premises outside the ambit of Class XI.
>
> For my part, I am quite unable to see that it does. Mr Collins, who has said everything that could be said on behalf of the respondent Secretary of State, has drawn attention to the fact that in the definition of a "shop" there are specifically excluded from the definition, *inter alia*, hotels and premises licensed for the sale of intoxicating liquor, which are there dealt with as separate entities. That confirms the provisional view I expressed that they are indeed different uses. But it does not follow from that, in my judgment, that an hotel providing sleeping accommodation in any way ceases to come within Class XI if it includes *as one of the incidents of the hotel use* [my emphasis] use of part of the hotel premises as a bar open to non-residents. Both bars and dining rooms are common features of hotels providing sleeping accommodation. I suppose the vast majority of hotels up and down the country include such facilities, and in many, many cases those facilities
> L are open to the general public and not confined to use by persons who are residing in the hotels.

I interpret that passage as meaning that the fact that you find a public house use, that is to say the provision of premises to which non-residents can resort for the purposes of drinking and socialising, does not mean you are necessarily outside Class XI. It is a question of whether, to use his words, "it is one of the incidents of the hotel use".

He goes on in the judgment to say: "For those reasons I have come to the conclusion that the Secretary of State, who does not appear to have addressed his mind and his decision letter to the
M issue raised under Class XI of the Uses Classes Order, came to a conclusion which cannot be supported in law", and he set aside the decision and sent it back to the Secretary of State for further consideration. As I understand it, what the learned Lord Justice was doing was saying to the Secretary of State, "Please consider whether, on the facts of this case, the hotel use included, as one of its incidents, the provision of non-resident drinking facilities".

It went back to the Secretary of State and he has remained of the opinion that he originally held, namely, that the enforcement order was right and his reasons are contained in paragraph 7 of his decision letter dated October 19 1979. In fairness to him I had

A better quote it in full. After setting out the relevant parts of the judgment of the Divisional Court, he says:

The Secretary of State has given particular consideration to the issue raised in relation to Article 3 and Class XI of the Town and Country Planning (Use Classes) Order 1972 and fully accepts that the use of a part of an hotel which provides sleeping accommodation as a bar open to non-residents as an incident of the hotel use would not remove the hotel from Class XI. The issue in the present case appears to be whether the use of the enlarged area as a non-residents' bar is in fact an incident of the hotel use of 25, 26 and 27 The Leas. The facts have been fully reconsidered. It is noted that, on the estimation of your clients' managing director, not more than 20 per cent to 30 per cent of the persons using this bar are hotel guests. It is also noted that the bar is advertised and operated in a manner more akin to a public house than an hotel bar: there are external advertisements reading "open to the
B public" and "Oak Hall Free House" and there are signs on the front windows of the room which state that it is a free house and which advertise different brands of beer. Your clients' general manager stated that this part of the premises was run on the same lines as a public house. The Secretary of State concludes that, as a matter of fact and degree, the use of the larger area which is the subject of the enforcement notice, together with the permitted area, as a non-residents' bar for the sale of intoxicating liquor and other drinks to the general public and the consumption of intoxicating liquor and other drinks by the general public is, having regard to the scale of its use by non-residents of the hotel, the way in which it is advertised and operated, with its distinctive character and appearance, and the fact that it can be readily isolated physically from the rest of the building, not an incident of the hotel use, but a separate one. It is further concluded that the introduction of that use constituted a material change of use of the planning unit to a mixed use as an hotel providing sleeping accommodation and as premises
C for the sale of intoxicating liquor and other drinks to the general public. There was therefore a breach of planning control and the appeal fails on ground (b).

For my part, I think that the Secretary of State, in that paragraph, has isolated precisely the point which he was asked to consider by this court, and that he has considered it. But I am quite unable to see how, directing himself correctly, it would be possible for him to decide that this was not an incident of the hotel use. A non-residents' bar by its nature has to be run as a public house. There is no particularly different way in which you treat resident users of a bar except that you may, perhaps, allow them to charge their drinks to their hotel account. However, I have never understood that
D putting drinks "on the slate" was unknown in a public house, and that really is all that is happening here.

The fact that it is a free house may well attract customers in larger numbers than would be the case if it were a tied house, but it does not follow from that that it ceases to be an hotel or the attracting of non-residents involves a separate composite use within the same planning unit as something in the nature of a public house.

Unless you have a very large number of bedrooms, or a very large number of people sleeping in those bedrooms, and in addition residents who are prone to make use of the drinking facilities to a very large extent, I really do not see why the Secretary of State should be in the least bit surprised that only 20 per cent to 30 per
E cent of the users of the public bar were residents.

The other factors encouraging the general public to come in and drink seem to me all necessary incidents of having a non-resident bar. There is no point in having a non-residents' bar if you do not invite people to use it.

The fact that it can be readily isolated physically from the rest of the building does not seem to me to be a relevant factor unless steps are taken to isolate it. That would be a different matter altogether, but unless and until that is done I cannot see on what basis the Secretary of State reached the conclusion that this was not an incident of use, as had been held.

Mr Brown did say that there is no challenge in the notice of
F appeal on *Wednesbury* principles. I do not, for my part, think it matters, since this point was fully argued, but if for the purposes of any subsequent proceedings he would rather have the notice of appeal amended, I have always personally taken the view that much the best time to amend a notice of appeal, or any pleadings, come to that, is when the case is concluded. For my part, I should be entirely happy to give leave for any necessary amendments to be made.

For those reasons, I fear that the Secretary of State has erred and I would once again set aside his determination and return it.

HODGSON J agreed.

The appeal was allowed with costs. Leave was given to appeal to the Court of Appeal.

QUEEN'S BENCH DIVISION G

July 15 1980

(Before Sir Douglas FRANK QC, sitting as a deputy judge of the division)

THORNVILLE PROPERTIES LTD v SECRETARY OF STATE FOR THE ENVIRONMENT AND ANOTHER

Estates Gazette April 11 1981
(1980) 258 EG 172

Town and Country Planning Act 1971 — Application under section 245 to quash decision of Secretary of State refusing permission for residential development — Refusal on the ground that the development would prejudice the preparation of the local plan — In fact the only real issue was when the development should begin — Duty of Secretary of State to give intelligible reasons, grappling with the issues raised — Secretary of State should have indicated what was in the local plan which might vitiate the conclusion of the inspector (rejected by the Secretary of State) and why — Secretary of State failed to give adequate reasons — Decision quashed H

This was an application by Thornville Properties Ltd under section 245 of the Town and Country Planning Act 1971 to quash the decision of the Secretary of State refusing, contrary to the J recommendation of the inspector, to allow residential development on an area of 6.4 acres at Gnosall, Staffs.

Michael Howard (instructed by Berrymans) appeared on behalf of the applicants; J G Laws (instructed by the Treasury Solicitor) represented the Secretary of State; the second respondents, Stafford Borough Council, were not represented and took no part in the proceedings.

Giving judgment, SIR DOUGLAS FRANK said: This is an application under section 245 of the Town and Country Planning Act 1971 to quash a decision by the Secretary of State refusing planning permission for the development of land at Gnosall for residential development. The application as originally made was for 9.52 K acres, but by the time the matter came before the Secretary of State's inspector it had, by consent, been reduced to an area of 6.4 acres. There were originally three grounds for refusal, but the only one relied upon by the council was that "the development of this site outside the village plan area would prejudice the preparation of the local plan for Gnosall and its surrounding area".

The inquiry was held by the inspector on November 9 1978, by which time the 6.4 acres had become included in the draft local plan. In reality the only question at issue at the inquiry was not whether the site should be allocated for residential development in the local plan but whether the development should begin no earlier L than 1986. The inspector in his conclusions found that the proposal would be an acceptable rounding-off of the existing development and that there were no valid objections to it. He did not deal expressly with the question of postponement to 1986 but necessarily implied that the case for it had not been made out. Accordingly he recommended that permission be granted. The material part of the letter conveying the decision of the Secretary of State is as follows:

The Secretary of State sees no reason to disagree with the inspector's conclusions relating to the suitability of the appeal site for residential development. However, he takes the view that the inspector has given insufficient consideration to the effect which the granting of planning per- M mission now might have on the preparation of the district plan. The district council have carried out public participation on the draft plan and have indicated their preference for the appeal site for future residential development. Nonetheless, the draft plan has to go through further statutory stages before a final plan is adopted; and modifications to the existing draft plan are possible at any stage. (The district council's intention as a result of past public participation, to introduce a measure of phasing into the draft plan's residential proposals is an example of changes which can arise during the statutory stages of the plan's preparation.) In the Secretary of State's view to grant planning permission now for development of the appeal site could prejudice the remaining statutory processes in the adoption of a local plan for the area, in that it could well finalise the residential proposals in the plan

A at this early stage in its preparation. No sufficient reason is seen which would justify prejudicing the preparation of the local plan in this way. Consequently, the Secretary of State is unable to accept the inspector's recommendation that the appeal be allowed.

As Mr Howard for the applicants pointed out, there is only one matter which it was alleged could result from the further procedures on the local plan and that is the question of phasing, which the inspector impliedly has concluded was not a valid reason for the refusal of permission. The first question then is whether the Secretary of State gave adequate reasons for disagreeing with the inspector's recommendation. It cannot be doubted that the Secretary of State was entitled to have regard to the draft local plan but, B on the other hand, he was also bound to have regard to the other material considerations highlighted in the inspector's reasons for allowing the development. Taken literally, the Secretary of State's reasons for refusing permission would apply to every planning appeal because, as yet, no local plan has been approved. Therefore the fact that the local plan here had not been adopted cannot be conclusive as the Secretary of State had decided by necessary implication. I should say in passing that in my view the local plan was not in an early stage of its preparation but in an advanced stage. All that remained was the local inquiry prior to the final consideration by the council. I understand that the inquiry has since taken place. In *French Kier Developments Ltd v Secretary of State for the* C *Environment* (1977) 244 EG 967 at p 974 Forbes J said that it was clear to him that, on the cases, the Secretary of State had got to give intelligible reasons for his decision and had got to grapple with the issues which were raised and deal with them (see Lord Denning MR in *Earl of Iveagh v Minister of Housing and Local Government* [1964] 1 QB 395). If he did not deal with the points that arise, he failed in his duty and the court had to order him to make good the omission.

It may be that the decision would be unassailable, if this case had been the same as in *Price Brothers (Rode Heath) Ltd v Secretary of State for the Environment* (1978) 252 EG 595, where the case for development turned on a possible allocation of the site in a local D plan. However, in the instant case the inspector found, and the Secretary of State did not dissent, that the development was such as should be permitted, irrespective of a local plan. It seems to me, therefore, that if the Secretary of State is relying on the local plan, he is not grappling with the issues unless he specifies what could be in the local plan to nullify the assumption that permission should be granted.

Applying the tests which have been pronounced by successive Secretaries of State, including the present incumbent, permission should never be refused unless there are positive reasons against it. Therefore, even if the local plan had been approved and the development of the appeal site had been phased, it would have been E incumbent on the Secretary of State to show reasons why the phasing should prevail. In my judgment the Secretary of State's approach was wrong when he said: "No sufficient reason is seen which would justify prejudicing the preparation of the local plan in this way." He should have said what was in the local plan which might vitiate the conclusion come to by the inspector and, more particularly, why. He should have specified in what way the local plan would be prejudiced and in failing to do so he has not given adequate reasons for his decision. Accordingly, the decision must be quashed and the appeal redetermined.

Application to quash Secretary of State's order granted with costs.

F

G

QUEEN'S BENCH DIVISION

December 5 1980

(Before Mr Justice WOOLF)

SOUTH OXFORDSHIRE DISTRICT COUNCIL v SECRETARY OF STATE FOR THE ENVIRONMENT AND ANOTHER

Estates Gazette May 23 1981

(1980) 258 EG 753

Town and Country Planning Act 1971 — Application by planning authority to quash decision of Secretary of State — Secretary of H **State had quashed an enforcement notice relating to a temporary builder's hut and had granted planning permission for bungalows on the site — A point of planning law of general application involved in the case — Planning permissions granted before April 1 1969 were deemed to be subject to a condition that development must be begun not later than the expiry of five years beginning with that date — – Whether as a general principle a time-expired planning permission was not a relevant consideration in deciding whether to grant a fresh permission — Judge ruled that such a time-expired permission may still be relevant or material, although not necessarily of great weight — It was part of the past history of the site — On the merits of the present application, the judge decided to quash the** J **Secretary of State's decision — Planning permissions for retention of the hut and for the two bungalows were quashed — The Secretary of State's reasoning appeared to be vitiated by a failure to appreciate the distinction between a situation where there was an existing valid planning permission and the situation where a former planning position was time-expired**

In these proceedings the appellants, South Oxfordshire District Council, applied under section 245 of the Town and Country Planning Act 1971 to quash a decision of the Secretary of State for the Environment, the first respondent, in favour of the second respondents, Faherty Brothers Ltd. The Secretary of State had quashed an K enforcement notice and allowed appeals against the refusal of planning permission.

D Latham (instructed by Sherwood & Co) appeared on behalf of the appellants; Simon Brown (instructed by the Treasury Solicitor) represented the first respondent; R M K Gray (instructed by Francis & Parkes, of Reading) represented the second respondents.

Giving judgment, WOOLF J said: In this application under section 245 of the Town and Country Planning Act 1971 the appellants, who are the planning authority for South Oxfordshire, are seeking to quash the decision of the Secretary of State contained in a letter dated December 18 1978 in which, in respect of three appeals, the Secretary of State decided to quash an enforcement notice which L the appellants had served on the second respondents, to grant conditional planning permission for the retention of the building which was the subject-matter of the enforcement notice, and to allow the second respondents' two appeals against the refusal of the appellants to grant them planning permission.

There is one point of planning law of general application raised on which both the planning authority and the Secretary of State indicate they would welcome guidance. The point arises out of the provisions which are now contained in Schedule 24 to the Town and Country Planning Act 1971. Paragraph 19 of that Schedule provides:

> every planning permission granted or deemed to have been granted before M April 1 1969 shall . . . (subject to exceptions which are not relevant here) if the development to which it relates had not been begun before the beginning of 1968, be deemed to have been granted subject to a condition that the development must be begun not later than the expiration of five years beginning with April 1 1969.

The effect of that provision is that unless the development is begun before April 1 1974 the planning permission can no longer be relied upon as permitting the development to which it referred. It re-enacts, as a transitional provision, the time-limit as to the validity of planning permissions first introduced by the Act of 1968. Similar provisions dealing with planning permissions granted under

A the Town and Country Planning Act 1971 are contained in sections 41 to 44 of that Act. In particular section 43 defines when a development is to be taken to be begun for the purpose of section 41 and section 42 and paragraph 19 of Schedule 24.

On behalf of the planning authority it is contended that where development is not commenced before the time-limit expires, on a fresh planning application the earlier planning permission is no longer a relevant consideration which it is permissible to take into account in deciding whether or not to grant the fresh application.

In deciding whether or not to grant planning permission, the planning authority is required by section 29(1) of the 1971 Act to "have regard to the provisions of the development plan . . . and to
B any other material considerations", and it is submitted that a time-expired planning permission cannot be a material consideration. In support of this submission Mr Latham relies upon the only previous relevant authority, which is *Peak Park Joint Planning Board v Secretary of State for the Environment and Kay* (April 27 1977), a report of which appears at [1979] JPL 618, the transcript of which is before me. The judgment is that of Sir Douglas Frank, whose opinion as to planning matters I treat with very great respect because of his unrivalled experience. In that case Sir Douglas was dealing with a decision by an inspector on behalf of the Secretary of State in which the inspector, in dealing with an appeal against the refusal of planning permission, purported to apply circular 17/69 of
C the Ministry of Housing and Local Government in deciding an application for planning permission where there was a previous time-expired permission. The circular dealt with the position where applications are made to, in effect, renew planning permissions before they became time-expired and stated:

> As a general rule such applications should be refused only where
> (a) there has been some material change of planning circumstances since the permission was granted, eg a change in planning policy for the area or in the relevant highway considerations,
> (b) continued failure to begin the development will contribute unacceptably to the uncertainty about the future pattern of development, or
> (c) the application is premature because the permission still has a reason-
> D able time to run.

As Sir Douglas rightly pointed out, the reliance upon the circular on the facts of that case was misconceived, because, unlike the circular, it did not concern a planning permission which was not time-expired. Sir Douglas did, however, go on to say at p 6 of the transcript:

> Of course nobody I would have thought would dispute that the existing planning permission is something which has to be taken into account. But the whole purpose of the amendment to the law introduced by section 38 of the 1968 Act, now sections 41 and 42 of the 1971 Act, is to as it were remove stale longstanding planning permissions from any consideration, to
> E prevent them inhibiting the planning authority in their decision-making process. I therefore think that paragraph is not a planning policy consideration. If anything it is some form of administrative policy unrelated to planning, and if it is unrelated to planning questions, then following the well-known principle enunciated in the *Pyx Granite Co v Minister of Housing and Local Government* [1960] AC 260 it is *ultra vires*. This view I find supported by a reference to section 29 of the Act of 1971 where it is provided that in dealing with a planning application the local planning authority shall have regard, so far as material to the application, to any other material considerations (that is other than the development plan). But as I have said the consideration must be material to planning, and I do not think that having regard to permissions which no longer have any effect is a planning consideration. And if the matter rested there on the question of
> F planning policy, then I would have had little hesitation in quashing the decision, but in my judgment it does not end there.

The case Sir Douglas was dealing with was a special one, because he dismissed the appeal, taking the view that the inspector, on the way the appeal was argued before him, had no alternative to dealing with the matter in the way in which he did. It may be, therefore, that Sir Douglas did not intend his remarks to be regarded as having general effect. If they were intended to be of general effect, I am bound to say that I have difficulty in agreeing with the views expressed.

The effect of the changes made by the 1968 Act, now embodied in section 41 and section 42 of the 1971 Act, is to get rid of out-of-date permissions and to allow a planning authority, if it wants to do G so, to prevent development by refusing permission in respect of development for which permission previously existed. A planning authority is thus in no way bound by a previous planning permission which has expired. However, the fact that it is in no way bound does not mean that it is forced wholly to disregard that pre-existing permission. That there was a pre-existing permission may still be a relevant or material circumstance which a planning authority is permitted to take into account, though it must do so properly, and, as it is unlikely to be of great moment, not give it more weight than appropriate. For example, it is not unreasonable for a planning authority to want to be consistent in its consideration of planning applications, and taking into account a planning permission which H has expired and considering whether there has been any change of circumstances on a fresh application may assist in achieving consistency. I would regard a planning permission which has expired as still being part of the planning history of the site. In my view, it is not without significance that the planning authority can impose a special time-limit which will then override the time-limit provided for by the Act. If it does so, by the time a fresh application is made there could have been a change of circumstances which made the special time-limit inappropriate. In such circumstances a developer should be entitled to ask the planning authority to take into account on a fresh application the fact that if the original planning authority had known that that would prove to be the position, that time-limit J would not have been imposed and the fresh application would not have been necessary. If it could be relevant to have regard to a planning permission which has expired because of a special time-limit, it should be permissible in the appropriate circumstances to have regard to a planning permission which has expired because of a statutory time-limit.

Before leaving the *Peak Park Joint Planning Board* case, it is right that I should say that I see no reason why the general policy set out in the circular should be regarded as *ultra vires* and I would therefore reserve the question as to whether or not the criticism of that policy set out in the *Peak Park Joint Planning Board* case is justified. K

It follows that I do not accept the planning authority's general contention as to the law. This, however, is not sufficient to dispose of the application. It is necessary to consider the specific decision of the Secretary of State on the facts of this case.

The enforcement notice related to a temporary builder's hut which the second respondents justified solely on the ground that it was needed for the erection of a bungalow which was the subject of a planning permission granted on February 8 1957. If that planning permission were still valid, then the enforcement notice would have to be quashed because the hut would be permitted development under class 4 of Schedule 1 to the Town and Country Planning General Development Order 1977. It was therefore necessary for L the Secretary of State to decide whether as a matter of fact the development in respect of which the permission had been granted on February 8 1957 had been begun before the end of March 1974. The second respondents contended that it had been so begun because trenches had been dug which fulfilled the requirements of section 43(1) and 2(b). While trenches had been dug, they had unfortunately not been for the bungalow to which the planning permission related but for a different bungalow which had been designed so that "it is possible to reorientate the trenches actually dug so that one wall more or less coincides with where the corresponding wall of the 1957 bungalow ought to be." The Secretary of State did not consider this was sufficient for the provisions of M section 43. He accordingly decided that the planning permission had expired on April 1 1974.

The Secretary of State went on to consider the two appeals from the decisions of the planning authority to refuse permission for bungalows on the same site, and the deemed planning application for permission to retain the building which is the subject of the enforcement notice. One of the bungalows which was the subject of the appeal was of the same design as the bungalow which had been the subject-matter of the earlier planning permission. The other differed.

In his decision letter the Secretary of State quoted what the

A inspector had concluded with regard to the planning merits of the section 36 appeals, which was as follows:

> On the planning merit of section 36 appeals the inspector has concluded: "Turning to the section 36 appeals, this is an area of largely undeveloped country, well away from any recognisable settlement. Although visually a bungalow here would not be unduly prominent, and when seen from any distance to the south would probably be dominated by the bigger house (Hadden House) standing above and to the north of it, it would certainly intensify such residential development as there is. Forgetting, temporarily, the history of the site it would be difficult to distinguish a dwelling here from other dwellings nearby which other people might apply to build. It could lead to other applications which could not justly be resisted. However, I conclude that there is already a valid planning permission on the site, and it
> B would therefore be inconsistent to reject the first section 36 appeal because the bungalow concerned is identical with the one for which I consider permission exists. But since this bungalow is in my view banal and unsuitable, every encouragement to build a more suitable dwelling should be given. The bungalow in the second section 36 appeal is, I consider, a major improvement on that for which permission exists. I consider, therefore, that both appeals should be allowed."

He recommended that permission should be granted. The Secretary of State then went on:

> 10. These conclusions are not accepted insofar as they relate to the validity of the planning permission which had already been given on the site in 1957. It is noted that the 1957 conditional planning permission was granted by the
> C former Wallingford Rural District Council acting on behalf of the former Berkshire County Council. It is further noted that following local government reorganisation measures of 1974 the present council took over the administration of planning functions in the former rural district area and adopted the policies of the former council. An application for a design of dwelling different from that approved in the 1957 permission was still outstanding and was not determined by the new council until the changeover date for the new council which coincided with the expiry date for planning permissions granted before the passing of the 1968 Act. The development plan for the area had not changed since the 1957 permission and at the time of these applications planning authorities remained which were taken over by the new council in 1974. The character of the area too remains what it was for many years, that is predominantly rural and agricul-
> D tural with the bulk of Hadden House standing on higher ground above the appeal site when viewed from the south. Taking into account the history of the case which is considered to be a vitally material consideration in this instance and bearing in mind that the character of the area has not changed in the intervening years since 1957, the inspector's conclusion that the bungalows "would not be unduly prominent when seen from any distance to the south and would probably be dominated by the bigger house (Hadden House) standing above and to the north of it" is accepted. In the special circumstances of this case the view is taken, in agreement with the inspector, that planning permission should be granted on these appeals.
>
> 11. The inspector's remarks on the design and the council's remarks recorded in the report about their preference for the "second" design of bungalow is however drawn to your client's attention.

E The first thing which is to be noted about the Secretary of State's decision is that he has granted a planning permission for a bungalow, the design of which the inspector concluded was "banal and unsuitable". It is understandable that the inspector should have recommended that permission should be granted for that bungalow, because he took the view that there was already an existing planning permission for such a bungalow so there would not be any adverse effect in granting further permission. However, the Secretary of State had already concluded there was no longer an existing planning permission and therefore there was not that justification for granting planning permission in what was largely undeveloped countryside. As the Secretary of State apparently did
F not disagree with the inspector's view of the design of that bungalow, and as the second respondents would have been quite content with planning permission for the second bungalow, I am bound to say that I am at a loss to understand why planning permission for a building of the original bungalow design was granted. Mr Simon Brown, with his usual candour, confessed that he was equally surprised. It does suggest that there was some error in reasoning of the person responsible for making the decision of the Secretary of State in this case.

I also find it difficult to understand the reference that the decision should be granted on the appeals in agreement with the view taken by the inspector. It is true that the inspector did recommend that planning permission should be granted but it is reasonably clear G that he did so only because of his erroneous conclusion that the original permission was still in existence. The reference by the inspector to largely undeveloped country and intensification of residential development and other applications seems clearly to indicate that the inspector was averse to the granting of planning permission but for the pre-existing permission which he regarded as still valid.

Next it is to be noted that the Secretary of State says that the history is "a vitally material consideration". The reference to history must be because of the expired planning permission; as I have indicated, such an expired planning permission can be a relevant consideration, but I find it very difficult to see how it could ever be H regarded in the context of this case as being "a vitally material consideration". While the weight to be given to a particular consideration is for the Secretary of State, such a conclusion indicates that the person responsible for making the decision on behalf of the Secretary of State either misdirected himself or he was acting perversely. In either event his decision was wrong in law. It is therefore not necessary for me to decide which is the appropriate label to apply. I suspect, however, that the person responsible for the decision has failed to appreciate the distinction between a situation where there is an existing planning permission which is still valid and the position where there was a former planning permission which is no longer valid because it became time-expired. J I say this because the very words "vitally material consideration" appear in the case of *Spackman v Secretary of State for the Environment* [1977] 1 All ER 257, at the end of the judgment of Willis J at p 261. In *Spackman* it was a previous planning permission which was still in existence which Willis J regarded as a "vitally material consideration".

I was referred to the case of *Spackman* by counsel on behalf of the second respondents because I was asked on behalf of the second respondents if the matter was going back to the Secretary of State to indicate that the Secretary of State's approach to the question of the trenches was wrong, bearing in mind that decision. I do not, I am afraid, agree. As far as the trenches are concerned, it seems to me K that the Secretary of State's decision was a perfectly proper one and, indeed, probably the only possible one, bearing in mind the facts of this case. *Spackman's* case is quite different because the trenches which were dug in the wrong position were intended for the house for which permission had been given and not a property of different design.

It follows that I quash the Secretary of State's decision to grant planning permission for both bungalows and also quash his decision to grant permission for the retention of the hut.

Judgment was given for the appellant with costs against the first respondent, the Secretary of State.

L

QUEEN'S BENCH DIVISION

January 16 1981

(Before Sir Douglas FRANK QC, sitting as a deputy judge of the division)

LADBROKE (RENTALS) LTD v SECRETARY OF STATE FOR THE ENVIRONMENT AND ANOTHER

Estates Gazette June 6 1981

(1981) 258 EG 973 M

Town and Country Planning Act 1971 — Application to quash inspector's decision on appeal under section 36 of the Act — Complaint was against conditions imposed by inspector on planning permission for use of premises as a gaming club and casino — Appellants, having themselves been refused a gaming licence, proposed to lease accommodation to a person who was likely to be granted one — Conditions as to parking of vehicles were conceded to be invalid — A condition prohibiting the use on Sundays held by the judge to be unsupported by evidence of probable disturbance to local residents — "Wednesbury" principle applied — Inspector's decision quashed

A This was an application by Ladbroke (Rentals) Ltd to quash a decision of an inspector appointed to consider an appeal by the appellants under section 36 of the Town and Country Planning Act 1971 from a refusal of planning permission by the second respondents, the Royal Borough of Kensington and Chelsea. The inspector appointed by the first respondent, the Secretary of State, decided to grant permission for the use in question, which was of part of the ground floor of the Park Tower Hotel, Knightsbridge, as a gaming club–casino, but subject to certain conditions. The complaint which was the reason for the present application was against the conditions.

R Sears QC and T Straker (instructed by Norton, Rose, Botterell
B & Roche) appeared on behalf of the applicants; Alan Cooper (instructed by the Treasury Solicitor) represented the first respondent; the second respondents, the planning authority, were not represented and took no part in the proceedings.

Giving judgment, SIR DOUGLAS FRANK said: This is an application to quash a decision by an inspector appointed by the first respondent to determine an appeal by the appellants made under section 36 of the Town and Country Planning Act 1971.

The application was for the continued use of part of the ground floor of the Park Tower Hotel, Knightsbridge, as a gaming club–casino, for which planning permission had been granted for a limited period expiring on September 30 1979. The second respon-
C dents refused permission for a further limited period on the grounds that "the proposed use is considered undesirable, and has in the past proved detrimental to the amenities enjoyed by the residents at the rear of the premises, by virtue of increased noise and added congestion".

The appellants have been refused a renewal of the licence granted to them under the Gaming Act 1968. Accordingly their intention is to lease the casino part of the premises, which are not now in use, to a person to whom such a licence has been or will be granted. I may say in passing that it was alleged that the inspector and the second respondents had been influenced by the adverse publicity connected with the court proceedings, but in my opinion
D that allegation is without foundation.

The inspector held a public inquiry into the appeal and on November 14 notified his decision to grant permission. However, the permission was granted subject to the following conditions:

1. So far as is reasonably practicable all vehicles attracted to the premises including vehicles of staff and others shall be parked within the car park of the Park Tower Hotel and a sufficient number of spaces shall be reserved at all times for this purpose.
2. Within 3 months of the date of this letter a scheme of operation to enable the above condition to be implemented shall be submitted to the local planning authority for approval. The use hereby permitted shall not commence until the scheme has been approved by the local planning authority,
E or if no agreement can be reached with the local planning authority, the method of operation has been determined by the Secretary of State.
3. The use hereby permitted shall be discontinued on or before a date not later than 4 years after final approval or determination of the scheme required by the above conditions at which all materials, goods and equipment associated with the use shall be removed from the site.
4. The use hereby permitted shall not operate between the hours of 2 am and 2 pm on weekdays or between the hours of midnight on Saturdays and 2 pm on Mondays.
5. The shop windows along the Knightsbridge frontage to the premises shall be retained and shall be available for display purposes at all times during the period of this permission.

The appellants' complaint in this court is that the inspector erred
F in imposing conditions nos 1, 2 and 4. I may say that counsel for the Secretary of State conceded that as the car park referred to in condition 1 is not under the control of the applicants that condition is invalid and it follows that condition 2 is also invalid. On those grounds alone the decision must be quashed.

I turn to the grounds upon which condition 4 is attacked. First, it is said that the inspector did not give proper, adequate or intelligible reasons for the imposition of the condition. It seems to me that, whatever else may be said about the decision letter, it is clear that what the inspector was purporting to do was to protect the local G residents from disturbance caused by the attraction of additional traffic into Lowndes Square nearby. That seems to me to be a proper, adequate and intelligible reason for the imposition of the condition. The second ground of attack gives me more cause for concern: It is said, citing *Hilliard v Secretary of State for the Environment* [1978] JPL 840, that there was no evidence on which the inspector could conclude that the imposition of the condition was necessary to protect the surrounding residents. I am bound to say that it would have been helpful if, in this case as indeed I think in all cases, the inspector had set out in clear terms his findings of fact and if he had clearly distinguished between evidence and argument. I find great difficulty in discerning whether the inspector concluded H that there had been any disturbance by noise in the past or whether there would be some disturbance in the future and, if so, on what evidence he relied for that conclusion. The witnesses called by the second respondents gave no direct evidence of the effect of the use of the casino. They seem to have relied exclusively on two letters, one of which is expressed in general terms and does not deal directly with the question of noise by traffic and the other does not directly connect the complaints expressed with the use of the casino. There was no oral evidence giving direct evidence on this question except by G Adrian Eve, the senior partner in a well known and highly respected firm of chartered surveyors. He conducted a survey for four nights when the casino was in use and found that the traffic J generated by the casino in relation to the traffic as a whole was minimal. The inspector in his report said:

> The evidence of noise and disturbance was mainly contained in letters from nearby residents and to some extent was circumstantial but nevertheless there would seem to be a case to answer. Your clients' evidence on this point was more direct but the survey was taken over a limited period and the incidents causing disturbance were probably more isolated rather than occurring on a regular basis.

I assume that when the inspector referred to evidence as being circumstantial, what he really meant to say was that it was hearsay. In fact there was no evidence, not even hearsay evidence, which could be taken as showing that the casino was responsible for any K disturbance. On the other hand, I can find nothing which justifies the inspector in doubting the results of Mr Eve's survey. In any event, I do not understand why, if it is permissible to allow the casino to operate up to 2 am on weekdays, it should not be so on Sundays, as there was certainly no evidence whatsoever enabling the inspector to draw a distinction between weekdays and Sundays. Although the inspector's reason is clear his reasoning is obscure and devoid of any evidential content. In my judgment there was no evidence on which the inspector could reasonably have come to the conclusion that the use of the casino was likely to cause such disturbance as to justify the imposition of this condition and accordingly, applying the judgment of Lord Greene MR in *Associated* L *Provincial Picture Houses Ltd v Wednesbury Corporation* [1948] 1 KB 223 and the judgment of Shaw LJ in the *Hilliard* case, the decision must be quashed.

It was also argued by Mr Sears that, having regard to the power in paragraph 24 of Schedule 2 to the Gaming Act 1968 vested in the licensing authority, for the express purpose of preventing disturbance or annoyance to other occupiers in the vicinity, to impose restrictions on the hours during which gaming will be permitted, the Secretary of State and therefore his appointed person were deprived of exercising a similar power. I have no doubt that as a matter of sensible administration that would be right, but I do not think that the court is entitled to imply that, because the licensing M authority is given this express power, the very wide powers given to the Secretary of State under the Town and Country Planning Act are thereby diminished. Accordingly I would not hold that the inspector had acted in excess of his powers on this ground.

In view of the foregoing, the decision is quashed and remitted for redetermination.

The decision was quashed with costs against the Secretary of State.

VENDOR AND PURCHASER

A **CHANCERY DIVISION**

June 11 1980

(Before Mr Justice NOURSE)

WALTERS AND OTHERS v ROBERTS

Estates Gazette June 6 1981

(1980) 258 EG 965

Application by purchaser for summary judgment for specific performance under Order 86 — Incidental agricultural holdings point determined — Case with a number of "bizarre features" including a purchaser under a subsequent conditional contract in actual occu-
B **pation — Land in question consisted of two parcels of sheep farming land in the Welsh mountains and the sale included the vendor's flock of sheep — A letter setting out what were held to be the terms of the contract provided that as two previous dates specified for completion had passed the completion date would have to be agreed and would have to be after the end of the lambing season — Rejecting a submission that the date for completion was left wholly open for further agreement and that the contract was thus void for uncertainty, the judge held that the date was to be one which covered the lambing season and a reasonable period for the removal of the sheep and lambs from the land — View expressed obiter that even if the date were left open for further agreement the law would imply a reason-**
C **able time in default of agreement — Submission considered that the purchaser in occupation under conditional contract was granted a licence which was transformed into a tenancy from year to year by virtue of section 2(1) of the Agricultural Holdings Act 1948 — Held that the licence given to the present occupier in this case was of a special and subordinate nature liable to be determined without notice and with no existence independent of the occupier's equitable interest under a contract to purchase — The application of section 2(1) would not result merely in a "necessary modification" but a radical transformation, so that it would not remain "recognisably the same agreement" in the words of Pearson LJ in Harrison-Broadley v Smith — Order for specific performance granted**

D

In this case the plaintiffs, John Walter Richard Walters and Maldwyn Morgan (the original purchasers) and Roy Miles Walkden, Mrs Shaun Walkden and Michael Booth (sub-purchasers) sought a summary judgment for specific performance under RSC, Order 86, against John David Roberts in respect of land in the parishes of Llanfair and Llandbedr in Gwent. The application for summary judgment was made in the action by the plaintiffs as purchasers for specific performance.

D Wood QC and D Neuberger (instructed by Macfarlanes) appeared on behalf of the plaintiffs; A W Simpson (instructed by Ward Bowie, agents for Gwyndaf-Williams & Roberts, of Port-
E madoc) represented the defendant.

Giving judgment, NOURSE J said: This is an application for summary judgment under Order 86 in a purchaser's specific performance action. The action is based on a contract dated March 25 1974 made between the defendant, John David Roberts, as vendor, and the original plaintiffs in the action, John Walter Richard Walters and Maldwyn Morgan, as purchasers. The property comprised in that contract was two parcels of sheep farming land in the Welsh mountains. The first parcel consists of a farmhouse, some farmland and a lake in the parishes of Llanfair and Llandbedr in Gwent, formerly Merioneth, and the second parcel of some adjoining farmland. F

The material terms of the contract were that the purchase price should be £100,000, that completion should be on September 30 1974 and that the defendant should give vacant possession on completion. The sale also included the defendant's flock of sheep. A 10 per cent deposit of £10,000 was paid. The agreement was expressed to be subject to certain special conditions and to the National Conditions of Sale, 18th ed, so far as the latter conditions were not inconsistent with the special conditions. The latter included as condition no 22 the familiar provision enabling either party to give a special notice to complete.

Mr Wood, who appears for the plaintiffs on this application, has G
described the case as having a number of bizarre features. Perhaps the most bizarre is the fact that the contract described the two parcels of land as having a combined acreage of 3,156 acres or thereabouts; whereas the plaintiffs later calculated it to be rather less than half that amount. They calculated it to be 1,515 acres or thereabouts. Not surprisingly, that caused difficulties in the completion of the contract at that stage and negotiations took place between the parties. In 1976 the defendant, the vendor, commenced an action for specific performance against the purchasers. That action still remains on the file, although it may well be that it will become a dead letter as a result of this application.

Going forward a little in time, I should mention that in April H
1979 the original plaintiffs, the purchasers, entered into an agreement to subsell the land to Roy Miles Walkden, his wife, Shaun Walkden, and a Michael Booth, and they also are plaintiffs in this action. There is no conflict of interest between the two sets of plaintiffs. I shall therefore refer to them as one.

I should say that on June 22 1977 the defendant entered into a second contract for the sale of the same property to a Mr Evans and his wife. That contract was expressed to be subject to the 19th edition of the National Conditions of Sale, which was by then the current edition, except so far as varied by the special conditions, one of which was in these terms:

J

(f) This contract is conditional upon a prior contract dated 28th day of March, 1974 and made between the vendor of the one part and Maldwyn Morgan and John Walter Richard Walters of the other part regarding the property hereby agreed to be sold being rescinded unless the Court otherwise directs. If the Court otherwise directs then it is agreed that the purchasers will recover the purchase price hereby agreed to be paid, but no more, and shall give up possession, if applicable, of the property hereby agreed to be sold.

It appears that shortly after the signing of that contract Mr Evans went into occupation of the property. He must have done that on the terms of condition no 8 of the National Conditions of Sale, 19th ed, to which I will have to return later. K

Meanwhile the negotiations between the defendant and the original purchasers were continuing. In February 1979 those negotiations started to take a shape which at least for the purposes of this application can be said to be recognisable. On February 23 1979 the plaintiffs' then solicitors, Careless & Co, wrote a letter to the defendant's solicitors which, after referring to a previous telephone conversation, said this: "We confirm that our clients are prepared to proceed with the purchase of the above property" — that is the property which I have already described — "upon the following

A amended terms." Those terms were then set out. First, the price was to be reduced to £50,000, but still to include an equivalent quantity and value of sheep as the original contract; secondly, the plaintiffs would agree to pay the defendant's reasonable conveyancing charges and the VAT on those charges; thirdly, each party would bear their own costs in connection with the defendant's action for specific performance to which I have referred. Paragraph 4 was in these terms: "The matter should be completed within 28 days." Fifthly, it was proposed that in so far as interest obtained on the deposit of £10,000 was concerned this should belong to the plaintiffs.

It appears that there was no answer to that letter and on March 9
B and 23 1979 the plaintiffs' then solicitors wrote two reminders. On April 19 there was an important telephone conversation between the plaintiffs' then solicitors and the defendant's solicitors. It is accepted by the defendant that the oral agreement which was then arrived at between the two solicitors was capable of being an enforceable agreement (no point is taken under section 40 of the Law of Property Act 1925) and that the terms of it are fully and accurately set out in a letter which the plaintiffs' then solicitors wrote to the defendant's solicitors on the following day. I will quote that letter almost in full. It is in these terms:

> We write to confirm our telephone conversation of yesterday when it was agreed that your client accepted the terms set out in our letter of February
> C 23 1979 with exception only that in regard to paragraph no 5 the interest earned on the deposit account should be divided equally between the vendor and purchaser.

I interpose to say that that no doubt was to take account of the fact that if the original purchase price of £100,000 was to be reduced to £50,000 the purchaser would have overpaid the deposit by 100 per cent.

Then the letter goes on as follows:

> The completion date specified in paragraph 4 is obviously not now appropriate. You explained the position in regard to possession in that someone was lambing a number of sheep there and could not move them on short notice. We appreciate this point and have explained it to our clients, and a
> D completion date will therefore have to be agreed. We will arrange for our clients to meet your clients to discuss the question of the sheep and to save time they can also discuss the question of possession.

And then the letter said that the plaintiffs' then solicitors were preparing the draft conveyance.

I have already referred to the first and the major problem which arose, which was the question of the acreage. After that letter of April 20 there was a further problem. There was some doubt as to whether the defendant owned the sheep, or at least whether he owned all the sheep, which had been included in the contract. That problem, however, was sorted out comparatively quickly. On June 12 1979 the solicitors agreed that the sum of £7,500 should be
E deducted from the purchase price as being referable to the value of the flock, and it was agreed that that sum should be paid into a stakeholder's account pending a resolution of the question of ownership, with the balance of £42,500 less the deposit of £10,000 being paid on completion.

For reasons which will appear later, it is not necessary for me to deal with later events, except to say that on September 18 1979 the plaintiffs' solicitors journeyed to Portmadoc and made an attempt to complete the purchase. On that occasion the defendant's solicitors had in their possession a conveyance of the land which had been signed and sealed by the defendant, but the defendant claims that it was held by his solicitors as an escrow of which one of the
F conditions was that a date for the completion of the contract should be agreed. In any event, the purchase was not completed then or afterwards and the writ in this action was issued on October 29 1979.

Pleadings were duly served on each side and the summons which is now before me was issued on February 29 1980. That summons claims the following relief: first, delivery of the conveyance already signed and sealed by the defendant on the footing that it was duly delivered by him and that it was not held as an escrow, as he alleges; and secondly (and necessarily in the alternative) specific performance of the agreement. I shall explain shortly what exactly is the agreement upon which the plaintiffs now rely.

The position immediately before the oral agreement of April 19 G
was this: The original date fixed for completion, September 30 1974, had long since passed and the new date proposed in the plaintiffs' solicitors' letter of February 23 1979 had also passed. But assuming for this purpose that the original agreement of March 25 1974 was still subsisting, nothing had occurred to prejudice the plaintiffs' right, or indeed the defendant's, to serve a special notice to complete under condition 22 of the National Conditions of Sale, nor any right which either party had to sue for specific performance of the agreement independently of the service of any such notice.

The effect of the oral agreement of April 19 1979 was to adopt the terms set out in the letter of February 23 with certain excep-
tions. That letter had proposed that the purchase should proceed on H
certain "amended terms"; that is to say that the original agreement should stand subject to such amendments. And it seems to me that the oral agreement of April 19 must be construed in the same way. In other words, on April 19 the parties agreed that the original agreement should stand, subject only to the amendments set out in the letter of February 23 and the further amendments then orally agreed and recorded in the letter of April 20. I therefore cannot accept the argument of Mr Simpson, who appears for the defendant, that the agreement of April 19 as he put it, pushed the original agreement off the map. In my judgment it did nothing of the kind. It preserved the original agreement so far as not amended.

Then the question arises: Was the original agreement amended J
in such a way as to leave the date for completion open for further agreement and thereby to render the whole agreement void for uncertainty? Although that would, to say the least, produce a startling result, Mr Simpson says that it was. He says that the agreement of April 19 as recorded in the letter of April 20 did import an agreement that completion was to take place on a date to be agreed. I must say at once that I do not think that that is a correct construction of the agreement of April 19. It is true that the letter of April 20 states that "A completion date will therefore have to be agreed"; but it seems to me that the parties were clearly not intending to leave the matter wide open for any agreement at which
they might choose to arrive. The material passage in the letter starts K
with this sentence: "The completion date specified in paragraph 4 is obviously not now appropriate". That was a fixed date which had passed some four weeks earlier, and the sentence is consistent with the notion that the parties still had another ascertainable date in mind. This is confirmed by the next sentence, which reads: "You explained the position in regard to possession in that someone was lambing a number of sheep there and could not move them on short notice". That suggests that the new date for completion was to be one which would cover the lambing season and a reasonable period for the removal of the sheep and the lambs thereafter. In my judgment that was the agreement at which the parties arrived on
April 19 and I can see no warrant for thinking that they intended L
that there should be any further latitude in the matter. But since that completion date, although ascertainable, was not at that date certain, it was obviously sensible that the parties should, if possible, make it so by meeting to agree a date. That, in my judgment, is the significance, and the only significance, of the words "a completion date will therefore have to be agreed", and the subsequent suggestion that the clients should meet together to discuss the questions of the sheep and possession.

On this view of the matter it is not necessary for me to decide what would have been the position if on its true construction the agreement of April 19 did import an agreement that completion
was to take place on a date to be agreed. But since the point was M
fully argued and is one of general importance, I propose to state briefly the view I have formed upon it.

I was referred to a number of authorities which touched on this question, but in only one of them did the agreement in terms provide that completion was to take place on a date to be agreed between the parties. That was *Gavaghan v Edwards* [1961] 2 QB, 220, where the county court judge had held on that ground that the agreement was not a completed and binding contract. However, the Court of Appeal, without calling on the respondent vendor who had served a cross notice on that point, affirmed the county court judge's judgment on another ground, and it may in the circum-

A stances be of some significance that Danckwerts LJ, who delivered the leading judgment, said at p 224 that it was unnecessary for the Court of Appeal to consider that point.

I think that what Danckwerts LJ had in mind in that case was that it might have been possible to go back to the date fixed by the material provision of the Law Society's Conditions of Sale. On consideration of the point I do not myself think that that can be done, at least with condition 5(1) of the National Conditions of Sale, which reads as follows: "The completion date shall be the day mentioned for that purpose in the contract or, if none, the first working day after the expiration of five weeks from the delivery of the abstract of title." In my judgment if the contract provides that B completion shall take place on a date to be agreed that date is the day mentioned for that purpose in the contract, so that condition 5(1) can have no further operation.

However, bearing in mind the invariable implication, in a case where the agreement is silent on the point, that completion will take place within a reasonable time, I strongly incline to the view that in a case where the date is left for future agreement the court would imply a reasonable time in default of agreement. It would indeed be very strange if an open contract was valid, but a far more sophisticated agreement invalid merely because it provided that completion was to take place on a date to be agreed. I have not been able to plumb the origins of the species of implication which is made in a C case where the agreement is silent as to completion, but they may well be bedded in strata of an era more remote and less frigid than that from which *The Moorcock* (1889) 14 PD 64 has evolved. But even if that is not so and the same principles apply, it seems to me to be pretty clear that they would produce the same result. After all, the date for completion is merely a part of the machinery for executing a contract whose essential terms are all agreed. Both parties must be taken to intend that completion will occur sooner or later, once the necessary investigation of title and the preparation of the conveyancing documents and so forth are done. On that footing it is not a very big step to assume that both parties, being reasonable men, would, if asked, unhesitatingly agree that in the D event of their being unable to agree the date it should be at the expiration of a reasonable period after the date of the agreement; cf the recent restatement of the principles of implication by the House of Lords in *Liverpool City Council v Irwin* [1977] AC 239. However, it is, as I have said, unnecessary for me to express a concluded view on this question, and perhaps it is better that I should not do so.

In the circumstances I hold that the effect of the oral agreement of April 19 1979 was that completion should take place on the expiration of a reasonable period after the end of the lambing season, the reasonable period being that required for the removal of the sheep and the lambs from the land. That period must have E expired on some date in 1979, and certainly before the issue of the writ on October 29 1979. No special notice to complete was served, but that is unnecessary because the defendant's arguments on this summons necessarily evince a refusal to complete. The plaintiffs are therefore entitled to have the agreement specifically performed unless a point which has been taken by Mr Simpson at a rather late stage, and perhaps as a result of something which I said yesterday, is a good one.

Mr Simpson points to the fact that Mr Evans is at present in occupation of the property under the terms of his conditional contract of June 22 1977. He says that the defendant may encounter very real difficulties in getting Mr Evans out of occupa- F tion so as to be able to give the plaintiffs vacant possession on completion of their contract, and he says that that is a sufficient reason for the court to refuse the plaintiffs a decree of specific performance. I should add that he does, if necessary, ask for an adjournment to file further evidence to show the difficulties which stand, or may stand, in the defendant's way in this respect.

In the first instance, in support of his argument on this point, Mr Simpson refers to special condition (f) of Mr Evans' contract. Mr Simpson says that it is not clear from that special condition that Mr Evans will, as between himself and the defendant, be obliged to get out if specific performance of the plaintiffs' contract is ordered. I reject that argument. Although the wording of special condition (f) is not perhaps as felicitous as one would have liked to G see, it seems to me to be clear that it does indeed oblige Mr Evans to get out if the plaintiffs' contract is ordered to be specifically performed. It that order is made the court "otherwise directs" for the purposes of special condition (f).

Then Mr Simpson takes a point which at first blush is of a more formidable nature. He says, correctly, that under condition 8(1) of the 19th edition of the National Conditions of Sale Mr Evans' status as an occupier pending the completion of his contract is that of a licensee, and he says that he is in occupation of the property for agricultural purposes. He then refers to section 2(1) of the Agricultural Holdings Act 1948, which can be abstracted, so far as material for present purposes, as follows: H

> Subject to the provisions of this section, where under an agreement made on or after March 1 1948, . . . a person is granted a licence to occupy land for use as agricultural land, and the circumstances are such that if his interest were a tenancy from year to year he would in respect of that land be the tenant of an agricultural holding, then . . . the agreement shall take effect, with the necessary modifications, as if it were an agreement for the letting of the land for a tenancy from year to year.

Mr Simpson says that that provision may have the effect of giving Mr Evans the protection of the Agricultural Holdings Act as against the defendant, and that the result might be that the defendant would not be able to give the plaintiffs vacant possession on completion. Although there is no authority on the point, he says that it is J one which might go either way, and that the defendant cannot therefore be required to embark on litigation to get Mr Evans out of occupation of the property. Therefore the court cannot, or should not, order specific performance of the plaintiffs' contract.

In this connection Mr Simpson referred me to a passage in the judgment of Megarry J in the case of *Wroth v Tyler* [1974] Ch 30 at p 50, where the learned judge said this:

> A vendor must do his best to obtain any necessary consent to the sale; if he has sold with vacant possession he must, if necessary, take proceedings to obtain possession from any person in possession who has no right to be there or whose right is determinable by the vendor, at all events if the vendor's right to possession is reasonably clear; but I do not think that the vendor will K usually be required to embark upon difficult or uncertain litigation in order to secure any requisite consent or obtain vacant possession. Where the outcome of any litigation depends upon disputed facts, difficult questions of law, or the exercise of a discretionary jurisdiction, then I think the court would be slow to make a decree of specific performance against the vendor which would require him to undertake such litigation.

A little later on the learned judge turned to consider the facts of that particular case, from which it appeared to him to be clear that the person concerned — in that case the vendor's wife — had demonstrated that she was not prepared to get out of possession short of the vendor taking proceedings to that end.

Accordingly, the position in the present case is that if the defen- L dant's right to possession is reasonably clear he must take proceedings to get Mr Evans out, should that course become necessary. But if it does become necessary and the defendant's right to possession depends on a difficult question of law then I ought to be slow to order specific performance of the plaintiffs' contract. So I must ask myself: Is the defendant's right to possession reasonably clear or does it depend on a difficult question of law? On the one hand, it can be said that the absence of authority on the point is an indication that the right to possession is not reasonably clear; on the other, that if the right does depend on a difficult question of law then it is most surprising that it does not appear to have been contested in any reported case in the 32 years or so since the M Agricultural Holdings Act was enacted.

It seems to me that the first thing I must do is to analyse the rights of the defendant and Mr Evans under their contract. It is true that under condition 8(1) of the National Conditions of Sale the quality of the purchaser's occupation is that of a licensee, but the licence is only granted as an adjunct to the purchaser's equitable interest in the land and in the expectation that that interest will in due course mature into full legal ownership. If the contract is discharged or rescinded the equitable interest comes to an end; likewise the licence, either under special condition (f) or condition 8(2) of the National Conditions of Sale — it matters not which. It follows that the licence is of a special and subordinate character liable to be

A determined without notice and with no existence independent of the purchaser's equitable interest in the land.

Next I must deal with such authority as there is which bears on the point. Mr Wood has referred me to the judgment of Pearson LJ in *Harrison-Broadley v Smith*, [1964] 1 WLR 456. There the facts were rather different. I need not refer to them, except to say that that was a case where the Court of Appeal held that section 2(1) could not apply to a licence granted by someone who remained in occupation with the licensee. At p 467 Pearson LJ started his judgment by saying this: "The main question here is: How should section 2 of the Agricultural Holdings Act 1948 be construed and applied in relation to the situation which has arisen in this case?".
B Then he called attention to the most material words in section 2(1) and proceeded as follows:

> It is to be observed that "the agreement" under which a person is granted a licence to occupy land for use as agricultural land is to "take effect, with the necessary modifications, as if it were an agreement for the letting of the land, for a tenancy from year to year". That which is to take effect is the original agreement, with the necessary modifications. It is not permissible to substitute for the original agreement a radically different agreement and make that take effect instead of the original agreement.

Then he referred to a passage in the judgment of Lord Evershed MR in *Goldsack v Shore* [1950] 1KB 708 and went on as follows:

> C Of course, it must be recognised that any modification of an agreement must be in a sense inconsistent with its own terms. But I understand that passage to mean that the agreement must remain, so to speak, recognisably the same agreement after the necessary modifications have been made. This section is not applicable to an agreement which is not capable of taking effect, with the necessary modifications, as an agreement for the letting of the land for a tenancy from year to year. The necessary modifications have to be distinguished from a transformation of the agreement into something radically different.

Mr Wood submits that if a licence for a purchaser to occupy pending completion of the purchase were to be transformed into a tenancy from year to year that would be a transformation into
D something radically different. I agree with that submission. It would be the transformation of a licence of a special and subordinate character liable to be determined without notice and with no existence independent of the purchaser's equitable interest in the land into an agreement for a tenancy determinable only on six months' notice and with an independent existence of its own. In my judgment that must be a transformation of one thing into another radically different. I should add that it would indeed be a most remarkable state of affairs if the Agricultural Holdings Act did catch an arrangement of this kind and afforded a protection which was clearly never intended.

Broadly speaking, the object of that Act is to extend the rights of
E a person who is contractually entitled to no more than the right to occupy land for agricultural purposes. It is not its object to extend the rights of a person who has contracted to purchase the land, who is let into occupation pursuant only to that contract and whose occupation was never intended to survive its extinction. I do not think it would be right for me to say that the point is one which is beyond argument, because, as Mr Simpson points out, the wording of the subsection itself does allow an argument to be advanced. Nevertheless, I do not myself have any hesitation in concluding that the defendant's right to possession as against Mr Evans is reasonably clear and that it does not depend on a difficult question of law.

F In those circumstances it is unnecessary for me to consider whether I should grant Mr Simpson an adjournment to file further evidence because I have assumed that Mr Evans might indeed adopt the attitude that he was not going to get out without an order for possession, and I have nevertheless concluded that the objection to a decree for specific performance is not a good one.

Mr Wood suggests the following course of action, which I propose to adopt. I propose to make an order for specific performance in terms which I will discuss with counsel. One thing which the order will certainly not do at this stage is to fix a date for completion. If, when the master has fixed that date, it turns out that there are any difficulties in the defendant's giving vacant possession on completion, then the matter should be referred to me for further consideration. If necessary, and if so requested by both parties, I will retain the case accordingly. That states in general terms the sort of order I propose to make. G

This makes it unnecessary for me to deal further with the plaintiffs' alternative plea for delivery of the executed conveyance. I will only say that on the evidence as it stands at present it would in my view be impossible for me to make such an order on an application for summary judgment. It is for this reason that I have found it unnecessary to deal in any detail with events subsequent to June 12 1979.

H

QUEEN'S BENCH DIVISION

December 22 1980

(Before Sir Douglas FRANK QC, sitting as a deputy judge of the division)

CEDAR TRANSPORT GROUP LTD v FIRST WYVERN PROPERTY TRUSTEES CO LTD AND OTHERS (MICHAEL RAYNER & PARTNERS, THIRD PARTY)

Estates Gazette June 13 1981
(1980) 258 EG 1077 J

Claim by purchasers of a warehouse building against vendors and architects employed by vendors for alleged breach of contract and breach of duty — Dispute arose from provisions in agreement to purchase by which vendors undertook to carry out and complete certain works including works necessitated by removal of heavy machinery and equipment used by previous occupiers — Claims under heads of defects, delay and vandalism examined in detail by judge, the claim in respect of defects being against the architects — Architects held to be acting as experts, not arbitrators, but no breach of duty established against them — Claims against vendors under heads of delay and of negligence resulting in vandalism likewise not established — Action fails K

In this action by Cedar Transport Group Ltd, in which they claimed damages for breach of contract and breach of duty, the first defendants were First Wyvern Property Trustees Co Ltd, the second defendants were Second Wyvern Property Trustees Co Ltd, and the third defendants were Michael Rayner & Partners, a firm of architects, who were also the third party.

J G Boggis (instructed by Grant, Saw & Sons) appeared on behalf of the plaintiffs; H Burnett (instructed by Baileys, Shaw & Gillett) appeared on behalf of the first and second defendants; P I Vallance (instructed by Russell, Jones & Walker) represented the third defendants (third party). L

Giving judgment, SIR DOUGLAS FRANK said: The plaintiffs carry on business as road-hauliers and warehousemen. The first and second defendants, who for the purposes of this case are indistinguishable and I shall refer to them jointly as "Wyvern", invest in and manage commercial properties for the benefit of a pension fund. The third defendants and third party are a firm of architects and during the material time were employed by Wyvern. The partners concerned with this matter mainly were Mr Michael Rayner AA Dip FRIBA and Mr Peter Canterbury Dip Arch ARIBA. I may say now that both impressed me as competent architects and as witnesses of integrity.

At St Mary Cray in the London Borough of Bromley there is an M
industrial estate fronting Cray Avenue, which contains five industrial or warehouse premises which are identified respectively as units 1-5. The road leading off Cray Avenue into the estate is called Lagoon Road and unit 1 fronts those two roads. In 1975 Wyvern owned unit 2 but had leased it to a manufacturing company called Commercial Plastics Ltd, who also carried on their manufacturing business in unit 1, of which they owned the freehold. In that year Commercial Plastics Ltd commenced negotiations with Wyvern which resulted in their surrendering their lease of unit 2 and selling their interest in unit 1 to Wyvern in the following year; the contract of sale was made on March 23 1976.

A The plaintiff company is for all practical purposes synonymous with its managing director, Mr A J Dyer. About the time that Wyvern were negotiating for unit 1, Mr Dyer was looking for a warehouse. For a number of reasons the area of St Mary Cray was particularly suitable and his attention was drawn to unit 1 by Mr R C M Gower, a chartered surveyor and estate agent, who was asked by Mr Dyer to find a warehouse in that area. Following inspections and negotiations the plaintiffs and Wyvern entered into an agreement to buy the freehold interest in part of unit 1 on July 26 1976. I shall refer to that part as unit 1 and to the part not acquired as unit 1A. The date specified in the agreement for completion was October 1 1976. In that agreement the vendors undertook to carry B out the works described in the schedule at their own cost in a good and workmanlike manner in compliance with all statutory and local requirements applicable thereto and in accordance with certain identified drawings. They further undertook to use their best endeavours to complete the works by the date fixed for completion, but the agreement provided that "such dates shall be extended if there shall be any delay in the execution of the works by reason of any cause whatsoever beyond the control of the vendors, the length of the extension is to be not less than the length of such delay and as shall be certified by the vendors' architects". It was also provided that the works should be deemed to be completed in accordance with the agreement upon the issue of a certificate to the vendors to C that effect by the vendors' architects. The works specified in the schedule relevant to this action were as follows:

1. Formation of a loading bay off Lagoon Road including the moving of the existing Seeboard sub-station.
2. The forming of a steel canopy covered with asbestos sheets with 15 % translucent sheeting falling to rain water pipe at the rear. . . .
5. Remove all existing machinery (except overhead gantry) and carry out necessary consequential repairs to floors, walls and surfaces the load bearing capacity of the floors thereafter to be not less than they were prior to the commencement of the works. . . .
7. Arrange (in so far as not already provided) connection to the main . . . electricity service. . . .

D

The machines used by Commercial Plastics Ltd were large and heavy, resting on a deep concrete floor and over pits constructed within the floor. By July 26 the company, in pursuance of an obligation under their contract of sale, had taken down some machinery and removed some from the premises. The removal work continued in the ensuing weeks. Wyvern employed J E Wiltshier & Co Ltd to carry out the repair works under the JCT standard form of building contract, which also applied to the rebuilding of unit 1A which was being carried out concurrently.

In September Mr Gower telephoned Mr R E W Atkins [partner of Bernard Thorpe & Partners, acting as estate agents for Wyvern] E making a number of general complaints. On September 28 Wyvern's solicitors sent the engrossment of the counterpart lease for sealing to the plaintiffs' solicitors. Further complaints were made in September; in particular it was said that the central heating was not working, that the roof was leaking and that the electrical system was defective. On October 4 the architects issued a certificate to the contractor stating that the works were found to be practically complete on October 1. The certificate stated that "the items outstanding were the erection of canopy, removal of switchgear room and filling of opening in side wall. All these outstanding items were delayed by conditions outside your control and we confirm that you have used your best endeavours to try and com- F plete all the works". on January 21 1977 Mr Gower sent a schedule of works to the architects which he said were required to be carried out before his client considered the property in a satisfactory condition for completion. Wyvern and the architects took the view that for the most part the terms of the agreement did not require them to carry out the work specified in the schedule. On March 8 the plaintiffs' solicitors issued a notice requiring completion within 28 days.

On March 31 1977 the architects certified that the works were substantially completed by October 1 1976 with the exception of the removal of the electricity sub-station and the renovation of the roadway which would be consequent on that. Concurrently, correspondence was still taking place regarding the plaintiffs' complaints. G On April 22 Wyvern's solicitors asked the plaintiffs' solicitors to make immediate arrangements to complete but the latter reiterated that a dispute existed, particularly concerning the leaking roof. On May 25 the architects sent a letter to Wyvern in which they said:

In case it is required we confirm that the length of the delay in completing was equal to the delay between 1st October 1976 and 21st April 1977 for reasons which were beyond your control.

Completion took place on May 27, the plaintiffs' solicitors recording that it was on the basis that there was no merger of their clients' rights under the contract, in particular those relating to the schedule of works, the construction of clause 7 and the vendors' duty in H relation to the property from exchange of contracts to completion.

On July 11 1977 the plaintiffs issued their writ in this action in which they claimed damages for breach of contract and breach of duty against Wyvern, and for breach of duty and care against the architects. The causes of action conveniently fall under three separate heads, namely, defects, delay and what has been given the rubric "vandalism" and I shall deal with each in that order.

Defects

Mr Boggis for the plaintiffs conceded that if the final certificate given by the architect was valid he has no claim against Wyvern under this heading and that his claim against the architects must be J on the footing that they were negligent in giving a certificate. He accepted for the purposes of this hearing that the certificate was valid and, therefore, contends that his remedy is against the architect.

The first question to be determined is whether the architects were acting as arbitrators and as such were immune from proceedings. A number of authorities were cited in argument, but I find it unnecessary to look beyond *Arenson v Arenson* [1977] AC 405 and in particular to the test adumbrated in the opinion of Lord Wheatley at p 428 in which he said:

the indicia are as follows: (a) there is a dispute or a difference between the parties which has been formulated in some way or another; (b) the dispute K or difference has been remitted by the parties to the person to resolve in such manner that he is called upon to exercise a judicial function; (c) where appropriate, the parties must have been provided with an opportunity to present evidence and/or submissions in support of their respective claims in the dispute; and (d) the parties have agreed to accept his decision.

Mr Vallance argued that his clients were acting as arbitrators and not mere certifiers. He said that by the time the certificate was given there was a formulated dispute with rival contentions. He contended that the function of the architect, in circumstances such as in this case, can change and that in reality the architect's function became that of an arbitrator. In my judgment the submission in this case fails to pass Lord Wheatley's summation of the relevant test in L at least two respects. First, I do not consider that on the facts the plaintiffs acted so that it can be said that they remitted their dispute to the architects in such a manner as to call upon them to exercise a judicial function. Secondly, the plaintiffs did not agree to accept the architects' decision either expressly or implicitly. Accordingly, I do not consider that the architects satisfy the requirements to be entitled to immunity for negligence,

The next question for determination is one of construction of the agreement, whether under paragraph 5 of the schedule the obligation to repair is consequential only upon machinery removed after the date of the agreement, namely July 26 1976. The argument is that it rests upon the footing that "existing" must mean M existing at the date of the agreement. Notwithstanding Mr Boggis' arguments that the meaning of the paragraph should be determined by the intention of the parties, I hold to the view that if the words in the paragraph are unambiguous in their plain and ordinary meaning and do not lead to an absurd result, then I am not entitled to take into account any other factors which may cause uncertainty where none exists. However, I think there is some force in Mr Boggis' contention that the paragraph, as it stands, is open to question as to which point in time the word "existing" is to be determined and that being so, I am entitled to look at the surrounding circumstances. The position as it was known to the parties was that Commercial

Plastics Ltd were in the process of removing their heavy machinery and that would necessitate the carrying out of some repairs. I find it inconceivable that the parties could have intended the repairs to be divided between before and after the agreement. Perhaps another way of arriving at the same conclusion is that to construe the word "existing" as prayed for by the defendants would lead to an absurd result. Accordingly, I hold on both grounds that the paragraph applies to all machinery moved by Commercial Plastics Ltd.

The next question is: what is meant by machinery? That is a word which has appeared in numerous statutes and has been subject to consideration by the courts over the years. However, I am content to take the meaning from the *Shorter Oxford Dictionary* and say that in the context of this case it means an apparatus for applying mechanical power and that the repair consequent on its removal means the restoration of the floors, walls and surfaces to a good condition by the renewal or mending of any damaged part. It is important to bear in mind that the duty is a limited one and is not to be compared with a repairing covenant in a lease. Clearly it was not contemplated that the vendors were to be responsible for all dilapidations, damage or defects resulting from the activities carried on by Commercial Plastics Ltd. I therefore first eliminate those items of claim which in my judgment do not fall within paragraph 5 of the schedule.

The building was purpose-built for the processes carried out on the premises and the machinery installed by Commercial Plastics Ltd. As part of the structure four large holes were incorporated within the first floor of the building and a further such hole made later. Over these holes stood machines used for mixing the material to be processed and from them the mixed material fell through the holes in hoppers below. The mixing machines and hoppers were removed by Commercial Plastics Ltd and the plaintiffs claim in this action the cost of filling in the holes. However, in my view that did not involve the carrying out of consequential repair because the holes are not consequent upon the removal of the machinery but are part of the structure of the premises. I turn now to the kitchen and laboratory from which sinks, cookers and other equipment were removed. I am not concerned with the question whether that equipment was wrongly removed but only whether the architects are responsible for making good the walls which became exposed. The short answer to this claim is the equipment that was removed was not machinery and I would add that even if it was, most of the claim relates to improvements and very little to repairs. I turn then to the ducting used for ventilation and heating. I first remark that I think it likely that this was removed by the plaintiffs. In a letter dated March 7 1977 their solicitors said: "The only works carried out by our clients consisted of the removal of certain ducting work which took approximately two hours to carry out." Mr Dyer said that referred to pipes on the ground floor, but curiously went on to say that although he sent two men to remove them, Mr Canterbury sent them away and no work was done. That statement is impossible to reconcile with the one in the solicitors' letter and the evidence of Mr Canterbury. Moreover, Mr Dyer did not identify with conviction any other pipes. Further Mr Canterbury was adamant that the removal of the ducting was not done by Wiltshiers. Be that as it may, in my judgment the ducting was not machinery and, accordingly, any works resulting from its removal do not fall within paragraph 5 of the schedule and that applies to a separate claim made in respect of the hole above a doorway.

I turn now to deal shortly with another category, namely that one where I am not satisfied that there was any evidence or sufficient evidence to show that the work was consequential upon the removal of machinery. This is alleged damage to a manhole and to a swing door and to various other miscellaneous items; the cause of the damage is not known and there is no ground for making an assumption. On the other hand, it is not seriously contested that an opening in the wall was damage and not completely made good and that the cost of making good was £30 as claimed.

Now for what has been called the hole in the wall. An area of brickwork, which I am satisfied measured about 12 ft by 12 ft, was removed from the wall fronting Lagoon Road. That work was done to facilitate the removal of the machinery and the evidence points to the fact that this area had originally been constructed so that it could be taken down and rebuilt for that purpose. It was alleged in the further and better particulars of the statement of claim that the original wall was 9 in thick but had been replaced to a thickness of only 4½ in covered with tiles. What is now said is that it was replaced by two layers of brick totalling 9 in, leaving a gap of about 6 in to be filled in on the inside. Mr Canterbury's evidence was that the reinstatement consisted of an 11-in cavity wall with tiling on the inside. He produced a day-work-sheet dated October 9 1976 which described work done in the previous week. Mr Dyer not only denies that this was the work done but further averred, and maintained his averment in the face of the oral and documentary evidence to the contrary, that the hole was not filled until February 1977. That day-work-sheet was signed by the foremen and by Mr Canterbury and was the subject of a claim for a day-works payment. It was sent to the architects under cover of a letter dated October 26 1976. It therefore follows that if Mr Dyer is to be believed, the day-work-sheet is a false document and a fraudulent claim and that Mr Canterbury certified work which had not been carried out. I have no doubt that the work was carried out as stated in the day-work-sheet, and that being so, the credibility of Mr Dyer generally must be questionable. It is conceded that the inside of the wall was tiled, although Mr Dyer did not think much of the type of tile used. Accordingly, I hold that this claim has not been made out. I find similarly in respect of the claims in respect of miscellaneous items quantified at £140.

It is also alleged that no electricity supply was connected to unit 1 until August 1977. In fact, it is clear that by December 1976 SEEBOARD had provided a cable to unit 1 and that by January 29 1977 that cable had been put through into the building by ducting which had by then been installed. From that time onwards the plaintiffs were in a position to ask for the supply to be made live and metred.

The next allegation is that the removal of the machinery by the use of pneumatic drills or like equipment caused extensive vibration to the roof of the premises so as to cause the roof to leak. The roof is flat with panels of glass cemented in, flush with the surface of the roof, which was asphalted. The inside of the roof was lined with hessian, apparently necessary for the processes carried on by Commercial Plastics Ltd. The plaintiffs' evidence that in the autumn of 1976 the roof leaked is not disputed. They also say that there was no evidence of leakage when they inspected the premises early in that year, although on occasions Mr Gower surveyed the inside of the building with binoculars but at no time did he go on the roof. Mr Dawson, the research and development manager employed at these premises by Commercial Plastics Ltd until he left in January 1976, said that the premises were wind and watertight when he left. He walked round the premises at least once daily and said that there was no evidence of any leakage. On the other hand, witnesses called for the defendants gave contrary evidence. Mr D R Graystone ARIBA, who was employed by the architects until June 1976, said that in March when he showed Mr Dyer and Mr Gower over the premises it was apparent that there was damp in the roof caused either by water ingress or condensation as some of the hessian had fallen away. Mr R E W Atkins, a partner in the firm who acted as estate agents for Wyvern, said that he had seen damp marks on the hessian and that it had fallen away in places. Photographs taken on July 30 1976 show some of the hessian lining hanging down from the roof. 1976 was the year of the very long drought and there was no rainfall during the whole of the summer. The question for determination is whether the plaintiff has discharged the onus of proof that vibration caused by pneumatic drills caused cement fillets round the window lights to crack and in some cases fall away, thereby allowing the ingress of rainwater. That pneumatic drills were used is not disputed, but there is no technical evidence that the vibration caused by them could be carried through the structure and cause the damage alleged. Accordingly, I am being asked in effect to conclude *post hoc ergo propter hoc* that because there was vibration the cracks must have been caused by it. However, this does not necessarily follow and it was incumbent upon the plaintiffs to lead technical evidence to that effect. This they have not done and having regard to the conflict of evidence regarding the state of the roof prior to the agreement and the

A possibility that the hot summer could have caused or contributed to any defect, in my judgment this claim fails. It is also significant that the works of repair which were carried out were to the external side of the roof whereas the evidence on which the plaintiffs rely is the internal damage.

Finally under this heading, I must refer shortly to the loading-bay, with which I shall deal in more delail later. It is alleged that it was left in an unusable state, particularly because of an unfilled trench. It is curious that although on May 30 1977 Mr Dyer took nearly 60 photographs of those parts of the premises in respect of which he has complaints and that one of them was of part of the loading-bay, there is no photograph illustrating that particular
B complaint. Mr Canterbury is adamant that the trench had been filled in and surfaced with concrete: it had been done by Wiltshiers and they had paid for it, as is evidenced by day-sheets produced. I have little doubt that Mr Canterbury's evidence is to be believed and that if there was an unfilled trench at the date of completion then it was as a result of work carried out by some authority for which the architects had no responsibility.

It follows from the forgoing that in my judgment the only defect, which was not repaired, for which Wyvern were contractually responsible was the plastering of the doorway, for which the claim is £30. In my view this is *de minimis* and in any event cannot amount to a failure to exercise such proper skill and judgment so as to
C constitute a breach of duty by the architects to the plaintiffs.

Delay

Leaving aside any question arising as to the finality of the certificate, the defendants' case is that the delay in the execution and completion of the works resulted from the removal of the electricity sub-station and was beyond their control. The existing electricity sub-station was within the building behind unit 1. The proposal was for it to be demolished, partly to provide for the loading-bay and partly for the rebuilding of unit 1A. The removal of that sub-station was a matter for the South Eastern Electricity Board (SEEBOARD) and the plaintiffs had started negotiations with
D them early in July and by July 12 had agreed a site and a price. The site selected was adjacent to what was to be a flank wall of the new building, unit 1A. I do not find it necessary to go into all the subsequent dealings with SEEBOARD. It suffices to say that they took an extraordinarily long time in dealing with conveyancing matters and that Wyvern and their solicitors did all they could to expedite matters, including the offering of an indemnity pending completion. That indemnity was sent on September 2 and even then the work was not put in hand because SEEBOARD were not prepared to do that until the new wall had been erected and the roof put over the new building. In the event it was not until March 10 that the new sub-station was commissioned, thereby enabling
E demolition of the old sub-station to proceed. It is said that the defendants could have done more to deal with this problem. In particular, first, that they should have suggested relocating the site of the sub-station, but the defendants say that would have involved starting again from scratch with the negotiations and with the conveyancing processes and consequently there might have been further delay. Secondly, it is said that the architects could have proposed a temporary construction but, as the documents show, that is precisely what Mr Rayner did but SEEBOARD did not pursue that suggestion. I have no difficulty in finding that Wyvern and the architects did everything that could reasonably be expected of them, as indeed it was in their interest so to do. Following the
F erection of the new sub-station, certain arrangements had to be made concerning a supply to another occupier on the estate and equipment had to be removed by SEEBOARD and Commercial Plastics Ltd. The demolition of the sub-station and the consequential works were carried out and completed by April 20. Mr Canterbury, allowing a day for the cement in the trenches to harden, granted his final certificate on April 21. I have no hesitation in concluding that the delay up to that date was by reason of causes beyond the control of Wyvern and the architects. Even if it were otherwise, I am far from satisfied that the plaintiffs suffered any damage by the delay. There are a number of unexplained inconsis- G
tencies with their claim, which is based on loss of rent on leasing offices and profit on the use of the warehouse. In fact as early as August 1976 Mr Dyer instructed Mr Gower to sell unit 1 at an asking price of £375,000, which would have shown about 100 per cent profit. It was not until July 30 1977 that Mr Dyer decided instead to let part of the premises for offices and use the rest as a warehouse. A large part of the claim relates to loss of rents of offices and yet it was not until August 30 1977 that the plaintiffs applied for planning permission for that use.

Mr Boggis submitted that in any event Wyvern's solicitors were wrong to charge interest from May 4 to completion, apparently on the ground that Wyvern did not have a valid certificate under clause H
7(b) at the time they called for completion. I can find nothing in the agreement for sale nor in the National Conditions of Sale which makes it obligatory to have a certificate in hand when calling for completion. Mr Boggis further argued that because the certificate given under clause 7(b) failed to supply a contractual completion date, that date remained October 1 and so Wyvern were liable for damages for failing to complete on that date. I find nothing in the agreement which requires a certificate for extending completion date other than a certificate which states the length of the delay resulting from reasons beyond the control of the vendors. It is true that the clause uses the words "not less than", but for my part I do not see how they could validly certify for a longer period. J

Vandalism

It is here alleged that the defendants owed the plaintiffs a duty to take reasonable care to preserve the premises during the period from contract to completion and that they failed in that duty in that the premises were subjected to extensive vandalism involving, *inter alia*, damage to and theft of the electrical wiring and other electrical installations. It is conceded by Wyvern that they were under a duty to use the care of a prudent owner, but they say that the onus is on the plaintiffs to prove not only the loss or damage but also that it resulted in Wyvern's failure of duty to take care. It seems to me that, K
although Wyvern had a primary duty in this matter, that duty was discharged by putting the building in the care of the architects and builders. The evidence points to such vandalism as occurred as being at the time when the builders were on site.

It seems to me on the authority of *Clarke v Ramuz* [1891] 2 QB 456 that the plaintiffs have to prove that Wyvern failed to take reasonable care of the property. It is conceded that the plaintiffs have no claim against the architects and therefore the architects are involved only as third parties if Wyvern are liable. Having appointed architects, who in turn contractually placed responsibility for security on Wiltshiers, it is difficult to see what more Wyvern could have done and nothing was put to Mrs De Casenove, L
Wyvern's managing director, saying what she should have done or not done. It is true that Mr Dyer said that Mrs De Casenove had undertaken to provide a full-time security guard, but that is entirely unsupported by other witnesses, including the plaintiffs' witnesses, and I do not accept it. As soon as Mrs Casenove was alerted that there was a security risk, namely a missing padlock, she took proper steps. I conclude that the plaintiffs have failed to discharge the onus of proving negligence. In any event, the major part of this claim relates to the rewiring of the electrical system and I have no doubt that that was not rendered necessary by such vandalism, if any, as occurred. Certainly some of the complaints related to equipment used in the supply of electricity to the machinery. Neither the M
electrician who carried out the work nor any other witness gave any evidence which would enable me to determine the extent of liability or to apportion the cost.

It follows from all the forgoing that this action fails.

Judgment was given for the first and second defendants, judgment also for the third party against the first and second defendants, and judgment for the third defendants. It was ordered that the plaintiffs pay the costs of the first and second defendants, the costs to include the costs of the third party proceedings.

LANDS TRIBUNAL

COMPENSATION

A
SMITH v STRATHCLYDE REGIONAL COUNCIL
(LTS/APP/3/111)

(LORD ELLIOTT, president of Lands Tribunal for Scotland, and WM HALL FRICS)

December 5 1980

Estates Gazette January 31 1981 and February 7 1981

(1980) 257 EG 391 & 501

Part of public house in Hamilton, Lanarks, acquired in 1973 — Compulsory purchase known to claimant to be pending from about 1972 — Various schemes for rehabilitating and extend-
B
ing remainder of building (the "rump") were discussed with acquiring authority but proved abortive, and "rump" demolished in 1976 — Compensation agreed payable for total extinguishment of business with valuation on before-and-after basis — Average gross turnover over three years prior to valuation date — YP of 0.92 applied — Value of "rump" as it stood in 1973 with new side wall built by authority — Disturbance claim — Whether expenditure incurred prior to notice to treat in furtherance of mitigation of loss is compensatable — Tribunal says "yes" — Only reasonable to take steps beforehand to ensure any temporary closure of business is as short as possible — Acquiring authority's submissions based on alleged significance of deemed notice to treat rejected — Bloom (Kosher) & Sons v Tower Hamlets not followed — Simply a question
C
of circumstances in each case whether particular losses or expenditure are naturally and reasonably incurred by reason of acquisition or are too remote and whether steps are in furtherance of duty to mitigate loss — Professional fees incurred both before and after notice to treat allowed — Loan interest payable owing to claimant's personal financial circumstances too remote and disallowed — Demolition costs for "rump" similarly disallowed.

D Hope QC and I G Mitchell (instructed by Ballantyne & Copland, of Edinburgh) appeared for the claimant; J Fleming, solicitor (chief executive's department, Strathclyde Regional Council) for the acquiring authority.

D
Giving the opinion of the Lands Tribunal for Scotland, THE PRESIDENT and MR HALL said: This reference is to determine the amount of compensation payable by Strathclyde Regional Council (the authority) to Andrew Fleming Smith (the claimant) for the acquisition of part of his property "The Golden Feather" public house at 1 Shawburn Street, Burnbank. The area of land acquired extends to 228.67 m² and was acquired under two compulsory purchase orders authorised under the Roads (Scotland) Act 1970. The first was the Burgh of Hamilton, 1 Shawburn Street (Hamilton) Compulsory Purchase Order 1972 confirmed by the Secretary of State on March 29 1972. This authorised the purchase of an area of 159.67 m². The second was the Burgh of Hamilton, Burnbank Road Hamilton Compulsory Purchase Order 1972 confirmed on
E
June 6 1972 which authorised the purchase of a further area of 69 m². The land acquired was needed for the realignment of the Hamilton to High Blantyre Road.

General vesting declarations were made on April 4 1973 and the land was vested in Hamilton Town Council on November 11 1973. In terms of paragraph 6 of Schedule 24 to the Town and Country Planning (Scotland) Act 1972 the date of the deemed notice to treat is April 4 1973 and the date of valuation is November 11 1973. Strathclyde Regional Council are statutory successors to Hamilton Town Council as roads authority.

On the two areas of land acquired stood part of the public house F
known as the Golden Feather. The remainder stood on other land and hence the complication of the present application. At the valuation date, the building was in part 12 and in part five years old. It was two storeys in height, built of brick which was rendered and it had a flat roof. The gable fronting Burnbank Road had a stone facing. The accommodation was as follows:

	Net floor area *sq ft*
Basement containing cellar with concrete floor	1553
Ground floor	
Public bar	784
Toilets and store	413 G
Games room/off-sales	425
Lounge bar	696
Upper floor	
Function room	1123
Toilets and kitchen	297
Eight bedrooms	1023

The subjects were situated between Peacock Cross and Burnbank Cross, Hamilton, but somewhat closer to the latter. At the valuation date, there was an adjacent area of housing with shops and containing seven public houses. But most of these have since been demolished and replaced by new local authority houses, some shops and a new public house under what is conceded to be a H
separate planning scheme. The Greenfield Social Club, which opened on March 19 1971, is at the opposite end of Shawburn Street.

In brief outline (we consider the details later) the prior history of the subjects is as follows:

The claimant built the Golden Feather in 1961 on part of an area of 0.859 acre. In 1967-68 he also built an extension for an hotel which included eight bedrooms and a function suite, but the bedrooms were never used apart from two bedrooms used as manager's accommodation. A provisional grant of an hotel licence was also obtained, but owing to ground and foundation problems the full extension to comply with the conditions of the licence was not built. J
Following annual renewals the provisional grant was later withdrawn in March 1973 prior to the valuation date.

The original building had an over-walls area of 5,175 sq ft, 1,725 on each of the two floors and 1,725 cellarage. The extension provided an additional 1,181 sq ft on the ground floor and 1,037 on the first floor. The total over-walls area at the valuation date in November 1973 was 5,668 sq ft excluding the cellarage. The ground adjacent to the public house and also in the claimant's ownership was used as a car park.

Only part of the building was demolished for the road widening, but the demolition extended beyond the limit of the compulsory
purchase order. In early 1974 an area of 2,147 sq ft plus part of the K
cellar was taken down and the authority, with the claimant's agreement, built a wall, tied in to the remaining part of the building with the necessary beams and lintel, to provide a structurally complete wind and watertight enclosure. The wall intersected the cellar, public bar and function room and removed the access to the remainder of the cellar and to the first floor. The intention at the time was that the portion of the building left standing — or "the rump" as it was called by the parties — would be adapted for use as it stood with an extension.

A During the period of negotiations the licence was renewed and, following the closing of the business, has been kept in force. In 1976 it was renewed on condition that the rump be demolished within six months.

Value of going concern

The basis of compensation was by the end of the hearing agreed. At the commencement of the hearing it was conceded for the acquiring authority that there was, at the valuation date, total extinguishment of business, so the valuations were presented on a "before and after" basis, ie by valuing the subjects before the acquisitions as a going concern and then deducting the value of the
B remaining rump. The claimant's valuer had in the alternative submitted that as the claimant had been unable to make use of the rump and eventually demolished it, there was no value left to be offset against the "before value". This basis was not, however, in the end of the day pursued or referred to in the closing submissions. We have, of course, to assess what a purchaser would have paid for the heritage *rebus sic stantibus* as at November 1973.

The valuations were prepared and spoken to in evidence by J S Traill, MSc FRICS MRTPI for the claimant and R I Brown ARICS, of Inland Revenue Valuation Office, for the authority. Their valuations are as set out below.

In addition to the £70,000 there is a disturbance claim sub-
C mitted by the claimant on the basis of a total loss of business of £9,866.98, which we deal with later.

Mr Traill and Mr Brown referred to comparable transactions in support of their valuations the details of which are set forth in the Appendix opposite.

Claimant's valuation

1. *Value before acquisition*			
Estimated average drawings £75,000			
D Multiplier	1	£75,000	
E *Add* for bedrooms			
1,023 sq ft @ £5		5,115	
		80,115	
		say	£80,000
2. *Value after acquisition*			
ground floor 1,607 sq ft @ £5		£8,035	
F first floor 1,607 sq ft @ £1.25			
		2,009	
		10,044	
		say	£10,000
		Compensation	£70,000

Acquiring authority's valuation

1. *Value before acquisition*			
Gross drawings			
1969-70	£46,759		
1970-71	£44,681		
1971-72	£46,674		
	£138,114		
Average for period	46,038		
Multiplier	0.87	£40,054	
Supporting calculation			
Average drawings as above	£46,038		
Deduct for drawings needed to pay for entertainers £600 × 4	2,400		
	43,638		
YP	0.9		
	£39,274		
Add for whatever potential there may have been in the bedrooms			
1,120 sq ft @ £4		4,480	
		44,534	
	say	44,500	
Deduct for movables	sum	2,000	
			£42,500
2. *Value after acquisition*			
Site	sum	£5,000	
Add for value of structure remaining including new gable provided as accommodation work 3,720 sq ft @ £2.50	say	9,000	
			14,000
Difference			28,500
Add for loss on forced sale of movables			1,500
			£30,000

APPENDIX G

Sales of public houses

Address	*Turnover (average)* £	*Price* £	*YP*	*Date*
Mr Brown				
In Hamilton				
Barleybree	69,000 (previous 3 years	52,500	0.77	October 1972
Birkmill Road	100,000 (previous 3 years)	97,000	0.98	October 1974
Victoria Bar Quarry Street	34,400 (previous 3 years)	21,000	0.61	October 1972 H
Modern properties elsewhere in Central Lanarkshire				
Sholto North Road Coatbridge	68,000 (previous 3 years)	59,519[1]	0.875	March 1973
Mr Traill				
Sholto North Road Coatbridge	71,691[3] (previous 2 years)	65,000[1]	0.91	October 1973
Clachan Bar Barrachnie	80,319 (1971-73)	75,000[2]	0.93	1976

NOTES 1. dispute about price paid. J
2. dispute about price of house included in sale.
3. if £68,000 year's purchase is 0.95.

The claimant, Andrew F Smith, spoke to the effect of the acquisition on his business. He said he had built the Golden Feather in 1961 and then made a substantial extension about 1967-68. He said that he had other business interests and had employed managers to run the Golden Feather although he had also assisted. The claimant decided in the winter of 1970 to advertise the business for sale and to sell it if a price in the region of £65,000 could be obtained. Ind
Coope were interested, but he said that they told him they were K
annoyed when they learnt on making inquiries from the local authority that the property was affected by road proposals.

The Burgh of Hamilton then indicated their interest in the whole property and the district valuer, on their behalf, wrote on May 18 1971 saying he had been asked to negotiate on their behalf. On July 30 1971 the town clerk himself wrote to Mr Smith's solicitors confirming that the property was affected by the proposed development. He informed them that the town council had approved a plan for the realignment of the road which required that at least part of the property would need to be demolished. The claimant's solicitors, Ballantyne & Copland, replied on September
15 1971 saying that the only practical course appeared to be to L
demolish the whole building and erect a new building on the remaining ground. This was Mr Smith's wish.

A meeting was then held on September 23 1971 at which it was proposed by the town clerk that the rump should be adapted. Detailed discussions took place between the authority's officials and the claimant's advisers on alternative schemes of adaptation, but in August 1972, the town clerk wrote to confirm that a compulsory purchase was to be made with entry on November 11 1973. The order was made and the claimant ceased trading on or about the same date.

The effect of all the delay and uncertainty, according to Mr
Smith, materially affected the running of his business — stock was M
reduced, decoration was not carried out, no forward booking for functions could be accepted and he had difficulty in obtaining a good manager.

There are several headings of dispute in the valuation of the going concern. It is agreed that licensed premises should generally be valued by applying a multiplier (or years' purchase) to average gross turnover over the three years prior to the date of valuation.

(i) *Turnover.* The accounts for the year ended October 16 1967-1972 were produced. The profit and loss account for the year 1972-1973 was not available as the books were destroyed in a fire caused by vandals.

A The turnover for the various years is as follows:

1967	£42,385	1970	£46,759
1968	£41,758	1971	£44,681
1969	£51,333	1972	£46,674

Mr Brown said that it was normal practice to use the figures three years prior to the valuation date. As the accounts for 1973 were not available he had used the last three years for which information was available. Mr Traill agreed with the practice but said that it was also necessary to analyse the accounts to ascertain if they are representative of the property and of assistance in determining the average gross turnover. He did not consider the accounts after 1970 were of
B much assistance because of the problem of supervision, lack of control, staff difficulties and the run-down of the business prior to closure. Mr Traill said that he considered the 1967-69 accounts were representative of the proper trading of the business and, using a straight-line projection, the average turnover for the period 1971-73 would be £71,556.

Turnover in licensed premises generally rose steadily in the 1970s owing to increased prices and Mr Traill had analysed the turnover of five public houses in Burnbank, which showed that the increase in turnover from 1969 to 1973 was an average of 1.64 times — the individual increases range from 1.47 to 1.81. The average turnover related to floor area of drinking space represented
C £23.86 per sq ft in 1973. The individual rates range from £13.43 to £31.96. The average rate per sq ft in 1969 was, according to Mr Traill's analysis, £15.52. Applying a rate per sq ft, slightly higher than the average for Burnbank to the ground floor and slightly less for the upper floor, Mr Traill calculated a turnover for 1973 of £73,000 as follows:

Ground floor		
Public bar 784 sq ft		
Games room 425 sq ft		
Lounge bar 696 sq ft	1,905 sq ft @ £26.50 =	£50,482
First floor		
D Function room	1,123 sq ft @ £20.00 =	£22,460
		£72,942
	say	£73,000

Mr Traill also considered the increase in the retail price index group II: alcoholic drink: and found that the December 1973 figure was 1.21 times the November 1969 figure. Taking the 1967, 1968 and 1969 turnovers and increasing them relative to increases in personal expenditure for the years 1971, 1972 and 1973 gave an average of £71,339. Having regard to all that information and assuming that the business had been efficiently managed, he esti-
E mated that the average turnover for the years 1971-73 should have been £75,000.

Mr Brown said that he was not aware of any reason associated with the scheme which would have resulted in any reduction of the drawings in the three years he had used to find the average. The adjacent area was being redeveloped as a clearance area, but this was not the scheme under which the subject property was acquired. He considered that the poor turnover record was due, firstly, to the closure and demolition of houses in the vicinity of Burnbank Cross over the period — although this should have been offset to some extent by the closure of other public houses towards the end of the period. Secondly, and in his opinion more material, the opening of
F the Greenfield Social Club on March 19 1971. This club, he said, had approximately 600 members and the original two-storey premises had been extended three times since 1971, the first of these extensions being in 1973. The accommodation now extends to over 15,000 sq ft and comprises a large hall for functions, lounge bar, cellar, kitchen and toilets on the ground floor with two lounge bars, function room, large games room and toilets on the upper floor. He understood that the gross turnover from the bars in March 1973 was over £100,000 and by 1979 almost £400,000.

Mr Smith disputed that the social club affected his business, as it was a different type of facility. He suggested that the club was 800 yds away from the Golden Feather.

We have considered the evidence and inspected the locus of the G
subject property and the Greenfield Social Club. Although the roads have changed it appears to us that the club is about 300 yds from where the Golden Feather was located near the garage. It is agreed that the Golden Feather's turnover has not followed the general pattern of turnover of public houses in the period 1969-1973. Between 1967 and 1969 the turnover increased in a way which one would have expected following the opening of the function suite, which, according to Mr Smith, was in 1968. However, in the year ended October 16 1970, the turnover had dropped and it was at that stage Mr Smith decided to sell. Ind Coope informed him early in 1971 that they had been told that the property was to be acquired. The district valuer wrote him on May 18 1971 saying he H
had been instructed to enter into negotiations. The Greenfield Social Club opened on March 19 1971. The turnover for the year ended October 16 1971 dropped by about £2,000. For the year ended October 16 1972, the actual turnover recovered to the 1971 level, but in real terms this was a drop relative to 1970.

As the road scheme could hardly have affected the 1971 figure the drop must be due to some other cause. Apart from management and the clearance of houses in Burnbank, the only other influence would appear to be the opening of the social club. The house clearance does not appear to have affected the other public houses in Burnbank over that period and Mr Smith did not appear to have management problems until 1972. The creation of a function suite J
with dining facilities does not appear to have materially enlarged the business and it would be this part of the going concern which would be most vulnerable to the competition of the social club. We accept that by 1972 the proposed acquisition would affect the running of the business so far as staff, management and forward bookings were concerned. In the normal way one would expect an increase in turnover in the public bar even if the business did not expand, because prices were increasing. We have come to the conclusion that a purchaser would consider the year 1971 as the most reliable to start from as it reflected the opening of the Greenfield Club, but he would then, in the absence of the acquisition, anticipate that the turnover would increase broadly in line with the K
indices of alcoholic drink expenditure as set out and adopted by Mr Traill. The effect of these calculations is therefore that the anticipated turnover would be the average of the following figures: in 1971 — £44,681, 1972 — £50,495, 1973 — £56,616, which produces an average figure of (say) £50,000.

Disregarding, in terms of section 16 of the Land Compensation (Scotland) Act 1963, the depreciation caused by the proposal to acquire, we conclude that a purchaser in November 1973 would estimate the turnover at £50,000.

(ii) *Multiplier.* Mr Traill applied a multiplier of 1.00 to his estimated drawings of £75,000 to give a value of £75,000 for the going concern including the property, the goodwill and fixtures and fit- L
tings. He considered this multiplier to be appropriate for the size, type and age of the building and the level of potential turnover adopted. He also had regard to the sale of the Sholto Bar, Coatbridge, and the Clachan Bar, Barrachnie, which he had respectively analysed at 0.91 and 0.93.

Mr Brown applied a multiplier of 0.87 to his calculated average turnover of £46,038 to give a value of £40,054 for the going concern. In arriving at the multiplier he had considered that a purchaser would have regard to the reduction of the drawings in real terms and to the fact that extraordinary expenditure was incurred in employing entertainers. He also took into account his opin-
ion that the building was of cheap construction, not well- M
maintained and in need of immediate repairs and renewals. The building, in his view, was over-large having regard to the competition and there was no evidence that any significant amount was paid in the open market for potential. He accepted that a multiplier of 1.00 would have been appropriate if the Golden Feather had been a good modern public house with a turnover at least holding its own in real money terms.

Mr Brown referred to the sale of two public houses in Hamilton and also to the Sholto Bar, Coatbridge, relied on by Mr Traill. The Barleybree in Hamilton sold in October 1972 at a multiplier of 0.77 of the average turnover of the previous three years; and again in

October 1974 at 0.98 of the turnover, which by that time had increased from £69,000 to £100,000. The Victoria Bar, which he referred to, was an older public house, small in area and having drawings of £34,000. It sold for £21,000 in 1972, which represented a multiplier of 0.61. Reference was also made to acquisition settlements for seven older public houses at or around Burnbank Cross. The valuation dates in six of the cases were much earlier than November 11 1973 and multipliers in these cases ranged from 0.9 to 0.76. One of the settlements in September 1974 produced a multiplier of 0.8.

Mr Traill did not have the opportunity of seeing the building at the valuation date. Mr Brown saw it and photographs were produced to the tribunal. Mr Brown agreed, however, that there were no structural defects in the property and that the disrepair due to vandalism could have been remedied. He also agreed that the question of appearance was subjective and that the stone facing, which did not appeal to him, had been a requirement of the planning authority.

Mr Brown said that his information about Mr Traill's comparisons of both the Sholto Bar and the Clachan Bar indicated that Mr Traill's facts were incorrect. The average turnover of the Sholto Bar had been taken only over two years instead of the customary three years; the price quoted of £65,000 was not correct according to his records — his information was that the price was £59,519 (apportioned at £29,519 for the property and £30,000 for the goodwill etc). On the adjusted figure he said the multiplier was 0.875. The price for the Clachan was paid in 1976 and he understood that the average turnover for 1973-75 was £88,000 — the price also included a flatted dwelling-house, which he understood had been agreed by the parties to be priced at £12,500 compared with the £6,000 apportioned by Mr Traill. The effect of these adjustments was to reduce the multiplier to 0.78. In connection with the Clachan, Mr Brown said that in the year after the sale the turnover was £114,118, which showed that the potential did not appear to be reflected in the price paid by the purchaser.

This evidence is difficult to reconcile but is perhaps not so important when Mr Brown concedes that he would have agreed a multiplier of 1.00 with the conditions already referred to. So far as the Sholto Bar is concerned, the apportionment of £29,519 to the property is unusual and it appears to us that the balance of the £65,000, is probably attributable to the apportioned price for the movables. If the £65,000 is related to the £68,000 turnover the multiplier is 0.95.

We have decided that the turnover and thus the value of the relevant interest was affected by the indication that the property was to be acquired by the Burgh of Hamilton. We therefore consider that Mr Brown's multiplier should be increased, seeing that he had reduced the multiplier because of the effective fall in turnover. Conversely, Mr Traill said that the multiplier tends to increase with turnover and he would have taken a lower multiplier on lower drawings. We find that the minor repairs and decoration would not substantially reduce the price which a purchaser would pay, as it is likely that he would probably redecorate in any event. We also find that the building was a modern one, conforming to the requirements of the planning authority and the building authority; but not with a high standard of internal finish. We do not accept Mr Brown's view that a purchaser would consider the building generally to be of cheap specification and unattractive and reduce his offer materially on that account.

Increasing Mr Brown's multiplier for condition, appearance and higher turnover and decreasing Mr Traill's for lower turnover and also his wrong assumption that the subjects had, at the valuation date, the capacity of being classified as an hotel, we determine the appropriate multiplier to be 0.92.

(iii) *Bedrooms.* The opinions of value expressed by Mr Traill and Mr Brown were £5,115 and £4,480 respectively. Both are experienced valuers but neither was able to produce any comparable evidence to support his figure and it would be difficult for them to do so. Mr Brown arrived at his figure by applying one-half of the average cost rate for domestic building. This is a difficult item to assess and we determine the value at £4,750.

(iv) *Movables.* In making a valuation under rule (2) as a going concern Mr Brown deducted a sum of £2,000 representing the movable items and assumed that a price of £500 would be received for these on a forced sale. He therefore added £1,500 for the loss on forced sale of these items. While this is strictly an item of disturbance, like Mr Brown we agree that it should be covered at this point. This was the only evidence before us on this aspect and was not challenged by the claimant. We therefore accept the figure of £1,500.

Value after acquisition

The values for the remaining part of the building and site proposed by Mr Traill and Mr Brown were £10,000 and £14,000 respectively. Again these were expressions of opinion not supported by evidence of direct comparable sales. Mr Brown produced details of three transactions of sites on which new licensed premises have been erected. These produced relatively high figures, the capitalisation of the ground rent in two of the cases being in excess of £20,000. He had valued this particular site at £5,000 and added £9,000 for the structure of the building, which represented a rate of £2.50 per sq ft of over-walls area. Mr Traill valued the building including the site at about the same rate but on a net-floor-area basis. We were impressed by the point made by counsel for the claimant that the relationship between Mr Brown's value of the whole as a going concern at £40,000 and the rump at £14,000 was out of proportion. We agree that this does seem too narrow a difference and, in all the circumstances, we determine that the value of the rump in November 1973 was £10,000.

Tribunal's summary

The compensation for the value of the property taken and the severance to the remainder with an adjustment for the movables is therefore £40,250 as follows:

Value before acquisition			
Turnover	£50,000		
Multiplier	0.92	£46,000	
Add bedrooms		4,750	
		50,750	
Deduct for movables		2,000	£48,750
Value after acquisition			10,000
Difference			38,750
Add for loss on forced sale of movables			1,500
Compensation for value of property and severance			£40,250

Disturbance

In addition to the above claim a disturbance claim is submitted under rule (6) as follows:

(a) Loss of profits	£3,856.00	
(b) Professional fees–		
(i) Robert C Brown, chartered quantity surveyors, Hamilton	£339.08	
(ii) Cullen, Lochhead & Brown, architects, Hamilton	£2,319.90	
(iii) James Rusk & Associates, chartered surveyors	£280.00	
(iv) Henry J Crone & Partners, chartered surveyors, Glasgow	£420.00	
(v) Ballantyne & Copland, solicitors	£750.00	
(c) Loan interest	£902.00	
(d) Demolition costs	£1,000.00	£9,866.98

Two legal points are raised by the solicitor for the acquiring authority which it is appropriate to deal with first before turning to the specific items of claim.

It was first submitted that as the entire loss of goodwill was being compensated under rule (2) by applying a multiplier to turnover there would normally be no further claim (as included in the above fees) in respect of the investigation of alternative premises. For

A otherwise double compensation would be involved. We were referred to the case of *Prestwick Hotels Ltd v Glasgow Corporation* 1975 SLT Notes 53 and also to the well-known dictum of Scott L J in *Horn v Sunderland Corporation* [1941] 2 KB 26 at p 42. In the *Prestwick Hotels* case compensation was based upon the total extinguishment of the hotel business, including all its goodwill, thus reflecting the expectation of future profits. In these circumstances the Court of Session disallowed, as other items of claim, the costs incurred in finding a different hotel in a different location and in negotiating its purchase as a going concern; as also a further sum to reflect the cost of the claimant's own time in searching for new premises. However, the Lord President observed (*supra* at p 54):
B "it must be emphasised that no suggestion was made that any expense was incurred in discovering the business could not be transferred elsewhere". While no element of double compensation is of course allowable, claimants may be forced to look for another house or to try to relocate their business and it is also their duty to mitigate their damage. In view of the above observation and also for reasons which we shall develop later, involving the duty to mitigate damage, we consider that expenditure incurred thereon may be recovered if proved to have been naturally and reasonably incurred because of the acquisition.

The next related point is whether, even so, any such expenditure, however reasonably incurred, is compensatable if incurred prior to
C the date of the notice to treat.

Counsel for the claimant submitted that such expenditure was an allowable item, provided it was not too remote and fell within the dictum of Romer L J in *Harvey v Crawley Development Corporation* (1957) 8 P & C R 141, where his Lordship stated (at p 148) "Any loss sustained by a dispossessed owner (at all events one who occupies his house) which flows from a compulsory acquisition may properly be regarded as the subject of compensation for disturbance, provided, first, that it is not too remote and, secondly, that it is the natural and reasonable consequence of the dispossession of the owner". Counsel also referred us to *Venables v Department of Agriculture for Scotland* 1932 SC 573.

D The solicitor for the acquiring authority, on the other hand, submitted that no expenditure, however reasonably incurred, prior to the date of the deemed notice to treat was allowable unless the subject of an express contract to that effect made with the acquiring authority beforehand. He also relied upon the decision of the English Lands Tribunal in *Bloom (Kosher) & Son Ltd v London Borough of Tower Hamlets* (1978) 247 EG 1091.

There was no actual notice to treat served in the present case for, as in the case of most Scottish compulsory acquisitions, this was effected by general vesting declaration made under section 278 of and Schedule 24 to the Town and Country Planning (Scotland) Act 1972. Paragraph 6 of that schedule provides that at the end of the
E period specified in the general vesting declaration the provisions of the Lands Clauses Acts shall apply as if, on the date on which the declaration was made, a notice to treat had been served on every person on whom, under section 17 of the Lands Clauses Consolidation (Scotland) Act 1845 the acquiring authority could have served such a notice. The relevant date in this case was April 4 1973.

The question now arises whether this date constitutes a dividing line on one side of which "disturbance or other matters" mentioned in rule (6) may be claimed; and on the other side of which they may not, even if the items of expenditure were "naturally and reasonably incurred" or "the natural thing for a dispossessed owner-occupier to do" — to quote other passages from the same page of
F Lord Romer's judgment in *Harvey v Crawley*.

Both Romer L J and Denning L J in *Harvey* relied on the Scottish case of *Venables*, which was one of the first cases in which the nature of a disturbance claim was reconsidered following the passing of the Acquisition of Land (Assessment of Compensation) Act 1919 in which the so-called rule (6) first appears.

In *Venables* the Court of Session emphasised that claims for disturbance do not arise *totidem verbis* from section 2(6) itself but are really based on the original provisions of the Lands Clauses Acts. Lord Justice Clerk Alness referred in particular to sections 17, 19, 71 and 114 of the Lands Clauses Consolidation (Scotland) Act 1845. These sections, which have their English counterparts, will be seen to refer variously to compensation for damage or for G any loss or injury that may be sustained. Hence rule (6) gives no new right but merely preserves the existing right to be compensated not only for the taking of the land but also for consequential loss caused by the compulsory acquisition. Lord Justice Clerk Alness said (at p 581): "The sound principle would seem to be that a person dispossessed should get compensation for all loss occasioned to him by reason of his dispossession". He also quoted with approval a dictum of Lord Kinnear in *Lanarkshire & Dunbartonshire Railway Co v Main* (1895) 22 R 912 at p 919: "It is a well-settled rule in the construction of the Lands Clauses Act that when lands have been taken in the exercise of powers of compulsory purchase, the owner or occupier, as the case may be, is entitled H not only to the market value of his interest but to full compensation for all the loss which he may sustain by being deprived of his land".

The tribunal do not consider that the reference to "full compensation" can be read as supporting the view that loss, injury or expense reasonably incurred prior to the date of a notice or deemed notice to treat is not to be compensated. On the contrary it is couched in language apt to embrace prior loss, provided always that it is naturally and reasonably incurred and can truly be described as loss or expenditure incurred through the claimant's being deprived of his land as opposed to something extraneous or due to some independent business decision of his own. *Harvey* followed *Venables* without criticism and we can find no contrary statement of J principle therein. Romer L J plainly regarded the main test to be that of remoteness, saying (at p 14) "(objection) might have been raised with regard to the expenditure which was incurred by Mrs Harvey in connection with the abortive purchase of a house which she was advised by her surveyor was not a reasonable residence for her". But he concluded by saying: "It is bad enough in itself for a person to be compulsorily dispossessed of his house, but it is worse still if he has himself to bear expenses of the kind which are in issue in the present case in finding another house in which to live".

In the light of that dictum, we cannot think that his Lordship, by using the words "in consequence of" rather than "by reason of" in a case in which all the expenditure incurred was after the notice to K treat can have intended that expenditure by a displaced householder should be disallowed if it preceded that date; nor can we see any reason why there should be such disallowance if the dispossessed claimant is to receive full compensation for all loss and injury sustained by reason of the compulsory acquisition.

At the time of the Lands Clauses Act compensation was, of course, required to be paid before entry could be taken and it would be seldom, if ever, that expenditure would be incurred prior to the service of the notice to treat.

The principle in assessing compensation is that of equivalence, described by Scott L J in *Horn v Sunderland Corporation supra* as: "The right to receive a money payment not less than the loss L imposed on him in the public interest, but on the other hand, no greater". Why then, more particularly since the decision of the House of Lords in *Birmingham Corporation v West Midland Baptist (Trust) Association* (1969) 211 EG 527 and 629 (HL), should the service of a notice to treat operate so as to prevent full compensation for loss or expenditure reasonably incurred? As appears from the modern textbooks dealing with the subject, the point is still in law an open one: see *Corfield and Carnwath on Compulsory Acquisition and Compensation*, p 351, and *Leach on Disturbance* 3rd ed p 31.

As already remarked, there was no actual service of a notice to treat in the present case. Nor is there in the case of a deemed M compulsory purchase following the issue of a blight notice. There is only a deemed notice to treat in terms of section 185(2) of the Town and Country Planning (Scotland) Act 1972. Under the equivalent Northern Ireland legislation the Lands Tribunal for Northern Ireland have held that expenditure incurred prior to the deemed contract to purchase in endeavouring to market the blighted property and find an alternative house was recoverable as compensation as being the natural and reasonable consequence of the blighting of the interest: see *McGonigal v Department of the Environment for Northern Ireland* [1976] RVR 56.

There is also no actual or even deemed notice to treat where an

A authority possessing compulsory purchase powers voluntarily purchases property without using those powers. Compensation may then require to be assessed by the tribunal under a voluntary reference as if for a compulsory purchase. In the case of *Sisters of Charity of the Order of St Vincent de Paul v Lanark Town Council* 1974 SLT (Lands Tr) 5 a voluntary reference was made to this tribunal in which it was contended that the completion of missives was the equivalent of a notice to treat prior to which no expenditure incurred was compensatable. This contention the tribunal rejected in principle. On the evidence, however, they found that the claimants' unsolicited attempt to sell their disused hospital to the authority at a time when the latter had no compulsory purchase in mind
B was an independent action on their part which made the relevant expenditure too remote.

We have direct evidence of the prevalent practice spoken to by the district valuer in this case of allowing pre-notice expenditure if not too remote, provided always that it is stipulated for in the course of negotiations and is the direct and reasonable consequence of the compulsory purchase. The same result, but more concealed this time, occurs when surveyors' fees based upon clause 5(a) of the RICS scale of professional charges are allowed as compensation. For under this scale the fee is an *ad valorem* one based upon the final determination of compensation due. The latter, however, cannot distinguish between fees incurred prior to or after the
C service of a notice to treat. We consider this practice to be in accordance with the law.

It thus appears that current valuation practice does not always accord with what is here contended to be a legal rather than a practice rule, namely that expenditure incurred prior to the date of the notice or deemed notice to treat fails to be excluded. Where there is an actual notice this merely signifies that the acquiring authority is taking up the compulsory powers confirmed by the Secretary of State under a compulsory purchase order which they have promoted. This is all part of an ongoing process which starts with the scheme behind the compulsory acquisition.

Under the general vesting declaration procedure ordinarily used
D in Scotland there is also only a small gap (usually 28 days) between the making of the vesting declaration when there is deemed service of a notice to treat and the expiry of the date mentioned therein when in terms of para 7 divestiture occurs and in terms of para 30 compensation is payable. In this case the gap was a longer one, from April to November 1973; but, even so, why should a claimant have to delay before starting to find alternative accommodation or to take other steps to ensure that total loss of goodwill does not occur? It is surely only reasonable to take steps beforehand to ensure that any temporary closure of business is as short as possible. In the present case the claimant was, as early as 1971, encouraged by the authority to consider preserving and adapting the rump for con-
E tinuance of business. While their efforts were abortive, it did not seem so at the time, even to the acquiring authority. It is, of course, a claimant's duty to mitigate his loss and so lessen the compensation otherwise payable.

The measure of loss on compulsory purchase has been equated to that in delict or tort, where the duty to mitigate damage also arises. Compulsory purchase cases, however, differ from delict in that pending loss, prior to the acquisition, can often be seen from afar; so anticipatory action to mitigate loss becomes more feasible. If successful this lessens the compensation which the acquiring authority would otherwise have to pay. Even if abortive, as in this case, such expenditure on investigating ways of mitigating the loss
F may exceptionally be recoverable under rule (6) if reasonable in the circumstances and proved to be referable to the compulsory purchase. As indicated in *Prestwick*, this may be in addition to a claim for the total extinguishment of the business which is eventually found to be unavoidable.

The main question, however, recurs, namely whether expenditure in furtherance of mitigation is only compensatable if incurred after the date of service of the notice to treat or whether the claimant is entitled to have prior expenditure incurred in endeavours to mitigate his loss also taken into consideration. We consider that he can.

In our opinion, following *Venables*, *Harvey* and the *Birmingham Corporation* cases, we therefore reject the legal submissions based G
upon the alleged significance of the deemed notice to treat. In our judgment it is simply a question of circumstances in each case whether particular losses or expenditure claimed as items of compensation under rule (6) were naturally and reasonably incurred by reason of the compulsory acquisition; or, in contrast, were too remote as being independently incurred. And, where questions of anticipatory mitigation of damage arise, whether the steps taken to avoid what would otherwise be greater loss were reasonable steps to take in all the circumstances.

With these preliminary observations we then turn to the various disputed fees now claimed.

Item (a) loss of profits. The first item of claim is for loss of profits. H
This was based on the assumption that the rump of the building was adapted and used for the continuation of the business. It was assumed that there would be a one-year building period and profits would be lost for that year. The authority argued that this item should be disallowed because they were conceding that the claim should be dealt with on the basis of total extinguishment of the business. In the end of the day, this appeared to be accepted by counsel for the claimant. We agree with the authority's contention and disallow this item of claim.

Item (b) professional fees. The next item in the disturbance claim is in respect of professional fees incurred by the claimant, most of which relates to the consideration of the adaptation of the rump. It J
was the authority who proposed that the rump be adapted, and a probable cost of £56,000 was calculated by R C Brown, the claimant's quantity surveyor. Mr Smith agreed reluctantly to discuss adaptation, and sketch plans which have been produced were prepared by the architects, Cullen, Lochhead & Brown, for an extension to, and adaptation of, the rump. Discussions took place about these, but by August 21 1972 site problems cast doubts on whether the extension to the rump could be built.

As already mentioned, the town clerk wrote to confirm that a compulsory purchase was to be made taking possession of the property on November 11 1973. He also set out the necessary K
permissions which Mr Smith would have to obtain in order to be able to have the building ready and move into it before that date. In the meantime, plans had been lodged with the licensing court for Mr Smith's own scheme of an entirely new single-storey building and were approved by them. A revised cost for adaptation was given by the quantity surveyors in October 1972 in the sum of £70,300.

It will be observed that the date of the deemed notice to treat when the authority made their general vesting declaration here intervenes on April 4 1973.

Further discussions appeared to have taken place during 1973, and in November 1973 following suggestions by the burgh architect L
a revised scheme of adaptation was prepared. The financial position at that time was complex. A payment to account of £18,900 had been made at the end of 1973, but it is said that this was insufficient to repay the loans secured over the property. No offer of compensation had been received from the authority and the claimant said he was unsure of his financial position. Agreement that only part of the building be demolished and a new gable be built by the authority was eventually reached. The architects were involved in the planning of this and also in the supervision of the building of the gable. Planning permission was granted on February 12 1974 for the reconstruction and extension of the rump. Demolition of all but the rump and the new gable was completed by April 1974. M

In July 1974, Ushers, who had been approached about financing the scheme, wrote saying that they were not prepared to lend because of the claimant's strained financial resources. Because of these problems the question of the authority's acquiring the whole was again canvassed; but the town council replied saying that they had always been willing to acquire the whole premises but such an acquisition would require to be unconditional, ie the town council would have been free to dispose of the site in any way they thought fit. The claimant in March 1975 obtained an offer of a loan from the Guardian Royal Exchange Insurance Group.

The reasons for the delays are confusing and it is pointless to try

A to apportion blame. It is pertinent, however, to note that it was Hamilton Town Council who were directly responsible for encouraging the investigation of — and hence the abortive expenditure in relation to — the abortive scheme for preservation and adaptation of the rump even to the extent of themselves building the gable wall; and this, of course, well before the date of the deemed notice to treat. It was the town council who were thus the authors of the attempt to continue trading in the rump. Mr Smith was never very keen on this idea, although he went along with it, and we consider all his expenditure thereon — even prior to April 4 1973 — is allowable. Mr Smith has been forced into these complications by being deprived of his property and the expenditure after B July 1971 in meeting the authority's suggestion on adaptation should be reimbursed to him. We would add that Mr Smith himself is not claiming for any expenditure in respect of his own professional scheme for an entirely new building.

We now turn to the details of the fee expenditure. In case the matter is taken further we have apportioned the amounts to show what was incurred before and after April 4 1973.

(i) *Robert C Brown, quantity surveyors.* This fee of £339.08 is specified as being solely for the extension proposal and we accept it. The fee for the period before April 1973 was £203 and after that £136.08.

(ii) *Cullen, Lochhead & Brown.* Lockhart W Hutson, formerly C senior partner of the firm, explained the account of £2,319.90. In November 1971 he was instructed to prepare a scheme for adaptation and extension of the rump. Sketch plans were prepared and used for discussion with the authority's officials. In October 1973 a revised scheme of adaptation and extension of the rump was suggested by Mr Wyler, burgh architect, and a scheme was prepared by his firm in relation and contrast to that. In January 1974 after consultation and meetings, also involving the new gable, drawings were prepared for submission to the licensing court, the Dean of Guild Court and the planning authority and approval by all three was obtained. On the probable cost, he had applied the RIBA scale for work to existing buildings up to the stage of scheme design. This D resulted in a percentage fee of £2,127.13, which with outlays and value added tax totalled £2,232.50. After April 4 1973 they had charged £32.40 for time involved in meetings etc and before that date £55 for preparing sketch plans and attending meetings.

The solicitor for the authority challenged the stage to which the scheme had been taken and argued that only stages A and B were necessary to investigate the feasibility of the proposal. Stage B is described in the scale of charges: a preliminary technical appraisal of a project sufficient to enable a client to decide whether and in what form to proceed, and making town planning inquiries or application for outline planning approval. If he were correct, the fee according to the scale should be on a time basis.

E We have studied the drawings of the 1973-1974 scheme, which we accept are to design stage (ie stage D of the scale). These, in our view, were necessary for submission to the various authorities for approval and we accept that, at that date, there was an intention to proceed with the adaptation and extension of the rump. We accordingly allow these fees.

(iii) *James Rusk & Associates, chartered surveyors.* It is an unusual circumstance which leads to this claim. Mr Rusk died during the negotiations and the claimant was advised that in such a complicated case he should seek the advice of a surveyor who was very experienced in licensed premises. He was recommended to employ Henry J Crone & Partners and we consider he was entitled to F choose his adviser. It is expenditure which is caused by the acquisition. We allow the £280, although as Mr Rusk was instructed in mid-1971 it was mostly incurred before April 4 1973. We apportion the account as £190 and £90.

As will be noted below, no claim for Messrs Crone is being made. In all the circumstances, we consider the fee reasonably incurred.

(iv) *Henry J Crone & Partners.* This firm was instructed in July 1974 following Mr Rusk's death, but Mr Crone also died before negotiations were completed. We were informed, however, that no account has been rendered and the claim was abandoned.

(v) *Ballantyne & Copland, solicitors.* Ian Livingstone, senior partner of the firm, gave evidence that he was consulted by Mr Smith in 1971 about the possible acquisition of the Golden Feather. He was G asked by the town clerk if he would attend all the meetings, which he did. He was also involved in the renewal of the licences, the loan requests and the consents for the buildings. Of the £746, £120 had been incurred before April 4 1973 and £626 since that date. The claim is slightly adjusted because of value added tax being involved on April 1 1973. £20 of the fee is in connection with planning permission for the new single-storey building, which it is agreed should be disallowed. We apportion the allowed claim of £724.80 as £120 and £604.80.

We summarise our award for professional fees as follows:

H

Item	*Claim*	*Award* Before April 4 1973	*Award* After April 4 1973
Robert C Brown	£339.08	£203.00	£136.08
Cullen Lochhead & Brown	£2,319.90	£55.00	£2,264.90
James Rusk & Associates	£280.00	£190.00	£90.00
Henry J Crone & Partners	£420.00	—	—
Ballantyne & Copland	£746.40	£120.00	£604.80
		£568.00	£3,095.78
	£4,105.38	£3,663.78	

Item (c) loan interest. In connection with the loan interest, we have set out the circumstances of the payment to account. The payment to account was said not to be enough to meet the liabilities and pay J off the business overdraft. As, however, the compensation is based on total extinguishment, it does not seem appropriate to allow this overdraft interest, which is payable owing to the personal financial circumstances of the claimant. If the compensation now determined had been paid at November 11 1973, then the overdraft could have been paid out of the realisation of compensation. The compensation is still due as at November 11 1973 and statutory interest is payable on the full amount less the payment to account from that date. We consider this claim too remote and disallow it.

Item (d) demolition cost. The demolition cost was said to be £1,000. The failure to use the rump of the building was due to a number of circumstances. Mr Smith since the beginning has said that he K wanted a cleared site to erect a new building, although he may have envisaged that the authority would have demolished the whole building at their expense. On the basis of compensation adopted in this case, it has been assumed at the valuation date that there was a market for the rump and the site. The actual demolition in 1976 was activated by the condition in the renewed licence for new premises, although Mr Smith had been wanting to demolish it for some time before. As with the professional expenditure on his own scheme for the new buildings, we consider that the related demolition costs for the rump are too remote from the acquisition and accordingly disallow them.

The total compensation is therefore determined at £43,914: L

Value of property and severance		£40,250.00	
Disturbance			
Quantity surveyor's fee	£339.08		
Architect's fee	£2,319.90		
Valuer's fee (Rusk)	280.00		
Solicitor's fee	724.80	3,663.78	
		£43,913.78	say £43,914

In addition to that, the acquiring authority will pay a surveyor's fee calculated in accordance with scale 5 of the RICS scales.

The acquiring authority will pay to the claimant his expenses of M the reference on the Court of Session scale. We certify the case as appropriate for the employment of counsel and certify Mr Traill and Mr Brown as expert witnesses.

A WILLIAMS v BRITISH GAS CORPORATION

(REF/227/1979)

(V G WELLINGS QC and C R MALLETT FRICS)

July 3 1980

Estates Gazette February 28 1981

(1980) 257 EG 833

Preliminary issue of law — Husband and wife joint owners of relevant property — Reference to tribunal by husband only — Husband undischarged bankrupt — Claim in respect of gas pipe laid across river upstream of property at Glan Conway, Gwynedd, B **allegedly polluting lakes by release of fine clay — Gas Corporation's contention that notice of reference disclosed no right to compensation — Tribunal holds that notice must be struck out unless wife joined as claimant — Notice defective in other respects, but tribunal considers it can be put right by amendment — Nearly three months allowed for joinder of wife and provision of further and better particulars — Point raised by claimant concerning adding EEC official as party to proceedings**

The claimant appeared in person; Gary Flather (instructed by Linklaters & Paines) appeared for the respondent

Giving their decision, MR WELLINGS and MR MALLETT said: The respondents, the British Gas Corporation, apply for determination C of points of law as preliminary issues. Originally, only one issue of law was ordered to be tried in this manner; namely, whether the reference, as the respondents contended, should be dismissed forthwith on the ground that "the applicant is an undischarged bankrupt and while the legal title to the land in question is vested in him and his wife on trust for sale, it is his wife and/or his trustee in bankruptcy who should be making any application".

At the hearing of the application, while Mr Flather, counsel for the corporation, submitted that the reference should be dismissed forthwith, the grounds on which he relied did not quite take that form. He contended that because Mr Williams and his wife are joint owners of the relevant property, he has no *locus standi* to institute a D reference to the Lands Tribunal unless his wife be joined as claimant with him; that, his wife not having been so joined, the reference was bad in law and should be dismissed.

Shortly before the hearing of the application the corporation's solicitors wrote to the Registrar alleging that the Lands Tribunal has no jurisdiction to hear the reference in the present case and stating that the works therein complained of by Mr Williams were not of themselves authorised by specific statute and that any claim for injurious affection by those works has its remedy at common law. As will be seen, at the hearing of the application, Mr Flather put that point a little differently.

In the notice of reference, the land and property (which we will E together call "the property") to which the reference relates is described as being all that land belonging to Penrhiw Isa, Glan Conway, Gwynedd, that lies in the valley alongside the two streams that merge shortly after they enter the property, and which have now been made into four lakes, inclusive of the lakes themselves and the dams that form them. A copy of the notice of reference is contained in the schedule hereto. It describes Mr Williams as being the claimant and gives particulars of his claim.

At the hearing of the application, Mr Williams conceded that he is an undischarged bankrupt, having been adjudicated bankrupt on August 1 1977, and that he and his wife are, and at all material times were, joint owners of the property.

F Mr Flather submitted that the notice of reference disclosed no right to compensation against the corporation. In the first place, he said, unless Mrs Williams were joined with him as claimant in the notice of reference, Mr Williams was not "a person entitled under any enactment" within the meaning of rule 16(1) of the Lands Tribunal Rules 1975. Secondly, since Mr Willliams was an undischarged bankrupt and no order for costs made against him would be enforceable, justice required that the proceedings should not continue unless his wife (who was not a bankrupt) were joined. We were referred to RSC Ord 15 rule 4. Mr Flather's third submission was that the notice of reference did not indicate, or sufficiently indicate, the nature of the question, dispute or case to be determined by the Lands Tribunal, the enactment(s) conferring jurisdic- G tion on it, the nature of compensation claimed or the estate or interest in respect of which compensation is claimed. Further, notwithstanding the mandatory requirements of rule 16(2) (ii) of the Lands Tribunal Rules 1975 and the warning to the same effect given in Note (3) to Form 4 Schedule 1 to the rules, the notice of reference, when sent to the Registrar, had not been accompanied by any copy of any order, direction, notice, decision, authorisation, or other document in consequence of which proceedings for the determination of the reference were instituted.

Mr Flather said that the corporation had placed a gas pipe across the relevant stream upstream of the property. If the owners of the property had any claim in respect of the pipe it lay at common law. H The notice of reference did not disclose that the Lands Tribunal had any jurisdiction. The provisions of the Gas Act 1972, the Town and Country Planning Act 1971 and the General Development Order 1977 referred to in the notice of reference, disclosed none. The notice of reference should be struck out. The corporation were not seeking to exclude Mr Williams' claim for all time. The corporation wished to know what his case was. There was nothing to prevent him from starting again if his wife were joined.

Mr Williams said that his wife was not willing to be joined with him as claimant. He was not willing to join her as a respondent. He described the claim which he seeks to make against the corporation in the Lands Tribunal as follows. J

The corporation had put a gas pipe across the river upstream of his property and polluted the lakes. The pollution took the form of fine clay which was released by the works of excavation carried out by the corporation in the course of constructing the pipeline. The pollution occurred on several occasions and amounted to several hundred tons in total. It killed the natural growth at the bottom of the lakes on which fish fed and reduced the water level. He was claiming compensation for depreciation in the value of the property in an amount required to put the property back into the condition which existed before the damage was done. The placing of the pipeline, he said, was authorised by statute and the damage could have been guarded against. The corporation was, he said, negligent. K They had allowed the works to become blocked with debris. He wished to put his claim under Part I of the Land Compensation Act 1973. The property qualified as an agricultural unit within the meaning of section 2 of that Act and he was claiming in respect of the interest of "an owner-occupier", also within the meaning of that section. Mr Williams referred to a cheque for £2,000 paid on August 2 1978 by Wales Gas to Mr and Mrs Williams' tenants for pollution arising from incidents up to July 21 1978 due to pipe-laying activities in the feeder stream. The corporation had given a written undertaking to the owners and tenants that if any further pollution occurred they would put it right. They had reneged on their undertaking. L As an owner he wanted the matter put right. No land had been acquired from himself and his wife or from their tenants. No rights had been acquired from or created against himself and his wife or their tenants. A right to lay and maintain a pipe across the river upstream of the property on the other land had, however, been created or acquired under the Gas Act 1972.

It is clear law that, although as between themselves joint tenants (and, therefore, joint owners) have separate rights, as against everyone else they are in the position of a single owner. There is absolute unity between them and together they form one person. Apart from equitable remedies *inter se*, one of the joint tenants cannot commence any proceedings without the aid of the other or M others. See *Megarry & Wade, Law of Real Property (4th ed)*, pp 391 *et seq*. See also *Cheshire, Modern Law of Real Property (12th ed)*, pp 211 *et seq*.

The claim which Mr Williams seeks to put forward in the Lands Tribunal, as described by him, is either a claim for compensation for injurious affection, no land being taken from him, under section 10, Compulsory Purchase Act 1965 or a claim for compensation for depreciation caused by use of public works under Part I of the Land Compensation Act 1973. In either case he has, in our judgment, no *locus standi* to commence or maintain the proceedings unless his wife be joined with him as claimant. If his claim is in the nature of a

A claim for compensation for injurious affection, one of the matters which he must establish is that, but for the statute authorising the works, the damage giving rise to the alleged depreciation in value of the property would be actionable in the courts. It would not be actionable in the courts at the suit of Mr Williams without the aid of Mrs Williams. Furthermore, of course, Mr Williams' beneficial interest in the property has vested in his trustee in bankruptcy, so that his interest at the present time in it is limited to that of a trustee for his wife. One of two trustees cannot act without the other.

For these reasons we hold that the notice of reference must be struck out unless Mrs Williams be joined as claimant with her husband. As will be seen, we shall allow time for that course to be
B taken.

A notice of reference in Form 4 in the schedule to the Lands Tribunal Rules 1975 is a kind of pleading, but it is a rudimentary pleading. Not much in the way of precision of allegation is required from a claimant either under the rules or in the form itself. In a suitable case, the lack of precision, where it exists, can be remedied by exercise of the tribunal's power to order further and better particulars of the notice of reference to be delivered.

The notice is defective in some respects. Quite apart from the non-joinder of Mrs Williams there is no express mention of compensation for injurious affection nor of compensation for depreciation caused by use of public works. The list of enactments confer-
C ring jurisdiction on the Lands Tribunal does not include a reference to section 10 of the Compulsory Purchase Act 1965 or to Part I of the Land Compensation Act 1973. Moreover, the compensation claim appears, at the least, to include a sum in respect of trade losses; such a claim would be inadmissible whether compensation is claimed under section 10 of the Act of 1965 or Part I of the Act of 1973.

However, we are not convinced that the notice of reference entirely fails to disclose a right to compensation, because section 6 of the Gas Act 1972 (referred to in the notice of reference) expressly imports Schedule 2 to that Act, which schedule states the powers of acquisition of the corporation and amends the law relat-
D ing to compensation for compulsory acquisition so as to assimilate procedure for compensation for a new right created under the schedule to that for compensation for compulsory acquisition of land.

It appears to us to be a reasonable interpretation of the notice of reference that it claims compensation for injurious affection, but without indicating whether land has been acquired from, or a right created over, land belonging to the claimant. The reader cannot therefore determine whether the notice of reference alleges that the claim is made under section 7 or section 10 of the Compulsory Purchase Act 1965.

It appears to us that all these defects can be put right by amend-
E ment and we propose to give to the claimants, upon the joinder of Mrs Williams being effected, the opportunity to make suitable amendments in the notice of reference. We may add that there were inconsistencies in the description which Mr Williams gave us of the nature of the case which he seeks to put forward. If the fine particles of clay, which caused the damage, were released during the process of excavation prior to installation of the gas pipe, the claim cannot be put on the basis of compensation for depreciation caused by the *use* of public works. In so far as Mr Williams alleges negligence against the corporation, that would not, so far as we know, have been authorised by any statute and no claim in respect of it could be made in the Lands Tribunal.

F Finally, we are not sure to what extent the existence of tenants and their occupation prevents Mr and Mrs Williams being able to say that they are owner-occupiers for the purpose of Part I of the Act of 1973. It would be in the interests of Mr and Mrs Williams if they took legal advice upon these matters.

We shall dismiss the reference unless not later than September 30 1980:

(1) Mr Williams' wife shall have been joined with him as claimant in the proceedings;
and
(2) Further and better particulars of the notice of reference dated November 13 1979 specifying:
(a) the nature of the question, dispute or case to be deter- G
mined by the Lands Tribunal and, more particularly, indicating whether the claim is in the nature of a claim for compensation for injurious affection (and, if so, specifying whether any land or right over land has been acquired from or created against the claimants) or is made under Part I of the Land Compensation Act 1973 or whether both such claims are made in the alternative the one to the other;
(b) The enactment or enactments conferring jurisdiction on the Lands Tribunal and, more particularly, indicating whether section 10 of the Compulsory Purchase Act 1965 or Part I of the Land Compensation Act 1973 is H
relied upon or, as the case may be, whether both such provisions are relied upon in the alternative;
(c) The nature and amount of the compensation claimed, and more particularly, indicating whether the claim is made in respect of depreciation of the freehold interest or the reversion upon a lease;
(d) All the facts and circumstances relied upon shall have been delivered to the corporation and to the Registrar of the Lands Tribunal:

AND

(3) A copy of the order, direction, notice, decision, authorisation or other document in consequence of which proceedings J
for the determination of the reference are instituted shall have been sent to the Registrar of the Lands Tribunal.

We reserve the question of the costs of the application until a later date.

We ought to mention certain other matters to which Mr Williams referred in argument on the application.

He complained that the order of the Registrar ordering the hearing on a preliminary point of law was erroneous in that it recited that no objection to such a hearing had been received from the claimant. The recital was technically correct in that by the date of the order, December 13 1979, no such objection had been received in the Lands Tribunal. An objection by Mr Williams to K
such a hearing was delayed in the post and not received in the Lands Tribunal until December 28 1979. The President did not accede to a request by Mr Williams that the order should be corrected. At the hearing of the application we assured Mr Williams that he would suffer no injustice by reason of his objection not having been taken account of in the making of the order. That assurance appeared to satisfy him.

Mr Williams contended that a sizeable part of the corporation's project for the installation of the gas pipeline had been paid for by the European Commission in Brussels and that the corporation was the agent in respect of the installation for the commission. He (Mr Williams) might seek to add the Director-General for Regional L
Policy in Brussels as a party to the proceedings in the Lands Tribunal at a later stage. A question affecting the jurisdiction of the Lands Tribunal might then, he said, arise. British law, he said, was subject to European law. It appears to us that whether or not Mr Williams seeks to join the director-general in the present proceedings or to take proceedings against him elsewhere is not a matter which affects our decision on the present application. Moreover, we note that article 222 of the European Economic Community Treaty provides that "this treaty shall in no way prejudice the rules in member states governing the system of property ownership".

At the beginning of the hearing Mr Williams objected to the presence of C R Mallett FRICS as a member of the tribunal M
appointed to hear the application on the ground that he was not "learned in the law". Wisely, he did not pursue the point. It is quite misconceived. It is sufficient to say that surveyor members of the Lands Tribunal are "persons who have had experience in the valuation of land" appointed by the Lord Chancellor "after consultation with the president of the Royal Institution of Chartered Surveyors". Their authority to sit is contained in sections 2 and 3 of the Lands Tribunal Act 1949.

Mr Williams asked that the proceedings on the application should be recorded, but we were unable to arrange for that to be done.

THE SCHEDULE

Lands Tribunal Act 1949

(Lands Tribunal Rules 1975)

To: The Registrar,
Lands Tribunal,
5 Chancery Lane,
London WC2A 1LX.

The description of land and property to which this reference relates is all of that land belonging to Penrhiw Isa, Glan Conway, Gwynedd that lies in the valley alongside the two streams that merge shortly after they enter the property, and which have now been made into four lakes, inclusive of the lakes themselves and the dams that form them.

I Richard Hugh Williams of Brynrhedyn Farm, Glan Conway, Colwyn Bay, Clwyd hereby apply for determination by the Lands Tribunal of the question of which particulars are set out below.

I wish to appear in person and all communications regarding this reference should be addressed to me at the address shown above.

Particulars

The name and address of the authority against whom I complain is the British Gas Corporation, Wales Region, Snelling House, Bute Terrace, Cardiff CF1 2UF.

I, Richard Hugh Williams am the claimant.

The nature of question, dispute and case to be determined by the Lands Tribunal is the damage to the lakes by pollution; loss of some 6,000 Rainbow Trout between 8 ounces and 6 pounds in weight; loss of income and goodwill and delay in the development of a scheme involving the building of a Fishing Centre and Restaurant and nine Fishing Lodges of quality construction, for which planning permission in detail has been granted.

The enactments conferring jurisdiction on the Lands Tribunal are the powers granted to the British Gas Corporation under Section 2(2)(e) and Section 6 of the Gas Act 1972 and the Town and Country Planning Act 1971 and General Development Order 1977. Schedule 1 Order 289, Article 3.

The nature of the compensation claimed is a fair and just payment to allow the site to be rehabilitated to cover the loss of fish, income and interest on non-productive capital, by me in a capacity as owner.

The party by whom this notice is signed proposes to call three expert witnesses to give evidence.

13th November 1979.

(Signed) R Hugh Williams

REYNOLDS v MANCHESTER CITY COUNCIL (REF/250/1979)

(R C WALMSLEY FRICS)

December 19 1980

Estates Gazette March 7 1981

(1980) 257 EG 939

Repair garage and house at Longsight, Manchester, acquired and demolished — Value of interest and disturbance — No weight attached to claimant's valuer's evidence — "Opinion evidence given at second-hand is not really evidence at all" — Tribunal's reservations about authority's valuer's evidence, which did not convey that he had seen premises before demolition — Difference on condition of premises — Higher rental value and YP adopted — Substantial difference on amount of compensation for "loss of profit" — Tribunal says compensatable loss more accurately described as "loss of ability to derive a future profit out of the particular premises from which [claimant] has been dispossessed" — 3-stage approach conventionally adopted by valuers set out — Reasons for acceptance of authority's valuer's calculation of average profit for period prior to cessation of trading — Factors affecting YP on adjusted annual profit — No substantial help to be derived from previous decisions or settlements because every case falls to be dealt with on its own merits — Minor adjustments made to authority's figure

S J Reynolds appeared for the claimant by leave of the tribunal; P B Keenan (instructed by the solicitor to Manchester City Council) represented the acquiring authority.

Giving his decision, MR WALMSLEY said: This is a reference to determine the compensation payable by the Manchester City Council to Robert G Reynolds on the acquisition of the latter's interest in premises at 97 Stanley Grove, Longsight, Manchester, about two miles south-east of the city-centre. The premises (now demolished) consisted of a large detached dwelling-house and a repair garage.

The business part of these premises comprised (a) a garage building with an inspection pit (b) an open-ended leanto with a second inspection pit (c) yard space in front of the house on which cars were displayed for sale (d) yard space at the back of the house and (e) part of the house itself. The cellars of the house were used as stores, as was also part of the ground floor, and another part of the ground floor was used for office/reception purposes. The living accommodation comprised four rooms and bathroom/wc upstairs, together with a kitchen on the ground floor (I shall be referring to this living accommodation as "the house").

The subject premises were included in the City of Manchester (Albert Grove, Longsight) Housing Compulsory Purchase Order 1972, made on December 18 1972, and confirmed by the Secretary of State for the Environment on October 31 1973. Notice to treat was served on December 21 1973. The claimant continued trading until August 31 1975. Possession was taken on November 1 1975. It is common ground that although the relevant date for my purpose is November 1 1975, the calculation of the compensation payable for trade disturbance should be related to the date of cessation of trading on August 31 1975.

At the hearing P Keenan of counsel appeared for the authority and called evidence from Michael Laurence Watts ARICS, a principal assistant valuer in the housing section of the City Estates and Valuation Department of the city council. The claimant was represented, by leave, by his father, Sidney James Reynolds, who gave evidence himself and also called evidence from George Issac Woolard FIAS, managing director of a company, Woolard Holdings Ltd, surveyors/assessors/valuers/industrial consultants. (I shall refer to S J Reynolds as "Mr Reynolds", and his son will be referred to as "the claimant".)

The claim put forward at the hearing was in the sum of £123,561 against which the authority proposed £38,370. During the course of the three-day hearing various adjustments were made to both these figures, and by the close of the hearing the amounts that were being respectively contended for had become:

For the claimant			
Interest in land	(adjusted from £50,630)	£42,760	
Disturbance	(adjusted from £72,931)	£78,870	£121,630
For the authority			
Interest in land	(adjusted from £10,500)	£11,600	
Disturbance	(adjusted from £27,870)	£27,370	£38,970

I will deal first with interest in land and then with disturbance.

Interest in land

The interest being acquired is the leasehold for a term of 999 years subject to an annual ground rent of £8. The respective valuations, by Mr Woolard for the claimant and by Mr Watts for the authority, were made up thus:

	Claimant		*Authority*	
Business premises	£34,410		£7,152	
House	8,450		4,500	
	£42,860		£11,652	
Less for ground rent	100		64	
			£11,588	
		£42,760		say £11,600

Inasmuch as both valuers had decided to value first the business premises and then the house I will take these separately. In respect of the business premises the respective calculations, slightly adjusted for ease of comparison, may be expressed thus:

A

Rental value			*Claimant*		*Authority*
Office/reception	255 sf	@ £1.20	£306	@ £0.70	£178
Garage	300 sf	@ £2.50	750	@ £0.50	150
Leanto, with pit	293 sf	@ £2.50	732	@ £0.25	73
Front yard (sales site)	218 sy	@ £3.60	785	@ £1.00	218
Stores (ground floor)	185 sf	@ £0.35	65	@ £0.40	74
Stores (basement)	646 sf	@ £0.35	226	@ £0.30	194
Rear yard (car-wash)	108 sy	@ £5.40	583	@ £0.50	54
			£3,447		£941
Less					
Repairs		Nil			
Insurance		£6	6	5% £47	47
Net rental value			£3,441		£894
YP			10		8
B			£34,410		£7,152

Mr Watts, by way of support for the components of his valuation above, relied both on his general experience and on a number of specific comparables, all of which are now demolished, but details of which were produced together with photographs. Mr Woolard, on the other hand, relied neither on comparables nor on his own experience: in his evidence-in-chief and also in cross-examination Mr Woolard was explicit not only that he had never seen the subject property prior to its demolition but also that the figures appearing in his valuation had been adopted after discussion with Mr C Reynolds; his own preference indeed would have been (a) to adopt lower rental values psf or psy for various of the constituent areas (b) to include some allowance for repairs and (c) to use some lower multiplier than 10 YP. This being the foundation of his valuation, I regret that I can attach no weight to Mr Woolard's evidence. Valuation is a matter of independent uninfluenced judgment. Opinion evidence given at second-hand is not really evidence at all.

I find myself thus left with the valuation produced by Mr Watts for the authority. In the normal course I should have been disposed to accept this as presented. Mr Watts has had 13 years' professional experience; he gave his evidence fairly and clearly; and the figures that go to the make-up of his valuation at £7,152 are not inconsistent with the analyses of his comparables. Nevertheless, I have D reservations. There was nothing in Mr Watts' evidence which conveyed to me that he had himself seen the subject premises prior to their demolition; he confirmed that he had had no hand in the earlier negotiations which had followed the lodging of the first claim in this matter, in August 1975; and it was not so long after that date that demolition took place.

I note that the photographs of 97 Stanley Grove produced for the authority had been taken on November 25 1975, some three weeks after the date when possession was taken, and by which time (on the evidence of Mr Reynolds) the contractors had moved in. Certainly the authority's photographs present an impression of disrepair and E neglect, in contrast to the photographs produced by Mr Reynolds; these had been taken earlier in 1975 and they show a trim and well-kept property.

The unchallenged evidence given by Mr Reynolds was that, after the property had been bought by his son in 1961, expenditure had been incurred in modernisation and improvements. An existing inspection pit in the garage had been brought up to Ministry of Transport ("MOT") standard; a second inspection pit, also up to MOT standard, had been constructed under the leanto; the surface of the back yard had been concreted, and graded for use as a car-wash; a wc in the back yard had been rebuilt for customers' use; the cellar walls in the house had been pointed and refaced to serve as stores.

F Having regard to these matters, I propose to adopt a somewhat higher rental value than that put forward by Mr Watts, coupled with a somewhat higher figure of years' purchase, as follows:

Rental value		£1,000
Less for repairs and insurance	5%	50
Net rental value		£950
	YP	8.5
Value of business premises		£8,075

In respect of the house Mr Woolard's figure of £8,450 represents: G

Rental value		£750	
Less Repairs	£60		
Insurance	14	74	
		£676	
	YP	12.5	£8,450

Mr Watts for his part had gone straight to his figure of £4,500 as being derivable from a settlement figure of £5,500 for a nearby substantial semi-detached house, 93 Stanley Grove, purchased H under the same statutory powers at a valuation date of January 1974.

There are various matters to be borne in mind in considering the helpfulness of this comparable:

(a) 93 Stanley Grove is now demolished. No floor plan or photograph of it was put before me. Neither Mr Woolard nor Mr Watts claimed to have seen it prior to demolition.

(b) One of Mr Reynolds' photographs shows the gable-end of the house no 95, which was semi-detached with no 93 and — per Mr Reynolds — was in a similar condition thereto. That con- J dition can be seen to be one of some disrepair.

(c) The £5,500 paid for no 93 was in respect of the whole house, whereas the value now under consideration for the subject house relates only to that part which was used for living accommodation.

(d) Unchallenged evidence was given by Mr Reynolds that the subject house had been substantially overhauled; lead piping was replaced with copper; new water tanks were installed; new windows put in; wiring renewed; a new bathroom suite installed. "I lived there for 14 years" said Mr Reynolds, "I did everything to that house. It was beautiful".

(e) In the compulsory purchase order the comparable is classified K as a "fit" house, whereas "House, 97 Stanley Grove" is shown (rather surprisingly) coloured pink on the order map as unfit for human habitation. It is agreed that the claimant is entitled to "full compulsory purchase value" as an owner-occupier, and I was presented only with evidence as to the proper amount of this value, without any separation of that amount between (i) the site value, payable under Part III of the Housing Act 1957 and (ii) the owner-occupier supplement, payable under Schedule 5 to the Housing Act 1969.

Taking into account these various matters I have decided to L allow, as the value of the house, the sum of £5,250.

Disturbance

Turning now to disturbance, the figures that were being contended for by the end of the hearing had become:

	Claimant	*Authority*
Loss of profit	£75,300	£23,800
Loss on stock	1,415	1,415
Loss on fixtures and fittings	1,950	1,950
Domestic disturbance	205	205	M
	£78,870	£27,370

It will be seen that the outstanding issue is as to the compensation payable for "loss of profit". This head of claim had been spoken to by Mr Woolard, being part of his calculation showing the original total figure of £123,561, but it became clear during the hearing not only that Mr Woolard's figures had come from Mr Reynolds but also that Mr Reynolds wished to give the evidence on this matter himself. And so he did.

A The make-up of the parties' respective figures under this head of claim may be set out thus:

		Claimant			*Authority*	
Annual profit			£20,452			£8,200
Deduct						
Rental value		£3,441			£894	
Interest on capital employed:						
Stock in trade	£1,610			£2,750		
Fixtures and fittings	2,100			2,100		
Cash	300			300		
	£4,010			£5,150		
	@ 7%	281	3,722	@ 10%	£515	1,409
Adjusted annual profit			£16,730			£6,791
B YP			4.5			3.5
			£75,285			£23,769
Capital loss		say	£75,300		say	£23,800

There are five elements in the above calculations to be considered, *viz*: (a) Annual profit, (b) Rental value, (c) Amount of capital employed, (d) Rate of interest on the capital employed, and (e) Years' purchase.

(a) Annual profit

C As can be seen, there is a dramatic difference of opinion as to the proper figure to be adopted for annual profit.

The approach adopted by Mr Reynolds is to estimate the profit that would have been earned over the three years after the cessation of trading. What Mr Watts has done is to take the average profit over the 41 months up to the cessation of trading.

In more detail, Mr Reynolds' approach goes like this:

(a) The profit for the five months from April 1 1975 to August 31 1975 is agreed to have been £5,422, which figure may be apportioned as to £1,506 being fees from MOT tests and as to £3,916 being income from sales/repairs.

D (b) As regards the MOT work, the test fee became increased in August 1975 from £1.70 to £2.10 (23.52 per cent) and became further increased in January 1977 from £2.10 to £4.10 (95.23 per cent). For the 17 months from September 1975 to December 31 1976 the monthly income from MOT work could therefore be assumed to increase by 23.52 per cent, from £301.20 to £372.04. And for the next 19 months, from January 1 1977 to August 31 1978, that monthly income could be assumed to increase by a further 95.23 per cent from £372.04 to £726.33. This indicates a total estimated profit from MOT work in the three-year period to August 31 1978 of £20,124.95.

(c) As regards the sales/repairs side of the business, the profit could be assumed to increase in line with inflation. Mr Reynolds put the rate of inflation during 1975/76 at 27 per cent, during 1976/77 at 18 per cent and during 1977/78 at 8 per cent. For the five months to August 31 1975 the E profit from sales/repairs had been running at an annual rate of £9,398.40. So the profit over the next three years could be expected to be:

Year to August 31	1976	£ 9,398.40 +27 per cent or	£11,935.95
do	do 1977	£11,935.95 +18 per cent or	£14,084.44
do	do 1978	£14,084.44 + 8 per cent or	£15,211.20
			£41,231.59

(d) **The total future profit over the three years from cessation of trading could therefore be put at, say, £61,357, or an average of £20,452 a year.**

F **The annual profit figure adopted by Mr Watts had been derived from the following (agreed) figures showing excess of income over expenditure:**

Year to April 5 1973	£5,964
do 1974	7,939
do 1975	8,695
5 months to August 31 1975	5,422
Total for 41 months' trading	£28.020

Average annual excess of income over expenditure £8,200

G I shall be accepting, with one adjustment, the approach adopted by Mr Watts, and I shall be rejecting the approach to annual profit propounded by Mr Reynolds. For the assistance of Mr Reynolds it is as well that I should explain why.

Although the loss which is under consideration is often referred to as "loss of profit", or as "loss of future profit" or even as "damage to goodwill", the claimant's compensatable loss is more accurately described as "loss of ability to derive a future profit out of the particular premises from which he has been dispossessed". The authority's scheme does nothing to impair the claimant's skills and abilities, which are released to be deployed elsewhere.

H Because of the lack of any external check on the figure finally arrived at under this head of claim, the assessment of loss is inherently a difficult task, as I indicated at the hearing. What the tribunal has found is that valuers, whether acting for claimants or for acquiring authorities, consistently use a form of calculation similar to one which is sometimes employed when assessing the value of a business changing hands in the open market. In the context of a "one-man business" such as this, that approach, as adjusted for use in the assessment of compensation on compulsory acquisition, consists of three stages:

J *Stage 1.* The ascertainment of a figure of historic profit, which is customarily (but not necessarily) taken as the average of the three previous years' trading.

Stage 2. The adjustment of this historic profit by making certain deductions, as for instance (a) an allowance for the rental value of the subject premises to the extent that this has not been fully charged in the accounts used to ascertain the average annual profit, and (b) an allowance for interest on the capital employed in the business. (With a "one-man business" it is customary not to make any deduction in respect of the notional wages earned by the proprietor.)

K *Stage 3.* The capitalisation of the adjusted annual profit by applying a multiplier in terms of years' purchase. This multiplier is intended to be such as will produce an end-figure representing, as fairly as may be, the value to the claimant of the loss of his ability to derive a future profit out of the premises from which he had been dispossessed.

The main reason why this conventional approach is used is because with any other method it is not easy to make the necessary deductions at stage 2 above.

L Mr Reynolds cited to me a number of Lands Tribunal decisions. In all those decisions the figures of years' purchase used by the tribunal are referable to a factual and historic profit, not to a future and conjectural profit. In one of the cited cases, *Tragett v Surrey Heath Borough Council* (1975) 237 EG 423, the claimants' witness had put forward a calculation based on a conjectural future profit, and this approach was rejected by the tribunal (V G Wellings QC) as being "artificial and having no relevance".

M It is against this background that I am accepting Mr Watts' calculation of annual profit, but subject to the adoption of an average of 36 months' trading instead of 41 months. No good reason was advanced by Mr Watts as to why the customary period of three years had not been used. The making of this adjustment has the effect of producing an annual profit figure of £8,500 instead of £8,200, thus:

7	months ended	April 5 1973	£3,479
12	do	April 5 1974	7,939
12	do	April 5 1975	8,695
5	do	August 31 1975	5,422
			£25,535

Average annual excess of income over expenditure say £8,500

A (b) *Rental value of the business premises*

The next element in the calculation is the deduction to be made from the annual profit to allow for the rental value of the business premises.

The calculations presented by the parties in respect of interest in land show, respectively, £3,441 (for the claimant) and £894 (for the authority). On my earlier finding as to the value of that interest the appropriate deduction becomes £950.

(c) *Amount of capital employed in the business*

As to the capital employed in the business: in respect of stock in trade it is suggested for the claimant that £1,610 is a sufficient B allowance. Mr Watts told me, however, that his allowance of £2,750 had come from figures previously supplied for the claimant, and I adopt that sum. It will be recollected that the loss on sale of this stock is agreed to have been £1,415.

The fixtures and fittings are agreed at £2,100. These realised only £150 on sale, thus resulting in a loss of £1,950 as set out earlier.

A token allowance of £300 is agreed for the cash employed.

This gives a total of £5,150 capital employed in the business.

(d) *Rate of interest on the capital employed in the business*

Mr Reynolds told me that he had adopted a 7 per cent rate of interest on the capital employed in the business because it was his C impression that at the relevant date this was the Bank Rate. As it happens, the Bank Rate fluctuated during 1975 between 12½ per cent and 13½ per cent. Inasmuch as Mr Watts has used a rate not disadvantageous to the claimant, of 10 per cent, I adopt that figure.

The effect of the findings (a) to (d) above is to produce a figure of adjusted annual profit of £7,035, thus:

Annual profit		£8,500
Less rental value	£950	
Interest on capital, 10 per cent of £5,150	515	1,465
Adjusted annual profit		£7,035

D

(e) *Years' purchase of the adjusted annual profit*

There remains the final stage of the calculation, that of deciding on a multiplier which, when applied to the adjusted annual profit, is to produce an end-figure that fairly represents the claimant's loss. No substantial help as to what is the appropriate mutiplier is to be derived from figures used in previous decisions of the tribunal or from evidence as to settlements effected in other cases, because every case falls to be dealt with on its own merits. It is clear that Mr Watts recognises this to be so, because although he referred to E various settlements in respect of "loss of profit" or "goodwill" relating to other car businesses in Manchester, negotiated narrowly within the range of 1½ to 2½ YP, he nevertheless proposes a multiplier of 3½ YP in the instant case.

In the present reference there are, it seems to me, three factors which are of particular significance, namely (i) the likelihood that future profits would have been at a higher level than past profits (ii) the extent to which the earning capacity of the business was attributable to the personal skill and ability of the claimant himself and (iii) the degree of interruption in the profit-earning capacity of the claimant resulting from the scheme.

F (i) *Likely level of future profits.* Mr Reynolds asks me to regard the profits of £5,422, earned during the last five months of trading and equivalent to an annual rate of profit of over £13,000, as being earned in normal conditions and as showing the growth of the business over the two preceding years when the profits had been only £8,695 and £7,939. I apprehend, however, that during those closing months special factors may well have been at work; indeed Mr Reynolds mentioned one of these himself — it appears that during one particular 10-day period no less than 11 cars had been sold, whereas the normal expectation was about one car a week.

Be this as it may, there was certainly a rising trend of profit, and that trend would have been reinforced by increases in MOT test fees which took place not only in 1975 and 1977, as noted, but also G subsequently in 1978, 1979 and 1980. This is not to say that such increases would have been wholly profit: the increases were linked with stringent requirements by the Ministry of Transport in respect of equipment and other matters. Although these statutory requirements applied in the first instance to *applications for appointment* as authorised examiners, those examiners who were already authorised (as was the claimant) were notified in May 1975 that they would become subject to the same requirements at a later date. That date was first specified to be June 30 1979 but was, in the event, postponed to January 1 1980.

(ii) *Personal skill of the claimant.* I entertain no doubt that the H profit-earning capacity of the business was in large measure attributable to the personal skill and ability of the claimant himself. He was assisted by his father but that assistance was, I gather, mainly on the car sales side. At the relevant date in 1975 the claimant was aged about 39 and his father was aged about 66.

(iii) *Interruption of profit-earning capacity.* In answer to a question from myself Mr Reynolds said that within a few months after being dispossessed his son had found himself running, with others, a similar car business in Bramhall, and that it was a business which also held an MOT authorisation.

The submission for the authority is that a multiplier of 3½ YP J applied to the adjusted annual profit produces a capital sum sufficient to compensate the claimant. The appropriate figure of adjusted annual profit I have found to be £7,035, and this produces @ 3½ YP a capital sum of, say, £24,650. There is nothing in the evidence before me which suggests that such a sum would be inadequate to compensate the claimant for the loss of his ability to derive a future profit out of his former premises at 97 Stanley Grove.

The total compensation payable is accordingly found to be £41,470, as follows:

K

Interest in land			
Business premises	£8,075		
House	5,250		
	£13,325		
Less for ground rent, say	75	£13,250	
Disturbance			
Loss of profit	£24,650		
Loss on stock	1,415		
Loss on fixtures and fittings	1,950		
Domestic disturbance	205	£28,220	£41,470

L

Mr Reynolds sought to contend that some additional payment was due by virtue of section 37 of the Land Compensation Act 1973, but it was pointed out to him that this statutory provision concerns itself only with persons without compensatable interests and is therefore irrelevant to the present reference.

The acquiring authority will pay the claimant the sum of £41,470 to which will be added a surveyor's fee based on scale 5 of the scales of professional charges of the Royal Institution of Chartered Surveyors. The acquiring authority will also pay the claimant his costs of this reference, such costs unless agreed to be taxed by the M Registrar of the Lands Tribunal on the High Court Scale.

I was told that a payment of £16,325 on account had been made on May 3 1976 together with accrued interest from November 1 1975, and that a home loss payment under section 29 of the Land Compensation Act 1973 had also been made.

A DIXON v ALLGOOD AND SECRETARY OF STATE FOR TRANSPORT

(REF/97/1980)

(Wm HALL FRICS)

January 28 1981

Estates Gazette March 28 1981

(1981) 257 EG 1268

Section 19 application for rent apportionment for land at Hexham — Agreement betweeen lessee and DV for £2 for aquired land and £50 for residue not acceptable to lessor — £2 broadly a nominal
B **figure in which DV would have concurred if suggested in negotiations — Approach different for formal apportionment — Necessary to examine different qualities of land — Largest proportion of rent applied to comparatively small area — Lessee's proposal for pro-rata basis rejected by tribunal — DV's figure of 56p accepted — Apportionment understood to have possible implications in relation to Leasehold Reform Act 1967 — No order on costs between lessor and lessee — Liberty to apply for costs against acquiring authority**

James Dixon, the lessee, appeared in person; J H Fryer Spedding, (instructed by Wilkinson, Marshall, Clayton & Gibson, of Newcastle upon Tyne) represented the lessor; the Secretary of
C State for Transport (as successor to the Secretary of State for the Environment) was not represented.

Giving his decision, MR HALL said: This reference to the tribunal was made by James Dixon, of Riverside Cottages, Acomb, Hexham, to apportion the rent between land acquired by the Secretary of State for the Environment and the residue of the land leased by him.

Mr Dixon was granted a lease dated December 31 1964 of an area of land extending to 6.11 acres, being part of the Hermitage Estate at Acomb, together with two derelict cottages on the land. He was obliged by the lease to rebuild the two cottages, which he has done. The lease is for a period of 51 years from 1964 and the
D rent for the whole period is £52 per year. At the commencement of the lease the 6.11 acres consisted of the derelict houses and garden ground, a former railway embankment, an area of the River Tyne including the abutments of the old railway bridge and an area of land bounding the river which had been part of the adjacent farm.

In May 1974 a notice to treat was served following the Hexham-Corbridge By-pass Compulsory Purchase Order (CNE No 3) 1973 for the acquisition of Mr Dixon's leasehold interest in land for the erection of a bridge. The area of land acquired was 1,429 m² and in addition there were other areas affected by easements for the overhead bridge deck and licences for working areas. Possession of the land acquired appears to have been taken at July 2
E 1974, which was adopted as the valuation date for compensation purposes.

Mr Dixon's surveyors had agreed with the district valuer, who was acting for the acquiring authority, that of the rent of £52, £2 should be apportioned to the land acquired and £50 to the remaining land and buildings. This apportionment, however, was not acceptable to the present lessor, L G Allgood, and agreement could not be reached. Hence the reference to the tribunal.

Section 19(1) and (2) of the Compulsory Purchase Act 1965 states:

(1) If part only of the land comprised in a lease for a term of years unexpired is required by the acquiring authority, the rent payable in
F respect of the land comprised in the lease shall be apportioned between the land so required and the residue of the land.

(2) The apportionment may be settled by agreement between the lessor and lessee of the land on the one part and the acquiring authority on the other part, and if the apportionment is not so settled by agreement between the parties, it shall be settled by the Lands Tribunal.

Valuation evidence

J M Clark MA FRICS, partner in J M Clark & Partners and agent to the estate since 1971, said, on behalf of the lessor, that from the records the area of 6.11 acres consisted of the following parts:

(a) the cottages and immediate garden ground extending to 0.761 of an acre.

(b) an area of about 1.472 acres which had been part of Bridge End Farm; the tenant had relinquished his tenancy of that area and in 1964 no abatement of rent was asked for. It was rough riverside grazing. G

(c) the remainder, which had been partly a railway embankment — then a cinder track and overgrown embankment.

He said that part of the land acquired was the bed of the River Tyne. This was not, in his view, in Mr Dixon's lease, but as it only extended to an area of 122 m² and did not materially affect the end-result he did not propose to adjust the area of 1,429 m² for his computation of the apportionment of rent. The cottages and land had been substantially improved by the lessee. Mr Clark had assumed that the garden ground which had been used by the former H
occupiers of the cottages would be appropriate for the cottages and this was included in the area of 0.761 of an acre.

In his opinion, the rental value of that part, which included the walls, was £50. The remaining 5.349 acres was, in his opinion, not useful land and he applied a rate of 40p per acre to that area. The 40p, he said, was reasonable in 1964. This gave a rental value of £2 for the land; the area acquired was 0.353 of an acre and as this was one-fifteenth of the 5.349 acres, the proportionate rent was 14p. In these circumstances, he considered no reduction of rent should be made. He did not accept that dividing the total area of 6.11 acres by 0.353 of an acre and applying that fraction to the £52 gave a result which accorded with valuation practice. He agreed that in 1974 the J
grassland extending to about 4½ acres could have been let for grazing and Mr Dixon had done "a remarkable job" in creating that.

Richard J Wilkie ARICS, deputy district valuer of the Valuation Office, Inland Revenue, Northumberland, said that his office had agreed £2 as the apportionment with the lessee at the time when the compensation was being settled. It was broadly a nominal figure to which he would have concurred if it was suggested to him in negotiations. In making a formal apportionment, however, his approach was different. He did not agree that a straight apportionment of area acquired to the whole area was correct. He accepted that it was necessary to examine the different qualities of the leased land and K
building and agreed that the allocation of 0.761 acre for the house and garden plot and 5.349 acres for the remainder was correct. The largest proportion of the rent would be on the 0.761 acre. He had been advised that the apportionment should be made having regard to the condition of the land at the valuation date adopted for compensation purposes.

He submitted a rental valuation as at July 2 1974 as follows:

Land — 6 acres @ £16 per acre (ie as amenity land being twice agricultural value)	say £100
Riverside Cottage West	say £250
Riverside Cottage East	say £200
	£550 pa

L

His apportionment calculation was:

Land $\frac{100}{550} \times £52 = £9.45$p

Area acquired 1,429 m² or 1,715 sq yds*

So apportionment $\frac{1,715}{6 \times 4,840} \times £9.45$

= say 56p

* Mr Dixon calculated this at 1,709 sq yds, which appears to be correct.

So on a strict statutory basis in accordance with section 19 of the Compulsory Purchase Act 1965, he submitted that the apportionment to the land acquired was 56p. M

Mr Dixon, the lessee, said that, in his view, the apportionment should be on a straight proportionate basis. The 1,709 sq yds was one-fifteenth of the total area of 6.11 acres and so one-fifteenth of £52 was approximatly £3. His surveyors had agreed £2 with the district valuer and as he considered £50 a round figure for the remainder he was prepared to accept the £2. He criticised Mr Wilkie's valuation as not reflecting the considerable capital expenditure which he had made both on the cottages and the land and also Mr Clark's assumption that the house plot should be 0.761 acre. He calculated the area of the cottage as 0.317 acre.

A J Fryer Spedding, for the lessor, said that there was little between Mr Clark's and Mr Wilkie's opinions and that both were in accordance with proper valuation practice. The legal basis as set out in the district valuer's calculation was accepted by him as the correct one and a broad basis either in negotiations or on Mr Dixon's method was not justified.

I have inspected the leased subjects. I agree that either at 1974 or when the lease was entered into the greater value would be in the cottage area and the land south of the cottages. The remainder, being a narrow strip of land, was the cinder track on the top of the embankment and sloping down to the river. The part acquired was at the beginning of the narrow strip.

B I do not accept Mr Dixon's proposal that the rent should be apportioned on a pro-rata basis, as the value to the lessee would vary with the use which could be made of the various parts of the land. Both valuers disagreed with a pro-rata apportionment. As there was no other contrary legal argument before me on the basis of apportionment adopted by Mr Wilkie, I accept it. The figure calculated by Mr Wilkie is not substantially different from that of Mr Clark, who appeared to apportion the rent relative to the original contract for the derelict cottages and the land. I accept that, on either basis, the apportionment to the area acquired will be nominal. The only valuation evidence I have before me once I have rejected the pro-rata basis is that of Mr Wilkie and Mr Clark. As Mr C Wilkie's approach is agreed to by the lessor, I accept his apportionment of 56p and so determine.

Discussion on costs

I heard a submission on costs and it was proposed by both parties that the Secretary of State should meet them. It was argued by the parties that he had initiated the acquisition and appeared to be the person to call the lessor and lessee before the tribunal to determine the amount of the apportionment. (*Slipper v Tottenham & Hampstead Junction Railway Co* [1867] LR 4 Eq 112). The tribunal, however, in terms of section 1 of the Land Compensation Act 1961 can have any question as to the apportionment of the rent payable D under the lease referred to it. This would appear to give the right to any one of three parties to make an application to the tribunal to apportion the rent. On the Secretary of State's behalf the district valuer had agreed the apportionment with the lessee. Mr Dixon made the application to the tribunal as he could not get the lessor to agree to that apportionment. In the circumstances, he was entitled to consider that apportionment might be upheld. The lessor was contending, through Mr Clark, that there should be no reduction in the rent. Mr Wilkie has explained his reasons for the £2, which Mr Dixon had not been aware of, but then gave his apportionment on a strict statutory basis. As I have accepted Mr Wilkie's figure, which is above the lessor's figure and below the lessee's figure, I have E come to the conclusion that there should be no order on costs as between the lessor and the lessee.

Although the Secretary of State is a party to the proceedings, an order dated June 18 1980 stayed all proceedings against him until determination of the appropriate apportionment of rent was made. That stay has not been removed and the Secretary of State has not been heard on the application for costs against him. He is entitled to be heard on it and if the lessor and/or lessee wish to pursue that application against him, they must apply on notice to him not later than February 28 1981 for an order removing the stay and seeking costs against the Secretary of State. Liberty to apply accordingly. If no such application for removal of the stay and for costs shall have F been made on or before the said date this decision shall operate as a final decision; otherwise it shall operate as an interim decision.

The amounts involved in the dispute are small, but I was led to understand that the apportionment may have implications in relation to the Leasehold Reform Act 1967.

PENNINE RACEWAY LTD v KIRKLEES METROPOLITAN COUNCIL G

(REF/45/1980)

(V G WELLINGS QC)

February 2 1981

Estates Gazette April 11 1981

(1981) 258 EG 174

Small airfield near Huddersfield used for drag motor racing under licence created by deed — Entitlement to claim compensation under s 164 of Town and Country Planning Act 1971 — "28-day permission" under article 3 of GDO withdrawn by direction under article H 4 — Claim resisted on ground that claimants not "a person interested in the land", words which Member "suspects" are shorthand for "a person having an interest in the land" — Held that licence not coupled with an interest — It was a revocable contractual licence which gave claimants no more than a personal privilege to use land for commercial purposes — Barriers and fencing erected were not fixtures and there was no grant of any interest in them — On issue whether letter from council officer had contained unequivocal representation that claimants entitled to make claim, held that council not precluded by estoppel or election from challenging entitlement — Claimants not entitled to claim compensation J

M E M Brooke (instructed by Booth & Co, of Leeds) appeared for the claimants; M J Green, solicitor, for the compensating authority.

Giving his decision, MR WELLINGS said: The claimants apply for determination of a preliminary point of law as to whether they are entitled to make a claim for compensation under section 164, as applied by section 165, of the Town and Country Planning Act 1971.

At the material time the claimants had a licence from the owner of a small airfield known as Crosland Moor Airfield, Crosland Moor, Huddersfield, West Yorkshire, to use the airfield for the purpose of drag motor racing. The frequency of use for that purpose K was such that the use was covered by the permission (the "28-day permission") contained in article 3 of the Town and Country Planning General Development Order 1973, and Class IV, paragraph 2 of the First Schedule thereto. On November 8 1974 the respondent council made a direction under article 4 of the Town and Country Planning General Development Order 1973 that the abovementioned permission should not apply to the use of the airfield for the purpose of motorcar or motorcycle racing. The direction was confirmed by the Secretary of State for the Environment on December 13 1974. The claimants made an appropriate application for planning permission which was refused. The claimants appealed to the Secretary of State, who dismissed the appeal. L

Sections 164 and 165

Section 165 of the Town and Country Planning Act 1971 provides that where planning permission for the development of land has been granted by a development order and that permission is withdrawn by the issue of directions under powers in that behalf conferred by the order, the provisions of section 164 of the Act are to apply. Section 164 provides that where planning permission is revoked or modified, then if, on a claim made to the local planning authority, it is shown that:

a person interested in the land:
(a) has incurred expenditure in carrying out work which is rendered abortive by the revocation or modification; or M
(b) has otherwise sustained loss or damage which is directly attributable to the revocation or modification,
the local planning authority shall pay to that person compensation in respect of that expenditure, loss or damage.

The preliminary issues are as follows:

(1) Whether the claimants were "a person interested in the land", that is to say, in the airfield;
(2) If not, whether the council had elected not to challenge the claimants' entitlement to compensation and/or had represented

that it had made such an election, and is now estopped from doing so.

The facts

The facts are as follows:

(1) Drag motor racing is a race between two motor vehicles over a measured distance (quarter of a mile) from a standing start. The course is straight. The maximum speed attained by the fastest car is 165 mph. The cars have parachute brakes. The sport is recognised by the governing bodies of motor sports in the United Kingdom and race meetings are supervised by the Royal Automobile Club, who impose various safety requirements in relation to race tracks. Race meetings take two days, usually over weekends.

(2) In or about February 1974, David Michael Whitham, the owner of the airfield, a Mr Wormald (his cousin) and Stephen Morty orally agreed that the company should have the sole right to promote drag racing on the airfield. They were shareholders in the company, but it was agreed between them that since Mr Whitham was providing the site and suffering a reduction in the land available for grazing and the inconvenience of permanent safety barriers and fencing, he should provide no capital. However, the three of them were to have equal shares in the profits of the venture.

(3) Other express terms of the oral agreement were as follows:

(a) that it should be the responsibility of the company to take all such measures, including the installation and maintenance of safety barriers, fences and spectator banks as were necessary for the safety of the public;

(b) that the meetings should be conducted in such a way that the RAC would approve of them;

(c) that on ceasing to use the airfield for drag racing the company were to remove such barriers, fences and other fixtures and fittings as it should have erected and reinstate the airfield;

(d) that Mr Whitham was to be paid £500 per meeting except during the first trading year, in which he was to receive his share of the profits only.

(4) Mr Whitham had bought the airfield in order to promote private flying and he was anxious that his plans to develop it for an aeroclub and air displays should not be prejudiced by the drag racing project.

(5) Mr Murty was anxious that the company should be allowed to go on holding drag racing meetings on the airfield for as long as they proved to be commercially viable and he received Mr Whitham's assurance to that effect.

(6) After the agreement had been entered into, the claimants erected along either side of the eastern section of the runway about 400 yds of safety barriers which were substantial and permanently installed. They cleared an area of land in the vicinity for pits and a car park for competitors and put tarmacadam on that area. They created a bank for spectators, moving about 3,000 m^3 of top soil in the process. They erected fencing consisting of two strands of barbed wire and two strands of nylon cable attached to existing posts to a height of 5 ft over a length of three-quarters of a mile.

(7) The claimants held a practice meeting on June 8-9 1974 observed by RAC stewards but without the public being present. At about that time a letter from Mr Whitham, as the owner authorising public meetings, was forwarded to the RAC, as they require.

(8) Thereafter the oral agreement was reduced into writing by solicitors acting jointly on behalf of Mr Whitham and the company. The written agreement was executed in the form of a deed on October 2 1974, a copy of which deed is contained in the appendix hereto. In addition to the oral terms, the deed contained an arbitration clause and a provision for the review of the £500 fee.

(9) No race meeting took place after the execution of the deed. Accordingly, the claimants held no more than two public race meetings on the airfield and those race meetings were held before the execution of the deed.

(10) During race meetings it was not practicable for the airfield or any part of it to be used for any other purpose except that about one quarter of it was used during meetings for grazing of animals. Between meetings Mr Whitham used the airfield for private flying. That part of the runway used for race meetings was too narrow for aircraft because of the barriers and fencing, but two-thirds of the runway was usable for aircraft. Mr Whitham did not use the pits area or competitors' car park because they could not be grazed and because they were in the wrong position for air displays or gymkhanas. Between race meetings those areas were not used by anyone.

(11) On December 15 1976 the claimants' solicitors wrote to the director of administration of the respondent council and made a claim for compensation under section 164 of the Act of 1971 as applied by section 165 of that Act. In the letter the solicitors mentioned that at a public inquiry they had given details of the expenditure which the claimants had been put to, totalling £3,200. With the letter they enclosed a copy of a statement listing alleged additional losses and stated: "We should be pleased to hear from you that you are prepared to consider this claim"; they also enclosed a copy of the deed "under which our clients occupied the land".

(12) The statement listing additional losses referred to a number of items totalling £39,942, including the sum of £3,200 mentioned at the public inquiry. The list included claims made on behalf of persons other than the company, including a claim for £4,000 on behalf of Mr Whitham on account of loss of revenue (eight meetings at £500 per meeting). Indeed of the 10 items in the list only five appear to have been made on behalf of the claimants.

Letters from council officers

(13) On December 22 1976 the director of administration replied to that letter, stating: "I accept that under section 164 of the Town and Country Planning Act 1971 (as applied by section 165) circumstances exist in which a claim for compensation may validly be made". The letter pointed out that the entitlement to claim compensation lies with persons "interested in the land" and that the claims made related to a number of persons and organisations. The letter went on to ask that it might be indicated whether each claim fell under paragraph (a) or (b) of section 164(1).

(14) On April 28 1978 the council's estate property manager wrote to the claimants' solicitors: "consideration has been given to the claim and I am of the opinion that the compensation in this case should be £16,100". The letter continued: "Please let me know if your clients are prepared to accept this amount, whereupon I will report the figure to the relevant Committee for approval".

(15) On May 15 1978 the claimants' solicitors replied to the last-mentioned letter: "We confirm our telephone conversation of the 12th May that our clients are prepared to accept your offer of compensation in the sum of £16,100 together with payment of our legal costs and disbursements to be agreed".

(16) On March 6 1979 the director of administration wrote to the claimants' solicitors stating that the relevant committee of the council had decided that the claim should be resisted in total and that the reason for this decision was that the council had been advised that the claimants were not persons interested in the land.

(17) The letter dated December 22 1976, in addition to asking the claimants to indicate under which paragraph of section 164(1) each claim fell, also asked the claimants to itemise the various amounts claimed by reference to the individual or organisation which had incurred the expenditure or occasioned the loss or damage in question. The claimants provided this information and incurred expense in so doing. They took no steps to reinstate the airfield because they had no capital with which to do it (they estimated the cost of so doing at about £12,500) and they hoped to be able to do so out of the compensation when paid.

In consequence of the letter dated December 22 1976 they believed that compensation would be paid.

The case was argued on the basis that the claimants' rights were governed by the deed.

Claimants' submissions

Mr Brooke, for the claimants, submitted that the expression "a person interested in the land" in section 164(1) of the Act of 1971 has a more generous meaning than "a person who has an interest in the land". He said that "a person interested in the land" meant a person who had some proprietorial or occupational interest in the

A land: see *Willingale (Inspector of Taxes) v Islington Green Investment Co* [1972] 1 WLR 1533, *per* Buckley LJ, at p 1538 (where an executor in whose name the deceased's shares had not been registered but who had the right to call for registration though he had no beneficial interest in the shares, was held to be "interested in" the shares). Mr Brooke drew attention to section 166(6) of the Act of 1971 which, by limiting claims under section 164 for compensation for depreciation in the value of an interest in land to the fee simple or a tenancy only, may suggest that "a person interested in the land" may have a wider meaning than "a person having an interest in the land". The claimants' right under the deed were those of licensees but, in Mr Brooke's submission, the words "a person
B interested in the land" were wide enough to embrace those rights.

Mr Brooke's alternative argument, in the event that the relevant words in section 164(1) referred to a person having an interest in the land, was that under the deed the claimants had a licence coupled with an interest. That interest was, he said, the right to install the fencing and the barriers as fixtures and to keep them there and the right to carry on a profit-making activity on the land. The claimants were interested in the fixtures while they were there as a safety barrier. While they were installed they were land. He referred to the judgment of Megarry J (as he then was) in *Hounslow London Borough Council v Twickenham Garden Developments Ltd* [1971] Ch 233, where all the earlier cases are most usefully
C reviewed. The claimants, he said, were just as much licensees coupled with an interest as they would have been if they had had the right to take crops off the land: he referred to *Webb v Paternoster* (1619) Popham 151 (79 ER 1250); *Wallis v Harrison* (1838) 4 M & W 539 (150 ER 1543).

In the further alternative, Mr Brooke submitted that the claimants had a contractual licence which was irrevocable, alternatively it was revocable only in such way as not to cause damage to the licensees: that meant not revocable within the first five years and then only on giving an indemnity against the obligation to reinstate the airfield in clause 5 of the deed or of waiving that obligation. There was to be implied a covenant against revocation. Such a licence,
D even though it did not confer exclusive possession of the airfield or any part of it was irrevocable. Where a licensor allows another to expend money on the land under an expectation created or encouraged by the licensor that he will be able to remain there, that raises an equity in the licensee such as to entitle him to stay. He has a licence coupled with an equity: see per Lord Denning MR in *Inwards v Baker* [1965] 2 QB 29 at pp 36, 37.

Mr Brooke went further: viewed as an irrevocable licence, the deed created an interest in land: see *Plimmer v City of Wellington* [1884] 9 App Cas 699 (Privy Council); the note to section 1(3) of the Law of Property Act 1925 in *Wolstenholme and Cherry's Conveyancing Statutes* (13th ed), vol I, p 48; *National Provincial Bank v*
E *Hastings Car Mart* [1965] AC 1175 at p 1239 *per* Lord Upjohn, where he said:

> "the cases of *Foster v Robinson* [1951] 1 KB 149 and *Errington v Errington and Woods* [1952] 1 KB 290 were much canvassed before your Lordships. In the latter case the licensees were in exclusive occupation upon the terms of paying off the mortgage instalments and after the matrimonial rupture the wife continued to do so. This, I would have thought, would have given the spouses an interest in the land, in accordance with a well-known line of authority starting with *Webb* v *Paternoster*, valid against all except a purchaser for value without notice".

Mr Brooke also referred to *DHN Food Distributors Ltd v Tower Hamlets London Borough Council* [1976] 1 WLR 852.

F

Point on estoppel

Finally, Mr Brooke submitted that the council was estopped from taking the point that the claimants were not a person interested in the land. The estoppel arose out of the doctrine of election. While the director of administration had no authority to bind the council as to the amount of compensation payable, but merely a power to recommend a sum, and while Mr Brooke could not rely upon the letters dated April 28 1978 and May 15 1978 as constituting a contract to pay compensation (because the obligation of the director was merely to make a recommendation), the director nevertheless had authority to negotiate on behalf of the council and to make an election as to which of two courses should be adopted. Those G courses were:

(1) to take the point that the claimants were not a person interested in the land; or
(2) to decide not to take the point.

Mr Brooke said that the director had shown by his letter dated December 22 1976 that he knew of the requirement that the claimant should be a person interested in the land but, nevertheless, had elected not to take that point. He had made an election and the council could not resile from it: see *Spencer Bower and Turner, Estoppel by Representation* (3rd ed) pp 151 and 315.

H

Arguments for council

Mr Green submitted that the relevant words in section 164(1) of the 1971 Act refer to interests in land in the strict sense. Further than that they refer to legal estates and legal interests only, that is to say, the kind of legal estates and legal interests referred to in subsections (1) and (2) of section 1 of the Law of Property Act 1925 (an estate in fee simple absolute in possession, a term of years absolute, a legal easement, a rentcharge, and a charge by way of legal mortgage). The estates and interests to which subsection (3) of section 1 of the Act of 1925 refers as taking effect as equitable interests, namely, all other estates, interests and charges in or over J land, are intended by section 164(1) of the 1971 Act to be excluded from consideration.

He found support for that view in section 52 of the 1971 Act which, after authorising a local planning authority to enter into an agreement "with any person interested in land in their area for the purpose of restricting or regulating the development or use of the land" goes on to say in subsection (2) that an agreement made under this section with any person interested in land may be enforced by the local planning authority against "persons deriving title under that person in respect of that land". That meant, he said, that the person in question must have an interest in land in the strict sense. The person referred to had to be some person capable of K entering into an agreement and binding his successors. The words "a person interested in the land" must have the same meaning in section 164 as in section 52. The deed gave the claimants no legal interest in the land and accordingly they could not qualify under section 164 (1).

Mr Green submitted further that the deed did not purport to grant any interest in land but merely promotion rights: a right to trade. The deed contained no specific requirement that the owner should make the airfield available; he was merely agreeing not to allow other persons to carry on the same trade on the airfield. The deed did not require the claimants to hold any particular number of meetings or indeed any meetings at all; if none were held the owner L would have no right to payment. These matters pointed to a licence merely rather than a licence coupled with an interest.

The deed contained no specific requirement that any work should be carried out; that was left to the judgment of the claimants depending upon the need to safeguard the public. It was immaterial to the owner whether works were required or not. The deed did not entitle the claimants to possession of any land. Their interest in the barriers and fencing lay in the right to remove them on cessation of their trading activities.

The true parallel with the present case lay in the principle of the decision in *Hill v Tupper* (1863) 2 H & C 121 (where a canal company by deed granted to the plaintiff the sole and exclusive M right or liberty of putting or using pleasure boats for hire on their canal, it was held that the grant did not create such an interest or estate in the plaintiff as to enable him to maintain an action in his own name against a person who disturbed his right, but operated as a licence merely).

Mr Green submitted that there was no estoppel because:

(1) No statement by an officer of the council or indeed by the council itself could grant a right to compensation where it does not exist under the statute;
(2) Officers of the council have no delegated power to agree and settle compensation claims;

A (3) The claimants had not acted to their detriment. On the contrary they were proceeding with their claim in a normal fashion. The non-restoration of the site was irrelevant; the lack of capital was the true cause;

(4) One of the persons on whose behalf a claim had been made by the claimants in their letter dated December 15 1976 was a person interested in the land, namely, the owner, Mr Whitham. Accordingly the letter dated December 22 1976 in stating that circumstances exist in which a claim for compensation may validly be made was correct; there was no unequivocal representation that the claimants were entitled to make a claim.

B **Tribunal's findings**

The words "a person interested in the land" appear in a number of statutory contexts outside the provisions of the Act of 1971: for example, section 35 of the Law of Property Act 1925 and section 84 (1) of that Act. In section 84 (2) of that Act the words are "any person interested", the words "in the land" being omitted. In *Willingale's* case, *supra*, the Court of Appeal appears to have treated the words "interested in any shares" as being wider in meaning than "having an interest in any shares". But in *Re Todd and the North Riding of Yorkshire Agricultural Executive Committee* [1921] 1 KB 281, the Court of Appeal appears to have treated the words "a person who is interested in any land . . ." in section 9
C (1) of the Corn Production Act 1917 as meaning a person having an interest in the strict sense. However, since the court was concerned with the rights of a tenant, the point may not have been directly in question.

There is much to be said for the view that in the Law of Property Act 1925 and the Town and Country Planning Act 1971, both of them statutes concerned with rights in on or over land, the words "a person interested in the land" are shorthand for "a person having an interest in the land" and that they therefore refer to interests in the land properly so-called. I suspect that that is the correct interpretation of the words as they appear in section 164(1) of the Act of 1971. I am not convinced that section 166 (6) requires a
D different interpretation; that subsection would operate no differently if the relevant words were "a person having an interest in the land". I do not accept the narrow meaning which Mr Green places on the words. In my opinion section 164(1) applies to all interests in land, whether legal or equitable. But the question in the present case is whether the relevant words have a wider meaning than that. I think not but, however wide their meaning may be, the rights created by the deed dated October 2 1974 do not, in my judgment, come within it.

Licence not coupled with interest

I hold that the licence is not coupled with an interest and do so for
E all or any of the reasons following:

(1) I am not satisfied that the barriers and fencing were truly fixtures, because it is not certain that the purpose of their annexation was to effect a permanent improvement of the land rather than to enable them to be better enjoyed as chattels;

(2) The deed does not prevent the claimants from removing the barriers and fencing at any time (at all events if race meetings are not held) and the claimants would seem to be interested in them as chattels only;

(3) For there to be a licence coupled with an interest there must have been a *grant* of the interest by the owner of the land; there has been no grant by Mr Whitham to the claimants of any interest in the
F barriers or fencing or anything else; in so far as the barriers and fencing are required by the deed at all they are required only in so far as they are necessary for the protection of the public; they are not specifically mentioned; there is no grant in respect of them;

(4) The deed on its proper construction constitutes a trading licence and nothing more.

I am unwilling to accept that the licence created by the deed is irrevocable or irrevocable during the first five years or irrevocable except upon conditions of the kind mentioned by Mr Brooke. The deed is concerned with a commercial transaction and, in my view, is revocable upon reasonable notice being given to the licensee: see *Winter Garden Theatre (London) v Millennium Productions* [1948] G
AC 173; *Martin-Baker Aircraft Co v Canadian Flight Equipment* [1955] 2 QB 556.

This is not a case of a licence coupled with an interest nor of a possessory licence nor of an irrevocable contractual licence. It is a case of a revocable contractual licence which gives to the claimants no more than a personal privilege to use the land for commercial purposes: see *Torbett v Faulkner* [1952] 2 TLR 659.

I am not persuaded that the council are estopped from taking the point that the claimants are not a person interested in the land within the meaning of section 164(1) of the Act of 1971. Having regard to the fact that the claims made by the claimants were made
on behalf of several persons or bodies, one of whom certainly had H
an interest in the land by reason of his ownership of it, I am very doubtful whether the director of administration's letter dated December 22 1976 contained an unequivocal representation that the claimants were entitled to make a claim. No doubt the claimants incurred expense in proceeding with their claim after the receipt by them of a copy of the letter, but I doubt whether it can be said that they changed their position in any way.

I reject the contention that the director had the authority of the council to make an election on its behalf or that any such election made by him is binding on the council in any way: the question whether the claimants were interested in the land was a question of
substance and not a technicality, matter of procedure or an irregu- J
larity: see per Lord Denning MR in *Wells v Minister of Housing and Local Government* [1967] 2 All ER 1041 at p 1044.

Finally, in my opinion, neither the doctrine of estoppel nor that of election can entitle the claimants to make a claim under section 164 of the Act of 1971 which the statute does not intend them to have.

For these reasons I decide the preliminary issues as follows:

(a) the claimants are not entitled to claim compensation under section 164 of the Town and Country Planning Act 1971;

(b) the respondent council are not precluded by estoppel or election from challenging the claimants' entitlement to claim compen-
sation under the said section. K

In my opinion, my decision on the preliminary issues disposes of the proceedings on the reference. In the circumstances, under rule 49 of the Lands Tribunal Rules 1975, I order that the argument on the preliminary issues shall be treated as the hearing of the case and accordingly I order that the reference herein be dismissed.

Mr Green said that the council would not ask for an order for costs. Accordingly there will be no such order.

APPENDIX

THIS AGREEMENT is made the ** day of One thousand nine hundred and
seventy four BETWEEN DAVID MICHAEL WHITHAM, or Crosland Moor Air- L
field Crosland Moor Huddersfield West Yorkshire (hereinafter called "the Owner") of the one part and PENNINE RACEWAY LIMITED (hereinafter called "the Company") of the other part.

WHEREAS it is agreed as follows:

1. The Owner hereby grants to the Company sole rights to promote motor and motor cycle events on the Airfield at Crosland Moor.

2. The Company shall carry out all works fixtures and fittings necessary for the safe-guard of the public at the Airfield at the Company's own expense and shall be fully responsible for all payments as to future maintenance of such fixtures and fittings or the removal reinstatement or addition to the same.

3. The Company covenants with the Owner to insure with an Insurance Company to be approved by the Owner against any claim against the Owner
in any way relating to public liability or otherwise in connection with the M
events.

4. The consideration to be paid to the Owner shall be the sum of Five hundred pounds per meeting except for the first full year which shall be, treated as a trial year and during this year the Owner shall be entitled to receive Thirty per cent of the net profits of each event from the Company.

5. Upon ceasing to hold Meetings as detailed above, the Company hereby covenants to return the Airfield to its previous condition and remove all fixtures and fittings barriers or otherwise at the Company's expense. This work shall be carried out to the satisfaction of the Owner and if dispute shall arise be referred to a single arbitrator to be appointed by the Surveyor for the Owner whose decision shall be final and binding to all parties.

6. The said sum of Five hundred pounds per meeting shall be reviewed every five years but shall never be less than the said sum of Five hundred pounds.

A
IN WITNESS, whereof the Common Seal of the Company is hereunto affixed and the Owner has hereunto set his hand and seal the day and year first before written.

SIGNED SEALED AND DELIVERED
by the said David Michael
Whitham in the presence of:

SIM AND ANOTHER v ABERDEEN CITY DISTRICT COUNCIL

(LTS/COMP/1979/13)

B
(Wm HALL FRICS, member of Lands Tribunal for Scotland)

February 18 1981

Estates Gazette May 2 1981
(1981) 258 EG 451

House in Aberdeen redevelopment area acquired — Claimants still in occupation — Alternative house bought by claimants in 1973, three years before actual compulsory purchase, had remained unoccupied and had fallen into disrepair — Sufficient indications in 1973 that acquisition was likely in near future — Disturbance claim for £2,700, being mainly for cost of repairs to alternative house — Such cost disallowed as too remote from the acquisition — But, following its recent decision in Smith v Strathclyde, tribunal allows £76 legal
C
expenses arising from acquisition of alternative house on basis that they were not too remote even though incurred prior to deemed notice to treat — If claimants had delayed acquiring another house legal expenses would have been included in disturbance estimate — Council valuer's allowances for expenses of removal, redirection of mail, carpets and curtains held fair and acceptable — Estimate for disconnecting and reconnecting gas cooker increased to cover possible alterations to accommodate it

Mrs Margaret Sim appeared in person on behalf of herself and her husband James Sim; Miss A K Boyd (solicitor in Department of Law and Administration, City of Aberdeen District Council) for
D
the acquiring authority.

Giving his decision, Mr HALL said: This hearing was a continuation of the reference to deal with the compensation payable to Mr and Mrs James Sim by the City of Aberdeen District Council on the acquisition of their dwelling-house at 27 South Constitution Street, Aberdeen. On April 29 1980 the tribunal issued a determination *ad interim* on the value of the dwelling-house at £4,000. At that time there was no evidence in regard to the compensation for disturbance or any other matter not directly based on the value of land and this was to be dealt with in due course when removal had taken place. In the event, Mr and Mrs Sim still occupy the house although there is an action pending in the Sheriff Court to enable the council
E
to obtain possession. The council, wishing to have the total compensation determined in the absence of any negotiations taking place on the disturbance element of the claim, have asked that the compensation under section 12(6) of the Land Compensation (Scotland) Act 1963 be dealt with now.

The claimants had submitted a claim of £2,700 under this head but have not indicated to the council how this sum is justified or made up. Accordingly, the district council have prepared an estimate of the compensation to which they consider the claimants are entitled and in fact have made an offer of £203.88 which, however, is not acceptable to the claimants.

F
"Crawley" test

Roger McKechnie ARICS, principal assistant estates surveyor, City of Aberdeen District Council, had inspected the house at 27 South Constitution Street and also the one at 47 Constitution Street which was purchased some time ago by the claimants. In preparing his estimate of compensation for disturbance, he had regard to the test laid down by Romer LJ in *Harvey v Crawley Development Corporation* (1957) 8 P & CR 141 at p 148: "any loss sustained by a dispossessed owner (at all events one who occupies his house) which flows from a compulsory acquisition may properly be regarded as a subject of compensation for disturbance, provided, first, that it is not too remote and, secondly, that it is the natural and reasonable consequence of the dispossession of the owner".

He had previous experience in negotiating and agreeing distur- G
bance claims in Aberdeen and had regard to these in preparing his estimate. He considered the following items allowable:

Removal expenses

He estimated that a reasonable sum for removal expenses was £70. In determining this figure he had regard to the costs of removal in other disturbance claims which he had dealt with, to the situation of the house at 27 South Constitution Street with its only access through a relatively long and narrow pend, to the fact that the distance by road from the reference house to the alternative house is approximately 150 yds and to the current rate of charge from a firm of removal contractors in Aberdeen at £18.97 per hour includ- H
ing VAT.

Redirection of mail

He said that his information was that the standard charge by the Post Office for redirection of mail for one year was £7.50. He considered one year to be a reasonable period for redirection and was prepared to allow that sum.

Costs of lifting, adapting and relaying carpets and linoleum

He had examined all the floor coverings in the house and allowed a sum of £85, which was made up of individual items which could be considered either as an allowance for lifting and refitting of linoleum where appropriate or as a contribution towards the J
replacement.

Cost of adaptation of curtains and tracks

He had inspected the curtains and allowed £30 towards the cost of adaptation of the curtains and tracks or, alternatively, towards the replacement of them.

Disconnection and reconnection of cooker

There was a gas cooker in the house. Mr McKechnie had it noted that an electric cooker was in the house and had ascertained that the charge by the North of Scotland Hydro Electric Board for disconnection and reconnection would be £11.38. When it was established that a gas cooker was in the house, he said that the cost for connec- K
tion would be about the same but conceded that there might be some additional costs in adaptation of the sink area in the alternative house to take the larger cooker.

Negotiations for acquisition

So far as he understood, the £2,700 proposed by the claimants was mainly in connection with the repairs which were needed at 47 Constitution Street. This house had been bought in September 1973 by the claimants at a price of £7,500, and in April 1974 the solicitors acting for the claimants wrote to the former town council indicating that their clients had purchased another property and were therefore in a position to dispose of 27 South Constitution L
Street. The council had previously given general instructions to their officials on October 16 1972 to proceed with negotiations for the acquisition of properties within the redevelopment area when they were offered to the council. In these circumstances, the council's then estates surveyor replied that the council were willing to enter into negotiations.

On March 8 1976 the district council instructed the estates surveyor to endeavour to negotiate terms and conditions for the acquisition of the properties not already in the ownership of the council affected by phase 6 of King Street/Beach Boulevard redevelopment area in which this property was situated. A resolution to make a compulsory purchase in respect of that area was passed by M
the council on September 6 1976. A compulsory purchase order was confirmed on November 24 1977 and a general vesting declaration executed with effect from July 21 1978.

Mr McKechnie considered that any repairs necessary to 47 Constitution Street did not arise directly from the acquisition, so that he proposed no allowance for such expenditure. So far as the legal expenses arising from the acquisition of 47 Constitution Street in 1973 [were concerned] he considered that these could not be met as they had been incurred prior to the deemed notice to treat.

Mrs Sim gave evidence but was unable to give any details of how the £2,700 was made up. She accepted that the removal expenses

A were probably fair but was unable to comment on the allowance for the carpets or for the curtains. She said that she had a gas cooker in 47 Constitution Street but it was an old one. She had bought the gas cooker in 27 South Constitution Street about 10 years ago but she conceded the cooker was secondhand at that time. It was too large to go into 47 Constitution Street and therefore she would either have to get an alternative one or carry out some alterations to allow the cooker to be installed. She produced the account from her solicitors for the purchase of 47 Constitution Street which amounted to £114.95, being £90 for the sale fee and £24.95 for VAT and outlay relative to the price of £7,500.

Mrs Sim said that the house adjacent to her had been purchased
B by the town council about 1972 and that tenants had been put into the house by the council. From her point of view, the tenants were unsatisfactory and, as she said, "made her life a misery". She said that they had received a letter from the town council in 1972 saying that it would be prepared to enter into negotiations to purchase the house if they were disposed to sell, and when 47 Constitution Street became available for sale in 1973 she and her husband decided to buy it. She confirmed that their solicitor wrote in April 1974 to the town council saying that they were prepared to sell their house at no 27, but at that time the price offered by the council was £800, which they were not prepared to accept. Negotiations broke down and no progress was made until the reference by the council to the tribunal
C in 1980. The tribunal determined £4,000 for the house. She said they had not moved into their alternative house at no 47 in case they prejudiced a sale of their existing house and in the meantime no 47 had fallen into a state of disrepair. She informed the tribunal that it was still unoccupied.

Miss A K Boyd, solicitor for the district council, submitted that Mr McKechnie's estimate of the compensation for disturbance should be accepted as being in accordance with the established principles for such compensation. She also submitted that in so far as the legal expenses of acquiring no 47 were concerned, they should not be allowed because they had been incurred prior to the date of the deemed notice to treat and she referred to the English
D Lands Tribunal decision in *Bloom (Kosher) & Son Ltd v London Borough of Tower Hamlets* (1978) 247 EG 1091. She appreciated, however, that the tribunal's decision in the case of *Smith v Strathclyde Regional Council* issued on December 5 1980 (257 EG 391, 501) was that expenditure incurred prior to the date of a service of notice to treat or a deemed notice to treat can be allowed where, in the circumstances of a case, particular expenditure was naturally and reasonably incurred by reason of the compulsory acquisition or was not too remote by being independently incurred. She said that if the tribunal was against her on the notice to treat aspect, it should find that this expenditure was too remote. There was no firm evidence that the claimants had moved because of the acquisition.
E Mrs Sim had said she intended to move because of the neighbours. The fact that the house had been bought three years before the resolution to compulsorily acquire was in itself a strong indication of remoteness. Further, if the tribunal still allowed the item the expenses should be modified, as they had been incurred for a house worth £7,500 in 1973, whereas the tribunal had determined £4,000 for the reference house as at July 1978. The expenses of £114.95 should be restricted. In particular, the scale fee of £90 should be reduced to £60.

Tribunal's conclusions

The tribunal had inspected no 27 South Constitution Street when
F determining the value and, having considered Mr McKechnie's estimate and heard Mrs Sim's views on the items, we have come to the conclusion that his allowances for removal, redirection of mail, carpets and curtains are fair and acceptable. He had made his estimates from his own observations but supported them by producing a number of settlements of disturbance compensation in Aberdeen, detailed figures of which were in keeping with his proposals.

So far as the gas cooker is concerned, the tribunal have no detailed information but, having regard to the estimate for removal and the possible alterations to accommodate it, determine a figure of £15 for that item.

Although the tribunal have no details of the expenditure of no 47 which Mrs Sim claimed in her figure of £2,700, Mr McKechnie's G
view is accepted that the repair work carried out on no 47 or the work which is still to be done is too remote from the acquisition and cannot be allowed.

On the legal expenses, the tribunal follow the recent decision in *Smith* and hold that the deemed notice to treat is not a legal bar to allowing such expenditure. The question is rather whether the expenditure incurred in 1973 is too remote from the acquisition.

The claimants took the opportunity of purchasing this house near to their existing house when it became available on the market and certainly at that time there were indications that the property would be acquired for redevelopment purposes. The facts that the town council had purchased the adjacent house, that the claimants' sol- H
icitor wrote the town council in 1974 saying they were prepared to sell and the instruction from the town council as early as 1972 to buy any houses which were offered to them were sufficient, in the tribunal's view, to indicate that an acquisition was likely to take place in the near future. If the claimants had delayed acquiring another house until after the deemed notice to treat, the authority would have had to meet the reasonable legal expenses and therefore it is not financially prejudiced by having to meet these expenses when incurred earlier. In fact, these expenses may well be less than they would have been, even in 1978. Mrs Sim was definite that the council had written them in 1972 or 1973, but she was unable to produce the letter. The district council say they have no record of J
such a letter being sent and challenged her statement. As the figures spoken to by Mr McKechnie are estimates of what the disturbance is likely to be, an estimate would have to be made of the legal expenses which the claimants would incur if they still had to buy another house.

In the case of *Harvey* already referred to, which involved the cost of finding other accommodation, Romer LJ said (at p 149) "The natural thing for a dispossessed owner-occupier to do is to buy another home of a comparable kind"; and he concluded by saying "It is bad enough on itself for a person to be compulsorily dispossessed of his home, but it is worse still if he has himself to bear expenses of the kind which are in issue in the present case in finding K
another house in which to live."

Adjusted legal expenses

In the particular circumstances of this case, the tribunal accordingly allow these legal expenses. The expenses were, however, for a house which is larger and more expensive than the house which has been compulsorily acquired and the alternative submission of the acquiring authority is accepted that some adjustment should therefore be made. The scale fee for a house price of £4,000 would be £60 instead of £90 and, with VAT added, the amount would be £66. In addition, there will be the recording dues and incidentals of approximately £10. In these circumstances, the sum of £76 will be L
allowed.

A summary of the determination is as follows:

Removal expenses	£70.00
Fee for redirection of mail for one year	7.50
Cost of lifting, adapting and relaying carpets and linoleum	85.00
Cost of adaptation of curtains	30.00
Disconnection and reconnection of gas cooker	15.00
Legal expenses in connection with 47 Constitution Street	76.00
	£283.50

M

The tribunal accordingly determine the claimants' compensation for disturbance to be £283.50.

The acquiring authority made a submission on expenses and asked that the two hearings should be dealt with separately. So far as the second hearing is concerned it was argued that the claimants had not submitted a detailed claim in accordance with section 11(2) of the Land Compensation (Scotland) Act 1963 and therefore in terms of subsection (1) the claimants should pay the expenses of the acquiring authority.

The lack of information, however, has not prevented the authority from being able to make a proper offer. Normally, the establishment of disturbance claims is best done when the actual costs of

A removal and of disturbance are known. In this case, however, the authority wished the total compensation to be determined prior to completing the action in the sheriff court to obtain possession of the house and therefore estimates had to be made. In these circumstances, the tribunal consider that the expenses of the reference should be viewed as a whole and find the acquiring authority liable to the claimants in their expenses in the fixed sum of £25.

MEPC HOUSING LTD v EAST SUSSEX COUNTY COUNCIL

B (REF /131/1980)

(R C WALMSLEY FRICS)

April 13 1981

Estates Gazette May 23 1981
(1981) 258 EG 759

Cleared site on Hastings seafront close to town centre — Confirmed purchase notice following refusal to renew planning permission for 6-storey block of flats — Whether site a "flat-site" or a "house-site" — Comparables — Negotiated settlement in Hastings relied on by authority valuer accepted as dependable evidence of market value — Claimant's valuer's reliance on sale of flat-site in Bex- C **hill — Adjustments made for "uplift factor", to reflect difference in quality between sites, and "date factor", to reflect differing dates of transactions — Council valuer's two valuations; one a residual valuation showing a negative site value affords tribunal no assistance — Relative levels of prices in Bexhill and Hastings — Tribunal unable to follow its inclination to accept council valuer's opinion that subject site would not today be purchased as flat-site because his valuation evidence pointed the other way — Award of flat-site value**

W D A Hicks (instructed by Nabarro Nathanson) appeared for the claimants; N C Nardecchia (instructed by the solicitor to the East D Sussex County Council) for the acquiring authority.

Giving his decision, MR WALMSLEY said: This is a reference to determine the purchase price to be paid by the East Sussex County Council ("the authority") to MEPC Housing Ltd ("the claimants") for the freehold interest in the cleared site of White Rock House, White Rock Road, Hastings, following the confirmation on October 6 1978 by the Secretary of State for the Environment of a purchase notice.

The subject site comprises 0.24 of an acre and is located at the junction of White Rock Road and White Rock Gardens in Hastings. It adjoins the White Rock Pleasure Grounds to the west. The surrounding area is predominantly residential and recreational in E character. The site has the following planning history:

(i) On February 27 1970 the Hastings County Borough Council granted outline planning permission for the demolition of an existing building on the site and the erection of a block of 12 flats 4 storeys high with garages or parking space for 13 cars.

(ii) On June 25 1970 the Hastings County Borough Council granted outline planning permission for the erection of a block of 16 flats comprising one flat in a semi-basement and 5 storeys above, each with 3 flats and garages and parking space for 16 cars.

F (iii) On October 14 1971 the Hastings County Borough Council granted outline planning permission for the erection of a block of 17 flats comprising 1 flat in a semi-basement and 5 storeys above, each with 3 flats and 1 penthouse, garage and parking space for 18 cars.

(iv) On July 10 1974 the Hastings Borough Council granted planning permission for the erection of 17 flats comprising 57 habitable rooms on 6 floors with basement and ground-floor car parking.

(v) On December 17 1976, MEPC Housing Ltd made application in the form of a letter to Hastings Borough Council for the renewal of the previous planning permission dated July 10 1974. On April 6 1977 the Hastings Borough Council refused to grant permission for the erection of 17 flats on 6 G floors with basement and ground-floor parking because the site lies within the band of interest for the Hastings Town Centre Relief Road.

White Rock House, the building formerly existing on the site, was demolished in or about 1971.

On March 31 1978 the claimants served a purchase notice on the Hastings Borough Council under section 180 of the Town and Country Planning Act 1971. On June 28 1978 the Hastings Borough Council served a counternotice saying they were unwilling to comply with the purchase notice because the East Sussex County Council should be substituted as the acquiring authority. On October 6 1978 the Secretary of State confirmed the purchase H notice but substituted the East Sussex County Council as the acquiring authority.

A notice to treat is deemed to have been served on October 6 1978. The authority not having entered the site, the date for the assessment of the purchase price is the date on which the tribunal completed its hearing of evidence as to value, namely March 4 1981.

At the hearing the claimants were represented by W Hicks, who called valuation evidence from Alan Vivian Maltby BSc FRICS, a partner in the Bexhill-on-Sea firm of Staines & Co. The authority were represented by N C Nardecchia, who called valuation evidence from John Lucas Wood BSc FRICS, a partner in the East- J bourne firm of Bracketts.

The substantive valuation issue is as to whether the subject site is a "flat-site" or a "house-site". The term "flat-site" in this context connotes a site the most profitable use for which is development by flats on more than three storeys ("high-rise" flats). And "house-site" connotes a site the most profitable use for which is development by houses or, possibly, by low-rise flats. Mr Maltby was firmly of the opinion that the subject site is a flat-site and his valuation is on that basis. Mr Wood, while being equally firmly of the opinion that the subject site is a house-site, had prepared alternative valuations.

K

Comparables

Although several other transactions became referred to during the hearing, it became clear that Mr Maltby was relying on only one comparable, and so was Mr Wood. The circumstances of their respective comparables are briefly as follows:

Land at the junction of Station Road and Cranfield Road, Bexhill ("the Bexhill E1 site")

Sold in January 1980 at a price of £125,000 plus the right for the vendor to receive free a top-floor flat when the development was completed. The planning permission on this site was for 30 flats. The terms of purchase had been agreed in the summer of 1979.

The price paid is agreed to represent £1,685 per habitable room ("phr"). L

Land at the junction of Cambridge Road and Holmesdale Gardens, Hastings ("the Methodist Church site")

Purchased in April 1978 by East Sussex County Council at a price of £31,250 pursuant to a purchase notice. There was a certificate issued under section 17 of the Land Compensation Act 1961 which defined the appropriate alternative development as 22 flats comprising 73 habitable rooms on 7 floors with basement car parking for 14 cars, and the price was negotiated on that basis.

The price was agreed in March 1978 and represents £428 phr.

I propose to deal straight away with a submission for the claimants regarding the weight to be attached to this latter comparable. This transaction, say the claimants, should be treated with reserve M and indeed is of no great help because it was a negotiated settlement and not an open market transaction.

It is true that the tribunal has indicated more than once that prices negotiated as settlements are not automatically accepted as dependable evidence of open market value. In *Xerri and Shanks v Enfield London Borough Council* (1967) 18 P & CR 117, for instance, the tribunal had this to say:

Mr Phillips distrusted these settlements. In his view an owner anxious to sell had agreed to the first purchase and once the district valuer had that precedent in his bag, the pass had been sold and even surveyors acting for owners or leaseholders not in a financial position to litigate could not hope to do much better for their clients.

A This tribunal has said on more than one occasion that the danger to which Mr Phillips drew attention is a real one, though it seldom carries with it any stigma of unfair conduct by a district valuer or his officers; but the tribunal has also said that if settlements of this kind are to be disregarded, there must be evidence clear and positive to prove that they were on too low a basis.

And in *Koch v Greater London Council* (1969) 209 EG 1503:

In his final speech Mr Yorke argued the unreliability of settlements made by small traders after negotiation with powerful acquiring authorities. The same problem has confronted the tribunal on several occasions when dealing with slum houses. I do not think I am exaggerating when I say that there is a large body of surveyors who are convinced that some authorities deliberately make settlements with weak claimants on favourable terms and B then use those settlements as precedents to compel surveyors acting for impecunious clients to accept terms which they believe to be too low: nor am I betraying any secret if I say that the members of the Tribunal are on their guard against this possibility and, as they have said more than once, much prefer to rely on open market transactions if at all possible.

In the instant comparable the vendors were not weak, and they were powerfully advised. Evidence was given by Terence Keith Padgham ARICS, now the authority's principal estates surveyor, that the vendors — the Trustees of the Central Methodist Church — had been advised by two (named) firms of surveyors, one based in London and the other in Hastings, both well-established and of high repute. In these circumstances I have no hesitation in accepting the Methodist Church site comparable as dependable evidence C of market value. This comparable is indeed very helpful because it stands only about 400 yds from the subject site and it carried a broadly similar (deemed) planning permission.

Value as a flat-site

I take first the evidence relevant to the flat-site presumption but interpose two preliminary comments:

(a) As a yardstick for comparison, the values of flat-sites seem to me more conveniently expressed in terms of £s per habitable room rather than in terms of £s per unit.

(b) When considering the value phr of the subject site as a flat-site, D I find myself unable to derive any assistance from the evidence produced by Mr Maltby regarding the prices paid for various house-sites.

Mr Maltby's valuation is £85,000, representing a little under £1,500 phr for the permitted development of 57 habitable rooms. Mr Wood put in two valuations as a flat-site, one at £46,500 and the other at minus £18,668. I will deal with Mr Wood's valuations before turning to Mr Maltby's valuation.

Mr Wood's £46,500 figure was based directly on the Methodist Church site comparable, and was arrived at by stages as follows:

(a) Price phr for the comparable, agreed in March 1978	£428
(b) Corresponding value for the subject site phr in March 1978 —£428 uplifted by a factor of 100:66	£650
(c) Value of the subject site phr in March 1981 —£650 uplifted by a factor of 125:100	£812.50
(d) Entirety value of the subject site in March 1981 — 57 habitable rooms @ £812.50, £46,312, say	£46,500

E

The starting point of the above valuation is non-contentious, being the price paid for the Methodist Church site £31,250 divided by 73 as the permitted number of habitable rooms. This starting figure of £428 is then adjusted to reflect the difference in quality between that site and the subject site. The subject site is a very good one, being on the sea front, close to the town centre (albeit approached therefrom up a steepish hill) and in pleasant surroundF ings. Mr Wood considers that the difference between the two sites is fairly reflected by his uplift of 100:66. Mr Maltby considered this inadequate, and suggested instead an uplift of 100:50. Having inspected, I adopt an uplift factor of 100:55, the application of which gives a value for the subject site in March 1978 of £778 phr. The next adjustment is to translate a March 1978 value into a March 1981 value. Mr Wood adopts for this purpose a factor of 125:100. Mr Maltby disagrees, and in his oral evidence suggested 130:100.

This question, of the appropriate adjustment for the date-factor, came up on several occasions during the hearing, in connection with the differing dates of a number of transactions. Towards the end of the hearing I invited both witnesses to take time during an G adjournment to prepare for me a schedule showing their respective opinions. On studying the schedules thereafter produced, I have decided to adopt the figures put in by Mr Wood: the figures appearing in Mr Maltby's schedule bear no relationship to the oral evidence he had given earlier. In the remainder of this decision, therefore, when an adjustment of a price or value is required to bring into account the date-factor, I shall be making use of the following table:

	Flat-site value	
Date	*Hastings*	*Bexhill*
March 1978	100	100
June 1978	105	108
October 1978	112	115
Summer 1979	125	128
March 1981	125	128

H

Taking the factor of 125:100 from the above table and applying this to the ascertained March 1978 value of £778 phr produces a March 1981 value for the subject site of £972.50 phr. And multiplying this last figure by 57, as the permitted number of habitable rooms, produces an entirety value for the subject site, on this approach, in the amount of £55,432, say £55,500.

J

Residual valuation

Mr Wood's second valuation was on a residual basis, as follows:

Realisation		
6 × 3-bed flats @ £33,500 =		£201,000
11 × 2-bed flats @ £26,000 =		286,000
		487,000
Add 17 basement car-parking spaces @, say, £1,000 each		17,000
		504,000
Building costs		
6 × 3-bed flats @, say, 900 sq ft = 5,400 sq ft @ £30 per sq ft =	£162,000	
11 × 2-bed flats @, say, 750 sq ft = 8,250 sq ft @ £30 per sq ft =	247,500	
17 basement parking spaces @ £3,000 per space =	51,000	
	460,500	
Interest for, say, 9 months @ 18%	62,168	522,668
Showing a negative site value of		£18,668

K

I can see no way in which a site valuation at £46,500 can be L reconciled with another valuation of the same site, and on the same development assumption, at minus £18,668. It was in the case of *First Garden City Ltd v The Letchworth Garden City Corporation* (1966) 200 EG 123 that the tribunal commented [at p 460] "once valuers are let loose upon residual valuations, however honest the valuers and reasoned their arguments, they can prove almost anything". In the later case of *Clinker & Ash Ltd v Southern Gas Board* (1967) 203 EG 735 this opinion about residual valuations prepared other than under the discipline of open market conditions was spelled out in these terms [at p 736]:

It is a striking and unusual feature of a residual valuation that the validity of a site value arrived at by this method is dependent not so much on the M accurate estimation of completed value and development cost, as on the achievement of a right balancing difference between these two. The achievement of this balance calls for delicate judgment but in open market conditions the fact that the residual method is (on the evidence) the one commonly, or even usually, used for the valuation of development sites, shows that it is potentially a precision valuation instrument. If there are two equally proficient valuers acting respectively for a willing vendor and a willing purchaser they would thus expect to agree on a price for the site in question, it being irrelevant for this purpose that one valuer may have arrived at the agreed site value by using figures for completed value and development cost differing substantially from those used by his colleague.

When a residual valuation is prepared for arbitration purposes, however, the conditions are very different; the valuation is then a calculation made *in*

> *vacuo*; and although there may be a *deemed* open market there are no external sanctions acting as an incentive to the achievement of the delicate balance which I have described, because there is in effect a captive purchaser and a captive vendor; thus there is no risk on the vendor's part of losing a sale by reason of the price advised by his expert being too high, nor is there any risk on the purchaser's part of missing a buy because the price advised by *his* expert is too low. Possibly as a side-effect of this absence of any external constraint, the natural tendency of the vendor's (or claimant's) valuer to adopt full figures when calculating developed value and conservative figures when calculating development cost almost invariably results (in the experience of the Tribunal) in his putting forward an undependably high opinion of site value. Similarly the natural tendency of the purchaser's (or authority's) valuer to adopt conservative figures when calculating developed value and full figures when calculating development cost almost always results in his putting forward an undependably low opinion of site value, and on occasion it may even throw up a minus site value. . . . It is against this background, and for this reason, that the Tribunal has reluctantly found itself compelled to reject the residual method when put forward as opinion evidence, unless there is no simpler method of valuation available.

Although some considerable time was spent at the hearing on the figures that make up Mr Wood's residual valuation, for my part I propose to leave that "valuation" on one side as affording no assistance.

Turning now to Mr Maltby's valuation at £85,000, his basic figure of a little under £1,500 phr was derived directly from the price of £1,685 phr realised for the Bexhill E1 comparable. In cross-examination Mr Maltby concurred that his firm had also negotiated the sale of two other flat-sites in Bexhill; and that these had been at prices which showed a much lower level of value than that shown by the E1 comparable. The circumstances of these two other Bexhill sites are briefly as follows:

Land at the junction of Cantelupe Road and Sea Road ("the Bexhill 2 site")
Sold in October 1978 at a price of £40,000. The planning permission on this site was for 17 flats, and the site is now developed with a block of high-rise flats. The terms of purchase had been agreed in June 1978.
The price paid represents £770 phr.

Land between Eversley Road and Wilton Place, site of the Marina Garage ("the Bexhill 3 site")
Sold in January 1979 at a price of £135,000 plus the right for the vendor to receive free a top-floor flat when the development was completed. The planning permission on this site was for 54 flats and a restaurant. The terms of purchase had been agreed in October 1978.
The price paid is agreed to represent not more than £829 phr.

Taking the sale prices of the Bexhill sites E1, 2 and 3 and adjusting these to bring into account the date-factor points to corresponding March 1981 values as follows:

Site E1	£1,685 phr
Site 2	£912 phr
Site 3	£923 phr

I accept a comment by Mr Wood that the E1 price is "exceptionally high". Indeed it seems unaccountably so, apart from the circumstance that this site was apparently bought by an outside operator who had not previously bought any land in Bexhill. Nevertheless the price paid for Site E1 stands as much a part of the market as the prices paid for Sites 2 and 3. Of these three sites I find the most helpful to be Site 3. Having inspected, my own impression is that a reasonably expected price in March 1981 for a flat-site of importance in Bexhill comparable to the importance of the subject site in Hastings would be of the order of £1,200 phr.

As to the relative level of prices in Bexhill as compared with Hastings, I received conflicting evidence; with Mr Wood suggesting a differential of 25 per cent in favour of Bexhill, and Mr Maltby suggesting no differential at all for a first-class position but 10 to 15 per cent for other positions. My own view is that a March 1981 value of £1,200 phr for a Bexhill flat-site of comparable importance would point to a value of £1,000 phr for the subject site. Adopting this latter figure and multiplying it by 57, as the permitted number of habitable rooms, produces an entirety value for the subject site, on this approach, in the amount of £57,000.

Looking at this figure of £57,000 alongside the figure of £55,000 arrived at by reference to the Methodist Church site comparable, I find the value of the subject site as a flat-site to be £56,250.

Value as a house-site

As already mentioned, Mr Maltby put forward no valuation of the site as a house-site. Mr Wood's valuation on a house-site basis was £30,000, representing a spot price of £7,500 per plot for an assumed development with four houses. This site price of £30,000, said Mr Wood, could be seen to be realistic because it would show the purchaser a profit of about 18 per cent on the development, thus:

Realisation		
4 houses @ £42,000		£170,000
Costs		
4 houses @ 1,250 sq ft each @ £20 psf	£100,000	
Finance for, say, 6 months @ 18%	9,000	
Fees etc 5%	5,000	
Site cost	30,000	144,000
	Profit (= 18% of costs)	£26,000

Towards the end of his evidence, however, Mr Wood told me that, having listened to the details of various sales of house-sites in or near Hastings and Eastbourne as given in evidence by Mr Maltby, he was prepared to increase his spot price from £7,500 per plot to £9,000 and to make an end-addition of £4,000 for the "hope value" of obtaining permission for five houses instead of four, thus producing a revised valuation figure of £40,000.

In support of his opinion that the subject site would not today be purchased as a flat-site, Mr Wood said that as a result of the escalation since 1978 of the cost of building high-rise flats there were at the present time a number of development sites along this length of the South Coast with an unimplemented planning permission for high-rise flats. These sites were either remaining undeveloped or were now being developed with houses.

While my inclination is to accept Mr Wood's opinion that in March 1981 the subject-site is a house-site and not a flat-site, I find myself unable to follow that inclination. This is because Mr Wood's valuation evidence points the other way. If the subject site is regarded as a flat-site Mr Wood puts its value at £46,500, whereas he puts its value as a house-site at no more than £40,000.

I find the purchase price to be paid by the authority to be £56,250. In addition, the authority will pay the claimants a surveyor's fee based on scale 5(a) of the Scales of Professional Charges of the Royal Institution of Chartered Surveyors.

Counsel for the claimants asks that interest on the purchase price be paid from the date of this award, on the grounds that the subject site has been standing blighted since April 1977, its only use since then having been as a car park by the Hastings Area Health Authority under a licence from the claimants and without any payment passing. This seems to me a reasonable request. Interest on the sum awarded will be paid by the authority at the statutory rate from the date of this decision.

One further matter should be mentioned. During the course of the hearing a letter was handed in dated February 6 1980, marked "Subject to Contract", signed by the County Estates Surveyor, East Sussex County Council, and addressed to Mr Maltby's firm. The letter reads:

> Further to your letter of 21st November 1979 and to my assistant's recent discussion with Mr Maltby at your office on 18th January last, I confirm that I am prepared to recommend the County Council to pay the sum of £50,000 (fifty thousand pounds) for your clients' unencumbered freehold interest in the above site, with vacant possession on completion, and free from Tithe Redemption Annuity and all other charges. This sum is offered on the understanding that it includes compensation for all items of claim whether specifically mentioned or otherwise.
> In addition, the County Council will be recommended to pay your clients' costs of deducing title and conveyance, together with a surveyor's fee based on Scale 5(a) of the RICS and ISVA Scale of Professional Charges.
> If you will confirm that your clients are prepared to proceed on this basis, notifying me of the name and address of their solicitors, I will instruct the County Secretary accordingly.

I was asked to treat the above letter as for information only: and no weight has been attached to the figure mentioned therein in arriving at the amount of my award.

A Having opened the sealed offer of compensation lodged by the acquiring authority I find that the amount thereof is marginally greater than the amount of my award. The acquiring authority will pay the claimants their costs of this reference up to the date of the sealed offer, February 23 1981, and the claimants will pay the costs of the acquiring authority as from that same date, such costs unless agreed to be taxed by the Registrar of the Lands Tribunal on the High Court Scale.

B

CHALK v MOLE VALLEY DISTRICT COUNCIL

(REF/238/1979)

(V G WELLINGS QC)

March 24 1981

Estates Gazette May 30 1981

(1981) 258 EG 867

C

Depreciation following s 51 discontinuance order on caravan site at Ashtead, Surrey — Issue of number of caravans on which compensation payable — Extent of land with deemed planning permission under s 17 of Caravan Sites and Control of Development Act 1960 — Plan attached to discontinuance order showed four caravans, each within a larger rectangle of land, on 2-acre area ("blue land") — Site licence application originally made for 0.864-acre area ("green land") which was within blue land and very little bigger than area enclosing rectangles — Site licence restricted number of caravans to four — If applicant had applied for variation of conditions, 12 caravans would have been allowed — Whether "green land" or rectangles of land qualified for compensation — Tribunal rejects authority's submission that order plan showed single site with individual parts linked by paths — Held that green land represented area in regular use by caravanners in conjunction with their caravans and constituted existing site for 1960 Act purposes — Award of £7,622 for 12 caravans

D

Giles Clarke (instructed by Cotton Gumersall & Palmer, of Epsom) appeared for the claimant, Arthur Leonard Chalk; D Smith (solicitor, Mole Valley DC) for the compensating authority.

Giving his decision, MR WELLINGS said: This reference concerns the amount of compensation to which the claimant is entitled, under section 170 of the Town and Country Planning Act 1971, for depreciation of the value of his interest in land used as a caravan site in consequence of an order, made under section 51 of the Act, requiring the discontinuance of that use. The order was dated July 6 1977 and was made by the compensating authority, the Mole
E Valley District Council. The order provided that it should take effect on the date of confirmation thereof by the Secretary of State for the Environment, which date was June 5 1978. The order stated:

> This order relates to the land known as Greenslade Caravan Site, Farm Lane, Ashtead, shown edged blue on the attached plan, part of which land, shown hatched and coloured red on the attached plan (hereinafter referred to as "the said land") is used as a caravan site.

The "blue land"

The land shown edged blue on the plan was an area of about 2 acres (I will call it "the blue land"). On the plan four rectangles were drawn. In the centre of each rectangle the outline in plan of a
F caravan, also rectangular in shape of course, was drawn; each rectangle represented the area covered by taking a measurement of 20 ft outwards at right angles from the front and back and each side of the caravan. Each rectangle was shown on the plan within a red outline and was hatched in red.

The caravan shown in the northernmost rectangle was shown as caravan no 1; it faced north and south, approximately, so that the longest sides of that rectangle faced west and east, approximately. The caravan shown as caravan no 4 and the longest sides of the rectangle appertaining thereto were shown as facing in the same directions; but the north-south axis of that caravan, as shown on the plan, was about 2 m to the west of the north-south axis of caravan no 1. Caravan no 3 is shown as lying to the south-east of caravan no G
4. Caravan no 3, however, faces west and east, so that its sides and the longest sides of the rectangle face north and south. Approximately one half (the eastern half) of that rectangle lies to the east of the eastern side of the rectangle for caravan no 4 (if prolonged southwards). Caravan no 2 is shown on the plan as lying to the east of caravan no 4 and facing south-west and north-east, so that its north-western side and the north-western side of its rectangle look towards caravan no 1. The north-eastern corner of that rectangle is not shown on the plan because it would otherwise lie outside the blue land and therefore would represent land outside the claimant's ownership.

The distances separating the four rectangles are not very great: H

(1) between nos 1 and 4: about 8 m;
(2) between nos 4 and 3: about 1.5 m;
(3) between nos 4 and 2: about 16.5 m;
(4) between nos 3 and 2: about 14.5 m;
(5) between nos 1 and 2: about 22 m.

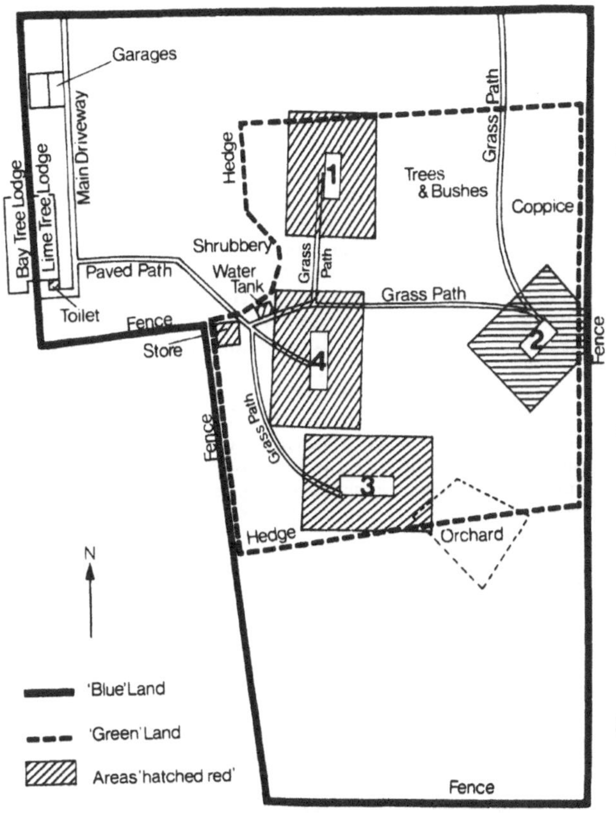

J

K

L

On the western side of the blue land a dwelling-house is shown. It is semi-detached, the western half, known as Bay Tree Lodge, being outside the blue land and the eastern half, known as Lime Tree Lodge, being within it. In the north-western corner of the blue land two adjoining garages are shown. The dwelling-houses and the garages are shown on the plan as uncoloured. A number of paths is shown on the land:

M

(1) A grass path proceeding northwards from caravan no 2 to the northern boundary of the blue land;
(2) A grass path proceeding westwards from caravan no 2 to the northern part of the rectangle for caravan no 4 and then westwards outside that rectangle, past the water tank, which is shown hatched red on the plan and as being partly within but mostly without that rectangle, and, then, proceeding to join up with another path, namely:
(3) The path made of paving stones (in whole or in part) proceeding north-westwards from caravan no 4 to join up with:
(4) The main driveway which links Lime Tree Lodge with the northern boundary of the blue land and with the garages;
(5) A grass path linking caravan no 1 with the path from caravan no 2;
(6) A grass path linking caravan no 3 with the paving stone path.

In addition to the features which I have so far mentioned, on the western side of the blue land there is shown on the plan a store outlined and hatched red: it is about 5 m from the western side of the rectangle for caravan no 4. Adjoining Lime Tree Lodge there is shown a toilet, outlined and hatched red. All the paths shown on the plan and the driveway are shown coloured red.

The order provided that:

> The use of the said land for the purposes set out in the Schedule hereto shall be discontinued within *twelve* months of the date of confirmation of this order and all sanitary accommodation, stand pipes, water tanks, footpaths, caravan pitches and store sheds, and other buildings or works connected with the use of the said land as a caravan site shall be permanently removed within the same period.

The purposes set out in the schedule to the order were:

> The use of the said land as a caravan site.

The "green land"

On March 6 1961, pursuant to the Caravan Sites and Control of Development Act 1960, the claimant's mother, who also was his predecessor in title, applied for a site licence. At that time the claimant was registered at HM Land Registry with title absolute in respect of the blue land, but the application for a site licence was nevertheless signed by his mother. However, nothing turns upon that. The application stated that the acreage of the site was 2 acres, but the plan attached to the application showed the site in question outlined in black and red. That area is the same as that shown on the plan attached to the agreed statement of facts and thereon outlined in green. I will refer to that area as "the green land".

The green land is part only of the blue land. The western boundary of the green land is the western boundary fence of the blue land, but at a point adjacent to the store where the fence turns westwards the green land turns eastwards and follows the curve of a shrubbery and then a hedge to its northern boundary much of which was also a hedge. The eastern boundary of the green land was a fence along the eastern boundary of the blue land. The southern boundary was along the line of a short hedge shown on the plan attached to the agreed statement of facts and the prolongation of that line eastwards. The green land is agreed to have an area of 0.864 acre, which area is stated on the plan attached to the application for a site licence.

Agreed facts

The agreed statement of facts contains the following statements:

(1) In the events which happened, the application for a site licence resulted in the deemed grant of planning permission under section 17 of the Act of 1960 for the use of all or part of the green land as a caravan site. By reason of an oversight no site licence was ever issued in response to this application.

(2) On December 21 1971 the claimant applied to the local authority for a site licence and on January 24 1974 such a licence was granted but was subject to a number of conditions, one of which restricted the number of caravans on the site to four.

(3) On March 3 1975 the claimant applied for a variation of that condition, but in his application he erroneously contended that the entirety of the blue land enjoyed deemed planning permission. That application was refused by the local authority.

(4) The claimant appealed to the magistrates' court under section 8 of the Act of 1960, but his appeal was abandoned when it became clear that the whole of the blue land did not enjoy deemed planning permission.

(5) The claimant did not make any new application for a variation of the conditions in the site licence because the council indicated that it was proposing to make a discontinuance order.

Agreed issues

The agreed statement of facts also contains the following statements:

(1) It is agreed that compensation should be paid by the compensating authority to the claimant following the order, and that the quantum of compensation should be determined by the number of caravans which the claimant could have stationed on the green land had the order not been made. It is further agreed that after allowing for full compliance with the conditions of any necessary site licence the amount of compensation should be based on a figure of £700 per caravan less the value of the site for agricultural purposes which is agreed at £900 per acre. Had the applicant reapplied for a variation of the conditions in the site licence he would have been allowed to station 12 caravans on the green land.

(2) The issue between the parties is the extent of land which enjoyed deemed planning permission under the Act of 1960. The compensating authority contend that this land is restricted to the land hatched and shaded red on the plan accompanying the discontinuance order. The claimant's contention is that the planning permission extended to all the green land. It is agreed that if the claimant's contention is correct 12 caravans could have been stationed on the land and that if the compensating authority is correct only four.

(3) The Lands Tribunal is asked to determine the quantum of compensation appropriate by determining the number of caravans for which compensation would be paid. It is agreed that if the applicant succeeds the compensation will be for 12 caravans and be £8,400 less £777.60, being the agreed value of agricultural land, namely £7,622.40, and that if the authority succeeds it will be for four caravans and be £2,800 less £777.60, namely £2,022.40.

Facts found

The claimant gave evidence. It was virtually unchallenged. I accept it and from it find the following facts:

(1) The plan accompanying the discontinuance order, in so far as it showed fences and hedges and the shrubbery and trees and bushes was substantially correct save in so far as the trees were in 1960 mere saplings. The area to the south-east of caravan no 3 shown as orchard was, however, not an orchard but consisted merely of four small fruit trees. The water tank and store were correctly shown.

(2) His mother bought the blue land in 1935 and it was about then that the hedges other than the southernmost hedge were put in.

(3) The blue land had been known to the claimant since 1935. The first caravan came on to the green land in 1941 and was stationed in the area shown on the plan accompanying the order (which I will call "the order plan") as the site of the water tank. Thereafter several caravans came and went. One would come and would go and be replaced by another. At the end of the second world war there were five caravans on the green land: one on the space shown on the order plan as the site of the water tank and others in approximately the same positions as the pitches shown on that plan. In 1943 he spent three weeks at Lime Tree Lodge: there were then four caravans, all in the same positions as shown on the order plan. In 1948 there was a caravan situated on the area marked "trees and bushes" on the order plan, namely, immediately to the east of the rectangle shown thereon for caravan no 1. From 1958 to 1959 there was a caravan on the coppice situate to the east of "trees and bushes" and to the north of the rectangle shown on the order plan for caravan no 2. Thereafter there were consistently four caravans on the green land, including throughout 1960. The largest number of caravans ever situated on the green land was five. The order plan correctly showed the positions of the caravans as they were in 1977. They had been in those positions for from five to seven years. Prior to that there had been other caravans, but not necessarily in the same positions but nevertheless approximately the same positions. Indeed since 1948 the caravans had been in approximately those positions.

(4) The 1960 application for a site licence was made out of time but with the express consent of the local authority. The application had been signed by his mother because both he and she erroneously had thought that she was the owner of the blue land. Had he known that, as the case was, he was the owner he would have signed the application. It was at his request that his mother had signed it.

(5) The section of the hedge along the northern boundary of the green land, at the point where it is intersected by the easternmost grass path, in the course of time became worn down by caravanners proceeding throught it and ultimately he removed that section of the hedge.

(6) The access for caravans coming on to the green land was, first, over the driveway to a point in the vicinity of the garages and, then, eastwards over the northern part of the blue land, not comprising the green land, and then southwards, through the hedge, in the vicinity of caravan no 1.

(7) His mother lived at Lime Tree Lodge until her death in 1972. Over the years, except during the war, he had lived either at Lime Tree Lodge or at Bay Tree Lodge, which property he had owned until 1978.

(8) He regarded the area of the blue land to the north of the green land as an amenity area for the caravanners.

(9) The short hedge on the southern boundary of the green land had been put in by caravanners.

A (10) The area of the blue land south of the green land had been, in part, cultivated by caravanners until 1977 and, in part, used by them for pigeon-shooting.

(11) The caravanners had always had complete freedom to use the green land and the blue land and, indeed, so long as he owned it, the land of Bay Tree Lodge.

Submissions

Mr Clarke submitted that the relevant unit of occupation was the blue land but that the relevant planning unit was the green land and the whole of it: see *Burdle v Secretary of State for the Environment* [1972] 1 WLR 1207. If the claimant had appealed the discontinuance order it would have been quashed because it did not refer B to any recognisable planning unit. The deemed planning permission obtained under section 17 of the Act of 1960 was unconditional and unlimited. It was immaterial that the 1960 application specified only four caravans. There was no power to cut down the claimant's existing use rights.

Mr Clarke referred to the meaning of "existing site" in section 13 of the Act of 1960 and of "caravan site" in section 1(4) of that Act. The question was what was the land on which caravans were stationed for human habitation and what was the land used in conjunction therewith? It was ludicrous to suggest that any land other than the green land qualified under the definitions. It was ludicrous to suggest that, if one stepped outside any of the four C rectangles shown on the order plan, one would not be on land on which a caravan was stationed. There was no evidence that any part of the green land had any use save as caravan pitches or as an adjunct thereto. Moreover, the positions of the caravans had varied from time to time. The concept of a caravan site was a concept of a use of land, not a concept with reference to the number of caravans on the land: see *Hartnell v Minister of Housing and Local Government* [1964] 2 QB 510, at p 540 *et seq, per* Danckwerts LJ, approved by the House of Lords [1965] AC 1134.

Mr Clarke referred also to *Biss v Smallburgh Rural District Council* [1965] Ch 335 and *East Barnet Urban District Council v British Transport Commission* [1962] 2 QB 484. The decisions in D *Williams-Denton v Watford Rural District Council* (1963) 15 P & CR 11 and *R v Axbridge Rural District Council ex p Wormald* [1964] 1 WLR 442 were distinguishable.

Mr Smith submitted that the order plan showed a single site the individual parts of which were linked by paths. It did not show a number of sites. He, however, did not seek to justify the clumsiness of the draftsmanship of the order. The 20-ft measurements used for the purpose of drawing the rectangles shown on the order plan were arbitrary but they were taken from the facts of the *Axbridge* case. They were justified by the fact that 20 ft was the distance required between caravans under the model conditions for caravan site licences.

E Mr Smith relied upon the decision in the *Axbridge* case and more particularly on the observations of Lord Denning MR in that case at p 445, where he said:

> In order that land should be "used in conjunction with land on which a caravan is so stationed" it must be used immediately and directly in conjunction with it.

Mr Smith submitted that on the facts the "existing site" within the meaning of the Act of 1960 was correctly represented on the order plan.

It is plain that parts of the blue land, outside the green land, used F for recreation by caravanners from time to time or for pigeon-shooting were not part of any existing site. But the claimant does not allege that they were. It appears to me that the green land was a very compact piece of land. It is very little bigger than the area which would result if a line were drawn enclosing all the four rectangles shown on the order plan. There is no reason and indeed no evidence to support a limitation of communication between one caravan and another to the particular paths shown on the order plan, most of which paths were grass paths merely.

It appears to me that the green land represents the area in regular use by the caravanners in conjunction with their caravans and that the green land constitutes the existing site for the purposes of the Act of 1960. Accordingly the claimant is entitled to an award in the G higher of the two agreed sums and I therefore award to him the sum of £7,622.40. I have not been asked to construe the discontinuance order and I make no observations about it. The compensating authority will pay to the claimant his costs of the proceedings, such costs if not agreed to be taxed by the Registrar of the Lands Tribunal on the High Court Scale.

(The sketch plan shows the features referred to by the tribunal, reproduced to assist in appreciation of the facts, and does not form part of the decision.)

H

CHILTMEAD LTD v READING BOROUGH COUNCIL

(REF/245 & 246/1979)

(C R MALLETT FRICS)

April 8 1981

Estates Gazette June 20 1981

(1981) 258 EG 1191

Forced sale of contents of electronic-equipment-dealing company's premises at Reading — Value of stock prior to forced sale — Proportion of stock valued by firm of machinery and plant valuers relying in part on company director's expertise — Some stock J valued by director without any assistance from firm of valuers — Claimants' valuation of open market retail value of stock before discount accepted by tribunal — Discount to allow for likely rate of disposal of stock — Whether proper discount applied or whether council valuer's valuation based on certified accounts to be preferred — Method of valuing stock by reference to average purchases "singularly inappropriate" to valuation of claimant company's stock — No foundation for belief that compensation for loss on forced sale cannot exceed loss of profits — Total deduction of 36.4 per cent made by claimants from full retail value — Tribunal makes further deduction of 15 per cent after weighing certain specified matters K

Michael Rich QC and C J Lockhart-Mummery (instructed by Clark & Son, of Reading) appeared for the claimants; Christopher Cochrane (instructed by the solicitor to Reading Borough Council) for the acquiring authority.

Giving his decision, MR MALLETT said: These references relate to the compulsory acquisition by the Reading Borough Council of premises at 7 to 11 Arthur Road and land and buildings adjoining 2 Arthur Road, Reading, occupied by Chiltmead Ltd. Compensation for the acquisition of the premises has been agreed and the only matter in dispute is the compensation payable in respect of the forced sale of the contents. It is agreed L that the sum of £98,364.03 represents the proceeds of the forced sale and the only issue concerns the value of the stock prior to the forced sale, which the claimant puts at £1,037,192 and the acquiring authority at £198,744.

"Lucky dip" procedure

The company deals in surplus new and secondhand electronic equipment, usually purchased in lots from government bodies or in quantity from companies and sold to firms or the general public in bundles or as individual items. Mr Stewart [director of the claimant company] has gained an expertise in this trade over a number of years. A particular source of supply was auction sales of surplus government equipment where large quantities of M items, insufficiently described or catalogued, were sold in lots. Mr Stewart used his skill to identify items of value in each lot in order to limit his bid. If successful, he acquired items of known value on which he could expect to make a profit on resale. He also acquired, at little or no cost, a miscellany of items some of which were immediately discarded as valueless and others which proved to be useful acquisitions. At times this "lucky dip" procedure produced items of high value which had been sold and purchased unknowingly at a nominal price.

In the main, items of little value were identified and disposed of at the point of purchase. Other items were identified im-

A mediately before or after being transferred to the premises at Arthur Road. Once the items were identified their real value was capable of being estimated. Nevertheless, Mr Stewart estimated the value of his stock for the purpose of the company accounts on the basis of the historic cost, which was substantially below the true value.

Mr Stewart specialised in trading comparatively high-priced electronic test gear such as oscilloscopes. Because of the rapid technological development in the electronic world much of the equipment was obsolescent although still very much in use and in demand. It was a feature of his business to acquire a considerable quantity of spares for the maintenance of these items before or B after sale. Stock items such as plug and socket terminals remained in constant demand despite changes and development in equipment. It was necessary at very frequent intervals to go through the stock and identify and dispose of those items which were no longer in demand and were of little value. In this way the value of his stock was upgraded over the years.

In addition to his trading business Mr Stewart had in the past manufactured certain equipment, but this had proved to be nonprofitable and had been abandoned. At the material time, however, he was developing a piece of equipment known as an "XY Plotter" and was about to start production on this new venture. For this purpose he had collected on the premises a C large number of valuable parts.

In brief it may be said that the premises at Arthur Road were not particularly well suited either for the storage, manufacture or sale of articles. The company employed 11 staff, including Mr and Mrs Stewart, four of whom were only part time. Part of the premises were occupied upon an annual basis on expiry of a five-year lease and part of them were held on an oral annual tenancy.

The premises were affected by the Reading County Borough Council (Albert Road, Arthur Road, Leopold Road, Orts Road) Compulsory Purchase Order 1973 confirmed by the Secretary of State on December 4 1974. On December 7 1976 Chiltmead were D informed by the council that their premises were affected by the redevelopment scheme and of the council's willingness to discuss the availability of a redevelopment site in another part of Reading so as to enable Chiltmead to relocate their business. Notice to treat and notice of entry in respect of both premises were served on November 14 1977.

Sale of stock "in site"

It is unnecessary here to go into the various alternative courses of action, and the alternative basis of compensation, that were discussed between the parties over the years, and the reasons why the lease of an alternative site never took place, because on E December 8 1978 the council made it known that they would require both premises on or about March 1 1979 and it would not be prudent to reply upon any extension of time after the date. The acquiring authority took possession of both premises on March 1 1979 and this is the agreed valuation date. The parties agreed that there was no practical alternative to selling off the stock *in situ* and at a meeting on December 13 1978 it was agreed that Henry Butcher & Co be appointed by the claimants to value the stock prior to the forced sale. Stock was sold over the period January 2 1979 to March 1 1979. The public sale started on February 7 1979 and from January 2 1979 certain goods were sold to dealers. Items remaining on the premises at the end of the sale F were sold to a scrap merchant and removed. The total figure of £98,364.03 was realised exclusive of value added tax.

It is accepted by the parties that this is the proper figure to deduct from the value of the stock in order to arrive at the appropriate compensation.

The team from Henry Butcher was not available to start the valuation of stock until January 2 1979 and they had to complete their work before the public sale started on February 7 1979. It was recognised by the parties that due to the time available a detailed schedule and valuation could only be made on a proportion of the stock. On the residue, apart from the specific high value items, the valuers should consider a representative sample, so that this might be applied to the valuation of the G whole of the residue. The parties also recognised that Henry Butcher & Co would in part be relying on Mr Stewart's expertise and that part of the stock would be valued by Mr Stewart without any assistance from Henry Butcher.

Valuations for claimants

R B Watkins FSVA, a senior valuer in Henry Butcher and in charge of the team of assistants who carried out the work, gave evidence as to how he had arrived at the valuation of part of the stock on an impartial basis as between a willing purchaser and a willing seller to an ingoing tenant as part of a going concern in the sum of £737,191.55. He explained how the items had been H counted, either individually or in batches, and he produced lists of the computer equipment, test equipment, and other stock which had been itemised individually. All the items were then priced at their estimated retail value by reference to Mr Stewart and a number of retail catalogues. Initially the items were priced by Mr Stewart, but they were able to check some of these prices against their own knowledge and experience, particularly in the case of the test equipment, and other prices were checked against the catalogues of retail prices. With all the counted items it was necessary to assume a median type and a median price. For example, there was a wide variety of capacitors with a price range varying between 2p and 450p each. A price of 15p each J had been adopted to arrive at a total value of £24,496.50.

In some cases it had been possible to check the valuation figures against the company's invoices, their retail advertisements and also the price at the forced sale.

Mr Watkins then applied discounts to each range of items to allow for the difference between the estimated retail value and the value to the claimant. These percentages varied between 20 per cent and 50 per cent. The 20 per cent discount was applied to items where the stock was orderly and readily disposable, a higher discount to stock in an unsorted state and in large quantities, and the highest discount in the case of "other stocks" which consisted of small parcels of varied items. The percentage K allowance was also varied to take into account items which were slow moving.

By this means the retail value of £1,179,569.95 was reduced to £737,191.55, an average reduction of 37.5 per cent. A summary of his valuation is set out in appendix I.

Mr Stewart gave evidence as to the nature of his business and as to his part in the valuation. He assisted Henry Butcher in the identification and pricing of many of the articles in the list of items which they valued. He also prepared his own list of the items included in his valuation of £450,000 and which did not form any part of Henry Butcher's valuation. The figures are set out in appendix II. L

Mr Stewart was not able in the time available to count the items, but where there was any doubt about the numbers he erred on the side of substantial underestimate. In applying a unit price to each group of items he was able to refer to retail prices in catalogues. All these details were submitted in evidence with two exceptions. There was a number of residual items which could not be included in any particular category and on which he had put a spot figure of £18,000. There was also a large unknown quantity of spares which could not be identified with any catalogue or manual and on which he had placed a value of £150,000.

On the matter of the forced sale Mr Stewart had kept detailed M accounts and was able to produce a day-by-day record of the amount of the discount allowed and the amount of sales in each category of items. This helped to identify the retail value of the items sold and the items which were readily disposable.

Mr Stewart also considered what deduction was appropriate from his estimated retail value and reduced his figure of £450,000 to £300,000, a deduction of $33\frac{1}{3}$ per cent.

The total of Henry Butcher's and Mr Stewart's valuations, amounting to £1,037,192, was included in the claim for compensation made by the firm of Haslam & Son, chartered surveyors, acting on behalf of the claimant. As has been said, the

A other items of claim have been agreed and the only matter outstanding is the compensation due to the loss on forced sale. No evidence was given by Haslam & Son.

There is no criticism by the acquiring authority of the method of valuation adopted by the claimant. Indeed M C Matthews ARICS, chief valuer to the Reading Borough Council, freely admitted that the valuation of the stock by means of a proper inventory and proper pricing was the best method. Criticism is in the execution of the method. Mr Watkins and Mr Stewart were closely cross-examined on how they had identified, counted and priced the various items of stock. No evidence in rebuttal was brought by the authority.

B Having heard the evidence I am of the opinion that the claimant's valuation of the open market retail value of the stock, before any discounts, in the sum of £1,692,570, has not been upset and I accept it.

Council's valuation based on certified accounts

The real issue is whether or not a proper discount has been applied to this figure in arriving at the sum of £1,037,192, or whether Mr Matthews' valuation based upon the certified accounts should be preferred.

When Mr Matthews was told of the figure of claim for loss on forced sale he thought it wholly unacceptable. He had previously C been informed at various times between October 1978 and the end of that year that the figure was likely to be in the region of £100,000, then £250,000 and later £400,000. Therefore he felt it necessary to fall back on a valuation based upon the accounts as being the next best alternative.

His valuation of the stock at £198,744 is shown in appendix III. He arrived at this valuation by extracting from the certified accounts of the company for the five years ended June 30 1978 the figures of gross purchases and adjusted these to terms of value in 1979. To do this he used index figures extracted from a publication produced by the Department of Industry and Trade dealing with scientific and industrial instruments and systems. D Taking an average for the five years of the figure so adjusted he arrived at an average of adjusted gross purchases of £49,686 which he multiplied by four years' purchase. In his experience in dealing with compensation where traders were involved it was normal to take two years' purchase, but to allow three years' purchase where there was a substantial variation in demand and supply. In allowing four years' purchase in the instant case he had erred very much on the generous side.

The method of valuation adopted by Mr Matthews came under severe criticism. A Kerevan, an experienced chartered accountant, appeared on behalf of the claimants and produced a model based on the accounts of the company from its formation in E 1967-68 to December 31 1978. On the advice of Mr Stewart he had assumed that the annual cost of sales expressed as a percentage of the opening stock and purchases was 15 per cent in the first four years, 8 per cent in the last four years and 10 per cent in the intervening three years. This model demonstrated that the historic cost of the closing stock as at December 31 1978 was £227,870. The accuracy of this figure depended solely on the accuracy of the estimates of 15 per cent, 8 per cent and 10 per cent made by Mr Stewart. Making the same assumptions, but adjusting historic costs by index figures extracted from the same published figures as those used by Mr Matthews, increased the figure of historic cost of £350,254.

F This model demonstrated that Mr Matthews' valuation was suspect at the very starting point. By taking only five years' accounts he had omitted all the purchases prior to June 30 1974 and all the purchases in the last six months of 1978. Having regard to the fact that only a small percentage of the stock was sold in each year this was a serious omission. In taking the average of five years' figures he had failed to take into account the substantial increase in the company's business over the years, the increase in the quantity and quality of the stock, and to some extent the effects of inflation.

Mr Kerevan was also able to demonstrate that the accounts for the five and a half years to December 31 1978 showed an average gross profit, expressed as a percentage of sales, in the region of G $76\frac{1}{2}$ per cent. Therefore with stock-in-hand costing £350,264, a profit margin of 76.5 per cent would justify a retail value of the stock in the region of one and a half million pounds.

These figures demonstrate the uncertainty of valuing stock by reference to average purchases. The method may be a convenient check to apply in the case of a small retail business where there is a clear relationship between wholesale and retail prices, and between the amount of stock purchased and sold in any year. Even so, a distinction must be made between, say, a greengrocer who turns over his stock almost daily and a wine merchant who might store his stock for a number of years. The method appears to be singularly inappropriate to the valuation of stock held by H the claimant company.

The acquiring authority sought support for their valuation by calling J Orme, an experienced chartered accountant, who gave evidence of valuations by reference to the estimated loss of the company's profits. Mr Orme studied the company's accounts including those for the year ended June 30 1979. He then estimated that the total loss of profits for the four years ending June 30 1982 would be in the region of £93,500. To assist him in his estimate he had access to the VAT returns of the company which continued a somewhat similar business in other premises after the end of the forced sale. By two separate methods he arrived at the view that the estimated loss of profits of the J company would be in the region of £90,000 to £100,000. It is fundamental to Mr Orme's approach that compensation for loss on forced sale cannot exceed the loss of profits, but having heard the evidence I can find no foundation for this belief.

I am left with the retail value of the stock as at February 6 1979 in the sum of £1,629,570 as the only starting point. From this must be found the value to the claimant. Both Mr Watkins and Mr Stewart have applied their minds to the problem of adjusting the retail value of each group of items having regard to their quantity and the likely speed of disposal in normal circumstances. A further allowance had been made where stock had not been properly sorted. Mr Watkins makes adjustments of K between 20 per cent and 50 per cent and this results in an overall deduction of 37.5 per cent on the items which he has valued. Mr Stewart applies a deduction of $33\frac{1}{3}$ per cent overall to the items that he has valued. These are matters of opinion given in evidence which was not seriously challenged. The deductions were arrived at by Mr Watkins and Mr Stewart separately on different dates, and the two resultant figures were added together and submitted in the claim presented by Haslam & Sons in the sum of £1,037,192, representing a total deduction of 36.4 per cent from the full retail value.

I think that this figure must be looked at in the context of the claimant's business, its past record of trading and what could L reasonably be anticipated for the future but for the advent of compulsory acquisition. The problem is to decide by how much the retail value should be discounted to allow for the likely rate of disposal of stock. From the evidence, the turnover of stock has been estimated to vary between 8 per cent and 15 per cent annually. From the accounts, the record of sales indicate an expanding business, except in the years of abortive manufacture, with an increase between 1977 and 1978 of about 14 per cent. The gross profit as a percentage of sales is in the region of 76.5 per cent. The profit margins are high and there is sufficient flexibility to increase the rate of sales by increased advertising or by discounted retail prices. If the "XY Plotter" project were a M success this would increase the turnover of stock. Nevertheless, it would take a considerable number of years to dispose of the stock in the normal course of this company's business. During that time some stock might appreciate in value in real terms, some might become redundant, and some might prove unsaleable.

These are the matters that must be weighed. It is not an easy task, but I conclude that the discount from retail value is not sufficient and I make a further deduction of 15 per cent, giving a figure of £881,613. From this should be deducted the agreed proceeds of the forced sale in the sum of £98,364, resulting in a figure of compensation for loss on forced sale of stock in the sum of £783,249.

Having heard submissions as to costs I order that the acquiring authority shall pay to the claimant its costs of these references such costs unless agreed to be taxed by the Registrar of the Lands Tribunal on the High Court Scale.

Appendix I

Mr Watkins' valuation
of part of stock as at February 6 1979

No	*Type*	*Retail price per unit*	*Discount*	*Discounted retail value*
787,320	Integrated circuits	@ 75p	50%	£295,245.00
994,240	Resistors	@ 3p	30%	20,879.90
6,474	Motors	@ £10	20%	51,792.00
1,355	Tubes	@ £38	20%	41,192.00
62,200	Valves	@ £1	30%	43,540.00
41,325	Plugs & sockets	@ £1	30%	28,927.50
233,300	Capacitors	@ 15p	30%	24,496.50
133,960	Semi-conductors	@ 25p	30%	23,443.00
	Computer equipment		20%	48,000.00
	Test equipment		20%	132,272.00
	Other stocks		40%	27,403.65
				£737,191.55

Appendix II

Mr Stewart's valuation
of part of stock as at February 6 1979

40,000	Inductors & tuning cans — various	£1 ea	£40,000
20,000	Filter assemblies	£1 ea	£20,000
40,000	Coil formers, ferrite cores and oscillator coils	£1 ea	£40,000
500	Transformers, iron core	£10 ea	£5,000
20,000	Variable capacitors	75p ea	£15,000
1,000	RF chokes	50p ea	£500
4,000	Trimmers — various	50p ea	£2,000
50	Crystal filter units	£50 ea	£2,500
250	Wave guide assemblies & connectors	£4 ea	£1,000
25,000	RF connectors — various (mostly used)	50p ea	£12,500
7,000	Transformers ferrite, line output transformers scan coils	£1 ea	£7,000
10,000	Connectors — edge/pin (some used)	50p ea	£5,000
75,000	Knobs, escutcheons, handles, graticules feet and general hardware	20p ea	£15,000
1,000	Valves — special spares	£3 ea	£3,000
1,000	Tubes, shields — secondhand spares	£10 ea	£10,000
50,000	Switches — lever, rotary, slide, toggle, rocker, micro	50p ea	£25,000
50,000	Potentiometers — various 1, 2, 3 & 4 ganged, variable shaft size, switched	50p ea	£25,000
20,000	Lamps — neon, filament	10p ea	£2,000
300	Tubes — nixie, plasma, numeric diaplay, bulbs	50p ea	£150
10,000	Fuses — various	5p ea	£500
2,000	Fuse holders — various	25p ea	£500
150	Relays — thermal — various	£1 ea	£150
200	Relays — special oscilloscope	£3 ea	£600
10,000	Valve bases — various — specials — relay	40p ea	£4,000
1,000	Meters — spares for various equipments	£5 ea	£5,000
1,000	Servo motors — various — 400HZ	£10 ea	£10,000
30	Power units — module assemblies	£10 ea	£300
15,000	Creed spares	40p ea	£6,000
4,000	Variac spares — brushes etc	40p ea	£1,600
2,000	Thermistors — spares for various equipments	£1 ea	£2,000
2,000	Microwave switch assemblies & diodes	40p ea	£800
75	Magnetrons, backward wave oscillators klystrons, travelling wave tubes	£50 ea	£3,750
80,000	Resistors	3p ea	£2,400
25,000	Capacitors	15p ea	£3,750
		Carried forward	£272,000
		Brought forward	£272,000
10,000	Semi conductors/integrated circuits	25p ea	£2,500
	Residue		£18,000
	Spares — not identifiable by manuals held		£150,000
			£442,500

Appendix III

Mr Matthews' valuation
Loss on forced sale of stock
Valuation based on adjusted accounts

Average adjusted gross purchases		£49,686	
Years' purchase		4	
			£198,744
Deduct			
Proceeds of forced sale shown in Chiltmead's claim			
Jan 2 1979 to Feb 6 1979	£11,050.95		
Feb 7 1979 to Feb 28 1979	£88,913.08		
	£99,964.03		
Less plant and machinery	£1,600.00		
Total proceeds (net of VAT)	£98,364.03		
Deduct from stock valuation			£98,364
Loss on forced sale of stock therefore			£100,380

TURRIS INVESTMENTS LTD v CENTRAL ELECTRICITY GENERATING BOARD

(REF/31/1980; REF/105/1981)

(J H Emlyn Jones FRICS)

May 20 1981

Estates Gazette June 27 1981
(1981) 258 EG 1303

Acquisition of rights for electric line and pylon over residential building land in course of development at Worsley, Greater Manchester — Valuation methods — Claimants' valuations on basis of losses on sale, with depreciation factors applied to brochure prices of houses built and to be built — Tribunal prefers authority's approach of separately valuing land taken and injurious affection — Claimants' figure for loss on completed houses accepted — Depreciation factors applied to remainder of land as building land — Depreciation factors were visual intrusion of pylon and cables, potential purchasers' fears and apprehensions, noise of corona discharge and interference from authority's access rights — Value of land per acre taken at £50,000 instead of authority's valuer's figure of £30,000 — Additional sums allowed for extra fencing, higher-than-average contractor's charges in moving heavy plant to avoid cables, and further loss in respect of plot on which pylon stands — £65,700 awarded against claim for £245,323 and authority's opinion of £39,684

A J C Hoggett (instructed by Berry Berry Cocker Smith & Co, of Manchester) appeared for the claimants; George Bartlett (Instructed by the solicitors to the Central Electricity Generating Board) for the acquiring authority.

Giving his decision, which is published by consent of the parties, MR EMLYN JONES said: These references, which were heard together by agreement, are concerned with the assessment of compensation payable to the claimants, Turris Investments Ltd, by the acquiring authority, the Central Electricity Generating Board, following the acquisition by the board of certain rights pursuant to a compulsory purchase order known as the Central Electricity Generating Board (Worsley) Compulsory Purchase Order 1978, and also arising from a deed of grant made on December 31 1980 between the claimants and the board.

A Under the compulsory purchase order the authority were empowered to acquire a right to place, repair and maintain an overhead electric line across the claimants' land. Subsequently, it was apparently felt that some further compensation might be due to the claimants, and the deed of grant was entered into in order to grant to the board the additional right to remove the electric line and works and also the right to fell or lop from time to time trees or coppice wood. The parties agree, however, that for the purposes of determining compensation due to the claimants no distinction is to be drawn between the two references, and the evidence on both sides was directed towards the establishment of a single payment of compensation in respect of the totality of the
B claim. The deed of grant is therefore to be deemed as a consequence of the compulsory purchase order, and the figure of compensation to be awarded will be made under the first reference, with a nil award under the reference by consent.

The subject land comprises an area of 10.80 acres, the site being known as Stablefold, Barton Road, Worsley, which is in the county of Greater Manchester. The land lies to the east of Barton Road and is bounded on the south by a local authority housing estate to which it has no access. To the north-east lies the Bridgewater Canal and to the north is an area of industrial development. On the west side of Barton Road opposite the entrance to the subject land is a depot of the North Western
C Water Authority. Beyond that some housing and open land and further to the west at a distance of about 500 yds is the M63 motorway.

The claimants acquired the land for £70,000 in July 1977 at which time it carried outline planning permission for 77 houses. At that time there was in existence a soil report which had been commissioned by the City of Salford. This report suggested that there were likely to be certain subsoil difficulties to be encountered in developing the land for residential purposes. The claimants, however, had obtained their own report which indicated that the difficulties were not as great as had appeared in the earlier report.

D In March 1978 detailed planning permission was granted to the claimants for 106 dwellings on the land, comprising 66 houses, 20 bungalows and 20 flats in five blocks, each of four flats. The sale of the land was subject to an existing wayleave agreement in favour of the board's predecessor which had been in operation since 1957. Under this agreement one pylon had been placed on the north-eastern corner of the land and the line of cables ran across the land in a south-westerly direction. Originally this line had carried two circuits, one on each side at 275 kV, but in 1968 one of the circuits was up-rated to 400 kV when it became part of the national grid. This agreement was subject to 12 months' notice and the claimants on acquiring the land duly served notice
E on November 7 1977 terminating the agreement.

The board subsequently published a draft compulsory purchase order on December 16 1978 and, following a public inquiry in September 1979, the order was confirmed by the Secretary of State on November 20 1979. The pylon and the cables remained in position and a formal notice of entry was served by the board on the claimants on December 31 1979, so that the deemed date of entry on to the land is January 14 1980. It is common ground that this is the date of valuation for the purpose of assessing compensation.

Three phases of development

F By that date the claimants had started the development of the subject land which they intended to carry through in three phases. Phase one was to consist of 24 dwellings and of these 16 were more or less complete. Of the 16, eight were directly under the line of the cables including the limit of swing and another two were partly within the same area. At the relevant date the estate road had been laid out and covered up to base coat for nearly the whole development, together with the sewers and ducting for other public services.

Certain difficulties had been encountered in the drainage of the site, and a pump-house which had been built on the northern boundary of the site had to be abandoned because of the penetration of subsoil water. A further pump-house on an alternative site was in the course of construction at the relevant G
date. A number of test bore holes were drilled but no unusual complications were revealed and the houses were built on concrete raft foundations which involved no increase in construction costs.

Since the relevant date work has started on the houses in phase two, and of the 45 dwellings in this phase 18 are wholly or partly under the line of the cables. More recently a start has also been made on phase three. There is one pylon standing on the land. It occupies approximately 200 sq yds, being 44 ft 6 in square at its base and 164 ft 9 in high. The spread of the cross arms is 44 ft and the minimum clearance of the conductors on the pylon is 89 ft. The height of cables above ground at mid-span is at the H
minimum 45 ft.

For the claimants Mr Hoggett called Derek Johnson FRICS, who is in private practice as a chartered surveyor in Walkden, which lies just to the north of Worsley, and is also a director of the claimant company. Mr Johnson put forward a valuation of the loss in value amounting to a total of £269,673, amended during the course of the hearing to £245,323. In his opinion the presence of the pylon and the overhead cables seriously reduced the value of the subject land. This was ripe building land with detailed planning permission for 106 houses, with the first phase of development nearly complete. Quite apart from the grant of the rights themselves in the form of the permanent easement, the J
existence of the pylon resulted in the loss of one building plot; prices obtained for the houses would be substantially reduced; and the costs to be incurred in carrying out the development and disposing of the properties would be substantially increased.

Depreciation factors

The factors which gave rise to this depreciation were in his opinion:

(a) Visual intrusion of the pylon and the cables, which he said were ugly and quite inappropriate in a residential area.
(b) The fears and apprehension expressed by potential purchasers, who were worried about interference with radio and television and fears of the cables coming down. K
(c) The noise of the corona discharge from the cables in certain climatic conditions.
(d) The right of entry without notice to the whole of the land.

In Mr Johnson's experience, these factors had been raised on many occasions by prospective purchasers and although he had tried to allay some of the apprehension with regard to television interference these matters were effectively used as part of the bargaining process.

The value of the land entered only indirectly into Mr Johnson's calculations, but in his opinion residential land in Worsley, which L
he considered to be one of the most attractive villages in Greater Manchester, was worth at the relevant date about £70,000 per acre if unserviced and £100,000 per acre if serviced. At the relevant date his company had actually spent £150,000 on the laying down of roads and services over the whole area of the site. In support of these figures of land values Mr Johnson referred to a number of transactions involving the sale of building land in the locality with which he had been directly concerned. The detailed breakdown of his total valuation was as follows:

A.	Loss on sale of houses in phase one	£28,450
B.	Loss on sale of houses in phase two	56,300	M
C.	Loss on sale of houses in phase three	26,500
	A further loss in phase three attributable to the closer proximity of the pylon	27,300
	The loss of one building plot on land taken up by the pylon	10,000
D.	Loss on sale of other properties outside the cable line but within 40m	22,950
E.	Loss on sale of remaining properties	50,150
F.	Increased costs of development	23,673
	Total	£245,323

A In order to arrive at the losses which form the basis of this claim, Mr Johnson produced schedules setting out details of actual sales, a list of prices appearing in a brochure published in February 1980 and estimates of figures which, in his opinion, would be obtained for the same houses outside the line of the cables. The houses directly affected in this way showed losses as aggregated in "A", "B" and "C". In the case of the properties in "D" Mr Johnson applied an overall depreciation factor of 7½ per cent and in "E" at an average of 2½ per cent. The breakdown of the figures under the heading "F" was as follows:

1. "In order to achieve better sales conversion from inquiries, properties had to be finished in each cul-de-sac to minimise effect of cables thus B involving extra finance costs." This he calculated as additional interest for a two-month period based on 50 per cent of sales value taken at 18 per cent producing a total of £17,223.
2. "Extra fencing and landscaping will be necessary around pylon to minimise effect" — £950.
3. "Contractors' prices higher than average due to avoidance of cables by large machinery" — £1,000.
4. "Extra professional fees in preparation of layout due to design of more types to fit beneath minimum clearance of cables" — £1,500.
5. "Extra advertising and details due to low conversion factor of inquiries to sales" — £3,000.

Mr Hoggett also called B M Cook FRICS, who practises from C offices in Swinton and Worsley and has lived in the area all his life. Mr Cook adopted the brochure prices provided by Mr Johnson and took an overall 10 per cent as representing the loss in the case of all houses under the line in all phases. In the case of the phase two houses, however, he deferred the loss for two years on the 10 per cent table and those in phase three for three years. In the case of the phase three houses, which were those nearest to the pylon, he estimated that the sale prices would be further reduced by 5 per cent because of the effect on visual amenity. On the nine properties included in Mr Johnson's "D" he applied a reduction of 5 per cent with a deferment in the case of the phase two houses for two years; and on the houses D included in "E" he took a reduction in value of 1½ per cent deferred for two years in the case of phase two and three years in the case of phase three. He also included additional losses, including £10,000 for the loss of one plot due to the existence of the pylon. His other items broadly corresponded with those put forward by Mr Johnson in "F" except that his claim for additional finance costs was about double that of Mr Johnson. He included nothing in respect of fencing around the pylon and thought that the extra advertising would result in an additional expenditure of £5,000 rather than £3,000 as suggested by Mr Johnson.

Mr Cook also gave evidence regarding some figures of E compensation which he had agreed in respect of three houses in Granary Lane, which lay to the west of the subject land on the other side of Barton Road but was affected by the same power line. He had there negotiated compensation varying between £400 and £950 for each of the three houses, but his understanding of the compensation was that this compensation represented the difference between the value of the properties affected by the earlier 275 kV line and as affected by the up-rating to 400 kV. The compensation worked out at 5 per cent, 6 per cent and 10 per cent respectively, of the sale price of the three houses.

For the authority, Mr Bartlett called W S Wrigley ARICS. Mr Wrigley adopted a different approach. His valuation of the F compensation payable was made up of two parts. Firstly, in respect of the land taken, he valued the 196 sq yds occupied by the pylon, which produced a figure of £1,200 arrived at on the basis of £30,000 per acre. Also under the heading of land taken he attributed an amount of £100 for the easement. His second main heading was injurious affection. This he took in three parts. He firstly calculated the compensation arising from the loss in value of the 10 properties completed and adopted the total sales value of £349,500, which was the amended brochure price put forward by Mr Johnson. To this he applied a depreciation factor of 7½ per cent.

With regard to the rest of the land Mr Wrigley took the view that there were too many imponderables in looking into the G future from the relevant date to justify the adoption of an estimated difference in value between the sale prices to be anticipated with the pylon and cables in position and those which might have been expected if there had been no pylon and cables. He therefore looked to the value of the land as building land and applied a factor of depreciation which, he said, would reflect the effect of the pylon and cables on the mind of a prospective purchaser of the land. In the case of 2.21 acres under the cables he again applied £30,000 per acre and calculated the depreciation due to the injurious affection as 10 per cent. For the rest of the land, valued at £30,000 per acre, he applied 2½ per cent.

The total arrived at in this way came to £39,684 and this was H the figure which Mr Wrigley put forward as his opinion of the figure of compensation due. In support of his valuation he relied on three comparable transactions where compensation had been paid by the board or its predecessor in respect of overhead lines at Bromley Cross, near Bolton, at Norden, near Rochdale and at Drywood Avenue, Worsley, which lay immediately to the east of the subject land and was affected by the same line. In support of the land values which he had adopted he referred to the price paid by Turris when they bought the subject land in 1976 but conceded that the values which he had adopted for the land might not properly reflect the servicing which had been carried out by the relevant date in the provision of roads and sewers. J

Tribunal's findings

I have inspected the subject land and looked at most of the comparables to which reference was made at the hearing. I should say first of all that I prefer the approach adopted by Mr Wrigley in taking separately the value of the land and the injurious affection. As Parker J said in *Jones v Pritchard* [1908] 1 Ch 630 at p 638:

> . . . the grant of an easement is prima facie also the grant of such ancillary rights as are reasonably necessary to its exercise or enjoyment. Thus the grantee of an easement for a watercourse through his neighbour's land may, when reasonably necessary, enter his neighbour's land for the K purpose of repairing, and may repair, such watercourse. On this principle each party in the present case may do such acts on the property of the other as are reasonably necessary to the continued enjoyment of the easement;

It is therefore proper in determining the amount of compensation to take into account not only the value of the interest acquired but also the effect of the cables and the pylon on the value of the whole of the subject land and also the effect of the interference with quiet enjoyment which is posed by the possibility of access without notice by the board on to the land.

The first matter which has to be considered is the value of the land as residential building land at the material date in January L 1980. The amount paid for the land by the claimants in 1977, following negotiations in 1976, works out at about £6,500 per acre. The claimants say, however, that this price was artificially reduced because the sale of the land was linked to another transaction where they paid £5,500 for a plot of land and subsequently spent a further £8,000 making it good. This was done in order to help the vendors out of some difficulty. In Mr Johnson's opinion, this liability means that about £15,000 should have been added to the cost of the land and that works out at rather less than £8,000 per acre for the whole of the subject land.

Mr Johnson was unable to give any estimate of the increase in land values between 1976 and 1980. Mr Wrigley thought that a M doubling of values would have taken place. It is also true that when the land was bought the vendors were clearly aware of the site difficulties which had been revealed by the earlier soil-survey report and they were unaware of the rather more favourable report that the claimants had themselves obtained. It is reasonable, therefore, to assume that the real value was rather higher than that revealed by the purchase price.

Another consideration is that at the material date the planning permission had been improved in that detailed permission for 106 houses had been obtained instead of outline permission for 77 houses. At the same time, most of the road work had been done

A and the soil sewers laid, as well as the ducting for other services. All these are factors which must be properly taken into account in seeking to arrive at the value of the land as residential building land at the relevant date.

Comparable land transactions

I also look at the evidence of comparable land transactions referred to in the expert evidence of the claimants. Of the six transactions referred to by Mr Johnson, five were in respect of small plots for one, two or four houses. I think this evidence is of limited value. The remaining case was, however, for a sale by auction in November 1978 of 8.15 acres of freehold land at
B Walkden, just to the north of Worsley, with detailed planning permission for 89 properties. The site included 0.55 acres of reservoir. The price paid in this case works out at about £39,000 per acre overall or £42,000 per acre if the reservoir is excluded from consideration. In Mr Johnson's opinion, Walkden is a less desirable neighbourhood than Worsley and I think this is to some extent correct, although the subject land is outside the attractive village centre at Worsley.

Having regard to the details of the various transactions, I have come to the conclusion that the proper value to be attached to the subject land at the relevant date, on the assumption that the pylon and cables then existing were to be removed, would be
C £50,000 per acre. On this basis the value of the land taken and the payment for the easement is calculated at £2,100. The £100 attributable to the easement was not expressly included in the claimants' valuation, but this figure as put forward by Mr Wrigley was not challenged.

Turning to the question of injurious affection, it is necessary to consider the extent of the injury. First of all, and, I think, most serious, is the visual impact of the pylon and the cables. Then there is the apprehension felt by prospective purchasers of the houses on the land about television reception, and the risk of the breaking of the cables; fears that are largely unwarranted, but nonetheless real. As to noise, I can do no better than to quote a
D passage from the report of the inspector following the public inquiry held in 1979, the terms of which were accepted and adopted by Mr Johnson in the course of his evidence:

> I would agree with the remarks made by Mr Johnson that the noise from the corona discharge could well be a nuisance, and as mentioned earlier this noise was plainly evident during the period of the inspection on a day which was reasonably fine and dry when one might expect the corona discharge to be at a low level. I understand that the days in which there is little or no noise to be heard from the corona discharge are somewhat few and far between. It would appear that complaints concerning corona discharge have not been received from the immediate area and this would include those houses on Drywood Avenue and The Moorings which are of recent construction, where people moving into the area
> E might have had cause for commenting on the noise that they could hear. It seems to me that people who have come to live in this area have simply got used to the corona discharge noise, just as presumably they have got used to the fairly high level of background noise emanating in the main from the M62 motorway. It seems most likely to me that people purchasing houses on the Turris Investment site would also become used to this noise. I agree with the Board that in a fully furnished house, even with the windows open on a fine day, the attenuation of this noise is likely to be fairly rapid, and in consequence less noticeable to people inside their houses. However it must be recognised that this noise exists and must have an affect upon the local amenities, just as the existence of the present line has an effect upon the visual amenity, and I note that the Board is offering to pay compensation in this matter.

F Finally, there is the interference likely to arise from access by representatives of the board to the site of the pylon, and more generally on to the estate.

In considering the effect of these injuries, I turn first to the houses which were more or less complete and ready for sale at the relevant date. In some cases, prospective purchasers had been found although no contracts had at that time been exchanged. All three valuers deal with this part of the claim by starting with an aggregate figure of £349,500 in respect of the 10 houses which is produced by Mr Johnson and represents the sum of the prices quoted in his sales brochure. The loss claimed by Mr Johnson amounts to £28,450, which is arrived at by a series of 10 separate calculations. Mr Cook takes an overall depreciation G
in value of 10 per cent, giving £34,950, and Mr Wrigley $7\frac{1}{2}$ per cent, giving £26,202. There are a number of unexplained items in Mr Johnson's calculations, but at this stage I accept his total figure of £28,450, which represents a depreciation factor of about 8 per cent.

Up to this point there is little difference between the figures produced by Mr Wrigley's method (as amended by me) and by Mr Johnson's method. When, however, we turn to the injurious affection on the rest of the land very substantial differences emerge. Mr Johnson and Mr Cook apply factors of depreciation to the brochure prices of the houses to be built and Mr Wrigley looks at the land as it existed at the relevant date and values it as H
building land with planning permission and as affected by the pylon and cables.

I think Mr Wrigley's method is more likely to lead to the correct result. I say that for two reasons — firstly, because the loss occasioned by a reduction in the sale price of houses is not one which was suffered at the material date and indeed has still not been suffered in respect of very many of the houses which have not yet been started. This is a factor which is recognised in principle by Mr Cook, who adopts the principle of deferment. I do not consider, however, that he goes far enough in converting what amounted to a conjectural future loss into an actual value at the material date. J

This leads me to the second major objection to Mr Johnson's method. As I have already indicated, the actual calculations which he produced appear to show up a number of anomalies. I have studied his schedules and brochure prices, prices which he describes as being "obtained under the cable" and "obtained outside". These latter figures are based entirely on assumptions; the prices actually realised do not always coincide with the prices quoted in the schedule and require adjustment because of differences in time and date during a period when house prices were rising. The overall impression from an examination of the schedules does not support the view that houses under the line are invariably sold for lower prices than those outside it. In K
short, there are so many imponderable factors in calculations based on these notional differences that in my opinion they are quite unreliable as a basis for estimating the depreciation in value of the land.

I therefore adopt Mr Wrigley's approach, but as I have already said I consider that he had underestimated the value of the land, which I take at £50,000 per acre. He considers that the land most seriously affected (excluding the land already developed at the material date) amounted to 2.21 acres, to which he applied a reduction of 10 per cent. In my opinion, the land most seriously affected extends to a rather larger area than that immediately under the cables and I prefer to adopt 3 acres at £50,000 an acre, L
depreciated by $12\frac{1}{2}$ per cent to give an amount of £18,750 for injurious affection on this part of the land.

The remaining land amounts to 6.60 acres to which Mr Wrigley applies a reduction of $2\frac{1}{2}$ per cent. Here I think he has underestimated the visual effect of the cables on the land further away from the line and I adopt 3 per cent, which comes to £9,900.

Additional costs

There remain the various items contained under heading "F" of Mr Johnson's valuation for which no allowance is made by Mr Wrigley. With regard to items 1, 4 and 5 I am not persuaded that M
anything is due under these headings which is not already reflected in the depreciation in the value of the land as building land. Certainly something should be added for the fencing. Mr Wrigley accepted this in principle but considered that the sum of £950 was excessive. I think £500 should be allowed on this account. I also think that the allowance in respect of additional costs incurred in bringing heavy plant on to and off the site is a proper item reflecting, according to Mr Johnson, the costs actually incurred.

With regard to the claim made in respect of loss of the plot due to the existence of the pylon, this is partly covered by the value

A attributed to the land taken on which the pylon stands. Nevertheless, I accept that there is some further loss here which qualifies for compensation and I award £5,000 under this heading. The total compensation, therefore, can be expressed as follows:

Land taken		
Land occupied by pylon — 196 sq yds		
or 0.04 acre at £50,000 per acre		£2,000
Easement over land crossed by cables		100
Injurious affection		
B To developed land with houses completed.		
Sales value = £349,500		
Depreciation @ 8% = £27,960	say	28,450
To land most seriously affected.		
Three acres at £50,000 = £150,000		
Depreciation @ 12½%		18,750
To remaining land.		
6.60 acres at £50,000 = £330,000		
Depreciation @ 3%		9,900
	Carried forward	£59,200

Brought forward	£59,200 C
Additional items — extra fencing	500
Additional contractors' charges	1,000
Additional loss in respect of plot lost	5,000
Total	£65,700

I therefore determine the compensation payable by the acquiring authority to the claimants in the sum of £65,700. The acquiring authority will also pay the proper legal costs of the claimants from the date of the notice to treat up to the date of the notice of reference together with a surveyor's fee based on scale 5(a) of the scales of professional charges of the Royal D Institution of Chartered Surveyors. Having opened the sealed offer of compensation lodged by the acquiring authority I find that the amount thereof is less than the amount of my award; accordingly the acquiring authority will pay the costs of the claimants of this reference such costs if not agreed to be taxed by the Registrar of the Lands Tribunal on the High Court Scale limited to one expert witness only, Mr Johnson, and one-quarter of the costs in respect of the factual evidence of Mr Cook.

LANDS TRIBUNAL

RATING

A CITY OF LONDON CORPORATION v WATNEYS LONDON LTD AND OTHERS;
SAME v TRUMAN LTD AND ANOTHER;
SAME v CHEF AND BREWER LTD AND ANOTHER
(LVC/318-321/1978)
(J H EMLYN JONES FRICS)
October 7 1980

Estates Gazette May 9 1981
(1980) 258 EG 561

B **Three public houses in City of London — Preliminary issue on disclosure of documents — Rule 40 application by City Corporation to require other parties to disclose inter alia audited trading accounts and details of sales and purchases — Corporation contends that assessments by "direct method" of valuation, agreed in present case between ratepayers and Inland Revenue licensed property valuer, did not produce correct GVs — In present context tribunal not prepared to hold that direct method of valuation is an incorrect method — Disclosure of VO's original valuations prepared for valuation list entries refused as not relevant to issues — Tribunal not prepared to find that other documents sought were irrelevant to questions of valuation — Tribunal's discre-** C **tion — Matters to be weighed in balance between parties in exercising that discretion — Profit and loss accounts would not be of real help — In public interest that a trader should be protected from disclosure in open court against his will of confidential accounts which could be of help to third parties not connected with proceedings — In present case in matter of accounts this consideration outweighs need in public interest for information to be available to assist in determination of correct GVs — Tribunal refuses disclosure of records of sales, deliveries and receipts at present stage of hearing but prepared to hear further argument if answers on forms of return in any way incomplete — Application for disclosure dismissed — Application for stay of proceedings for case to be stated** D **refused — Member considers he had no power to state a case at present stage, since he had given not a "decision" on a substantive issue but a "ruling" on a procedural matter arising during hearing**

B A Marder QC and J M Sullivan (instructed by the Solicitor for the City of London) appeared for the appellant rating authority, the City Corporation; G E Moriarty QC and Susan Hamilton (instructed by the solicitor to Watney Mann & Truman Holdings Ltd) for the first respondents in each of the three appeals and (instructed by Crossman, Block & Keith) for the second respondents, Goodhews Ltd, in the first appeal; R G Alderman, representing the Solicitor of Inland Revenue, for the respondent valuation officer, R Burrows ARICS, in all the appeals.

E Giving his decision, MR EMLYN JONES said: These appeals from the Greater London Central Valuation Court are made by the Corporation of the City of London and relate to the assessments for rating purposes of three public-houses within the City.

Watneys London Ltd and Goodhews Ltd as owners and occupiers respectively are the first and second respondents in the case of the "Sir Christopher Wren" public-house; Truman Ltd as owners and occupiers are the first respondents in the case of "The Cock Tavern"; and Chef and Brewer Ltd as occupiers are the first respondents in the case of "The Magogs" public-house. Additionally, the valuation officer for the City of London, R Burrows, is a F respondent in each appeal.

The issues in the three appeals appear to be the same and for convenience at this stage I will deal first with the circumstances surrounding the case of the "Sir Christopher Wren" public-house. This is situated at 17-19 Cathedral Place, London EC4, and was entered in the valuation list for the City of London on April 1 1973 with an assessment of gross value £14,500, rateable value £12,055.

LVC's decisions

Proposals were made on behalf of the first and second respondents on two occasions — the first on August 13 1973 and the second on March 21 1975. When the matter came before the local G valuation court on January 25 1978 it appears that altered values were put to the court as agreed between valuers acting for the respondents and the licensed property valuer acting on behalf of the valuation officer. The agreed figures were, in respect of the first proposal, gross value £9,500, rateable value £7,888, and in respect of the second proposal, gross value £8,000, rateable value £6,638. The Corporation of the City of London were not a party to this agreement but the figures were approved by the local valuation court, who in the course of their written decision made the following remarks:

> The court doubts that the "direct method" always effectively produces the correct assessments of licensed premises in the City of London in relation to H other types of property. However, in the absence of any firm evidence to the contrary, the court accepts the assessments agreed by the appellants and the Licensed Property Valuer.

The City Corporation being dissatisfied with this decision appeal to the Lands Tribunal.

The figures in respect of the other two public-houses were as follows:

"The Cock Tavern", 22 Fleet Street, London EC4 — assessment in the valuation list — gross value £10,750, rateable value £8,930; assessment agreed between the ratepayers and the valuation officer as confirmed by the local valuation court at the same hear- J ing — gross value £5,250, rateable value £4,347.

"The Magogs", 8 Russia Row, London EC2 — original assessment in the valuation list — gross value £6,450, rateable value £5,347; assessment confirmed by the local valuation court on February 14 1978 — gross value £3,900, rateable value £3,222.

It will be observed that the third case was dealt with at a different local valuation court and in the course of that decision the court said:

> The onus of proof is on the Appellant [at that stage of the proceedings that was the ratepayer company] whose method of valuation has been approved by a number of authorities referred to us in evidence. It showed an approach in line with the valuations of a number of other Public Houses in the Cities of K London and Westminster where agreement had been reached or Valuation Court decisions had been made.
>
> We found Mr Payne's evidence of value [for the City Corporation] insufficiently weighty to displace that of the Appellant and we directed a GV of £3,900, RV £3,222.

In the statement of case lodged by the appellant corporation the grounds of appeal in each case are set out in almost identical terms:

> The appellants now appeal on the grounds that the assessment determined by the Local Valuation Court is incorrect and insufficient; and seek an

A Order restoring the assessment in the Valuation List dated 1/4/73.

The appellants will submit:

(a) the rent appropriate to the hereditament supports the assessment originally entered in the Valuation List dated April 1 1973 and the Local Valuation Court did not or did not sufficiently consider the rental evidence against the assessment agreed between the . . . respondents.

(b) the direct method is not appropriate for assessments of licensed premises in the City of London [adding here in the case of the "Sir Christopher Wren" "in particular where there is a substantial catering element in the trade"] and the Local Valuation Court misdirected itself in not considering the alternative method of valuation, namely, the rental method proposed by the Appellants.

B In the case of the "Sir Christopher Wren" the appellants add:

(c) the Local Valuation Court did not or did not sufficiently take into account the substantial catering element in the premises, and the profits on mixers and soft drinks.

"Direct method" of valuation

For reasons which will presently become apparent, I have not yet heard evidence from the parties to the appeals, but the issues, as I understand it, can be summarised in the following way. The valuers acting for the brewers and the occupiers of these three public-houses and the licensed property valuer who acts on behalf of the valuation officer in these matters have both prepared and agreed
C values arrived at on what is known as the "direct method". That method is referred to by Thompson J in *Watney Mann Ltd v Langley* [1966] 1 QB 457 at p 463 as follows:

> . . . it appears again to be common ground that a method known as the direct method is now generally, if not indeed universally, used by those who have to value licensed premises for rating purposes. That method has been devised since the decision of the Court of Appeal in the case of *Robinson Bros (Brewers) Ltd* v *Houghton and Chester-le-Street Assessment Committee*, which swept away as misconceived some well-established beliefs and practices. The judgment of Scott LJ in that case was approved by the House of Lords on appeal and has been since, and is before me, accepted as the modern Bible on the subject and the authority for many of the principles I
> D have already cited from Ryde. That method is as follows: Take the premises in the physical condition in which they are and suppose them to be vacant and to let. Estimate the capability for trade of that house expressed as a quantity. Multiply by a money figure for the brewer's profit per unit of quantity. To the money figure thus obtained add the estimated tied rent that the brewer could expect to be paid by a tenant. From the total sum thus arrived at estimate and deduct the amount the brewer would want to keep for himself, and the balance resulting will be the rent he would be prepared to pay in order to obtain a tenancy of the house.

The City Corporation claim that this method does not produce the correct gross values, certainly in the case of public-houses within the City of London and the application of this method in
E their submission leads to widespread incorrectness and more particularly to under-assessment. Among the documents lodged by the appellants annexed to the statement of case on the "Sir Christopher Wren" are two valuations and a computation prepared by the corporation's expert witness, E C Payne. In the first valuation he says:

in the absence of full accounts valued by applying a percentage to the estimated maintainable gross takings:

Estimated maintainable yearly gross takings at 1972-73 prices:	£165,000
Take as gross value 11¾%	= £19,387
F say gross value	£19,250

The second valuation Mr Payne described as a check valuation using the rental basis by applying prices per m² derived from devaluation of gross values of neighbouring properties. Adopting this method Mr Payne arrives at a gross value of £16,000. His computation is an adjustment of the rent at which the property was let in 1963 converted to terms of gross value by adding an annual equivalent of estimated expenditure and an adjustment in respect of an assumed tenant's responsibility for internal repairs. In this way he arrives at a figure of £12,910. Mr Payne, however, in a footnote points out that he had not seen the lease and that the computation might require adjustment if more complete information became available as to the precise nature of the lessees' G
liability for repairs, insurance, etc.

At the start of the proceedings, Mr Marder for the appellants made an application under rule 40 of the Lands Tribunal Rules 1975 that the tribunal should require the other parties to the proceedings to disclose to the tribunal certain documents before the tribunal should proceed to a hearing of the substantive issues. All parties agreed that this application should be dealt with as a preliminary issue and an affidavit was admitted in support of the application. I do not propose to set out its contents at any length. It was made by E C Payne ARICS FRVA, who is the valuer advising the City Corporation on licensed property matters.

The affidavit seeks to cast doubt on the direct approach method H
of valuation for licensed premises and puts forward the view that, in order that the gross value may be properly determined, copies of audited trading accounts or, as a minimum, details of all sales and purchases broken down under different headings of intoxicating liquors, catering and other miscellaneous sources should properly be made available. Mr Payne suggests that in making a valuation (as he has done) based on a "percentage bid of estimated maintainable yearly gross takings of a hypothetical tenant at 1972-73 prices" it is customary and desirable to have details of actual accounts available as a starting point. Produced with the affidavit were graphs which purported to show observed consumption of draught beer at two of the public-houses made by him on various visits over a two-year J
period, which indicated to him that the application of the direct approach method to the quantities stated by the brewers would produce an incorrect answer. Also produced was a copy of a proposal for the alteration of the valuation list dated March 16 1978 and made by the valuation officer in respect of other premises within the City Corporation area. The proposal was for a reduction in assessment in the 1973 valuation list from gross value £9,000 to gross value £4,500 and for a change in description from restaurant and premises to public-house and premises.

Documents applied for

The documents which form the subject of this application are set K
out in the affidavit in the following terms:

A *In respect of the Sir Christopher Wren*

(1) The first respondents [Watneys London Ltd]
 (a) A full record of all sales to the premises of beer wine and spirits
 (b) A full record of all deliveries to the premises of beer wine and spirits
 (c) Returns to the licensed property valuer

(2) The second respondents [Goodhews Ltd]
 (a) Full profit and loss accounts for the years 1971-72 and 1972-73
 (b) A full record of all receipts including catering and ancillary sources, eg soft drink sales, cigarettes and gaming machines
 (c) Returns to the licensed property valuer L
 (d) A full record of all deliveries from all sources
 (e) VAT returns for the years 1972-73

(3) The third respondent [the valuation officer]
 (a) The returns from the first and second respondents
 (b) The valuations which led to the present list entry

B *In respect of the Cock Tavern*

(1) The first respondents [Truman Ltd]
 (a) Full profit and loss accounts for the years 1971-72 and 1972-73
 (b) A full record of all receipts including catering and ancillary sources, eg soft drink sales, cigarettes and gaming machines
 (c) Returns to the licensed property valuer
 (d) A full record of all deliveries from all sources M
 (e) VAT returns for the years 1972-73
 (f) The amount of premium paid by the first respondents to H A Forte and another on the assignment of the lease in October 1959 and details of any dispensation of waiver granted in relation to clause 1(18) of the lease.

(2) The second respondent [the valuation officer]
 (a) The returns from the first respondents
 (b) The valuations which led to the present list entry

C *In respect of the Magogs Public House*

(1) The first respondents [Chef and Brewer Ltd]
 (a) Full profit and loss accounts for the years 1971-72 and 1972-73

A (b) A full record of all receipts including catering and ancillary sources, eg soft drink sales, cigarettes and gaming machines
(c) Returns to the licensed property valuer
(d) A full record of all deliveries from all sources
(e) VAT returns for the years 1972-73

(2) The second respondent [the valuation officer]
(a) The returns from the first respondents
(b) The valuations which led to the present list entry

During the course of the hearing the application was withdrawn in respect of some of these items. In the first place the valuation officer agreed to make available copies of the returns asked for
B under A(3)(a), B(2)(a), C(2)(a). Consequently, the copies of these returns asked for at A(1)(c), A(2)(c), B(1)(c) and C(1)(c) are no longer asked for. Similarly, the amount of premium and lease details asked for under B(1)(f) together with certain additional information concerning a rent review in respect of the "Sir Christopher Wren" public-house were also to be made available, so that these items no longer form part of the application. I should perhaps say at this stage, however, that if the applications on these items had proceeded I should have had little hesitation in allowing them. The remaining items to be excluded were the applications in respect of VAT returns — it having been clearly established that at the relevant date VAT was not payable.

C The application is made under rule 40 of the Lands Tribunal Rules 1975, which provides as follows:

> A party to proceedings shall deliver to the registrar on his request any document or other information which the Tribunal may require and which it is in the power of that party to deliver and shall afford to every other party to the proceedings an opportunity to inspect those documents (or copies of them) and to take copies:
> Provided that nothing in this rule shall be deemed to require any information to be disclosed contrary to the public interest.

Corporation's submissions

Mr Marder submitted that the issues raised in the appeals are
D clearly set out in the pleadings, the relevant passages of which I have already quoted. The information sought, he says, is clearly relevant to those issues and is therefore compellable subject only to the proviso under rule 40 relating to the public interest. On this, suggests Mr Marder, the onus rests with the respondents to show why disclosure should not be ordered. The proceedings are analogous to those relating to an order for discovery and, in Mr Marder's submission, the tribunal should have regard to what he describes as a certain loosening of the protection provided by the courts during recent years.

He referred me to *Cartwright v Sculcoates Union* [1900] AC 150, which he conceded might still be good law in a consideration of a
E roving investigation but submitted that doubt must be cast upon it by more recent cases. He relied in particular on *D v National Society for the Prevention of Cruelty to Children* [1978] AC 171. Although in this case protection to a source of information was extended in the House of Lords, Mr Marder referred to passages in the speeches of Lord Diplock, Lord Hailsham, Lord Simon and Lord Edmund-Davies all of which applied some limits in the test of deciding what constituted the public interest; in particular a passage in the speech of Lord Edmund-Davies on p 243:

> . . . it would be unthinkable to vest the judiciary with a power to exclude in its discretion evidence relevant to the issues in civil proceedings merely
> F because one side wants it kept out and the judge thinks that its disclosure is likely to prove embarrassing. In other words, the exclusion of relevant evidence always calls for clear justification.

Mr Marder also referred to *Science Research Council v Nassé* and *Leyland Cars (BL Cars Ltd) v Vyas* [1979] 3 WLR 762 and *Bankers Trust Co v Shapiro and Others* in the Court of Appeal (*Times* Law Report, June 5 1980). In the *Science Research Council* case the House of Lords held that the ultimate test was whether discovery was necessary for disposing fairly of the proceedings and in order to decide whether it was necessary the tribunal should inspect the documents. In the present case Mr Marder did not consider that the application could satisfactorily be dealt with in this way.

Respondents' objections G

The application was vigorously resisted by Mr Moriarty for the ratepayer respondents and by Mr Alderman for the valuation officer.

Mr Alderman's objections related solely to the remaining requirement which concerned the valuation officer — namely that the valuations which formed the basis of the assessments originally entered in the valuation list should be disclosed. In Mr Alderman's submission these valuations were irrelevant, and in any case the valuer who did prepare them had since died. They could in no way be taken as an admission by the valuation witness whom he proposed to call.

For the other respondents Mr Moriarty did not admit the rele- H
vance of any of the information which was now sought. He relied on *Watney Mann v Langley* as indicating the sort of information which the valuation officer could properly obtain. He submitted that the possession of actual figures of gross receipts would not help the tribunal in the interpretation of Mr Payne's valuations. They obviously had no relevance to valuation 2; nor would they advance valuation 1 because the question of the appropriate percentage would still remain at large.

In Mr Moriarty's submission, the forms of return which the valuation officer was authorised to serve under section 82 of the General Rate Act 1967 indicated the nature of the information which was properly to be made available; and the use to which that J
information was to be put was closely regulated by statute under section 83 of the Act. In any case the disclosure of Goodhews' accounts, being the tenants operating the "Sir Christopher Wren", to Watneys London Ltd, being their landlords, would represent a substantial breach of confidentiality and would be greatly oppressive. If the City Corporation were held to be entitled to put before the tribunal the information asked for, the same opportunity would be available to any aggrieved person who made a proposal to increase the rating assessment of any hereditament entered in the valuation list. This could not be right. Finally, said Mr Moriarty, certified statements of barrelage and of catering receipts had been provided at the local valuation court hearing and were available for K
the appellants. They were not entitled to anything else.

The issues raised in this application are clearly of great importance and formidable arguments have been presented on either side. I have thought it appropriate, therefore, to set out my reasons at some length for dealing with the application before proceeding to deal with the substantive issues raised in the appeals.

First of all there are two matters of general importance which I want to mention. It is suggested by Mr Marder that one of the issues in the appeals is the question of the correct method of valuation to be applied in valuing these three public-houses. In this I fear he is going to be disappointed, since from the information already before L
me it seems unlikely that I shall make any such finding.

The question before the tribunal, as it was before the local valuation court, is to give such directions with respect to the manner in which the hereditament in question is to be treated in the valuation list as appear to be necessary to give effect to the contention of the appellants in so far as that contention appears to be well founded — see section 76(5) of the General Rate Act 1967. The entry in the valuation list contains a description of the hereditament and in the present case figures for gross value and rateable value. In the course of arriving at those values it may well be true that certain methods of valuation are more helpful than others and observations to that effect may properly be made by a valuation court or by the M
tribunal; but in the context of the present appeals I am not prepared to hold that the direct method of valuation, which has certainly been adopted for more than 40 years and recognised by Thompson J in *Watney Mann v Langley*, is an incorrect method of valuation.

At the same time I think it is quite open to the appellants to bring forward valuations arrived at on another basis. Indeed, in the reported cases there are examples of valuations of licensed premises carried out on the profits basis, when the accounts have been produced by ratepayers of their own volition.

I observe also that Mr Payne himself puts forward two alternative valuations and a computation attached to the statement of case;

A whichever method is adopted, however, the question must arise as to whether it has been correctly applied.

Mr Payne's valuations arrive at different answers. Valuation 1 produces a gross value of £19,250, valuation 2 a gross value of £16,000 and computation 3 a rent in terms of gross value of £12,910. It is true that the third figure does not purport to be a value at the relevant date, but of the other two valuations one thing is apparent — that they cannot both be right. This is not to say that the method employed is of necessity the wrong method but merely that at least one of the methods has been wrongly applied. In considering the application, therefore, I am not concerned with deciding which method is a right method and which a wrong method of B valuation. What concerns me is whether the gross value determined by the local valuation court is insufficient and if it is to be increased, by how much.

The second general observation which I want to make is this. The application is made under the Lands Tribunal Rules and is not an application for a discovery order under the Rules of the Supreme Court. It seems to me, nevertheless, that in determining the application I should be guided by the same considerations *mutatis mutandis* which would have been taken into account by the courts in considering applications for discovery; and I understand that this is an approach which is accepted by counsel.

C **Review of authorities**

The first authority to which I turn is that of *Cartwright v Sculcoates Union*. That was a case where the valuation of a public-house had been referred to an arbitrator who, in reaching his decision, had rejected evidence relating to weekly takings which the assessing authority had sought to put before him by cross-examination of the tenant of the public-house. The guardians appealed and when the matter reached the House of Lords the House declined to interfere with the valuation on that account. On p 153 Lord Macnaghten said:

> What your Lordships have to consider is whether the learned arbitrator proceeded on the lines on which the Act of Parliament indicates that he D ought to proceed. Now what has the arbitrator done? He has found that in this particular case there does not exist the ordinary basis of computation. And he has excluded any inquiry into profits. In that I think he was perfectly right, not that the profits if they could be ascertained would not be an element in arriving at the rent which a tenant might reasonably be expected to pay, but that any inquiry into profits should be avoided if possible, because it would be oppressive. There is nothing that a tradesman so much dislikes as any inquiry into his profits. In some cases it cannot be avoided, and then according to the decided cases the inquiry may be made; but in this case there has been no inquiry into profits. What the learned arbitrator has done is to take into consideration the amount of business which this public house was doing. Was he wrong in that? Surely the very first thing that a tenant who was going to offer for a house of this sort would do would be to consider (roughly if he could not do it accurately) what amount of business E the house commanded. It appears to me that the volume of business done in a public-house, as apparent to the man in the street — if I may use such an expression — is the very first thing that a tenant proposing to make an offer for such a house would take into consideration. That really is all that the arbitrator has done. That is clearly one of the circumstances which would influence persons bargaining about the rent of such a house as this.

At p 159 Lord Davey said:

> If you are to take into account the fact that the premises command a trade, you must surely ask what trade. Is it a large trade, or is it a small trade? And I do not know myself any better test of what trade they may be expected to command than the trade which they actually do command. It is not that you rate the profits, it is not that you rate the man's skill and judgment or F discretion in the mode of carrying on the business, but you have to ascertain what sort of a trade the hypothetical tenant, as he is called, may reasonably expect to be able to carry on on those premises as an element in determining the rent he would be willing to offer.

From that authority I deduce that a valuation is not to be disturbed merely because details of actual profits have not been obtained so long as there is information about the volume of trade which enables a valuation to be prepared.

Information "reasonably required"

I turn to the case of *Watney Mann v Langley*. This case concerned the rights of a valuation officer to obtain information from an occupier, owner or lessee of any hereditament in his area under G powers contained in section 58 of the Local Government Act 1948 as amended. The equivalent provisions are now contained in section 82 of the General Rate Act 1967. The Act empowered him to call for a return "containing such particulars as may be reasonably required for the purpose of enabling him accurately to compile the list or to decide whether or not to make or as the case may be to object to a proposal".

The brewers were asked to "... state quantity of the following goods supplied by you to the property for retail sale there for each of the last three years ending on the date to which your accounts are usually made up; beers in casks and kegs; bottled and canned beer; wines and spirits, cider." H

Thompson J held that the expression "reasonably required" as used in the statute means "reasonably necessary", that is to say necessary to a prudent and reasonable valuation officer for the purpose of compiling an accurate valuation list. He held further that if reasonably required meant reasonably demanded, nevertheless, on balancing the advantage to the valuation officer of having figures of actual trade against the hardship and disadvantage to the plaintiffs in compelling them to provide them, it was on balance reasonable for the valuation officer to demand them. The question asked on the form of return was therefore not *ultra vires* and illegal.

In that case it was conceded by the brewers that the information sought was relevant and admissible and the learned judge recog- J nised that the general tenor of development in the decisions was towards allowing the valuation officer to make use of factual information rather than of confining him to generalities and estimates.

It is in my opinion significant, however, that in the *Watney Mann* case the information for which the valuation officer contended and which the court confirmed was reasonably required or reasonably necessary was the information regarding the volume of trade as expressed by deliveries to the hereditament.

Finally, I turn to *Science Research Council v Nassé* and *Leyland Cars v Vyas* where the question at issue was the discovery of confidential reports in proceedings before an industrial tribunal and the Employment Appeal Tribunal. At p 769 of the report Lord K Wilberforce in a reference to the issues in the appeals said:

> These are, broadly, two: First, is there in relation to confidential documents, or any relevant class of confidential documents, any immunity from disclosure? Second, if not, how should the tribunal exercise its discretion as to discovery in relation to confidential documents in this field? (Here and elsewhere I use the word "tribunal" so as to include, where appropriate, a county court.)
>
> On these points my conclusions are as follows:
>
> 1. There is no principle of public interest immunity, as that expression was developed from *Conway v Rimmer* [1968] AC 910, protecting such confidential documents as those with which these appeals are concerned. That such an immunity exists, or ought to be declared by this House to exist, was L the main contention of Leyland. It was not argued for by the SRC; indeed that body argued against it.
>
> 2. There is no principle in English law by which documents are protected from discovery by reason of confidentiality alone. But there is no reason why, in the exercise of its discretion to order discovery, the tribunal should not have regard to the fact that documents are confidential, and that to order disclosure would involve a breach of confidence. In the employment field, the tribunal may have regard to the sensitivity of particular types of confidential information, to the extent to which the interests of third parties (including their employees on whom confidential reports have been made, as well as persons reporting) may be affected by disclosure, to the interest which both employees and employers may have in preserving the confidentiality of personal reports, and to any wider interest which may be seen to M exist in preserving the confidentiality of systems of personal assessments.
>
> 3. As a corollary to the above, it should be added that relevance alone, though a necessary ingredient, does not provide an automatic sufficient test for ordering discovery. The tribunal always has a discretion. That relevance alone is enough was, in my belief, the position ultimately taken by counsel for Mrs Nassé thus entitling the complainant to discovery subject only to protective measures (sealing up, etc). This I am unable to accept.
>
> 4. The ultimate test in discrimination (as in other) proceedings is whether discovery is necessary for disposing fairly of the proceedings. If it is, then discovery must be ordered notwithstanding confidentiality. But where the court is impressed with the need to preserve confidentiality in a particular case, it will consider carefully whether the necessary information has been or can be obtained by other means, not involving a breach of confidence.

> 5. In order to reach a conclusion whether discovery is necessary notwithstanding confidentiality the tribunal should inspect the documents. It will naturally consider whether justice can be done by special measures such as "covering up", substituting anonymous references for specific names, or, in rare cases, hearing in camera.

As I have already indicated, the course suggested in paragraph 5 referred to by Lord Wilberforce is not urged upon me in the present case.

I have already said that the information contained in the forms of return in the present case and the supplementary information regarding rents are to be considered as relevant and reasonably necessary for a proper determination of the disputed gross values. There is no longer any dispute about these items.

Tribunal's rulings

With regard to the original valuations, however, which were prepared in connection with the entries originally appearing in the valuation list, I do not consider that these are in any way relevant to the issues and I am not prepared to order disclosure. It seems to me that the details to be revealed from these valuations would tell me no more than what is already apparent — that the earlier figures are now thought to be wrong by the licensed property valuer and should be reduced to the figures put before the local valuation court. It seems to me to be immaterial whether in now putting forward that valuation he adopts different figures of trade or different multipliers or indeed a different method of valuation.

I am not prepared to find in the present case that any of the other documents which were the subject of the application are irrevelant to questions of valuation; but that is not enough to dispose of the matter. As Lord Wilberforce says, relevance alone, although a necessary ingredient, does not provide an automatic sufficient test for ordering disclosure. There remains with the tribunal a discretion and in the exercise of that discretion it seems to me that I am required to weigh the balance of advantage between the parties; regarding on the one hand whether a requirement to produce the documents is oppressive, how far it involves making public information which would be injurious, by exposure to trade competitors or to a landlord, and how far the documents sought are readily available. On the other hand I must have regard to the degree of relevance and to my own opinion on the extent to which if at all they would be necessary or helpful. On this side of the scales, I should say that I attach little weight to Mr Payne's personal observations of beer consumption or to the valuation officer's proposal relating to other premises.

At the end of the day I do not consider that the profit and loss accounts would be of any real help to me even if they are in a sufficiently identifiable form. It is suggested that any such accounts, in the case of "The Cock Tavern" and "The Magogs", must inevitably include a number of items which are the result of arbitrary apportionments. As it is put by Mr Moriarty, I would be no further forward with figures of gross receipts because the question of the percentage to be applied would still be at large and not readily susceptible of support from any other source. I think as a general principle Parliament has provided that the valuatioh officer shall have access to that information which he reasonably requires.

The line of authority through *Cartwright v Sculcoates Union* and *Watney Mann v Langley* indicates that the volume of trade is adequate to provide a basis for valuation. I am not prepared to extend this further to include the disclosure of profit and loss accounts. If the tribunal were to decide to order the production of accounts at the request of the rating authority I can see no reason why in principle they should not make a similar order at the request of a third party who qualified as an aggrieved person. (See *Arsenal Football Club v Smith and Another* [1977] 2 All ER 267.) It is clearly in the public interest that information should be available to assist in the determination of correct gross values. But it is also in the public interest — as was recognised in *Cartwright v Sculcoates Union* — that a trader should be protected from the disclosure in open court against his will of confidential accounts which could be of assistance to third parties not connected with the proceedings. In my opinion, in the matter of the accounts asked for in the present case, the latter consideration outweighs the former. In this context it is perhaps significant that under section 86 of the General Rate Act the valuation officer is given power of entry on to premises whereas no such powers are granted to rating authorities or other aggrieved persons.

There remains the request for full records of sales, deliveries and receipts. I must say that I am in some doubt as to what is covered by these items. Certainly I think that they are expressed in terms which are far too wide, but some of this information is information which can properly be required. At first sight, it is the sort of information which is being made available by the production of the forms of return. I am, therefore, not prepared to order disclosure under this heading at this stage. If, however, a detailed examination of the forms of return shows that the answers are in any way incomplete, I would certainly be prepared to hear further argument. With these qualifications, the present application is dismissed.

At this stage Mr Marder applies for a stay of proceedings and indicates that the Registrar is to be asked for a case to be stated on my ruling. My initial reaction is one of some surprise in view of the fact that at the pre-trial review which I held on November 20 1979 no question arose of any application under rule 40 and I was told that no difficulty was anticipated in the agreement of facts for the purpose of disposing of the appeals.

Nevertheless, the matter has now been raised and I must consider whether at this stage it would be proper for me to grant the application for the stay. This in turn depends on whether or not a case is to be stated. Subsection (4) of section 3 of the Lands Tribunal Act 1949 indicates that a decision of the Lands Tribunal should be final, subject only to the proviso that anybody aggrieved by the decision as being erroneous in point of law may require the tribunal to state and sign a case for the decision of the court.

I appreciate that it is possible to argue that the ruling which I have given can be described as a decision, but if the right of appeal arises every time the tribunal gives a ruling, for example on the admissibility of evidence, then the consequence could be that there might be any number of appeals during the same case. Moreover by Order 61 rule 1(2) of the Rules of the Supreme Court a case stated must state the facts on which the decision was based; and that would seem to recognise that the word "decision" as used in the Act means the decision on the substantive issue (in this case the gross value of the hereditament) which the tribunal is required to determine.

I have reached the conclusion that what I have given is not a decision within the meaning of the Act but a ruling on a procedural matter arising during the course of the hearing.

I think that the balance of convenience lies in this view. It seems sensible that in any case the Court of Appeal should have before it all the facts of the case and should decide all questions of law which may arise from those facts at the same time. If it were otherwise, a litigant could raise a series of separate appeals over a period of time in respect of the same case. I hasten to add that in the present case there is no suggestion that the City Corporation have any such intention nor indeed can they benefit by any delay in the determination of these appeals.

Accordingly, I am of the opinion that I have no power to state a case at this stage and I am not, therefore, prepared to order a stay of proceedings on that account.